Return on or before the
last date stamped below

Kingston College
Kingston Hall Road
Kingston upon Thames
KT1 2AQ

Introduction to
Finite and Spectral Element Methods using MATLAB®

Introduction to
Finite and Spectral
Element Methods
using MATLAB®

C. Pozrikidis

University of California
San Diego, USA

Chapman & Hall/CRC
Taylor & Francis Group
Boca Raton London New York Singapore

Published in 2005 by
Chapman & Hall/CRC
Taylor & Francis Group
6000 Broken Sound Parkway NW, Suite 300
Boca Raton, FL 33487-2742

Library of Congress Cataloging-in-Publication Data

Pozrikidis, C.
   Introduction to finite and spectral element methods using MATLAB / C. Pozrikidis.
      p. cm.
   Includes bibliographical references and index.
   ISBN 1-58488-529-7
   1. Finite element method--Data processing. 2. MATLAB. I. Title.

TA347.F5P7 2005
620'.0042'0151825--dc22

2005041402

Taylor & Francis Group
is the Academic Division of T&F Informa plc.

Visit the Taylor & Francis Web site at
http://www.taylorandfrancis.com

and the CRC Press Web site at
http://www.crcpress.com

Preface

Five general classes of numerical methods are available for solving ordinary and partial differential equations encountered in the various branches of science and engineering: finite-difference, finite-volume, finite element, boundary element, spectral and pseudo-spectral methods. The relation and relative merits of these methods are briefly discussed in the "Frequently Asked Questions" section preceding Chapter 1.

An important advantage of the finite element method, and the main reason for its popularity among academics and commercial developers, is the ability to handle solution domains with arbitrary geometry. Another advantage is the ability to produce solutions to problems governed by linear as well as nonlinear differential equations. Moreover, the finite element method enjoys a firm theoretical foundation that is mostly free of *ad hoc* schemes and heuristic numerical approximations. Although this may be a mixed blessing by being somewhat restrictive, it does inspire confidence in the physical relevance of the solution.

A current search of books in print reveals over two hundred items bearing in their title some variation of the general theme "Finite Element Method." Many of these books are devoted to special topics, such as heat transfer, computational fluid dynamics (CFD), structural mechanics, and stress analysis in elasticity, while other books are written from the point of view of the applied mathematician or numerical analyst with emphasis on error analysis and numerical accuracy. Many excellent texts are suitable for a second reading, while others do a superb job in explaining the fundamentals but fall short in describing the development and practical implementation of algorithms for non-elementary problems.

The purpose of this text is to offer a venue for the rapid learning of the theoretical foundation and practical implementation of the finite element method and its companion spectral element method. The discussion has the form of a self-contained course that introduces the fundamentals on a need-to-know basis and emphasizes the development of algorithms and the computer implementation of the essential procedures. The audience of interest includes students in science and engineering, practicing scientists and engineers, computational scientists, applied mathematicians, and scientific computing enthusiasts.

Consistent with the introductory nature of this text and its intended usage as a practical primer on finite element and spectral element methods, error analysis is altogether ignored, and only the fundamental procedures and their implementation are discussed in sufficient detail. Specialized topics, such as Lagrangian formulations, free-boundary problems, infinite elements, and discontinuous Galerkin methods, are briefly mentioned in Appendix F entitled "Glossary," which is meant to complement the subject index.

This text has been written to be used for self-study and is suitable as a textbook in a variety of courses in science and engineering. Since scientists and engineers of any discipline are familiar with the fundamental concepts of heat and mass transfer governed by the convection–diffusion equation, the main discourse relies on this prototype. Once the basic concepts have been explained and algorithms have been developed, problems in solid mechanics, fluid mechanics, and structural mechanics are discussed as extensions of the basic approach.

The importance of gaining simultaneous hands-on experience while learning the finite element method, or any other numerical method, cannot be overemphasized. To achieve this goal, the text is accompanied by a library of user-defined MATLAB functions and complete finite and spectral element codes composing the software library FEMLIB. The main codes of FSELIB are tabulated after the Contents. Nearly all functions and complete codes are listed in the text, and only a few lookup tables, ancillary graphics functions, and slightly modified codes are listed in abbreviated form or have been omitted in the interest of space.

The owner of this book can freely download and use the library subject to the conditions of the GNU public license from the web site:

$$\texttt{http://dehesa.freeshell.org/FSELIB}$$

For instructional reasons and to reduce the overhead time necessary for learning how to run the codes, the library is almost completely free of .dat files related to data structures. All necessary parameters are defined in the main code, and finite element grids are generated by automatic triangulation determined by the level of refinement for specific geometries. With this book as a user guide, the reader will be able to immediately run the codes as given, and graphically display solutions to a variety of elementary and advanced problems.

The MATLAB language was chosen primarily because of its ability to integrate numerical computation and computer graphics visualization, and also because of its popularity among students and professionals. For convenience, a brief MATLAB primer is included in Appendix G. Translation of a MATLAB code to another computer language is both straightforward and highly recommended. For clarity of exposition and to facilitate this translation, hidden operations embedded in the intrinsic MATLAB functions are intentionally avoided as much as possible in the FSELIB codes. Thus, two vectors are added explicitly component by component rather than implicitly by issuing a symbolic vector addition, and the entries of a matrix are often initialized and manipulated in a double loop running over the indices, even though this makes for a longer code.

Further information, a list of errata, links to finite element recourses, and updates of the FSELIB library are maintained at the book web site:

$$\texttt{http://dehesa.freeshell.org/FSEM}$$

Comments and corrections from readers are most welcome and will be communicated to the audience with due credit through the book web site.

I owe a great deal of gratitude to Todd Porteous for providing a safe harbor for the files, and to Mark Blyth and Haoxiang Luo for insightful comments on the manuscript leading to a number of improvements.

C. Pozrikidis
San Diego, California

Contents

FSELIB *software library*

The FSELIB software library accompanying this book contains miscellaneous functions and complete finite and spectral element codes written in MATLAB, including modules for domain discretization, system assembly and solution, and graphics visualization. The codes are arranged in directories corresponding to the book chapters and appendices. The owner of this book can download the library freely from the Internet site:

http://dehesa.freeshell.org/FSELIB

The software resides in the public domain and should be used strictly under the terms of the GNU General Public License, as stated on the GNU Internet page cited below.

The following tables list selected FSELIB finite and spectral element codes for problems in one and two dimensions, corresponding to Chapters 1–5 and appendices. A complete list of the FSELIB functions and codes can be found in the first part of the subject index.

Chapter 1: The finite element method in one dimension

Directory: 01

Code:	Problem:	Element type:
beam	Cantilever beam bending	linear
scdl	Steady convection–diffusion	linear
sdl	Steady diffusion	linear
sdq	Steady diffusion, midpoint interior node	quadratic
sdqb	Steady diffusion, arbitrary interior node	quadratic
sdqc	Steady diffusion, condensed formulation	quadratic
udl	Unsteady diffusion	linear

Chapter 2: High-order and spectral elements in one dimension

Directory: 02

Code:	Problem:	Element type:
sds	Steady diffusion	spectral
sde	Steady diffusion	evenly spaced
sdsm	Steady diffusion, modal expansion	spectral
sdsc	Steady diffusion, condensed formulation	spectral
uds_cn	Unsteady diffusion, Crank-Nicolson method	spectral
uds_fe	Unsteady diffusion, forward Euler method	spectral

Chapter 3: The finite element method in two dimensions

Directory: 03

Code:	*Problem:*	*Element type:*
hlm3_n	Helmholtz's equation in a disk-like domain, with the Neumann boundary condition	3-node triangles
lapl3_d	Laplace's equation in a disk-like domain, with the Dirichlet boundary condition	3-node triangles
lapl3_dn	Laplace's equation in a disk-like domain, with Dirichlet and Neumann boundary conditions	3-node triangles
lapl3_dn_sqr	Laplace's equation in a square domain, with Dirichlet and Neumann boundary conditions	3-node triangles
scd3_d	Steady convection–diffusion in a disk-like domain, with the Neumann boundary condition	3-node triangles

Chapter 4: Quadratic and spectral elements in two dimensions

Directory: 04-05

Code:	*Problem:*	*Element type:*
lap16_d	Laplace's equation in a disk-like domain, with the Dirichlet boundary condition	6-node triangles
lap16_d_L	Laplace's equation in an L-shaped domain, with the Dirichlet boundary condition	6-node triangles
lap16_d_rc	Laplace's equation in a rectangular domain with a circular hole, with the Dirichlet boundary condition	6-node triangles
lap16_d_sc	Laplace's equation in a square domain with a circular hole, with the Dirichlet boundary condition	6-node triangles
lap16_d_ss	Laplace's equation in a square domain with a square hole, with the Dirichlet boundary condition	6-node triangles
scd6_d	Steady convection–diffusion in a disk-like domain, with the Neumann boundary condition	6-node triangles
scd6_d_rc	Steady convection–diffusion in a rectangular domain with a circular hole, with the Neumann boundary condition	6-node triangles

Chapter 5: Applications in solid and fluid mechanics

Directory: 04-05

Code:	*Problem:*	*Element type:*
bend_HCT	Bending of a clamped plate based on the biharmonic equation	HCT triangles
cvt6	Stokes flow in a rectangular cavity	6-node triangles
membrane	In-plane deformation of a membrane patch under a constant body force in plane stress analysis	6-node triangles
psa6	Plane stress analysis in a rectangular domain possibly with a circular hole	6-node triangles

Appendix C: Linear solvers

Directory: AC

Code:	*Problem:*
gel	Solution of a linear system by Gauss elimination
cg	Solution of a symmetric system by the method of conjugate gradients

Frequently Asked Questions

- *What is the finite element method (FEM)?*

The finite element method is a numerical method for solving partial differential equations encountered in the various branches of mathematical physics and engineering. Examples include Laplace's equation, Poisson's equation, Helmholtz's equation, the convection–diffusion equation, the equations of potential and viscous flow, the equations of electrostatics and electromagnetics, and the equations of elastostatics and elastodynamics.

- *When was the finite element method conceived?*

The finite element method was developed in the mid-1950s for problems in stress analysis under the auspices of linear elasticity. Since then, the method has been generalized and applied to solve a broad range of differential equations, with applications ranging from fluid mechanics to structural dynamics.

- *What is the Galerkin finite element method (GFEM)?*

The GFEM is a particular and most popular implementation of the FEM, in which algebraic equations are derived from the governing differential equations by a process called the Galerkin projection.

- *What are the advantages of the finite element method?*

The most significant practical advantage is the ability to handle solution domains with arbitrary geometry. Another important advantage is that transforming the governing differential equations to a system of algebraic equations is performed in a way that is both theoretically sound and free of *ad hoc* schemes and heuristic numerical approximations. Moreover, the finite element method is built on a rigorous theoretical foundation. Specifically, for a certain class of differential equations, it can be shown that the finite element method is equivalent to a properly-posed functional minimization method.

- *What is the origin of the terminology "finite element?"*

In the finite element method, the solution domain is discretized into elementary units called finite elements. For example, in the case of a two-dimensional domain, the finite elements can be triangles or quadrilateral elements. The discretization is typically unstructured, meaning that new elements may be added or removed without affecting an existing element structure, and without requiring a global element and node relabeling.

• *Is there a restriction on the type of differential equation that the finite element method can handle?*

In principle, the answer is negative. In practice, the finite element method works best for diffusion-dominated problems, and has been criticized for its inability to handle convection–dominated problems occurring, for example, in high-speed flows. However, modifications of the basic procedure can be made to overcome this limitation and improve the performance of the algorithms.

• *How does the finite element compare with the finite difference method (FDM)?*

In the finite difference method, a grid is introduced, the differential equation is applied at a grid node, and the derivatives are approximated with finite differences to obtain a system of algebraic equations. Because grid nodes must lie at boundaries where conditions are specified, the finite difference method is restricted to solution domains with simple geometry, or else requires the use of cumbersome boundary-fitted coordinates and artificial body forces for smearing out the boundary location.

• *How does the finite element method compare with the finite volume method (FVM)?*

In the finite volume method, the solution domain is also discretized into elementary units called finite volumes. The differential equation is then integrated over the individual volumes, and the divergence theorem is applied to derive equilibrium equations. In the numerical implementation, solution values are defined at the vertices, faces, or centers of the individual volumes, and undefined values are computed by neighbor averaging. Although the finite volume method is also able to handle domains with arbitrary geometrical complexity, the required *ad-hoc* averaging puts it at a disadvantage.

• *How does the FEM compare with the boundary element method (BEM)?*

Because the boundary element method (BEM) requires discretizing only the boundaries of a solution domain, it is significantly superior (e.g., [44]). In contrast, the finite element method requires discretizing the whole of the solution domain, including the boundaries. For example, in three dimensions, the BEM employs surface elements, whereas the FEM employs volume elements. However, the BEM is primarily applicable to linear differential equations with constant coefficients. Its implementation to more general types of differential equations is both cumbersome and computationally demanding.

- *How does the FEM compare with the spectral and pseudo-spectral method?*

In one class of spectral and pseudo-spectral methods, the solution is expanded in a series of orthogonal basis functions, the expansion is substituted in the differential equation, and the coefficients of the expansion are computed by projection or collocation. These methods are suitable for solution domains with simple geometry.

- *What should one know before one is able to understand the theoretical foundation of the FEM?*

The basic concepts are discussed in a self-contained manner in this book. Prerequisites are college-level calculus, numerical methods, and a general familiarity with computer programming.

- *What should one know before one is able to write a FEM code?*

Prerequisites are general-purpose numerical methods, including numerical linear algebra, function interpolation, and function integration. All necessary topics are discussed in this text, and summaries are given in appendices. Familiarity with a computer programming language is another essential prerequisite.

- *What is the spectral element method?*

The spectral element method is an advanced implementation of the finite element method in which the solution over each element is expressed in terms of *a priori* unknown values at carefully selected spectral nodes. The advantage of the spectral element method is that stable solution algorithms and high accuracy can be achieved with a low number of elements under a broad range of conditions.

- *How can one keep up with new developments in the finite and spectral element method?*

Several Internet sites provide current information on various aspects of the finite and spectral element methods. Links are provided at the book web site:

http://dehesa.freeshell.org/FSEM

The finite element method in one dimension

1

In the first chapter, we illustrate the fundamental concepts underlying the finite element method in one dimension by discussing the numerical solution of several elementary ordinary and partial differential equations, including the steady diffusion equation in the presence of a distributed source, the unsteady diffusion equation, the convection equation, and the inclusive convection–diffusion equation.

We shall begin by discussing the simplest possible implementation of the finite element method, where the solution domain is divided into a number of intervals called finite elements, and the solution itself is approximated with a linear function over each element. The union of the linear element functions yields a continuous, piecewise linear function. Implementations for quadratic and high-order element functions will arise by a straightforward extension of the basic approach. In the most general implementation, the solution over each element is approximated with a polynomial of arbitrary degree defined by an appropriate group of interpolation nodes or expansion modes. Demanding high interpolation accuracy and requiring robust and efficient solution algorithms, subject to the available degrees of freedom, leads us to the spectral element method discussed in Chapter 2.

1.1 Steady diffusion with linear elements

Consider steady-state heat conduction through a rod of length, L, in the presence of a distributed source of heat due, for example, to a chemical reaction, as illustrated in Figure 1.1.1(a). We begin by introducing the x axis along the rod, and assume that the heat flux along the rod is given by Fick's law,

$$q(x) = -k \, \frac{\mathrm{d}f}{\mathrm{d}x}, \tag{1.1.1}$$

where k is the thermal conductivity of the rod material, and $f(x)$ is the temperature. A heat balance over a section of the rod with infinitesimal length, δx, requires

$$q(x) - q(x + \delta x) + s(x) \, \delta x = 0, \tag{1.1.2}$$

and $s(x)$ is the rate of heat production per unit length of the rod; if $s(x)$ is negative, heat escapes from the rod across the cylindrical surface. Substituting Fick's law in (1.1.2), and dividing by δx, we find

$$-\frac{k}{\delta x}\left[\left(\frac{\mathrm{d}f}{\mathrm{d}x}\right)_x - \left(\frac{\mathrm{d}f}{\mathrm{d}x}\right)_{x+\delta x}\right] + s(x) = 0. \tag{1.1.3}$$

Taking the limit, $\delta x \to 0$, we find that the temperature distribution along the rod satisfies the steady heat conduction equation

$$k\,\frac{\mathrm{d}^2 f}{\mathrm{d}x^2} + s(x) = 0, \tag{1.1.4}$$

which is a second-order ordinary differential equation (ODE). We shall assume that the boundary conditions specify:

- The heat flux at the left end of the rod located at $x = 0$,

$$q_0 \equiv -k\,\left(\frac{\mathrm{d}f}{\mathrm{d}x}\right)_{x=0}, \tag{1.1.5}$$

  where q_0 is a given constant.

- The temperature at the right end of the rod located at $x = L$,

$$f(x = L) \equiv f_L. \tag{1.1.6}$$

Equation (1.1.5) expresses a *Neumann* boundary condition, that is, a condition on the derivative of the unknown function, and equation (1.1.6) expresses a *Dirichlet* boundary condition, that is, a condition on the value of the unknown function. Other possible boundary conditions will be discussed later in this chapter.

1.1.1   Linear element interpolation

We begin implementing the finite element method by dividing the solution domain, $0 \le x \le L$, into N_E intervals, called finite elements, defined by the element end-nodes, x_i, for $i = 1, 2, \ldots, N_E + 1$, where $x_1 = 0$ and $x_{N_E+1} = L$, as shown in Figure 1.1.1(a). Next, we approximate the temperature distribution over the individual elements with linear functions whose union yields a continuous, piecewise linear function, drawn with the dashed polygonal line in Figure 1.1.1(a).

To derive the mathematical representation of the polygonal approximation, we introduce the piecewise linear, tent-line, global interpolation functions, $\phi_i(x)$, $i = 1, 2, \ldots, N_E + 1$, where the ith function is supported over the interval $[x_{i-1}, x_{i+1}]$, as illustrated in Figure 1.1.1(a). The ith global interpolation function takes the value of unity at the ith nodes, drops linearly to zero at the adjacent nodes numbered $i - 1$ and $i + 1$, and remains zero outside the

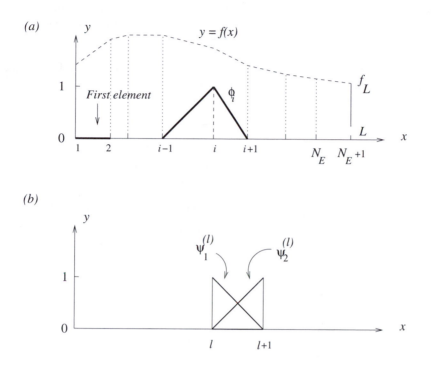

Figure 1.1.1 Steady-state heat conduction through a rod extending from $x = 0$ to L. The temperature distribution is computed by solving equation (1.1.4) subject to the boundary conditions (1.1.5) and (1.1.6) using a finite element method with linear interpolation functions. The function ϕ_i depicted in (a) is the global interpolation function of the ith node, and the functions $\psi_1^{(l)}$ and $\psi_2^{(l)}$ depicted in (b) are the local interpolation functions of the lth element.

supporting interval $[x_{i-1}, x_{i+1}]$. By construction then, the global interpolation functions satisfy the cardinal interpolation property

$$\phi_i(x_j) = \delta_{ij}, \tag{1.1.7}$$

where δ_{ij} is Kronecker's delta representing the identity matrix,

$$\delta_{ij} = \begin{cases} 1 & \text{if } i = j \\ 0 & \text{if } i \neq j \end{cases}. \tag{1.1.8}$$

The piecewise linear approximation of the solution is represented by the global finite element expansion

$$f(x) = \sum_{j=1}^{N_E+1} f_j \, \phi_j(x), \tag{1.1.9}$$

where

$$f_j \equiv f(x = x_j) \tag{1.1.10}$$

is the *a priori* unknown value of the solution at the jth node. The Dirichlet boundary condition (1.1.6) at the right end demands

$$f_{N_E+1} = f_L. \tag{1.1.11}$$

It is clear that the piecewise linear approximation cannot possibly be made to satisfy a second-order differential equation: taking the second derivative of the right-hand side of (1.1.9), we are faced with the zero function interrupted by a sequence of Dirac delta functions centered at the element nodes, as will be discussed in Section 1.1.7. However, while this is undoubtedly true, the difficulty will be bypassed by seeking a solution of the *weak formulation*, which arises from the Galerkin projection of the differential equation, as discussed in Section 1.1.3. When this is done, the order of the differential equation is effectively lowered by one unit, and the condition on the existence of the second derivative, which is implicit in the statement of the second-order differential equation, is replaced by a condition on the existence of the first derivative, which is satisfied by the piecewise linear approximation.

A continuous function with discontinuous first derivative at discrete points, such as the function described by the global finite element expansion (1.1.9), is called a C^0 function. A continuous function with continuous first derivative and discontinuous second derivative at discrete points, is called a C^1 function. An infinitely differentiable function is called a C^∞ function. We shall see later in this chapter that finite element solutions typically employ C^0 or C^1 functions, as the need arises, depending on the order of the differential equations.

1.1.2   Element grading

Two important features of the finite element method are: (a) the ability to ensure adequate spatial resolution by varying the element size manually or adaptively, and (b) ease for accommodating irregular geometrical shapes. In the case of the one-dimensional problem presently considered, because the solution domain is a line, only the first feature is relevant.

FSELIB function `elm_line1`, listed in the text, discretizes an interval of the x axis into a graded mesh of N elements, so that the element length increases or decreases from left to right in a geometrical fashion. Specifically, if Δx_1 is the size of the first element, and Δx_N is the size of the last element, then the ratio

$$\frac{\Delta x_N}{\Delta x_1} \equiv r \tag{1.1.12}$$

has a specified value, denoted as *ratio* in the code.

```
function xe = elm_line1 (x1,x2,n,ratio)

%=================================================
% Unsymmetrical discretization of a line segment
% into a graded mesh of n elements
% subtended between the left end-point x1
% and right end-point x2
%
% The variable "ratio" is the ratio
% of the length of the last
% element to the length of the first element.
%
% xe: element end-nodes
%=================================================

%-------------
% one element
%-------------

if(n==1)
   xe(1) = x1;
   xe(2) = x2;
   return;
end

%--------------
% many elements
%--------------

if(ratio==1)
   alpha = 1.0;
   factor = 1.0/n;
else
   texp = 1/(n-1);
   alpha = ratio^texp;
   factor = (1.0-alpha)/(1.0-alpha^n);
end

deltax = (x2-x1) * factor;    % length of the first element

xe(1) = x1;                   % first point

for i=2:n+1
   xe(i)  = xe(i-1)+deltax;
   deltax = deltax*alpha;
end

%-----
% done
%-----

return;
```

Function elm_line1: Unsymmetrical discretization of a line segment into a graded mesh of elements.

For a geometrical sequence, $\Delta x_2 = \alpha \, \Delta x_1$, where α is the element-size expansion coefficient, and by recursion,

$$\Delta x_k = \alpha^{k-1} \, \Delta x_1, \tag{1.1.13}$$

where $k = 1, 2, \ldots, N$. By definition then,

$$r = \alpha^{N-1}, \tag{1.1.14}$$

and the total length of the interval is

$$L \equiv X_2 - X_1 = (1 + \alpha + \alpha^2 + \ldots + \alpha^{N-1}) \, \Delta x_1$$
$$= \frac{1 - \alpha^N}{1 - \alpha} \, \Delta x_1, \tag{1.1.15}$$

where X_1 and X_2 are the left and right interval end-points. Rearranging (1.1.14), we obtain an expression for the expansion coefficient in terms of the ratio, r, and number of elements, N,

$$\alpha = r^{1/(N-1)}. \tag{1.1.16}$$

Rearranging (1.1.15), we obtain an expression for the size of the first element,

$$\Delta x_1 = L \, \frac{1 - \alpha}{1 - \alpha^N}. \tag{1.1.17}$$

Special provisions are made to accommodate the special cases $N = 1$ and $r = 1$, the latter corresponding to evenly spaced elements.

FSELIB script `elm_line1_dr`, listed in the text, drives the FSELIB function `elm_line1` to perform the element discretization and produce a graph of the element end-points. The top frame in Figure 1.1.2 shows an element distribution generated by this script.

FSELIB functions `elm_line2` and `elm_line3`, listed in the text, discretize an interval into a graded mesh of elements, so that the element distribution is symmetric with respect to the interval mid-point. Function `elm_line2` discretizes the interval into an even number of elements, and function `elm_line3` discretizes the interval into an odd number of elements. In both cases, the element length increases or decreases in a geometrical fashion by a specified factor from the left end-point to the interval mid-point. The ratio of the size of either one of the two elements joining at the interval mid-point to the size of the first element is denoted as *ratio* in function `elm_line2`. The ratio of the size of the element centered at the interval mid-point to the size of the first element is denoted as *ratio* in the function `elm_line3`. The middle and lower frames of Figure 1.1.2 show element distributions generated using the corresponding FSELIB driver codes `elm_line2_dr` and `elm_line3_dr` (not listed in the text).

```
%===================================================
% Code: elm_line1_dr
%
% Driver of the discretization function: elm_line1
%===================================================

%-----------
% input data
%-----------

x1=0.0; x2=1.0;   %   end points
n=10;             %   number of elements
ratio=0.1;        %   element geometrical ratio

%-----------
% discretize
%-----------

xe = elm_line1 (x1,x2,n,ratio);

%-----
% plot
%-----

ye = zeros(n+1,1);
plot (xe, ye,'-o');

%-----
% done
%-----
```

Script elm_line1_dr: Driver for discretizing a line segment into a graded mesh of elements based on the FSELIB function elm_line1.

1.1.3   Galerkin projection

Following the general blueprint of the Galerkin finite element method (GFEM), we multiply both sides of the governing differential equation (1.1.4) by each one of the global interpolation functions ϕ_i, for $i = 1, 2, \ldots, N_E$, and integrate the product over the solution domain to obtain the Galerkin projection,

$$\int_0^L \phi_i(x) \left[k \frac{\mathrm{d}^2 f}{\mathrm{d}x^2} + s(x) \right] \mathrm{d}x = 0. \tag{1.1.18}$$

Because the Dirichlet boundary condition (1.1.6) is imposed at the right end of the solution domain, the last global interpolation function, corresponding to $i = N_E + 1$, has been excluded from the Galerkin projection. In Section 1.1.6, we shall see that the projection of this function provides us with a way of computing the unknown flux at the Dirichlet end.

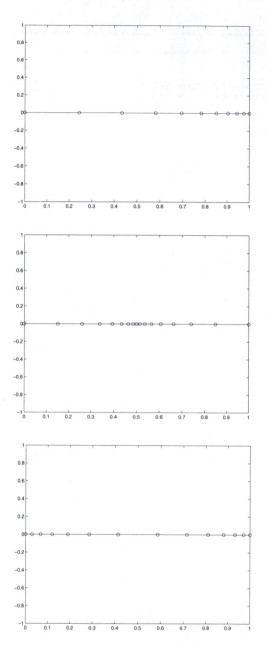

Figure 1.1.2 Graded element distributions generated by the FSELIB functions
elm_line1 (top), elm_line2 (middle), and elm_line3 (bottom).

```
function xe = elm_line2 (x1,x2,n,ratio);

%===========================================================
% Symmetric discretization of a line segment into a
% graded mesh of n elements subtended
% between the left end-point x1
% and the right end-point x2
%
% n is 1 or even: 2,4,6,...
%
% The variable "ratio" is the ratio of the size
% of the mid-elements to the size of the first element
%===========================================================

%------------
% one element
%------------

if(n==1)
   xe(1) = x1;
   xe(2) = x2;
   return;
end

%-------------
% two elements
%-------------

if(n==2)
   xe(1) = x1;
   xe(2) = 0.5*(x1+x2);
   xe(3) = x2;
   return;
end

%--------------
% many elements
%--------------

xe = zeros(n+1,1);
nh = n/2;
xh = 0.5*(x1+x2);    % mid-point

if(ratio==1)
   alpha = 1.0; factor = 1.0/nh;
else
   texp = 1.0/(nh-1.0);
   alpha = ratio^texp;
   factor = (1.0-alpha)/(1.0-alpha^nh);
end

deltax = (xh-x1) * factor;    % length of first element
```

Function elm_line2: Continuing $\longrightarrow$

```
%-----------
% discretize
%-----------

xe(1) = x1;                          % first point

for i=2:nh+1
   xe(i)  = xe(i-1)+deltax;
   deltax = deltax*alpha;
end

deltax = deltax/alpha;

for i=nh+2:n+1;
   xe(i)  = xe(i-1)+deltax;
   deltax = deltax/alpha;
end

%-----
% done
%-----

return;
```

Function elm_line2: ($\longrightarrow$ Continued.) Symmetric discretization of a line segment into a graded mesh with an *even* number of elements.

In the next step, we use the rules of product differentiation to recast the integral on the left-hand side of (1.1.18) into the form

$$\int_0^L \left[k \frac{\mathrm{d}}{\mathrm{d}x} \left(\phi_i \frac{\mathrm{d}f}{\mathrm{d}x} \right) - k \frac{\mathrm{d}\phi_i}{\mathrm{d}x} \frac{\mathrm{d}f}{\mathrm{d}x} + \phi_i \, s \right] \mathrm{d}x = 0, \qquad (1.1.19)$$

and apply the fundamental theorem of calculus to evaluate the integral of the first term on the left-hand side, obtaining

$$-\phi_i(x=0) \, k \left(\frac{\mathrm{d}f}{\mathrm{d}x} \right)_{x=0} + \phi_i(x=L) \, k \left(\frac{\mathrm{d}f}{\mathrm{d}x} \right)_{x=L}$$

$$-k \int_0^L \frac{\mathrm{d}\phi_i}{\mathrm{d}x} \frac{\mathrm{d}f}{\mathrm{d}x} \, \mathrm{d}x + \int_0^L \phi_i \, s \, \mathrm{d}x = 0. \qquad (1.1.20)$$

Finally, we use the cardinal interpolation property (1.1.7) to evaluate $\phi_i(x=0)$ and $\phi_i(x=L)$, and obtain

$$-\delta_{i1} \, k \left(\frac{\mathrm{d}f}{\mathrm{d}x} \right)_{x=0} + \delta_{i,N_E+1} \, k \left(\frac{\mathrm{d}f}{\mathrm{d}x} \right)_{x=L}$$

$$-k \int_0^L \frac{\mathrm{d}\phi_i}{\mathrm{d}x} \frac{\mathrm{d}f}{\mathrm{d}x} \, \mathrm{d}x + \int_0^L \phi_i \, s \, \mathrm{d}x = 0, \qquad (1.1.21)$$

```
function xe = elm_line3 (x1,x2,n,ratio);

%==========================================================
% Symmetric discretization of a line segment into a
% graded mesh of n elements subtended
% between the left end-point x1
% and the right end-point x2
%
% n is odd: 1,3,5,...
%
% The variable "ratio" is the ratio of the size
% of the mid-element to the size of the first element
%==========================================================

%------------
% one element
%------------

if(n==1)
   xe(1) = x1;
   xe(2) = x2;
   return;
end

%--------------
% many elements
%--------------

if(ratio==1)
   alpha = 1.0;
   factor = 1.0/n;
else
   texp = 2.0/(n-1.0);
   alpha = ratio^texp;
   tmp1 = (n+1.0)/2.0;
   tmp2 = (n-1.0)/2.0;
   factor = (1.0-alpha)/(2.0-alpha^tmp1-alpha^tmp2);
end

deltax = (x2-x1) * factor;   % length of first element

%-----------
% discretize
%-----------

xe(1) = x1;     % first point

for i=2:(n+3)/2
   xe(i)  = xe(i-1)+deltax;
   deltax = deltax*alpha;
end
```

Function elm_line3: Continuing $\longrightarrow$

```
deltax = deltax/(alpha^2);

for i=(n+5)/2:n+1
   xe(i) = xe(i-1)+deltax;
   deltax = deltax/alpha;
end

%-----
% done
%-----

return;
```

Function elm_line3: ($\longrightarrow$ Continued.) Symmetric discretization of a line segment into a graded mesh with an *odd* number of elements.

where δ_{ij} is Kronecker's delta. Because we have specified $i = 1, 2, \ldots, N_E$, the second term on the left-hand side of (1.1.21) makes a zero contribution. Implementing the boundary condition (1.1.5) in the first term on the left-hand side, we derive the Galerkin finite element equations

$$\delta_{i1} q_0 - k \int_0^L \frac{d\phi_i}{dx} \frac{df}{dx} \, dx + \int_0^L \phi_i \, s \, dx = 0, \tag{1.1.22}$$

which can be rearranged into the preferred form

$$\int_0^L \frac{d\phi_i}{dx} \frac{df}{dx} \, dx = \frac{q_0}{k} \delta_{i1} + \frac{1}{k} \int_0^L \phi_i \, s \, dx, \tag{1.1.23}$$

for $i = 1, 2, \ldots, N_E$.

1.1.4   Formulation of a linear algebraic system

In the next step, we substitute the finite element expansion (1.1.9) in the left-hand side of (1.1.23). Moreover, assuming that the source term is a continuous function, we introduce the analogous expansion

$$s(x) \simeq \sum_{j=1}^{N_E+1} s_j \, \phi_j(x), \tag{1.1.24}$$

where $s_j \equiv s(x = x_j)$. Rearranging the resulting equation, we derive a system of N_E linear equations for the N_E unknown values f_j,

$$\sum_{j=1}^{N_E} \left(\int_0^L \frac{\mathrm{d}\phi_i}{\mathrm{d}x} \frac{\mathrm{d}\phi_j}{\mathrm{d}x} \, \mathrm{d}x \right) f_j = - \left(\int_0^L \frac{\mathrm{d}\phi_i}{\mathrm{d}x} \frac{\mathrm{d}\phi_{N+1}}{\mathrm{d}x} \, \mathrm{d}x \right) f_L$$

$$\text{(1.1.25)}$$

$$+ \frac{q_0}{k} \delta_{i1} + \frac{1}{k} \sum_{j=1}^{N_E+1} \left(\int_0^L \phi_i \, \phi_j \, \mathrm{d}x \right) s_j,$$

where $i = 1, 2, \ldots, N_E$. Note that, because the Dirichlet boundary condition is specified at the right end, $x = L$, the terms multiplying the corresponding value, f_L, have been transferred to the right-hand side.

Detailed consideration of the global interpolation functions depicted in Figure 1.1.1(*a*) followed by analytical computation of the integrals on the left- and right-hand sides of (1.1.25), yields the integration formulas

$$D_{ij} \equiv \int_0^L \frac{\mathrm{d}\phi_i}{\mathrm{d}x} \frac{\mathrm{d}\phi_j}{\mathrm{d}x} \, \mathrm{d}x = \begin{cases} \frac{1}{h_1} & \text{if } i = j = 1 \\[2mm] \frac{1}{h_{i-1}} + \frac{1}{h_i} & \text{if } i = j \neq 1 \text{ and } N_E + 1 \\[2mm] \frac{1}{h_N} & \text{if } i = j = N_E + 1 \\[2mm] -\frac{1}{h_i} & \text{if } j = i + 1 \\[2mm] -\frac{1}{h_{i-1}} & \text{if } j = j - 1 \\[2mm] 0 & \text{otherwise} \end{cases} , \quad \text{(1.1.26)}$$

and

$$M_{ij} \equiv \int_0^L \phi_i \, \phi_j \, \mathrm{d}x = \begin{cases} \frac{h_1}{3} & \text{if } i = j = 1 \\[2mm] \frac{h_{i-1}}{3} + \frac{h_i}{3} & \text{if } i = j \neq 1 \text{ and } N_E + 1 \\[2mm] \frac{h_N}{3} & \text{if } i = j = N_E + 1 \\[2mm] \frac{h_i}{6} & \text{if } j = i + 1 \\[2mm] \frac{h_{i-1}}{6} & \text{if } j = j - 1 \\[2mm] 0 & \text{otherwise} \end{cases} , \quad \text{(1.1.27)}$$

where

$$h_i \equiv x_{i+1} - x_i \qquad \text{(1.1.28)}$$

is the size of the *i*th element. Substituting these values in (1.1.25), we derive the linear system

$$\mathbf{D} \cdot \mathbf{f} = \mathbf{b}, \qquad \text{(1.1.29)}$$

where:

- **f** is the vector of the unknown temperatures at the nodes,

$$\mathbf{f} \equiv \begin{bmatrix} f_1 \\ f_2 \\ \vdots \\ f_{N_E-1} \\ f_{N_E} \end{bmatrix}. \qquad (1.1.30)$$

- **D** is the $N_E \times N_E$ global *diffusion* or *conductivity* matrix, sometimes also called the *Laplacian* matrix, given by

$$\mathbf{D} \equiv \begin{bmatrix} \frac{1}{h_1} & -\frac{1}{h_1} & 0 & 0 & \cdots \\ -\frac{1}{h_1} & \frac{1}{h_1} + \frac{1}{h_2} & -\frac{1}{h_2} & 0 & \cdots \\ \cdots & \cdots & \cdots & \cdots & \cdots \\ 0 & 0 & \cdots & \cdots & \cdots \\ 0 & 0 & \cdots & \cdots & \cdots \end{bmatrix} \qquad (1.1.31)$$

$$\begin{bmatrix} \cdots & 0 & 0 & 0 \\ \cdots & 0 & 0 & 0 \\ \cdots & \cdots & \cdots & \cdots \\ 0 & -\frac{1}{h_{N_E-2}} & \frac{1}{h_{N_E-2}} + \frac{1}{h_{N_E-1}} & -\frac{1}{h_{N_E-1}} \\ 0 & 0 & -\frac{1}{h_{N_E-1}} & \frac{1}{h_{N_E-1}} + \frac{1}{h_{N_E}} \end{bmatrix}.$$

In the remainder of this text, we shall refer to this matrix as the global diffusion matrix.

- The right-hand side of the linear system is given by

$$\mathbf{b} \equiv \mathbf{c} + \frac{1}{k}\,\widetilde{\mathbf{M}} \cdot \mathbf{s}, \qquad (1.1.32)$$

where the nearly-null N_E-dimensional vector **c** encapsulates the boundary conditions,

$$\mathbf{c} \equiv \begin{bmatrix} q_0/k \\ 0 \\ \vdots \\ 0 \\ f_L/h_{N_E} \end{bmatrix}, \qquad (1.1.33)$$

$\widetilde{\mathbf{M}}$ is the $N_E \times (N_E + 1)$ *bordered mass matrix* given by

$$\widetilde{\mathbf{M}} \equiv \begin{bmatrix} \frac{h_1}{3} & \frac{h_1}{6} & 0 & 0 & \dots \\ \frac{h_1}{6} & \frac{h_1}{3} + \frac{h_2}{3} & \frac{h_2}{6} & 0 & \dots \\ \dots & \dots & \dots & \dots & \dots \\ 0 & 0 & \dots & \dots & \dots \\ 0 & 0 & \dots & \dots & \dots \end{bmatrix}$$ (1.1.34)

$$\begin{bmatrix} \dots & 0 & 0 & 0 & 0 \\ \dots & 0 & 0 & 0 & 0 \\ \dots & \dots & \dots & \dots & 0 \\ 0 & \frac{h_{N_E-2}}{6} & \frac{h_{N_E-2}}{3} + \frac{h_{N_E-1}}{3} & \frac{h_{N_E-1}}{6} & 0 \\ 0 & 0 & \frac{h_{N_E-1}}{6} & \frac{h_{N_E-1}}{3} + \frac{h_{N_E}}{3} & \frac{h_{N_E}}{6} \end{bmatrix},$$

and the $(N_E + 1)$-dimensional vector, $\mathbf{s}$, contains the nodal values of the source,

$$\mathbf{s} \equiv \begin{bmatrix} s_1 \\ s_2 \\ \vdots \\ s_{N_E} \\ s_{N_E+1} \end{bmatrix}.$$ (1.1.35)

Note that the global diffusion matrix has units of inverse length, whereas the global mass matrix has units of length.

Explicitly, the right-hand side of the linear system (1.1.29) is given by

$$\mathbf{b} = \mathbf{c} + \frac{1}{k} \begin{bmatrix} \frac{h_1}{3} s_1 + \frac{h_1}{6} s_2 \\ \frac{h_1}{6} s_1 + \left(\frac{h_1}{3} + \frac{h_2}{3}\right) s_2 + \frac{h_2}{6} s_3 \\ \vdots \\ \frac{h_{N_E-2}}{6} s_{N_E-2} + \left(\frac{h_{N_E-2}}{3} + \frac{h_{N_E-1}}{3}\right) s_{N_E-1} + \frac{h_{N_E-1}}{6} s_{N_E} \\ \frac{h_{N_E-1}}{6} s_{N_E-1} + \left(\frac{h_{N_E-1}}{3} + \frac{h_{N_E}}{3}\right) s_{N_E} + \frac{h_N}{6} s_{N_E+1} \end{bmatrix}.$$ (1.1.36)

Because the coefficient matrix of the linear system (1.1.29) is tridiagonal, the solution can be found efficiently using the Thomas algorithm discussed in Section 1.1.10. "Tridiagonal" derives from the Greek words $\tau\rho\iota\alpha$, which means "three," and $\delta\iota\alpha\gamma\omega\nu\iota\sigma$, which means "diagonal."

It is interesting to observe that the first equation of the linear system (1.1.29) takes the form

$$-\frac{f_2 - f_1}{h_1} = \frac{q_0}{k} + \frac{h_1}{6} \left(2\, s_1 + s_2\right). \tag{1.1.37}$$

The fraction on the left-hand side of (1.1.37) expresses the forward difference approximation of the slope $\mathrm{d}f/\mathrm{d}x = -q_0/k$ at the left end, $x = 0$, while the second term on the right-hand side contributes a correction based on the left-end and adjacent nodal values of the source.

1.1.5   Relation to the finite difference method

When the elements are evenly spaced, $h_1 = h_2 = \cdots = h_{N_E} \equiv h$, the global diffusion matrix takes the simple form

$$\mathbf{D} = \frac{1}{h}
\begin{bmatrix}
1 & -1 & 0 & 0 & \dots & 0 & 0 & 0 \\
-1 & 2 & -1 & 0 & \dots & 0 & 0 & 0 \\
0 & -1 & 2 & -1 & 0 & \dots & 0 & 0 \\
\dots & \dots & \dots & \dots & \dots & \dots & \dots & \dots \\
0 & \dots & \dots & 0 & -1 & 2 & -1 & 0 \\
0 & 0 & \dots & \dots & 0 & -1 & 2 & -1 \\
0 & 0 & \dots & \dots & 0 & 0 & -1 & 2
\end{bmatrix}, \tag{1.1.38}$$

and the bordered mass matrix takes the form

$$\widetilde{\mathbf{M}} \equiv h
\begin{bmatrix}
\frac{1}{3} & \frac{1}{6} & 0 & 0 & \dots & 0 & 0 & 0 & 0 \\
\frac{1}{6} & \frac{2}{3} & \frac{1}{6} & 0 & \dots & 0 & 0 & 0 & 0 \\
0 & \frac{1}{6} & \frac{2}{3} & \frac{1}{6} & 0 & \dots & 0 & 0 & 0 \\
\dots & \dots & \dots & \dots & \dots & \dots & \dots & \dots & 0 \\
0 & 0 & \dots & 0 & \frac{1}{6} & \frac{2}{3} & \frac{1}{6} & 0 & 0 \\
0 & 0 & \dots & \dots & 0 & \frac{1}{6} & \frac{2}{3} & \frac{1}{6} & 0 \\
0 & 0 & \dots & \dots & 0 & 0 & \frac{1}{6} & \frac{2}{3} & \frac{1}{6}
\end{bmatrix}. \tag{1.1.39}$$

Apart from the first and last rows, the matrix on the right-hand side of (1.1.38) is recognized as the negative of the central differentiation matrix for the second derivative associated with the finite difference formula

$$\frac{\mathrm{d}^2 f}{\mathrm{d}x^2} \simeq \frac{f_{i-1} - 2 f_i + f_{i+1}}{h^2}. \tag{1.1.40}$$

Thus, the finite element method with linear and uniform elements is equivalent to the finite difference method implemented with the second-order central difference approximation for the interior nodes. However, the extended mass matrix (1.1.39) enforces a neighbor averaging of the source term on either side of the ith node, thereby smearing out the local contribution.

1.1.6 Flux at the Dirichlet end

To compute the flux at the right end of the solution domain where the Dirichlet boundary condition (1.1.6) is imposed,

$$q_L \equiv -k \left(\frac{df}{dx} \right)_{x=L}, \tag{1.1.41}$$

we apply the Galerkin projection expressed by equation (1.1.21) for $i = N_E + 1$, and obtain

$$-q_L - k \int_0^L \frac{d\phi_{N_E+1}}{dx} \frac{df}{dx} \, dx + \int_0^L \phi_{N_E+1} \, s \, dx = 0. \tag{1.1.42}$$

Substituting expansions (1.1.9) and (1.1.24) and rearranging, we find

$$q_L = -k \left(\int_{x_{N_E}}^{x_{N_E+1}} \frac{d\phi_{N_E+1}}{dx} \frac{d\phi_{N_E}}{dx} \, dx \right) f_{N_E}$$

$$- \left(\int_{x_{N_E}}^{x_{N_E+1}} \frac{d\phi_{N_E+1}}{dx} \frac{d\phi_{N_E+1}}{dx} \, dx \right) f_{N_E+1} \tag{1.1.43}$$

$$+ \left(\int_{x_{N_E}}^{x_{N_E+1}} \phi_{N_E+1} \, \phi_{N_E} \, dx \right) s_{N_E} + \left(\int_{x_{N_E}}^{x_{N_E+1}} \phi_{N_E+1} \, \phi_{N_E+1} \, dx \right) s_{N_E+1}.$$

Finally, we perform the integrations using expressions (1.1.26) and (1.1.27), and find

$$q_L = -k \frac{f_{N_E+1} - f_{N_E}}{h_{N_E}} + \frac{h_{N_E}}{6} (s_{N_E} + 2 \, s_{N_E+1}), \tag{1.1.44}$$

which is the counterpart of the corresponding expression for the flux at the Neumann end displayed in (1.1.37). Formula (1.1.44) allows us to compute the flux at the Dirichlet end, after the solution of the linear system (1.1.36) has been completed.

It is instructive to consider the predictions of (1.1.44) for a special case where the source function, s, is constant, the flux at the Neumann end is zero, $q_0 = 0$, and the boundary value at the Dirichlet end is $f_L = -sL^2/(2k)$. The exact solution is readily found to be the quadratic function, $f(x) = -s\, x^2/(2\, k)$. Denoting, for simplicity $h_{N_E} = h$, and substituting in (1.1.44) $f_{N_E+1} = f_L = -s\, L^2/(2\, k)$ and $f_{N_E} = -s\, (L-h)^2/(2\, k)$, we find

$$q_L = s \frac{L^2 - (L-h)^2}{2\, h} + \frac{s\, h}{2} = s\, L, \tag{1.1.45}$$

which is precisely the exact solution.

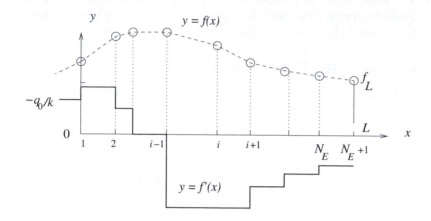

Figure 1.1.3 Graphs of the piecewise linear approximation (dashed line) and its first derivative (staircase bold line).

1.1.7 Galerkin finite element equations by way of Dirac's delta function

It is illuminating to rederive the Galerkin finite element equations compiled in the linear system (1.1.29) directly from the primary Galerkin projection (1.1.18). For this purpose, we consider the graphs of the piecewise linear global finite element expansion and its piecewise constant first derivative, drawn with the broken and bold solid lines, respectively, in Figure 1.1.3. Both graphs have been extended so that $f(x)$ is a linear function with slope $df/dx = -q_0/k$ beyond the left end of the solution domain, $x < 0$.

Dirac delta function

As a preliminary, we introduce the Dirac delta function in one dimension, denoted by $\delta(x - x_0)$, physically representing a unit impulse applied at a point, x_0. The induced field is distinguished by the following properties:

1. $\delta(x - x_0)$ vanishes everywhere, except at the singular point $x = x_0$ where it becomes infinite.

2. The integral of $\delta(x - x_0)$ with respect to x over an interval I that contains the point x_0 is equal to unity,

$$\int_I \delta(x - x_0) \, dx = 1. \tag{1.1.46}$$

This property reveals that the delta function with argument of length has units of inverse length.

3. The integral of the product of an arbitrary function $q(x)$ and the delta function over an interval I that contains the point x_0 is equal to value of the function at the singular point,

$$\int_D \delta(x - x_0)\, q(x)\, \mathrm{d}x = q(x_0). \qquad (1.1.47)$$

The integral of the product of an arbitrary function $q(x)$ and the delta function over an interval I that does *not* contain the point x_0 is equal to zero.

4. If a function $q(x)$ undergoes a discontinuity from the left value q_- to the right value q_+ at the point x_0, then the derivative at that point can be expressed in terms of the delta function as

$$q'(x_0) = (q_+ - q_-)\, \delta(x - x_0) + ..., \qquad (1.1.48)$$

where the dots denote nonsingular terms.

5. In formal mathematics, $\delta(x - x_0)$ arises from the family of test functions

$$g(x - x_0) = \left(\frac{\lambda}{\pi L^2}\right)^{1/2} \exp\left[-\lambda \frac{(x - x_0)^2}{L^2}\right], \qquad (1.1.49)$$

in the limit as the dimensionless parameter λ tends to infinity, where L is an arbitrary length.

Figure 1.1.3 shows graphs of the dimensionless functions $G \equiv g\,L$ plotted against the dimensionless distance $X \equiv (x - x_0)/L$ for $\lambda = 1$ (dashed line), 2, 5, 10, 20, 50, 100, and 500 (dotted line). The maximum height of each graph is inversely proportional to its width, so that the area underneath each graph is equal to unity,

$$\int_{-\infty}^{\infty} g(x)\, \mathrm{d}x = \left(\frac{\lambda}{\pi L^2}\right)^{1/2} \int_{-\infty}^{\infty} \exp(-\lambda \frac{x^2}{L^2})\, \mathrm{d}x = 1. \qquad (1.1.50)$$

To prove this identity, we recall the definite integral associated with the error function,

$$\frac{2}{\sqrt{\pi}} \int_0^{\infty} \exp(-w^2)\, \mathrm{d}w = 1, \qquad (1.1.51)$$

and substitute $w = \lambda^{1/2} x/L$.

Galerkin projection

Using the fourth of the aforementioned properties, we find that, in terms of the delta function, the second derivative of the global finite element expansion admits the generalized representation

$$\frac{\mathrm{d}^2 f}{\mathrm{d}x^2} = \sum_{j=1}^{N_E} (m_j - m_{j-1})\, \delta(x - x_j), \qquad (1.1.52)$$

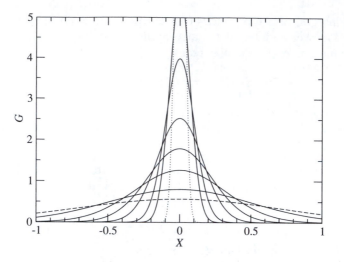

Figure 1.1.4 A family of dimensionless test functions $G \equiv gL$ described by (1.1.49), plotted against the dimensionless distance $X \equiv (x - x_0)/L$ for $\lambda = 1$ (dashed line), 2, 5, 10, 20, 50, 100, and 500 (dotted line). In the limit as the dimensionless parameter λ tends to infinity, we recover Dirac's delta function in one dimension.

for $0 \leq x < L$, where

$$m_j = \frac{f_{j+1} - f_j}{h_j} \tag{1.1.53}$$

is the slope of the approximate solution over the jth element, with the understanding that $m_0 = -q_0/k$, as required by the Neumann boundary condition.

Substituting (1.1.52) in the ith Galerkin projection (1.1.18), we obtain

$$\sum_{j=1}^{N_E} (m_j - m_{j-1}) \int_0^L \phi_i(x)\, \delta(x - x_j)\, \mathrm{d}x + \int_0^L \phi_i(x)\, s(x)\, \mathrm{d}x = 0. \tag{1.1.54}$$

Using the distinguishing properties of the Dirac delta function to evaluate the first integral, we find

$$\sum_{j=1}^{N_E} \phi_i(x_j)\, (m_j - m_{j-1}) + \int_0^L \phi_i(x)\, s(x)\, \mathrm{d}x = 0, \tag{1.1.55}$$

or

$$\sum_{j=1}^{N_E} \delta_{ij}\, (m_j - m_{j-1}) + \int_0^L \phi_i(x)\, s(x)\mathrm{d}x = 0, \tag{1.1.56}$$

yielding

$$m_i - m_{i-1} + \int_0^L \phi_i(x)\, s(x)\, \mathrm{d}x = 0. \tag{1.1.57}$$

Substituting the definition of the slopes m_i from (1.1.53), we recover precisely the linear system (1.1.29).

The analysis in this section demonstrates that integration by parts is not a mandatory step following the Galerkin projection and should be regarded as a venue for bypassing the potentially cumbersome device of Dirac's delta function. Some authors erroneously state that the critical step leading from the strong to the weak formulation is the integration by parts; in fact, the defining step is the Galerkin projection.

1.1.8 Element diffusion and mass matrices

In generating the coefficient matrix and right-hand side of the linear system (1.1.29), we have used the *global* integration formulas (1.1.26) and (1.1.27). In the practical implementation of the finite element method, the integrals arising from the Galerkin projection are assembled from corresponding integrals over the elements in a systematic way that facilitates the bookkeeping.

To demonstrate the process, we consider the Galerkin projection expressed by equation (1.1.23), introduce expansions (1.1.9) and (1.1.24), and replace the integrals over the solution domain $[0, L]$ with sums of integrals over the individual elements. The result is

$$k \sum_{j=1}^{N_E+1} \left(\sum_{l=1}^{N_E} \int_{E_l} \frac{\mathrm{d}\phi_i}{\mathrm{d}x} \frac{\mathrm{d}\phi_j}{\mathrm{d}x}\, \mathrm{d}x \right) f_j \tag{1.1.58}$$

$$= q_0\, \delta_{j1} + \sum_{j=1}^{N_E+1} \left(\sum_{l=1}^{N_E} \int_{E_l} \phi_i\, \phi_j\, \mathrm{d}x \right) s_j,$$

for $i = 1, 2, \ldots, N_E$, where E_l stands for the lth element subtended between the x_l and x_{l+1} nodes, as depicted in Figure 1.1.1(*b*).

A key observation is that, because the support of the global interpolation functions is compact, the diffusion integral on the left-hand side of (1.1.58) may be recast into the form

$$\int_{E_l} \frac{\mathrm{d}\phi_i}{\mathrm{d}x} \frac{\mathrm{d}\phi_j}{\mathrm{d}x}\, \mathrm{d}x = \delta_{il} \int_{E_l} \frac{\mathrm{d}\phi_l}{\mathrm{d}x} \frac{\mathrm{d}\phi_j}{\mathrm{d}x}\, \mathrm{d}x + \delta_{i,l+1} \int_{E_l} \frac{\mathrm{d}\phi_{l+1}}{\mathrm{d}x} \frac{\mathrm{d}\phi_j}{\mathrm{d}x}\, \mathrm{d}x$$

$$= \delta_{il} \left(\delta_{jl} \int_{E_l} \frac{\mathrm{d}\phi_l}{\mathrm{d}x} \frac{\mathrm{d}\phi_l}{\mathrm{d}x}\, \mathrm{d}x + \delta_{j,l+1} \int_{E_l} \frac{\mathrm{d}\phi_l}{\mathrm{d}x} \frac{\mathrm{d}\phi_{l+1}}{\mathrm{d}x}\, \mathrm{d}x \right) \tag{1.1.59}$$

$$+ \delta_{i,l+1} \left(\delta_{jl} \int_{E_l} \frac{\mathrm{d}\phi_{l+1}}{\mathrm{d}x} \frac{\mathrm{d}\phi_l}{\mathrm{d}x}\, \mathrm{d}x + \delta_{j,l+1} \int_{E_l} \frac{\mathrm{d}\phi_{l+1}}{\mathrm{d}x} \frac{\mathrm{d}\phi_{l+1}}{\mathrm{d}x}\, \mathrm{d}x \right),$$

where δ_{im} is Kronecker's delta. In compact notation,

$$\int_{E_l} \frac{\mathrm{d}\phi_i}{\mathrm{d}x} \frac{\mathrm{d}\phi_j}{\mathrm{d}x} \, \mathrm{d}x = \delta_{il}\delta_{jl} A_{11}^{(l)} + \delta_{il}\delta_{j,l+1} A_{12}^{(l)}$$

$$+ \delta_{i,l+1}\delta_{jl} A_{21}^{(l)} + \delta_{i,l+1}\delta_{j,l+1} A_{22}^{(l)}, \qquad (1.1.60)$$

where $A_{ij}^{(l)}$ is the lth-element diffusion matrix with components

$$A_{11}^{(l)} \equiv \int_{E_l} \frac{\mathrm{d}\phi_l}{\mathrm{d}x} \frac{\mathrm{d}\phi_l}{\mathrm{d}x} \, \mathrm{d}x = \int_{E_l} \frac{\mathrm{d}\psi_1^{(l)}}{\mathrm{d}x} \frac{\mathrm{d}\psi_1^{(l)}}{\mathrm{d}x} \, \mathrm{d}x,$$

$$A_{12}^{(l)} = A_{21}^{(l)} \equiv \int_{E_l} \frac{\mathrm{d}\phi_l}{\mathrm{d}x} \frac{\mathrm{d}\phi_{l+1}}{\mathrm{d}x} \, \mathrm{d}x = \int_{E_l} \frac{\mathrm{d}\psi_1^{(l)}}{\mathrm{d}x} \frac{\mathrm{d}\psi_2^{(l)}}{\mathrm{d}x} \, \mathrm{d}x, \qquad (1.1.61)$$

$$A_{22}^{(l)} \equiv \int_{E_l} \frac{\mathrm{d}\phi_{l+1}}{\mathrm{d}x} \frac{\mathrm{d}\phi_{l+1}}{\mathrm{d}x} \, \mathrm{d}x = \int_{E_l} \frac{\mathrm{d}\psi_2^{(l)}}{\mathrm{d}x} \frac{\mathrm{d}\psi_2^{(l)}}{\mathrm{d}x} \, \mathrm{d}x,$$

and $\psi_1^{(l)}$, $\psi_2^{(l)}$ are the first and second element-node interpolation functions, as shown in Figure 1.1.1(b). Carrying out the integrations, we find

$$\mathbf{A}^{(l)} = \begin{bmatrix} A_{11}^{(l)} & A_{12}^{(l)} \\ A_{21}^{(l)} & A_{22}^{(l)} \end{bmatrix} = \frac{1}{h_l} \begin{bmatrix} 1 & -1 \\ -1 & 1 \end{bmatrix}, \qquad (1.1.62)$$

where $h_l \equiv x_{l+1} - x_l$ is the size of the lth element.

Working in a similar manner with the integral on the right-hand side of (1.1.58), we find

$$\int_{E_l} \phi_i \, \phi_j \, \mathrm{d}x = \delta_{il} \, \delta_{jl} B_{11}^{(l)} + \delta_{il} \, \delta_{j,l+1} B_{12}^{(l)}$$

$$+ \delta_{i,l+1} \, \delta_{jl} B_{21}^{(l)} + \delta_{i,l+1} \, \delta_{j,l+1} B_{22}^{(l)}, \qquad (1.1.63)$$

where $B_{ij}^{(l)}$ is the lth-element mass matrix with components

$$B_{11}^{(l)} \equiv \int_{E_l} \phi_l \, \phi_l \, \mathrm{d}x = \int_{E_l} \psi_1^{(l)} \psi_1^{(l)} \mathrm{d}x,$$

$$B_{12}^{(l)} = B_{21}^{(l)} \equiv \int_{E_l} \phi_l \, \phi_{l+1} \, \mathrm{d}x = \int_{E_l} \psi_1^{(l)} \psi_2^{(l)} \mathrm{d}x, \qquad (1.1.64)$$

$$B_{22}^{(l)} \equiv \int_{E_l} \phi_{l+1} \, \phi_{l+1} \, \mathrm{d}x = \int_{E_l} \psi_2^{(l)} \psi_2^{(l)} \mathrm{d}x.$$

Carrying out the integrations, we find

$$\mathbf{B}^{(l)} = \begin{bmatrix} B_{11}^{(l)} & B_{12}^{(l)} \\ B_{21}^{(l)} & B_{22}^{(l)} \end{bmatrix} = \frac{h_l}{6} \begin{bmatrix} 2 & 1 \\ 1 & 2 \end{bmatrix}. \tag{1.1.65}$$

Substituting (1.1.60) and (1.1.63) in (1.1.58), interchanging the summation order with respect to j and l, and using the distinguishing properties of Kronecker's delta to simplify the resulting expression, we find

$$\sum_{l=1}^{N_E} \left[\delta_{il} \left(A_{11}^{(l)} f_l + A_{12}^{(l)} f_{l+1} \right) + \delta_{i,l+1} \left(A_{21}^{(l)} f_l + A_{22}^{(l)} f_{l+1} \right) \right]$$

$$= \frac{q_0}{k} \delta_{i1} + \frac{1}{k} \sum_{l=1}^{N_E} \left[\delta_{il} \left(B_{11}^{(l)} s_l + B_{12}^{(l)} s_{l+1} \right) \right.$$

$$\left. + \delta_{i,l+1} \left(B_{21}^{(l)} s_l + B_{22}^{(l)} s_{l+1} \right) \right]. \tag{1.1.66}$$

Cursory inspection of (1.1.66) reveals an important property: the lth element contributes only to the Galerkin-projection equations for $i = l$ and $l+1$. Thus, the first element corresponding to $l = 1$ contributes to the first and second equations associated with the projections of ϕ_1 and ϕ_2, and the second element corresponding to $l = 2$ contributes to the second and third equations associated with the projections of ϕ_2 and ϕ_3. Similar contributions are made by subsequent elements. Accordingly, the linear system can be assembled by scanning the elements one by one, while making appropriate additive contributions to the coefficient matrices on the left- and right-hand sides.

The linear system that arises by applying the finite element Galerkin equation (1.1.66) for $i = 1, 2, \ldots, N_E$, is, in fact, identical to (1.1.29). The only new feature is that the global diffusion matrix and bordered mass matrix are expressed in terms of the individual element matrices. Specifically, the global diffusion matrix takes the form

$$\mathbf{D} = \begin{bmatrix} A_{11}^{(1)} & A_{12}^{(1)} & 0 & 0 & \dots \\ A_{21}^{(1)} & A_{22}^{(1)} + A_{11}^{(2)} & A_{12}^{(2)} & 0 & \dots \\ 0 & A_{21}^{(2)} & A_{22}^{(2)} + A_{11}^{(3)} & A_{12}^{(3)} & \dots \\ \dots & \dots & \dots & \dots & \dots \\ 0 & 0 & 0 & \dots & \dots \\ 0 & 0 & 0 & \dots & \dots \\ 0 & 0 & 0 & \dots & \dots \end{bmatrix} \longrightarrow$$

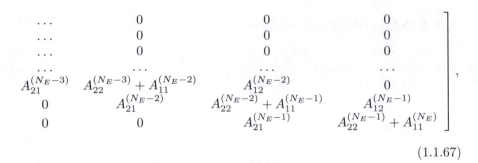

$$
\begin{bmatrix}
\cdots & 0 & 0 & 0 \\
\cdots & 0 & 0 & 0 \\
\cdots & 0 & 0 & 0 \\
\cdots & \cdots & \cdots & \cdots \\
A_{21}^{(N_E-3)} & A_{22}^{(N_E-3)} + A_{11}^{(N_E-2)} & A_{12}^{(N_E-2)} & 0 \\
0 & A_{21}^{(N_E-2)} & A_{22}^{(N_E-2)} + A_{11}^{(N_E-1)} & A_{12}^{(N_E-1)} \\
0 & 0 & A_{21}^{(N_E-1)} & A_{22}^{(N_E-1)} + A_{11}^{(N_E)}
\end{bmatrix},
\tag{1.1.67}
$$

and the bordered mass matrix takes the form

$$
\widetilde{\mathbf{M}} =
\begin{bmatrix}
B_{11}^{(1)} & B_{12}^{(1)} & 0 & 0 & \cdots \\
B_{21}^{(1)} & B_{22}^{(1)} + B_{11}^{(2)} & B_{12}^{(2)} & 0 & \cdots \\
0 & B_{21}^{(2)} & B_{22}^{(2)} + B_{11}^{(3)} & B_{12}^{(3)} & \cdots \\
\cdots & \cdots & \cdots & \cdots & \cdots \\
0 & 0 & 0 & \cdots & \cdots \\
0 & 0 & 0 & \cdots & \cdots
\end{bmatrix}
\tag{1.1.68}
$$

$$
\begin{bmatrix}
\cdots & 0 & 0 & 0 \\
\cdots & \cdots & \cdots & \cdots \\
\cdots & 0 & 0 & 0 \\
\cdots & B_{12}^{(N_E-2)} & 0 & 0 \\
\cdots & B_{22}^{(N_E-2)} + B_{11}^{(N_E-1)} & B_{12}^{(N_E-1)} & 0 \\
0 & B_{21}^{(N_E-1)} & B_{22}^{(N_E-1)} + B_{11}^{(N_E)} & B_{12}^{(N_E)}
\end{bmatrix}.
$$

The vector $\mathbf{c}$ defined in (1.1.33) is given by

$$
\mathbf{c} =
\begin{bmatrix}
q_0/k \\
0 \\
\vdots \\
0 \\
-A_{12}^{(N_E)} f_L
\end{bmatrix}.
\tag{1.1.69}
$$

Singular nature of the element diffusion matrix

It is instructive to observe that the determinant of the element diffusion matrix shown in (1.1.62) is equal to zero, which indicates that the matrix is singular. The reason is that, in the one-element implementation with the Neumann boundary condition at both ends and in the absence of a source term, the solution can be found only up to an arbitrary constant if the boundary fluxes are equal in magnitude and opposite in sign, and does not exist otherwise. Physically, steady state can be established only when the inward flux balances the outward flux to prevent accumulation. The absence of a unique solution implies a singular coefficient matrix.

1.1.9 Algorithm for assembling the linear system

To develop an algorithm for assembling the linear system (1.1.29), we make a correspondence between *element nodes* and *unique global nodes*. This is done by introducing the *connectivity matrix* $c_{l,j}$, where $l = 1, 2, \ldots, N_E$, and $j = 1, 2$, defined such that:

- $c_{l,1} = l$ is the global label of the first node of the lth element.

- $c_{l,2} = l + 1$ is the global label of the second node of the lth element.

In terms of the connectivity matrix, the global diffusion matrix, $\mathbf{D}$, and right-hand side, $\mathbf{b}$, of (1.1.29) can be assembled according to Algorithm 1.1.1.

1.1.10 Thomas algorithm for tridiagonal systems

The tridiagonal nature of the global diffusion matrix, $\mathbf{D}$, displayed in (1.1.31) and (1.1.67), allows us to solve the linear system efficiently using the legendary Thomas algorithm. To formalize the algorithm in general terms, we consider the $N \times N$ linear system

$$\mathbf{T} \cdot \mathbf{x} = \mathbf{r} \tag{1.1.70}$$

for the unknown vector $\mathbf{x}$, where the vector $\mathbf{r}$ is specified. The matrix $\mathbf{T}$ has the tridiagonal form

$$\mathbf{T} = \begin{bmatrix} a_1 & b_1 & 0 & 0 & 0 & \ldots & 0 & 0 & 0 \\ c_2 & a_2 & b_2 & 0 & 0 & \ldots & 0 & 0 & 0 \\ 0 & c_3 & a_3 & b_3 & 0 & \ldots & 0 & 0 & 0 \\ \ldots & \ldots & \ldots & \ldots & \ldots & \ldots & \ldots & \ldots & \ldots \\ 0 & 0 & 0 & 0 & 0 & \ldots & c_{N-1} & a_{N-1} & b_{N-1} \\ 0 & 0 & 0 & 0 & 0 & \ldots & 0 & c_N & a_N \end{bmatrix}, \tag{1.1.71}$$

where a_i, b_i, and c_i are specified matrix components.

Thomas's algorithm proceeds in two stages. In the first stage, the tridiagonal system (1.1.70) is reduced to the bidiagonal system

$$\mathbf{B} \cdot \mathbf{x} = \mathbf{y}, \tag{1.1.72}$$

involving the bidiagonal coefficient matrix

$$\mathbf{B} = \begin{bmatrix} 1 & d_1 & 0 & 0 & 0 & \ldots & 0 & 0 & 0 \\ 0 & 1 & d_2 & 0 & 0 & \ldots & 0 & 0 & 0 \\ 0 & 0 & 1 & d_3 & 0 & \ldots & 0 & 0 & 0 \\ \ldots & \ldots & \ldots & \ldots & \ldots & \ldots & \ldots & \ldots & \ldots \\ 0 & 0 & 0 & 0 & 0 & \ldots & 0 & 1 & d_{N-1} \\ 0 & 0 & 0 & 0 & 0 & \ldots & 0 & 0 & 1 \end{bmatrix}. \tag{1.1.73}$$

In the second stage, the bidiagonal system (1.1.72) is solved by backward substitution, which involves solving the last equation for the last unknown, x_N,

Do $l = 1, 2, \ldots, N_E$ *Run over the elements to define the connectivity matrix*
     $c(l, 1) = l$
     $c(l, 2) = l + 1$
End Do

Do $i = 1, 2, \ldots, N_E$ *Initialize to zero*
     $b_i = 0.0$
     Do $j = 1, 2, \ldots, N_E$
          $D_{ij} = 0.0$
     End Do
End Do

$b_1 = q_0/k$ *Neumann boundary condition*

Do $l = 1, 2, \ldots, N_E - 1$ *Run over the first N_E-1 elements*
     Compute the element matrices $\mathbf{A}^{(l)}$ and $\mathbf{B}^{(l)}$
          using (1.1.62) and (1.1.65)
     $i_1 = c(l, 1)$
     $i_2 = c(l, 2)$
     $D_{i_1,i_1} = D_{i_1,i_1} + A_{11}^{(l)}$
     $D_{i_1,i_2} = D_{i_1,i_2} + A_{12}^{(l)}$
     $D_{i_2,i_1} = D_{i_2,i_1} + A_{21}^{(l)}$
     $D_{i_2,i_2} = D_{i_2,i_2} + A_{22}^{(l)}$
     $b_{i_1} = b_{i_1} + (B_{11}^{(l)} \times s_{i_1} + B_{12}^{(l)} \times s_{i_2})/k$
     $b_{i_2} = b_{i_2} + (B_{21}^{(l)} \times s_{i_1} + B_{22}^{(l)} \times s_{i_2})/k$
End Do

Compute the last-element matrix entries
     $A_{11}^{(N_E)}$, $A_{12}^{(N_E)}$, $B_{11}^{(N_E)}$, and $B_{12}^{(N_E)}$,
          using (1.1.62) and (1.1.65)

$D_{N_E,N_E} = D_{N_E,N_E} + A_{11}^{(N_E)}$ *Process the last element*

$b_{N_E} = b_{N_E} + (B_{11}^{(N_E)} \times s_{N_E} + B_{12}^{(N_E)} \times s_{N_E+1})/k - A_{12}^{(N_E)} \times f_L$

Algorithm 1.1.1 Assembly of the finite element system, $\mathbf{D} \cdot \mathbf{f} = \mathbf{b}$, for linear
     elements.

Reduction to bidiagonal:

$$\begin{bmatrix} d_1 \\ y_1 \end{bmatrix} = \frac{1}{a_1} \begin{bmatrix} b_1 \\ r_1 \end{bmatrix}$$

Do $i = 1, 2, \ldots, N-1$

$$\begin{bmatrix} d_{i+1} \\ y_{i+1} \end{bmatrix} = \frac{1}{a_{i+1} - c_{i+1} d_i} \begin{bmatrix} b_{i+1} \\ r_{i+1} - c_{i+1} y_i \end{bmatrix}$$

End Do

Backward substitution:

$$x_N = y_N$$

Do $i = N-1, N-2, \ldots, 1$
$$x_i = y_i - d_i x_{i+1}$$
End Do

Algorithm 1.1.2 Thomas algorithm for solving the tridiagonal system of linear equations (1.1.70).

and then moving upward to compute the rest of the unknowns in a sequential fashion. The combined procedure is implemented in Algorithm 1.1.2. The FSELIB function **thomas** that implements the algorithm is listed in the text.

Thomas's algorithm is a special implementation of the inclusive method of Gauss elimination, which is applicable to general linear systems, as discussed in Section C.1 of Appendix C. The Thomas algorithm is designed to bypass idle operations with zeros.

1.1.11 Finite element code

We are in a position now to combine the computational modules developed previously in this section into an integrated finite element code that performs three main tasks: domain discretization, assembly of the linear system, and solution of the linear system.

To economize the computer memory, we store the diagonal, super-diagonal, and sub-diagonal elements of the global diffusion matrix in three vectors,

$$a_i^t \equiv D_{i,i}, \qquad b_i^t \equiv D_{i,i+1}, \qquad c_i^t \equiv D_{i-1,i}, \tag{1.1.74}$$

```
function x = thomas (n,a,b,c,r)

%=================================================
%
% Thomas algorithm for solving a tridiagonal system
%
% n:     system size
% a,b,c: diagonal, superdiagonal,
%        and subdiagonal elements
% r:     right-hand side
%
%=================================================

%--------
% prepare
%--------

na = n-1;

%-----------------------------
% reduction to upper bidiagonal
%-----------------------------

d(1) = b(1)/a(1);

y(1) = r(1)/a(1);

for i=1:na
   i1 = i+1;
   den = a(i1)-c(i1)*d(i);
   d(i1) = b(i1)/den;
   y(i1) = (r(i1)-c(i1)*y(i))/den;
end

%------------------
% back substitution
%------------------

x(n) = y(n);

for i=na:-1:1
   x(i)= y(i)-d(i)*x(i+1);
end

%-----
% done
%-----

return;
```

Function thomas: A function for solving a tridiagonal system of algebraic
equations by the Thomas algorithm. A modification of this algorithm
solves a pentadiagonal system of equations.

Do $i = 1, 2, \ldots, N_E$ *Initialize the coefficients of the tridiagonal system*
     $a_i^t = 0.0, \qquad b_i^t = 0.0, \qquad c_i^t = 0.0 \qquad$ *and right-hand side to zero*
     $b_i = 0.0$
End Do

$b_1 = q_0/k$ *Neumann boundary condition*

Do $l = 1, 2, \ldots, N_E - 1$ *Run over the first N_E-1 elements*
     Compute the element matrices $\mathbf{A}^{(l)}$ and $\mathbf{B}^{(l)}$ using (1.1.62) and (1.1.65)
     $a_l^t = a_l^t + A_{11}^{(l)}, \qquad b_l^t = b_l^t + A_{12}^{(l)},$
     $c_{l+1}^t = c_{l+1}^t + A_{21}^{(l)}, \qquad a_{l+1}^t = a_{l+1}^t + A_{22}^{(l)}$
     $b_{i_1} = b_{i_1} + (B_{11}^{(l)} \times s_l + B_{12}^{(l)} \times s_{l+1})/k$
     $b_{i_2} = b_{i_2} + (B_{21}^{(l)} \times s_l + B_{22}^{(l)} \times s_{l+1})/k$
End Do

Compute the last-element matrix components $A_{11}^{(N_E)}$, $A_{12}^{(N_E)}$, $B_{11}^{(N_E)}$, and $B_{12}^{(N_E)}$, using (1.1.62) and (1.1.65)

$a_{N_E}^t = a_{N_E}^t + A_{11}^{(N_E)} \qquad$ *Process the last element*
$b_{N_E} = b_{N_E} + (B_{11}^{(N_E)} \times s_{N_E} + B_{12}^{(N_E)} \times s_{N_E+1})/k - A_{12}^{(N_E)} \times f_L$

Algorithm 1.1.3 Compact assembly of the tridiagonal system for linear ele-
    ments, $\mathbf{D} \cdot \mathbf{f} = \mathbf{b}$, where a^t, b^t, and c^t are the components of the tridiago-
    nal global diffusion matrix, $\mathbf{D}$.

corresponding to the tridiagonal form shown in (1.1.71). With this notation, the assembly Algorithm 1.1.1 simplifies to the compact Algorithm 1.1.3, which is implemented in the FSELIB function sdl_sys, listed in the text; "sdl" is an acronym for **steady diffusion with linear** elements.

FSELIB code sdl, listed in the text, implements the complete finite element code in six modules:

- *Data input.*

- *Finite element discretization.*

- *Specification of the source function, s.*

- *Assembly of the linear system.*

- *Solution of the linear system.*

- *Plotting of the solution.*

```matlab
function [at,bt,ct,b] = sdl_sys (ne,xe,q0,fL,k,s)

%=====================================================
% Compact assembly of the tridiagonal linear system
% for steady diffusion with linear elements (sdl)
%=====================================================

%-------------
% element size
%-------------

for l=1:ne
  h(l) = xe(l+1)-xe(l);
end

%---------------------------------
% initialize the tridiagonal matrix
%---------------------------------

at = zeros(ne,1);
bt = zeros(ne,1);
ct = zeros(ne,1);

%-----------------------------
% initialize the right-hand side
%-----------------------------

b = zeros(ne,1);

b(1) = q0/k;

%---------------------------------
% loop over the first ne-1 elements
%---------------------------------

for l=1:ne-1

   A11 = 1/h(l); A12 =-A11;
   A21 = A12;    A22 = A11;

   B11 = h(l)/3.0; B12 = 0.5*B11;
   B21 = B12;      B22 = B11;

   at(l) = at(l) + A11;
   bt(l) = bt(l) + A12;

   ct(l+1) = ct(l+1) + A21;
   at(l+1) = at(l+1) + A22;

   b(l)   = b(l)   + (B11*s(l) + B12*s(l+1))/k;
   b(l+1) = b(l+1) + (B21*s(l) + B22*s(l+1))/k;

end
```

Function sdl_sys: Continuing $\longrightarrow$

```
%-----------------------------
% the last element is special
%-----------------------------

A11 = 1/h(ne); A12 =-A11;

B11 = h(ne)/3.0; B12 = 0.5*B11;

at(ne) = at(ne) + A11;

b(ne) = b(ne) + (B11*s(ne) + B12*s(ne+1))/k - A12*fL;

%-----
% done
%-----

return;
```

Function sdl_sys: ($\longrightarrow$ Continued.) Compact assembly of the tridiagonal linear system for one-dimensional diffusion with linear elements.

Figure 1.1.5, shows the graphics output generated by the code for the frivolous source function

$$s(x) = s_0 \ \exp(-5 \ \frac{x^2}{L^2}), \qquad (1.1.75)$$

where s_0 is a constant, and $0 \leq x \leq L$.

Calculations for a fixed value of the element stretch ratio, *ratio*=5, and number of elements,

$$N_E = 8, \quad 16, \quad 32, \quad 64, \quad 128, \quad 256, \qquad (1.1.76)$$

yield the left-end value

$$f(0) = 1.9746, \quad 1.9662, \quad 1.9644, \quad 1.9640, \quad 1.9639, \quad 1.9639. \quad (1.1.77)$$

Another set of calculations with uniformly spaced elements, *ratio*=1, yield

$$f(0) = 1.9510, \quad 1.9606, \quad 1.9631, \quad 1.9637, \quad 1.9638, \quad 1.9639. \quad (1.1.78)$$

These results clearly indicate that $f(0) = 1.9639$ is the exact value accurate to the fourth decimal place, achieved with 128 or 256 elements. In the case of evenly spaced elements, the numerical error, $e(0) \equiv f(0) - 1.9639$, is

$$e(0) = -0.0129, \quad -0.0032, \quad -0.0008, \quad -0.0002,$$
$$-0.0001, \quad 0.0000. \qquad (1.1.79)$$

```
%====================================
% Code sdl
%
% Steady one-dimensional diffusion
% with linear elements
%====================================

%-----------
% input data
%-----------

L=1.0;
k=1.0; q0=-1.0; fL=0.0;
ne=16; ratio=5.0;

%----------------
% grid generation
%----------------

xe = elm_line1 (0,L,ne,ratio);

%-------------------
% specify the source
%-------------------

for i=1:ne+1
  s(i) = 10.0*exp(-5.0*xe(i)^2/L^2);
end

%-----------------
% compact assembly
%-----------------

[at,bt,ct,b] = sdl_sys (ne,xe,q0,fL,k,s);

%--------------
% linear solver
%--------------

f = thomas (ne,at,bt,ct,b);

f(ne+1) = fL;

%-----
% plot
%-----

plot(xe, f,'-o'); xlabel('x'); ylabel('f');

%-----
% done
%-----
```

Code sdl: Finite element code for steady one-dimensional diffusion with linear elements.

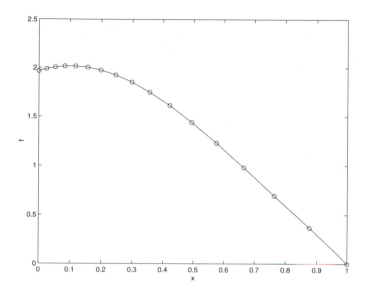

Figure 1.1.5 Graphics output of the FSELIB code `sdl` for steady diffusion with sixteen linear elements, $N_E = 16$, element stretch ratio equal to five, $L = 1.0$, $k = 1.0$, $q_0 = -1.0$, $f(L) = 0.0$, and $s_0 = 10.0$.

These data reveal that, when the number of elements is doubled, the numerical error is reduced approximately by a factor of four, which means that the error is proportional to h^2, where h is the element size. Thus, the finite element method with linear elements for the steady diffusion problem with a smooth source term is second-order accurate in the element size. This result can be confirmed by performing a formal analysis, keeping track of the error introduced by the various numerical approximations (e.g., [30]).

PROBLEMS

1.1.1 *Grid generation.*

(*a*) Explain in narrative form the discretization algorithms implemented in the FSELIB functions `elm_line2` and `elm_line3`.

(*b*) Execute codes `elm_line1_dr`, `elm_line2_dr` and `elm_line3_dr` to generate and display element distributions of your choice.

1.1.2 *Flux at the Dirichlet end.*

Consider the implementation of the finite element method with linear elements, as discussed in the text. Show that the flux at the Dirichlet end is given by

$$q_L \equiv -k \left(\frac{df}{dx} \right)_{x=L} = -k \left(f_{N_E} A_{21}^{(N_E)} + f_{N_E+1} A_{22}^{(N_E)} \right)$$
$$+ s_{N_E} B_{21}^{(N_E)} + s_{N_E+1} B_{22}^{(N_E)}. \qquad (1.1.80)$$

1.1.3 *Element diffusion and mass matrices.*

Perform the integrations to derive the element diffusion and mass matrices shown in (1.1.62) and (1.1.65).

1.1.4 *Code* sdl.

Execute the FSELIB code sdl with the same boundary conditions but a different source function of your choice, and discuss the results of your computation.

1.1.5 *Convection (Robin or mixed) boundary condition.*

A convection boundary condition at the left end, also called a Robin or mixed boundary condition, requires that the flux is related to the local function value by the linear relation

$$q_0 \equiv -k \left(\frac{df}{dx} \right)_{x=0} = h_T (f_0 - f_\infty), \qquad (1.1.81)$$

where h_T is a heat transfer coefficient, and f_∞ is a specified left-end ambient temperature. This boundary condition can be implemented in Algorithm 1.1.3 by replacing the fifth line expressing the Neumann boundary condition, with the two lines:

$$b_1 = h_T / (k \, f_\infty)$$
$$a_1^t = h_T / k \,.$$

Modify accordingly the FSELIB function sdl_sys, execute the modified code for $f_\infty = f_L$, and discuss the effect of the heat transfer coefficient, h_T.

1.1.6 *One-dimensional Helmholtz equation.*

Consider the Helmholtz equation in one dimension,

$$\frac{d^2 f}{dx^2} + \alpha f = 0, \qquad (1.1.82)$$

where α is a positive or negative constant coefficient. The solution is to be found over the interval $[0, L]$, subject to the boundary conditions stated in (1.1.4).

(*a*) Derive the system of linear equations arising from the Galerkin finite element formulation.

(*b*) Write a function hlm_sys that assembles the linear system in the spirit of the function sdl_sys discussed in the text, and a finite element code hlm that is analogous to sdl. Run the code hlm for a set of parameter values of your choice, and discuss the results of your computation.

1.2   Variational formulation and weighted residuals

For a certain class of differential equations, including the steady diffusion equation discussed in Section 1.1, the Galerkin projection method can be shown to be equivalent to Ritz's implementation of Rayleigh's variational formulation discussed in this section. Moreover, the Galerkin finite element method itself can be regarded as a special implementation of the method of weighted residuals. These complementary viewpoints accredit the finite element method and offer an alternative point of departure for building the mathematical framework. One advantage of the variational approach is that various physical effects can be incorporated readily in terms of energy functions that may be cumbersome to implement in the governing differential equations.

In some disciplines of computational mechanics, the weak formulation based on the governing differential equations discussed in Section 1.1 is altogether abandoned, and the finite element method is developed exclusively in the context of the variational formulation. In fact, some texts convey the impression that the finite element method arises strictly in the context of the variational formulation. The truth is that the method can be applied to a broader class of equations, even when a variational formulation is not available.

1.2.1   Rayleigh-Ritz method for homogeneous Dirichlet boundary conditions

To illustrate the connection with the variational formulation, we consider a linear differential operator, $L < \bullet >$, operating on the space of functions $w(x)$ defined in the interval $[a, b]$, subject to the homogeneous Dirichlet boundary conditions

$$w(x = a) = 0, \qquad w(x = b) = 0. \tag{1.2.1}$$

To simplify the notation, we define the inner product of two functions $w_1(x)$ and $w_2(x)$ as a curly bracket integral operation,

$$\left\{ w_1, \, w_2 \right\} \equiv \int_a^b w_1(x) \, w_2(x) \, \mathrm{d}x. \tag{1.2.2}$$

We further assume that the linear operator is self-adjoint, meaning that

$$\left\{ L < w_1 >, \, w_2 \right\} = \left\{ L < w_2 >, \, w_1 \right\}. \tag{1.2.3}$$

An example of a self-adjoint operator is the Sturm-Liouville second-order operator

$$L < \bullet > = \frac{\mathrm{d}}{\mathrm{d}x} \left(p(x) \frac{\mathrm{d} \bullet}{\mathrm{d}x} \right) - r(x) \, \bullet \quad , \tag{1.2.4}$$

where $p(x)$ and $r(x)$ are specified functions, and $\bullet$ stands for an appropriate function. To demonstrate that this operator is self-adjoint, we must show that

$$\int_a^b \left[\frac{\mathrm{d}}{\mathrm{d}x} \left(p(x) \frac{\mathrm{d}\omega_1}{\mathrm{d}x} \right) - r(x)\,\omega_1 \right] \omega_2 \, \mathrm{d}x$$
$$= \int_a^b \left[\frac{\mathrm{d}}{\mathrm{d}x} \left(p(x) \frac{\mathrm{d}\omega_2}{\mathrm{d}x} \right) - r(x)\,\omega_2 \right] \omega_1 \, \mathrm{d}x, \qquad (1.2.5)$$

for any two suitable functions, ω_1 and ω_2 that satisfy the homogeneous boundary conditions, $\omega_1(a) = 0$, $\omega_1(b) = 0$, and $\omega_2(a) = 0$, $\omega_2(b) = 0$. The proof follows readily by integrating by parts the first term inside the integral on either side.

Next, we introduce a source function, $s(x)$, and define the quadratic Rayleigh functional

$$\mathcal{F} < \omega(x) > \equiv \frac{1}{2} \Big\{ L < \omega(x) >, \, \omega(x) \Big\} + \Big\{ s(x), \, \omega(x) \Big\}, \qquad (1.2.6)$$

which maps a given function, $\omega(x)$, to a number. It can be shown that the minimum or maximum value of the functional $\mathcal{F} < \omega(x) >$ is achieved for a function $f(x)$ that satisfies the linear differential equation

$$L < f(x) > + s(x) = 0, \qquad (1.2.7)$$

subject to the boundary conditions $f(a) = 0$ and $f(b) = 0$. Stated differently, minimizing or maximizing the functional $\mathcal{F} < \omega(x) >$ over a class of admissible functions that satisfy the boundary conditions $\omega(a) = 0$ and $\omega(b) = 0$ is equivalent to solving the differential equation (1.2.7) subject to the homogeneous Dirichlet boundary condition. In this sense, the differential equation (1.2.7) is the *Euler-Lagrange equation* of the functional (1.2.6).

To prove the equivalence, we perturb the solution, $f(x)$, to $f(x) + \epsilon\,v(x)$, where $v(x)$ is an appropriate disturbance function, also called a variation, required to satisfy the homogeneous Dirichlet boundary condition $v(a) = 0$ and $v(b) = 0$, and ϵ is a dimensionless number whose magnitude is much less than unity. Next, we consider the variation

$$\delta\mathcal{F} \equiv \mathcal{F} < f(x) + \epsilon\,v(x) > - \mathcal{F} < f(x) >, \qquad (1.2.8)$$

and use the definition of $\mathcal{F}$ to write

$$\mathcal{F} < f(x) + \epsilon\,v(x) > = \frac{1}{2} \Big\{ L < f(x) > + \epsilon\,L < v(x) >, \, f(x) + \epsilon\,v(x) \Big\}$$
$$+ \Big\{ s(x), \, f(x) + \epsilon\,v(x) \Big\}. \qquad (1.2.9)$$

Expanding the products inside the brackets, using the property

$$\Big\{ L < v(x) >, \, f(x) \Big\} = \Big\{ L < f(x) >, \, v(x) \Big\}, \qquad (1.2.10)$$

and simplifying, we obtain

$$\delta \mathcal{F} = \epsilon \left\{ L < f(x) > +s(x),\ v(x) \right\} + \frac{\epsilon^2}{2} \left\{ L < v(x) >, v(x) \right\}, \qquad (1.2.11)$$

which shows that, if the first term on the right-hand side is zero for every x, then the functional is locally stationary and the minimum or maximum is achieved when $\omega(x) = f(x)$.

Ritz's method

In Ritz's method, the admissible functions $\omega(x)$ are expanded in a truncated series of a suitable set of basis functions, $\varphi_i(x)$, that satisfy the homogeneous boundary conditions $\varphi_i(a) = 0$ and $\varphi_i(b) = 0$, as

$$\omega(x) = \sum_{i=1}^{N} c_i \, \varphi_i(x), \qquad (1.2.12)$$

where N is a specified truncation level. Substituting this expansion in the Rayleigh functional, we find

$$\mathcal{F} < \omega(x) > \equiv \mathcal{G}(c_1, \ldots, c_N), \qquad (1.2.13)$$

where $\mathcal{G}$ is a quadratic function of the N coefficients, c_i. To compute these coefficients, we require the extremum condition

$$\frac{\partial \mathcal{G}}{\partial c_i} = 0, \qquad (1.2.14)$$

for $i = 1, 2, \ldots, N$, and solve the resulting system of linear algebraic equations to produce the solution, $f(x)$.

Weighted residuals and weak formulation

Expression (1.2.11) shows that solving the differential equation (1.2.7) is equivalent to computing a function $f(x)$ that satisfies

$$\left\{ L < f(x) > +s(x), \quad v(x) \right\} = 0, \qquad (1.2.15)$$

for any acceptable test function $v(x)$, interpreted as a disturbance or variation. The left-hand side of (1.2.15) is a weighted residual of the differential equation. Replacing the differential equation with the integrated form (1.2.15) is the starting point in a class of approximate methods, known as methods of weighted residuals. A solution procedure based on (1.2.15) is recognized as a *weak formulation*, whereas a solution procedure based on the primary differential equation is recognized as the *strong formulation*. The weak formulation carries the risk of introducing spurious components in the process of integration, playing the role of eigenfunctions.

In the Galerkin implementation of the method of weighted residuals, the solution is expanded as shown in (1.2.12), and the test functions, $v(x)$, are identified with each one of the basis functions, φ_i. When the operator, L, satisfies the conditions stated earlier in this section, the Galerkin method becomes equivalent to Ritz's method based on the Rayleigh variational formulation. In the Galerkin finite element method, both the basis functions, $\varphi_i(x)$, and the test functions, $v(x)$, are identified with the finite element global interpolation functions, ϕ_i, discussed in Section 1.1, as will be demonstrated later in this section by an example.

Rayleigh functional for one-dimensional diffusion

As an application, we note that the one-dimensional diffusion operator,

$$L < \bullet >= k \frac{\mathrm{d}^2 \bullet}{\mathrm{d}x^2}, \tag{1.2.16}$$

is a Sturm-Liouville operator with $p(x) = k$ and $r(x) = 0$. The variational formulation states that a solution of the differential equation

$$k \frac{\mathrm{d}^2 f}{\mathrm{d}x^2} + s(x) = 0, \tag{1.2.17}$$

subject to the homogeneous boundary conditions $f(a) = 0$ and $f(b) = 0$, maximizes the quadratic functional

$$\mathcal{F} < \omega(x) >\equiv \frac{k}{2} \int_a^b \frac{\mathrm{d}^2 \omega}{\mathrm{d}x^2} \, \omega(x) \, \mathrm{d}x + \int_a^b s(x) \, \omega(x) \, \mathrm{d}x, \tag{1.2.18}$$

over the set of all functions that satisfy the boundary-conditions $\omega(a) = 0$ and $\omega(b) = 0$. Integrating by parts the first term on the right-hand side of (1.2.18), and enforcing the homogeneous boundary conditions, we obtain the alternative expression

$$\mathcal{F} < \omega(x) >\equiv -\frac{k}{2} \int_a^b \left(\frac{\mathrm{d}\omega}{\mathrm{d}x}\right)^2 \mathrm{d}x + \int_a^b s(x) \, \omega(x) \, \mathrm{d}x. \tag{1.2.19}$$

Physically, the two integrals on the right-hand side express the thermal energy associated with the diffusive flux and the source. Maximizing this functional is equivalent to solving the steady diffusion equation (1.2.17) over the interval $[a, b]$, subject to the homogeneous Dirichlet conditions $f(a) = 0$ and $f(b) = 0$. In the absence of a source term, $s(x) = 0$, the functional is negative, and the maximum value of zero is clearly achieved for $\omega(x) = f(x) = 0$.

1.2.2   Rayleigh-Ritz method for inhomogeneous Dirichlet boundary conditions

Consider now the steady diffusion equation (1.2.17), to be solved in the interval $[a, b]$ subject to the inhomogeneous Dirichlet conditions

$$f(a) = f_a, \qquad f(b) = f_b. \tag{1.2.20}$$

It can be shown that solving this problem is still equivalent to maximizing the functional given in (1.2.19) over all continuous functions, $w(x)$, that satisfy the specified Dirichlet boundary conditions, $w(a) = f_a$ and $w(b) = f_b$. Integrating by parts the first term on the right-hand side of (1.2.19), we find

$$\mathcal{F} < w(x) >= \frac{k}{2} \left[\int_a^b \frac{\mathrm{d}^2 w}{\mathrm{d}x^2} \, w \, \mathrm{d}x - \left(w \, \frac{\mathrm{d}w}{\mathrm{d}x} \right)_a^b \right] + \int_a^b s(x) \, w(x) \, \mathrm{d}x. \quad (1.2.21)$$

When homogeneous Dirichlet boundary conditions are specified, $w(a) = 0$ and $w(b) = 0$, the second term on the right-hand side does not appear.

To prove the equivalence, we perturb the solution of the differential equation from $f(x)$ to $f(x) + \epsilon \, v(x)$, and compute the difference in $\mathcal{F}$ based on (1.2.19),

$$\delta \mathcal{F} \equiv \mathcal{F} < f(x) + \epsilon \, v(x) > - \mathcal{F} < f(x) >$$

$$= -\epsilon \, k \int_a^b \frac{\mathrm{d}f}{\mathrm{d}x} \frac{\mathrm{d}v}{\mathrm{d}x} \, \mathrm{d}x + \epsilon \int_a^b s(x) \, v(x) \, \mathrm{d}x + O(\epsilon^2) \quad (1.2.22)$$

$$= \epsilon \, k \int_a^b \frac{\mathrm{d}^2 f}{\mathrm{d}x^2} \, v \, \mathrm{d}x - \epsilon \, k \left(v \, \frac{\mathrm{d}f}{\mathrm{d}x} \right)_a^b + \epsilon \int_a^b s(x) \, v(x) \, \mathrm{d}x + O(\epsilon^2).$$

Applying the boundary conditions $v(a) = 0$ and $v(b) = 0$, we find

$$\delta \mathcal{F} = \epsilon \left[\int_a^b \left(k \, \frac{\mathrm{d}^2 f}{\mathrm{d}x^2} + s(x) \right) v(x) \, \mathrm{d}x \right] + O(\epsilon^2). \quad (1.2.23)$$

If the function $f(x)$ satisfies (1.2.17) subject to the aforementioned boundary conditions, then the first variation vanishes and the functional is stationary.

Alternatively, we may compute the difference in $\mathcal{F}$ based on (1.2.21), finding

$$\delta \mathcal{F} = \epsilon \frac{k}{2} \left[\int_a^b \left(\frac{\mathrm{d}^2 f}{\mathrm{d}x^2} v + \frac{\mathrm{d}^2 v}{\mathrm{d}x^2} f \right) \mathrm{d}x - \left(f \, \frac{\mathrm{d}v}{\mathrm{d}x} \right)_a^b \right]$$

$$+ \epsilon \int_a^b s(x) \, v(x) \, \mathrm{d}x + O(\epsilon^2). \quad (1.2.24)$$

Using Green's second identity

$$\int_a^b \frac{\mathrm{d}^2 v}{\mathrm{d}x^2} \, f \, \mathrm{d}x = \int_a^b \frac{\mathrm{d}^2 f}{\mathrm{d}x^2} \, v \, \mathrm{d}x + \left(f \, \frac{\mathrm{d}v}{\mathrm{d}x} - v \, \frac{\mathrm{d}f}{\mathrm{d}x} \right)_a^b, \quad (1.2.25)$$

and applying the boundary conditions, we recover precisely (1.2.23).

Rayleigh-Ritz method

In Ritz's method, the expansion (1.2.12) is endowed with a leading term, $\varphi_0(x)$, that satisfies the stipulated boundary conditions, $\varphi_0(a) = f_a$ and $\varphi_0(b) = f_b$,

$$w(x) = \varphi_0(x) + \sum_{i=1}^N c_i \, \varphi_i(x). \quad (1.2.26)$$

The remaining basis functions, φ_i, satisfy the homogeneous boundary conditions, $\varphi_i(a) = 0$ and $\varphi_i(b) = 0$, for $i = 1, 2, \ldots, N$. Substituting this expansion in the Rayleigh functional, we derive the quadratic algebraic form (1.2.13), and then compute the coefficients, c_i, by requiring the extremum condition (1.2.14).

1.2.3 Rayleigh-Ritz method for Dirichlet/Neumann boundary conditions

In the next case study, we consider the steady diffusion equation (1.2.17), to be solved in the interval $[a, b]$, subject to the Dirichlet boundary condition at the left end and the Neumann boundary condition at the right end,

$$f(a) = f_a, \qquad -k \left(\frac{\mathrm{d}f}{\mathrm{d}x} \right)_{x=b} = q_b, \qquad (1.2.27)$$

where the values f_a and q_b are specified. It can be shown that computing the solution to this problem is equivalent to maximizing the functional

$$\mathcal{F} < \omega(x) > \equiv -\frac{k}{2} \int_a^b \left(\frac{\mathrm{d}\omega}{\mathrm{d}x} \right)^2 \mathrm{d}x \qquad (1.2.28)$$

$$+ \int_a^b s(x)\, \omega(x)\, \mathrm{d}x - \omega(x = b)\, q_b,$$

over all continuous functions, $\omega(x)$, that satisfy the left-end boundary condition alone, $\omega(a) = f_a$. The satisfaction of the Neumann boundary condition is enforced by the last term on the right-hand side of (1.2.28).

Integrating by parts the first term on the right-hand side of (1.2.28), we derive the equivalent form

$$\mathcal{F} < \omega(x) > = \frac{k}{2} \left[\int_a^b \frac{\mathrm{d}^2\omega}{\mathrm{d}x^2}\, \omega\, \mathrm{d}x - \left(\omega\, \frac{\mathrm{d}\omega}{\mathrm{d}x} \right)_a^b \right]$$

$$+ \int_a^b s(x)\, \omega(x)\, \mathrm{d}x - \omega(x = b)\, q_b, \qquad (1.2.29)$$

or

$$\mathcal{F} < \omega(x) > = \frac{k}{2} \int_a^b \frac{\mathrm{d}^2\omega}{\mathrm{d}x^2}\, \omega\, \mathrm{d}x + \int_a^b s(x)\, \omega(x)\, \mathrm{d}x$$

$$+ \frac{k}{2} f_a \left(\frac{\mathrm{d}\omega}{\mathrm{d}x} \right)_{x=a} - \frac{1}{2} \omega(x = b)\, q_b. \qquad (1.2.30)$$

When homogeneous boundary conditions are prescribed, $f_a = 0$ and $q_b = 0$, the last two terms on the right-hand side do not appear.

To prove the equivalence, we introduce the perturbation $f(x) + \epsilon\, v(x)$, where the disturbance function $v(x)$ satisfies the homogeneous left-end Dirichlet boundary condition, $v(a) = 0$, and compute the variation based on (1.2.28),

$$\delta\mathcal{F} = -\epsilon\,k\int_a^b \frac{\mathrm{d}f}{\mathrm{d}x}\frac{\mathrm{d}v}{\mathrm{d}x}\,\mathrm{d}x + \epsilon\int_a^b s(x)\,v(x)\,\mathrm{d}x - \epsilon\,v(x=b)\,q_b + O(\epsilon^2)$$

$$= \epsilon\left[k\int_a^b \frac{\mathrm{d}^2 f}{\mathrm{d}x^2}\,v\,\mathrm{d}x - k\left(v\,\frac{\mathrm{d}f}{\mathrm{d}x}\right)_a^b + \int_a^b s(x)\,v(x)\,\mathrm{d}x\right.$$

$$\left. -\,v(x=b)\,q_b\right] + O(\epsilon^2). \qquad (1.2.31)$$

Enforcing the boundary conditions, we derive precisely (1.2.23). Thus, if the function $f(x)$ satisfies (1.2.17) subject to the aforementioned boundary conditions, the first variation vanishes and the functional is stationary.

Alternatively, we may compute the change in $\mathcal{F}$ based on (1.2.30), finding

$$\delta\mathcal{F} = \epsilon\,\frac{k}{2}\left[\int_a^b \left(\frac{\mathrm{d}^2 f}{\mathrm{d}x^2}\,v + \frac{\mathrm{d}^2 v}{\mathrm{d}x^2}\,f\right)\mathrm{d}x - \left(f\,\frac{\mathrm{d}v}{\mathrm{d}x} + v\,\frac{\mathrm{d}f}{\mathrm{d}x}\right)_a^b\right]$$

$$+\epsilon\int_a^b s(x)\,v(x)\,\mathrm{d}x - \epsilon\,v(x=b)\,q_b + O(\epsilon^2). \qquad (1.2.32)$$

Using Green's second identity (1.2.25) to simplify the right-hand side, we derive (1.2.23).

Rayleigh-Ritz method

To implement Ritz's method, we introduce expansion (1.2.26), repeated here for convenience,

$$w(x) = \varphi_0(x) + \sum_{i=1}^N c_i\,\varphi_i(x), \qquad (1.2.33)$$

where the leading term, $\varphi_0(x)$, satisfies the specified Dirichlet condition at the left end, $\varphi_0(a) = f_a$, and the rest of the basis functions, $\varphi_i(x)$, satisfy the homogeneous Dirichlet boundary condition, $\varphi_i(a) = 0$, for $i = 1, 2, \ldots, N$. Substituting this expansion in the Rayleigh functional, we derive the algebraic form (1.2.13), and then require the extremum condition (1.2.14).

As an example, we consider diffusion in the absence of a source, $s = 0$, whereupon the Rayleigh functional reduces to

$$\mathcal{F} < w(x) >\equiv -\frac{k}{2}\int_a^b \left(\frac{\mathrm{d}w}{\mathrm{d}x}\right)^2 \mathrm{d}x - w(x=b)\,q_b. \qquad (1.2.34)$$

Identifying $\varphi_0(x)$ with a constant function, $\varphi_0(x) = f_a$, truncating the expansion at the very first term, $N = 1$, and choosing $\varphi_1(x) = x - a$, we obtain

$$w(x) = f_a + c_1\,(x - a). \qquad (1.2.35)$$

A straightforward computation yields

$$\mathcal{F} < w(x) > \equiv \mathcal{G}(c_1) = -\frac{k}{2} L c_1^2 - (f_a + c_1 L) q_b, \qquad (1.2.36)$$

where $L = b - a$ is the length of the solution domain. Requiring $\partial \mathcal{G}/\partial c_1 = -k L c_1 - L q_b = 0$, yields $c_1 = -q_b/k$, which can be substituted in (1.2.35) to produce the exact solution,

$$f(x) = f_a - \frac{q_b}{k} (x - a). \qquad (1.2.37)$$

1.2.4   Rayleigh-Ritz method for Neumann/Dirichlet boundary conditions

As a complement to the problem discussed in the preceding section, we consider the steady diffusion equation (1.2.17), to be solved in the interval $[a, b]$ subject to the Neumann boundary condition at the left end and the Dirichlet boundary condition at the right end,

$$-k \left(\frac{\mathrm{d}f}{\mathrm{d}x} \right)_{x=b} = q_a, \qquad f(b) = f_b, \qquad (1.2.38)$$

where the values q_a and f_b are specified. Working as in the preceding section, we find that computing the solution is equivalent to maximizing the functional

$$\mathcal{F} < w(x) > \equiv -\frac{k}{2} \int_a^b \left(\frac{\mathrm{d}w}{\mathrm{d}x} \right)^2 \mathrm{d}x$$

$$+ \int_a^b s(x) \, w(x) \, \mathrm{d}x + w(x = a) \, q_a, \qquad (1.2.39)$$

over all continuous functions that satisfy the right-end boundary condition, $w(b) = f_b$. The satisfaction of the Neumann boundary condition at the left end is enforced by the last term on the right-hand side of (1.2.39).

To implement Ritz's method, we introduce expansion (1.2.33), where the leading term, $\varphi_0(x)$, satisfies the specified Dirichlet condition at the right end, $\varphi_0(b) = f_b$, and the rest of the basis functions, $\varphi_i(x)$, satisfy homogeneous Dirichlet boundary condition, $\varphi_i(b) = 0$, for for $i = 1, 2, \ldots, N$. Substituting this expansion in the Rayleigh functional, we derive the algebraic form (1.2.13), and then require the extremum condition (1.2.14).

To demonstrate the relation between the Rayleigh-Ritz and the Galerkin finite element method, we consider the steady diffusion problem discussed in Section 1.1, and set $a = 0$ and $b = L$. Next, we express the basis function $\varphi_0(x)$ in terms of the last-node global interpolation function, as

$$\varphi_0(x) = f_L \, \phi_{N_E+1}(x), \qquad (1.2.40)$$

and identify further basis functions with the previous node interpolation functions,

$$\varphi_i(x) = \phi_i(x), \qquad (1.2.41)$$

for $i = 1, 2, \ldots, N_E$, to obtain

$$w(x) = f_L \, \phi_{N_E+1}(x) + \sum_{i=1}^{N_E} w_i \, \phi_i(x),$$

(1.2.42)

$$w'(x) = f_L \, \phi'_{N_E+1}(x) + \sum_{i=1}^{N_E} w_i \, \phi'_i(x),$$

where a prime denotes a derivative with respect to x, and we have denoted $w_i \equiv w(x_i)$. In addition, we express the source term in terms of the global interpolation functions as

$$s(x) = \sum_{i=1}^{N_E+1} s_i \, \phi_i(x),$$

where $s_i \equiv s(x_i)$. Substituting these expansions in the Rayleigh functional (1.2.39), we find

$$\mathcal{F} < w(x) > \equiv \mathcal{G}(w_1, w_2, \ldots, w_{N_E})$$

(1.2.43)

$$= -\frac{k}{2} \sum_{i=1}^{N_E+1} \left(w_i \sum_{j=1}^{N_E+1} D_{ij} \, w_j \right) + \sum_{i=1}^{N_E+1} \left(w_i \sum_{j=1}^{N_E+1} M_{ij} \, s_j \right) + w_1 \, q_0,$$

where D_{ij} is the global diffusion matrix defined in (1.1.26), M_{ij} is the global mass matrix defined in (1.1.27), and $w_{N_E+1} = f_L$. In vector notation,

$$\mathcal{G}(\boldsymbol{w}) = -\frac{k}{2} \, \boldsymbol{w} \cdot \mathbf{D} \cdot \boldsymbol{w} + \boldsymbol{w} \cdot \mathbf{M} \cdot \mathbf{s} + w_1 \, q_0.$$

(1.2.44)

To maximize the functional we require

$$\frac{\partial \mathcal{F}}{\partial w_i} = 0,$$

(1.2.45)

for $i = 1, 2, \ldots, N_E$, set $w_i = f_i$, and derive precisely the linear system (1.1.29).

PROBLEMS

1.2.1 *Mixed boundary condition.*

Develop the variational formulation for the steady diffusion equation subject to the Dirichlet boundary condition at the left end, $f(a) = f_a$, and the mixed, convection, or Robin boundary condition at the right end,

$$q_b \equiv -k \left(\frac{\mathrm{d}f}{\mathrm{d}x} \right)_{x=b} = h_T \left(f(x = b) - f_\infty \right),$$

(1.2.46)

where h_T is a heat transfer coefficient, and f_∞ is a specified ambient temperature.

1.2.2 *Variational formulation for Helmholtz' equation.*

Consider the one-dimensional Helmholtz equation,

$$\frac{\mathrm{d}^2 f}{\mathrm{d}x^2} + \alpha\, f = 0, \tag{1.2.47}$$

where α is a constant, to be solved in the interval $[a, b]$ subject to the Dirichlet boundary conditions $f(a) = f_a$ and $f(b) = f_b$.

Show that solving this problem is equivalent to maximizing the Rayleigh functional

$$\mathcal{F} < \omega(x) > \equiv -\frac{1}{2} \int_a^b \left(\frac{\mathrm{d}\omega}{\mathrm{d}x}\right)^2 \mathrm{d}x + \frac{\alpha}{2} \int_a^b \omega^2(x)\, \mathrm{d}x, \tag{1.2.48}$$

over all continuous functions $\omega(x)$ that satisfy the specified Dirichlet boundary conditions, $\omega(x = a) = f_a$ and $\omega(x = b) = f_b$.

1.3   Steady diffusion with quadratic elements

One way to improve the accuracy of the finite element solution is to use a higher number of elements, which is equivalent to reducing the element size, h. An alternative is to approximate the solution over the individual elements with quadratic functions, which amounts to increasing the polynomial order, p. The first method is classified as an h-refinement, and the second method is classified as a p-refinement, as will be discussed in detail in Chapter 2. A combination of the two methods yield the hp refinement, not to be confused with the electronics manufacturer.

Element and global nodes

The quadratic polynomial is defined by three coefficients and requires three interpolation nodes over each element. To ensure that the solution composed of the union of quadratics is continuous over the entire solution domain, we use two *shared* element end-nodes, labeled as element nodes 1 and 3, and one *interior* element node, labeled as element node 2, as illustrated in Figure 1.3.1(a). Adding to the N_E interior nodes the $N_E + 1$ shared nodes, we obtain a total number of

$$N_G = 2\, N_E + 1 \tag{1.3.1}$$

unique global nodes, numbered sequentially from 1 at the left end of the solution domain located at $x = 0$, to N_G at the right end of the solution domain located

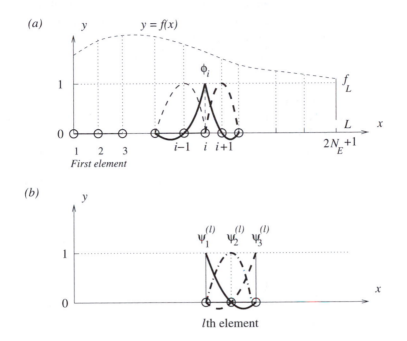

Figure 1.3.1 Quadratic finite element interpolation in one dimension involving two shared element end-nodes and one interior element node. The functions drawn with the heavy lines in (*a*) are the global interpolation functions; the solid line corresponds to the *i*th node, which is an element end-node, and the broken lines correspond to the $i \pm 1$ nodes, which are interior element nodes. The functions $\psi_1^{(l)}$, $\psi_2^{(l)}$, and $\psi_3^{(l)}$ depicted in (*b*) are the three local interpolation functions of the *l*th element.

at $x = L$. The global labels of the first end-node, interior node, and second end-node of the *l*th element are, respectively,

$$
\begin{aligned}
c_{l,1} &= 2\,(l-1) + 1 = 2l - 1, \\
c_{l,2} &= 2\,(l-1) + 2 = 2l, \\
c_{l,3} &= 2\,(l-1) + 3 = 2l + 1,
\end{aligned}
\tag{1.3.2}
$$

where $c_{l,j}$ is a connectivity matrix, for $l = 1, 2, \ldots, N_E$, and $j = 1, 2, 3$.

Interpolation functions

Next, we introduce the parabolic interpolation functions of the *l*th-element, $\psi_1^{(l)}$, $\psi_2^{(l)}$, and $\psi_3^{(l)}$, as shown in Figure 1.3.1(*b*). The first function, $\psi_1^{(l)}$, takes the value of unity at the first element node, and the value of zero at the second and third element nodes; the second function, $\psi_2^{(l)}$, takes the value of unity at the second element node, and the value of zero at the first and third element nodes; and the third function, $\psi_3^{(l)}$, takes the value of unity at the third element node, and the value of zero at the first and second element nodes. Specific expressions for these functions will be derived later in this section.

The union of the end-node element interpolation functions corresponding to adjacent elements, and the interior node interpolation functions themselves, yield *global* interpolation tent-like functions, $\phi_i(x)$. The global interpolation function of the ith global node, which happens to be an element end-node, is drawn with the solid line in Figure 1.3.1(a). The global interpolation functions of the $i \pm 1$ global nodes, which happen to be interior element nodes, are drawn with broken lines in Figure 1.3.1(a). Note that the global interpolation function associated with an end-node is supported by two elements, whereas that associated with an interior node is supported by one element alone.

The piecewise quadratic representation of the finite element solution is described by the expansion

$$f(x) = \sum_{j=1}^{N_G} f_j \, \phi_j(x), \tag{1.3.3}$$

where $f_j \equiv f(x_j)$. The left-end Dirichlet boundary condition (1.1.6) requires $f_{N_G} = f_L$.

Galerkin finite element equations

To compute the $2N_E$ unknown nodal values, f_j, for $j = 1, 2, \ldots, 2N_E$, we work as in Section 1.1, and derive the counterpart of the jth Galerkin projection equation (1.1.58),

$$k \sum_{j=1}^{N_G} \left(\sum_{l=1}^{N_E} \int_{E_l} \frac{\mathrm{d}\phi_i}{\mathrm{d}x} \frac{\mathrm{d}\phi_j}{\mathrm{d}x} \, \mathrm{d}x \right) f_j$$

$$= q_0 \, \delta_{j1} + \sum_{j=1}^{N_G} \left(\sum_{l=1}^{N_E} \int_{E_l} \phi_i \, \phi_j \, \mathrm{d}x \right) s_j, \tag{1.3.4}$$

for $i = 1, 2, \ldots, 2N_E$, where E_l stands for the lth element. As in the case of linear elements, because the Dirichlet boundary condition is specified at the right end of the solution domain, the last node is excluded from the projection. The counterpart of expression (1.1.60) for the diffusion matrix is

$$\int_{E_l} \frac{\mathrm{d}\phi_i}{\mathrm{d}x} \frac{\mathrm{d}\phi_j}{\mathrm{d}x} \, \mathrm{d}x = \delta_{i,c_{l,1}} \delta_{j,c_{l,1}} \, A_{11}^{(l)} + \delta_{i,c_{l,1}} \delta_{j,c_{l,2}} \, A_{12}^{(l)} + \delta_{i,c_{l,1}} \delta_{j,c_{l,3}} \, A_{13}^{(l)}$$

$$+ \delta_{i,c_{l,2}} \delta_{j,c_{l,1}} \, A_{21}^{(l)} + \delta_{i,c_{l,2}} \delta_{j,c_{l,2}} \, A_{22}^{(l)} + \delta_{i,c_{l,2}} \delta_{j,c_{l,3}} \, A_{23}^{(l)}$$

$$+ \delta_{i,c_{l,3}} \delta_{j,c_{l,1}} \, A_{31}^{(l)} + \delta_{i,c_{l,3}} \delta_{j,c_{l,2}} \, A_{32}^{(l)} + \delta_{i,c_{l,3}} \delta_{j,c_{l,3}} \, A_{33}^{(l)}, \tag{1.3.5}$$

where $\mathbf{A}^{(l)}$ is the 3×3 element diffusion matrix,

$$A_{11}^{(l)} \equiv \int_{E_l} \frac{d\phi_{c_{l,1}}}{dx} \frac{d\phi_{c_{l,1}}}{dx} \, dx = \int_{E_l} \frac{d\psi_1^{(l)}}{dx} \frac{d\psi_1^{(l)}}{dx} \, dx,$$

$$A_{12}^{(l)} = A_{21}^{(l)} \equiv \int_{E_l} \frac{d\phi_{c_{l,1}}}{dx} \frac{d\phi_{c_{l,2}}}{dx} \, dx = \int_{E_l} \frac{d\psi_1^{(l)}}{dx} \frac{d\psi_2^{(l)}}{dx} \, dx,$$

$$A_{13}^{(l)} = A_{31}^{(l)} \equiv \int_{E_l} \frac{d\phi_{c_{l,1}}}{dx} \frac{d\phi_{c_{l,3}}}{dx} \, dx = \int_{E_l} \frac{d\psi_1^{(l)}}{dx} \frac{d\psi_3^{(l)}}{dx} \, dx,$$

$$A_{22}^{(l)} \equiv \int_{E_l} \frac{d\phi_{c_{l,2}}}{dx} \frac{d\phi_{c_{l,2}}}{dx} \, dx = \int_{E_l} \frac{d\psi_2^{(l)}}{dx} \frac{d\psi_2^{(l)}}{dx} \, dx, \qquad (1.3.6)$$

$$A_{23}^{(l)} = A_{32}^{(l)} \equiv \int_{E_l} \frac{d\phi_{c_{l,2}}}{dx} \frac{d\phi_{c_{l,3}}}{dx} \, dx = \int_{E_l} \frac{d\psi_2^{(l)}}{dx} \frac{d\psi_3^{(l)}}{dx} \, dx,$$

$$A_{33}^{(l)} \equiv \int_{E_l} \frac{d\phi_{c_{l,3}}}{dx} \frac{d\phi_{c_{l,3}}}{dx} \, dx = \int_{E_l} \frac{d\psi_3^{(l)}}{dx} \frac{d\psi_3^{(l)}}{dx} \, dx.$$

For the physical reason discussed at the end of Section 1.1.8, this matrix is singular irrespective of the location of the interior node. The counterpart of expression (1.1.63) for the mass matrix is

$$\int_{E_l} \phi_i \, \phi_j \, dx = \delta_{i,c_{l,1}} \delta_{j,c_{l,1}} \, B_{11}^{(l)} + \delta_{i,c_{l,1}} \delta_{j,c_{l,2}} \, B_{12}^{(l)} + \delta_{i,c_{l,1}} \delta_{j,c_{l,3}} \, B_{13}^{(l)}$$

$$+ \delta_{i,c_{l,2}} \delta_{j,c_{l,1}} \, B_{21}^{(l)} + \delta_{i,c_{l,2}} \delta_{j,c_{l,2}} \, B_{22}^{(l)} + \delta_{i,c_{l,2}} \delta_{j,c_{l,3}} \, B_{23}^{(l)}$$

$$+ \delta_{i,c_{l,3}} \delta_{j,c_{l,1}} \, B_{31}^{(l)} + \delta_{i,c_{l,3}} \delta_{j,c_{l,2}} \, B_{32}^{(l)} + \delta_{i,c_{l,3}} \delta_{j,c_{l,3}} \, B_{33}^{(l)}, \qquad (1.3.7)$$

where $\mathbf{B}^{(l)}$ is the 3×3 element mass matrix,

$$B_{11}^{(l)} \equiv \int_{E_l} \phi_{c_{l,1}} \, \phi_{c_{l,1}} \, dx = \int_{E_l} \psi_1^{(l)} \, \psi_1^{(l)} \, dx,$$

$$B_{12}^{(l)} = B_{21}^{(l)} \equiv \int_{E_l} \phi_{c_{l,1}} \, \phi_{c_{l,2}} \, dx = \int_{E_l} \psi_1^{(l)} \, \psi_2^{(l)} \, dx,$$

$$B_{13}^{(l)} = B_{31}^{(l)} \equiv \int_{E_l} \phi_{c_{l,1}} \, \phi_{c_{l,3}} \, dx = \int_{E_l} \psi_1^{(l)} \, \psi_3^{(l)} \, dx,$$

$$B_{22}^{(l)} \equiv \int_{E_l} \phi_{c_{l,2}} \, \phi_{c_{l,2}} \, dx = \int_{E_l} \psi_2^{(l)} \, \psi_2^{(l)} \, dx, \qquad (1.3.8)$$

$$B_{23}^{(l)} = B_{32}^{(l)} \equiv \int_{E_l} \phi_{c_{l,3}} \, \phi_{c_{l,2}} \, dx = \int_{E_l} \psi_2^{(l)} \, \psi_3^{(l)} \, dx,$$

$$B_{33}^{(l)} \equiv \int_{E_l} \phi_{c_{l,3}} \, \phi_{c_{l,3}} \, dx = \int_{E_l} \psi_3^{(l)} \, \psi_3^{(l)} \, dx.$$

The Galerkin finite element implementation results in the linear system

$$\mathbf{D} \cdot \mathbf{f} = \mathbf{b}, \tag{1.3.9}$$

where the $2N_E$-dimensional solution vector, $\mathbf{f}$, encapsulates the unknown temperatures at all nodes, except for the last node where the Dirichlet boundary condition is specified,

$$\mathbf{f} \equiv \begin{bmatrix} f_1 \\ f_2 \\ \vdots \\ f_{2N_E-1} \\ f_{2N_E} \end{bmatrix}. \tag{1.3.10}$$

The northwestern portion of the $2N_E \times 2N_E$ pentadiagonal global diffusion matrix takes the form

$$\mathbf{D} = \begin{bmatrix} A_{11}^{(1)} & A_{12}^{(1)} & A_{13}^{(1)} & 0 & \cdots & \cdots & \cdots \\ A_{21}^{(1)} & A_{22}^{(1)} & A_{23}^{(1)} & 0 & 0 & \cdots & \cdots \\ A_{31}^{(1)} & A_{32}^{(1)} & A_{33}^{(1)} + A_{11}^{(2)} & A_{12}^{(2)} & A_{13}^{(2)} & 0 & \cdots \\ 0 & 0 & A_{21}^{(2)} & A_{22}^{(2)} & A_{23}^{(2)} & 0 & \cdots \\ 0 & 0 & A_{31}^{(2)} & A_{32}^{(2)} & A_{33}^{(2)} + A_{11}^{(3)} & A_{12}^{(3)} & \cdots \\ \cdots & \cdots & \cdots & \cdots & \cdots & \cdots & \cdots \\ 0 & 0 & 0 & \cdots & \cdots & \cdots & \cdots \\ 0 & 0 & 0 & \cdots & \cdots & \cdots & \cdots \end{bmatrix}, \tag{1.3.11}$$

which shows that $\mathbf{D}$ is composed of partially overlapping 3×3 diagonal blocks. "Pentadiagonal" derives from the Greek words $\pi\epsilon\nu\tau\epsilon$, which means "five," and $\delta\iota\alpha\gamma\omega\nu\iota o\varsigma$, which means "diagonal." The global diffusion matrix and right-hand side can be generated according to Algorithm 1.3.1, which is a generalization of Algorithm 1.1.1 for linear elements.

The right-hand side of the linear system (1.3.9) is given by

$$\mathbf{b} \equiv \mathbf{c} + \frac{1}{k} \widetilde{\mathbf{M}} \cdot \mathbf{s}, \tag{1.3.12}$$

where the $2N_E$-dimensional vector, $\mathbf{c}$, encapsulates the boundary conditions,

$$\mathbf{c} \equiv \begin{bmatrix} q_0/k \\ 0 \\ \vdots \\ 0 \\ f_L/h_{N_E} \end{bmatrix}, \tag{1.3.13}$$

Do $l = 1, 2, \ldots, N_E$ *Run over the elements to generate the connectivity matrix*
$\quad c(l, 1) = 2l - 1$
$\quad c(l, 2) = 2l$
$\quad c(l, 3) = 2l + 1$
End Do

Do $i = 1, 2, \ldots, 2\,N_E$ *Initialize the global diffusion matrix*
$\quad b_i = 0.0$ *and right-hand side to zero*
$\quad$ Do $j = 1, 2, \ldots, 2\,N_E$
$\quad\quad D_{ij} = 0.0$
$\quad$ End Do
End Do

$b_1 = q_0/k$ *Neumann boundary condition*

Do $l = 1, 2, \ldots, N_E$ *Run over all elements*

Compute the matrices $\mathbf{A}^{(l)}$ and $\mathbf{B}^{(l)}$ using (1.3.6) and (1.3.8)

$i_1 = c(l, 1), \qquad i_2 = c(l, 2), \qquad i_3 = c(l, 3)$

$$D_{i_1,i_1} = D_{i_1,i_1} + A_{11}^{(l)}, \qquad D_{i_1,i_2} = D_{i_1,i_2} + A_{12}^{(l)}, \qquad D_{i_1,i_3} = D_{i_1,i_3} + A_{13}^{(l)}$$
$$D_{i_2,i_1} = D_{i_2,i_1} + A_{21}^{(l)}, \qquad D_{i_2,i_2} = D_{i_2,i_2} + A_{22}^{(l)}, \qquad D_{i_2,i_3} = D_{i_2,i_3} + A_{23}^{(l)}$$
$$D_{i_3,i_1} = D_{i_3,i_1} + A_{31}^{(l)}, \qquad D_{i_3,i_2} = D_{i_3,i_2} + A_{32}^{(l)}, \qquad D_{i_3,i_3} = D_{i_3,i_3} + A_{33}^{(l)}$$

$$b_{i_1} = b_{i_1} + (B_{11}^{(l)} \times s_{i_1} + B_{12}^{(l)} \times s_{i_2} + B_{13}^{(l)} \times s_{i_3})/k$$
$$b_{i_2} = b_{i_2} + (B_{21}^{(l)} \times s_{i_1} + B_{22}^{(l)} \times s_{i_2} + B_{23}^{(l)} \times s_{i_3})/k$$
$$b_{i_3} = b_{i_3} + (B_{31}^{(l)} \times s_{i_1} + B_{32}^{(l)} \times s_{i_2} + B_{33}^{(l)} \times s_{i_3})/k$$

End Do

$b_{2N_E-1} = b_{2N_E-1} - A_{13}^{(N_E)} \times f_L$ *Process the last node*

$b_{2N_E} = b_{2N_E} - A_{23}^{(N_E)} \times f_L$

Algorithm 1.3.1 Assembly of the linear system $\mathbf{D} \cdot \mathbf{f} = \mathbf{b}$ for the quadratic elements illustrated in Figure 1.3.1.

the $(N_E + 1)$-dimensional vector, $\mathbf{s}$, contains the nodal values of the source,

$$\mathbf{s} \equiv \begin{bmatrix} s_1 \\ s_2 \\ \vdots \\ s_{2N_E} \\ s_{2N_E+1} \end{bmatrix}, \tag{1.3.14}$$

and $\widetilde{\mathbf{M}}$ is the $2N_E \times (2N_E + 1)$ bordered mass matrix,

$$\widetilde{\mathbf{M}} = \begin{bmatrix} B_{11}^{(1)} & B_{12}^{(1)} & B_{13}^{(1)} & 0 & \cdots & \cdots & \cdots \\ B_{21}^{(1)} & B_{22}^{(1)} & B_{23}^{(1)} & 0 & 0 & \cdots & \cdots \\ B_{31}^{(1)} & B_{32}^{(1)} & B_{33}^{(1)} + B_{11}^{(2)} & B_{12}^{(2)} & B_{13}^{(2)} & 0 & \cdots \\ 0 & 0 & B_{21}^{(2)} & B_{22}^{(2)} & B_{23}^{(2)} & 0 & \cdots \\ 0 & 0 & B_{31}^{(2)} & B_{32}^{(2)} & B_{33}^{(2)} + B_{11}^{(3)} & B_{12}^{(3)} & \cdots \\ \cdots & \cdots & \cdots & \cdots & \cdots & \cdots & \cdots \\ 0 & 0 & 0 & \cdots & \cdots & \cdots & \cdots \\ 0 & 0 & 0 & \cdots & \cdots & \cdots & \cdots \end{bmatrix}, \tag{1.3.15}$$

The entries of the element diffusion and mass matrices and the finite element solution itself depend on the precise location of the element interior nodes.

Because the global diffusion matrix $\mathbf{D}$ displayed in (1.3.11) is pentadiagonal, the solution of the linear system can be found efficiently using a modification of Thomas's algorithm for tridiagonal systems discussed in Section 1.1.10.

1.3.1   Thomas algorithm for pentadiagonal systems

To formalize the method in general terms, we consider the $N \times N$ linear system

$$\mathbf{P} \cdot \mathbf{x} = \mathbf{r} \tag{1.3.16}$$

for the unknown vector $\mathbf{x}$, where the vector $\mathbf{r}$ is specified, and the matrix $\mathbf{P}$ has the pentadiagonal form

$$\mathbf{P} = \begin{bmatrix} a_1 & b_1 & c_1 & 0 & 0 & \cdots & 0 & 0 & 0 \\ d_2 & a_2 & b_2 & c_2 & 0 & \cdots & 0 & 0 & 0 \\ e_3 & d_3 & a_3 & b_3 & c_3 & \cdots & 0 & 0 & 0 \\ \cdots & \cdots & \cdots & \cdots & \cdots & \cdots & \cdots & \cdots & \cdots \\ 0 & 0 & 0 & 0 & 0 & \cdots & d_{N-1} & a_{N-1} & b_{N-1} \\ 0 & 0 & 0 & 0 & 0 & \cdots & e_N & d_N & a_N \end{bmatrix}. \tag{1.3.17}$$

The modified Thomas algorithm works in three stages. In the first stage, the pentadiagonal system (1.3.16) is reduced to the quadradiagonal system

$$\mathbf{Q} \cdot \mathbf{x} = \mathbf{r}', \tag{1.3.18}$$

involving the quadradiagonal coefficient matrix

$$
\mathbf{Q} =
\begin{bmatrix}
a_1' & b_1' & c_1' & 0 & 0 & \cdots & 0 & 0 & 0 \\
d_2' & a_2' & b_2' & c_2' & 0 & \cdots & 0 & 0 & 0 \\
0 & 0 & d_3' & a_3' & b_3' & \cdots & 0 & 0 & 0 \\
\cdots & \cdots & \cdots & \cdots & \cdots & \cdots & \cdots & \cdots & \cdots \\
0 & 0 & 0 & 0 & 0 & \cdots & d_{N-1}' & a_{N-1}' & b_{N-1}' \\
0 & 0 & 0 & 0 & 0 & \cdots & 0 & d_N' & a_N'
\end{bmatrix}.
\tag{1.3.19}
$$

In the second stage, system (1.3.18) is reduced to the upper tridiagonal system

$$
\mathbf{T} \cdot \mathbf{x} = \mathbf{r}'',
\tag{1.3.20}
$$

involving the upper tridiagonal coefficient matrix

$$
\mathbf{T} =
\begin{bmatrix}
1 & b_1'' & c_1'' & 0 & 0 & \cdots & 0 & 0 & 0 \\
0 & 1 & b_2'' & c_2'' & 0 & \cdots & 0 & 0 & 0 \\
0 & 0 & 1 & b_3'' & c_3'' & \cdots & 0 & 0 & 0 \\
\cdots & \cdots & \cdots & \cdots & \cdots & \cdots & \cdots & \cdots & \cdots \\
0 & 0 & 0 & 0 & 0 & \cdots & 0 & 1 & b_{N-1}'' \\
0 & 0 & 0 & 0 & 0 & \cdots & 0 & 0 & 1
\end{bmatrix}.
\tag{1.3.21}
$$

In the third stage, the upper tridiagonal system (1.3.20) is solved by backward substitution, which involves solving the last equation for the last unknown, x_N, and then moving upward to compute the rest of the unknowns in a sequential fashion. FSELIB function `penta`, listed in the text, implements the integrated process outlined in Algorithm 1.3.2.

1.3.2 Element matrices

To simplify the notation, we denote the lth-element nodes by

$$
x_{c_{l,1}} = P_1^{(l)}, \qquad x_{c_{l,2}} = P_3^{(l)}, \qquad x_{c_{l,3}} = P_3^{(l)}.
\tag{1.3.22}
$$

To compute the element diffusion and mass matrices, it is convenient to map each element from the x axis to the standard interval $[-1, 1]$ of the parametric ξ axis, so that the first point, $P_1^{(1)}$, is mapped to $\xi = -1$, and the third point, $P_3^{(l)}$, is mapped to $\xi = 1$, as shown in Figure 1.3.2. The mapping of an arbitrary point from the x to the ξ axis is mediated by the linear function

$$
x = \frac{1}{2} \left(P_1^{(l)} + P_3^{(l)} \right) + \frac{1}{2} \left(P_3^{(l)} - P_1^{(l)} \right) \xi.
\tag{1.3.23}
$$

Accordingly,

$$
\mathrm{d}x = \frac{1}{2} h_l \, \mathrm{d}\xi,
\tag{1.3.24}
$$

where $h_l = P_3^{(l)} - P_1^{(l)}$ is the element size.

Reduction to quadra-diagonal :

Do $i = 1, 2, \ldots, N - 2$
$\quad c_i' = c_i$
End Do
$c_{N-1} = 0.0, \qquad c_N = 0.0$ $\qquad\qquad$ $\leftarrow$ *Initialize for convenience*
$a_1' = a_1, \quad b_1' = b_1, \quad r_1' = r_1$
$d_2' = d_2, \quad a_2' = a_2, \quad b_2' = b_2, \quad r_2' = r_2$

Do $i = 2, 3, \ldots, N - 1$

$$\begin{bmatrix} a_{i+1}' \\ b_{i+1}' \\ d_{i+1}' \\ r_{i+1}' \end{bmatrix} = \begin{bmatrix} a_{i+1} \\ b_{i+1} \\ d_{i+1} \\ r_{i+1} \end{bmatrix} - \frac{e_{i+1}}{d_i'} \begin{bmatrix} b_i' \\ c_i' \\ a_i' \\ r_i' \end{bmatrix}$$

End Do

Reduction to upper tridiagonal:

$$\begin{bmatrix} b_1'' \\ c_1'' \\ r_1'' \end{bmatrix} = \frac{1}{a_1'} \begin{bmatrix} b_1'' \\ c_1'' \\ r_1'' \end{bmatrix}$$

Do $i = 1, 2, \ldots, N - 1$

$$\begin{bmatrix} b_{i+1}'' \\ c_{i+1}'' \\ r_{i+1}'' \end{bmatrix} = \frac{1}{a_{i+1}' - d_{i+1}' b_i'} \begin{bmatrix} b_{i+1} - d_{i+1}' c_i'' \\ c_{i+1}' \\ r_{i+1}' - d_{i+1}' r_i'' \end{bmatrix}$$

End Do

Backward substitution:

$x_N = r_N''$
$x_{N-1} = r_{N-1}'' - b_{N-1}'' x_N$

Do $i = N - 2, N - 1, \ldots, 1$
$\quad x_i = r_i'' - b_i'' x_{i+1} - c_i'' x_{i+2}$
End Do

Algorithm 1.3.2 Modified Thomas algorithm for solving the pentadiagonal
$\quad$ system (1.3.16).

```
function x = penta(n,a,b,c,d,e,r)

%=========================================
% Modified Thomas algorithm for solving
% a pentadiagonal system
%
% n:            system size
% a,b,c,d,e: diagonal, super, super-super,
%            sub, and sub-sub elements
% r:            right-hand side
%=========================================

na = n-1;
nb = n-2;

%-------------------------------
% reduction to quadra-diagonal
%-------------------------------

for i=1:nb
  c1(i) = c(i);
end

c1(na) = 0.0; c1(n) = 0.0;

a1(1) = a(1); b1(1) = b(1); r1(1) = r(1);
d1(2) = d(2); a1(2) = a(2); b1(2) = b(2); r1(2) = r(2);

for i=2:na
  i1 = i+1;
  w = e(i1)/d1(i);
  a1(i1) = a(i1) - w*b1(i);
  b1(i1) = b(i1) - w*c1(i);
  d1(i1) = d(i1) - w*a1(i);
  r1(i1) = r(i1) - w*r1(i);
end

%-------------------------------
% generate the system T x = r2
%-------------------------------
```

Function penta: Continuing $\longrightarrow$

Mid-point interior node

When the interior node is situated at the element mid-point corresponding to $\xi = 0$,

$$P_2^{(l)} = \frac{1}{2} \left(P_1^{(l)} + P_3^{(l)} \right). \tag{1.3.25}$$

In Chapter 2, we shall see that the midway position is optimal in the sense of delivering the highest possible interpolation accuracy for the available degrees of freedom.

```
b2(1) = b1(1)/a1(1);
c2(1) = c1(1)/a1(1);
r2(1) = r1(1)/a1(1);

for i=1:na
  i1 = i+1;
  den =  a1(i1)-d1(i1)*b2(i);
  b2(i1) = (b1(i1)-d1(i1)*c2(i))/den;
  c2(i1) =   c1(i1)/den;
  r2(i1) = (r1(i1)-d1(i1)*r2(i))/den;
end

%------------------
% back substitution
%------------------

x(n) = r2(n);

x(n-1) = r2(n-1)-b2(n-1)*x(n);

for i=nb:-1:1    % step of -1
   x(i) = r2(i)-b2(i)*x(i+1)-c2(i)*x(i+2);
end

%-----
% done
%-----

return;
```

Function penta: ($\longrightarrow$ Continued.) A FSELIB function for solving a pentadiagonal system of algebraic equations by the modified Thomas Algorithm 1.3.2.

In terms of the canonical variable ξ, the element interpolation functions are readily found to be

$$\psi_1^{(l)}(\xi) = \frac{1}{2}\,\xi\,(\xi - 1), \qquad \psi_2^{(l)}(\xi) = 1 - \xi^2,$$

$$\psi_3^{(l)}(\xi) = \frac{1}{2}\,\xi\,(\xi + 1). \tag{1.3.26}$$

Substituting the first interpolation function in the first equation of (1.3.6), and using (1.3.24), we find

$$A_{11}^{(l)} = \int_{E_l} \frac{\mathrm{d}\psi_1^{(l)}}{\mathrm{d}x}\,\frac{\mathrm{d}\psi_1^{(l)}}{\mathrm{d}x}\,\mathrm{d}x = \frac{1}{2}\,h_l \int_{-1}^{1} \frac{2}{h_l}\frac{\mathrm{d}\psi_1^{(l)}}{\mathrm{d}\xi}\,\frac{2}{h_l}\frac{\mathrm{d}\psi_1^{(l)}}{\mathrm{d}\xi}\,\mathrm{d}\xi$$

$$= \frac{1}{2\,h_l} \int_{-1}^{1} (2\,\xi - 1)^2\,\mathrm{d}\xi = \frac{1}{h_l}\,\frac{7}{3}. \tag{1.3.27}$$

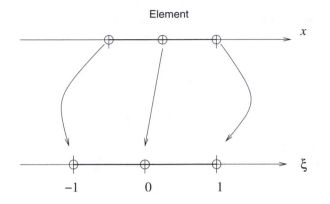

Element

Figure 1.3.2 Mapping of an element from the x axis to the standard interval $[-1, 1]$ of the parametric ξ axis. In this case, the mid-point interior node is mapped to the origin.

Working in a similar fashion with the rest of the elements, we derive the element diffusion matrix

$$\mathbf{A}^{(l)} = \frac{1}{3\,h_l} \begin{bmatrix} 7 & -8 & 1 \\ -8 & 16 & -8 \\ 1 & -8 & 7 \end{bmatrix}. \tag{1.3.28}$$

A straightforward computation shows that the determinant of this matrix is zero, which indicates that the matrix is singular (see also Problem 1.3.1(a)). The physical reason was discussed at the end of Section 1.1.8.

The components of the element mass matrix are given by

$$B_{ij}^{(l)} = \int_{E_l} \psi_i^{(l)} \, \psi_j^{(l)} \, \mathrm{d}x = \frac{1}{2}\,h_l \int_{-1}^{1} \psi_i^{(l)} \, \psi_j^{(l)} \, \mathrm{d}\xi. \tag{1.3.29}$$

Performing the integrations, we find

$$\mathbf{B}^{(l)} = \frac{h_l}{30} \begin{bmatrix} 4 & 2 & -1 \\ 2 & 16 & 2 \\ -1 & 2 & 4 \end{bmatrix}. \tag{1.3.30}$$

Note that the sum of all elements of this matrix is equal to the element size, h_l.

Arbitrary interior node

When the interior node is mapped to the point $\xi = \beta$, where $-1 < \beta < 1$,

$$P_2(l) = \frac{1}{2}\,(P_1^{(l)} + P_3^{(l)}) + \frac{1}{2}\,(P_3^{(l)} - P_1^{(l)})\,\beta. \tag{1.3.31}$$

The element interpolation functions are found to be

$$\psi_1^{(l)}(\xi) = \frac{1}{2\,(1+\beta)}\,(\xi - \beta)\,(\xi - 1),$$

$$\psi_2^{(l)}(\xi) = \frac{1 - \xi^2}{1 - \beta^2}, \qquad\qquad (1.3.32)$$

$$\psi_3^{(l)}(\xi) = \frac{1}{2\,(1-\beta)}\,(\xi - \beta)\,(\xi + 1),$$

and the components of the element diffusion matrix are found to be

$$A_{11}^{(l)} = \frac{1}{3\,h_l}\,\frac{1}{(1+\beta)^2}\,\Big(4 + 3\,(1+\beta)^2\Big),$$

$$A_{12}^{(l)} = A_{21}^{(l)} = -\frac{1}{3\,h_l}\,\frac{8}{(1+\beta)(1-\beta^2)},$$

$$A_{13}^{(l)} = A_{31}^{(l)} = \frac{1}{3\,h_l}\,\frac{1}{(1-\beta^2)}\,\Big(4 - 3\,(1-\beta^2)\Big),$$

$$A_{22}^{(l)} = \frac{1}{3\,h_l}\,\frac{16}{(1-\beta^2)^2},$$

$$A_{23}^{(l)} = A_{32}^{(l)} = -\frac{1}{3\,h_l}\,\frac{8}{(1-\beta)(1-\beta^2)}, \qquad\qquad (1.3.33)$$

$$A_{33}^{(l)} = \frac{1}{3\,h_l}\,\frac{1}{(1-\beta)^2}\,\Big(4 + 3\,(1-\beta)^2\Big).$$

When $\beta = 0$, we recover the previous expressions for the mid-point interior node.

To avoid a lengthy algebra, the element mass matrix can be computed numerically using the four-point Lobatto integration quadrature discussed in Section 2.3.2. Alternatively, both the diffusion and element matrices can be computed from those corresponding to the mid-point interior node using the transformation rules discussed in Section 2.1.3. Both methods are implemented in the code sdqb discussed in the next section.

1.3.3   Finite element code

We may now combine the elementary modules developed previously in this section into an integrated finite element code that performs three main functions: domain discretization, assembly of the linear system, and solution of the linear system. To economize the computer memory, we store the non-zero diagonal components of the global diffusion matrix in five vectors,

$$e_i^p \equiv D_{i,i-2}, \qquad d_i^p \equiv D_{i,i-1}, \qquad a_i^p \equiv D_{i,i},$$

$$b_i^p \equiv D_{i,i+1}, \qquad c_i^p \equiv D_{i,i+2}, \qquad\qquad (1.3.34)$$

corresponding to the pentadiagonal form shown in (1.3.17).

```
function [ap,bp,cp,dp,ep,b] = sdq_sys (ne,xe,q0,fL,k,s)

%=======================================================
%
% Compact assembly of the pentadiagonal linear system
% for one-dimensional steady diffusion
% with quadratic elements (sdq)
%
%=======================================================

%-------------
% element size
%-------------

for l=1:ne
  h(l) = xe(l+1)-xe(l);
end

%----------------------------
% number of unique global nodes
%----------------------------

ng = 2*ne+1;

%------------------------------------
% initialize the pentadiagonal matrix
% and right-hand side
%------------------------------------

ap = zeros(ng,1); bp = zeros(ng,1); cp = zeros(ng,1);
dp = zeros(ng,1); ep = zeros(ng,1);

b = zeros(ng,1); b(1) = q0/k;
```

Function sdq_sys: Continuing $\longrightarrow$

FSELIB function sdq_sys, listed in the text, implements the assembly Algorithm 1.3.1, for mid-point interior nodes; "sdq" is an acronym for steady diffusion with quadratic elements. FSELIB code sdq, listed in the text, implements the corresponding complete finite element code in the following familiar modules:

- *Data input.*

- *Finite element discretization.*

- *Specification of the source function.*

- *Assembly of the linear system.*

- *Solution of the linear system.*

- *Plotting of the solution.*

```
%-----------------------
% loop over all elements
%-----------------------

for l=1:ne

  cf = 1.0/(3.0*h(l));

  A11 = 7.0*cf; A12 = -8.0*cf; A13 = cf;
  A21 = A12;    A22 = 16.0*cf; A23 = A12;
  A31 = A13;    A32 = A23;     A33 = A11;

  cf = h(l)/30.0;

  B11 = 4.0*cf; B12 =  2.0*cf; B13 = -1.0*cf;
  B21 = B12;    B22 = 16.0*cf; B23 = B12;
  B31 = B13;    B32 = B23;     B33 = B11;

  cl1 = 2*l-1; cl2 = 2*l; cl3 = 2*l+1;

  ap(cl1) = ap(cl1) + A11;
  bp(cl1) = bp(cl1) + A12;
  cp(cl1) = cp(cl1) + A13;

  dp(cl2) = dp(cl2) + A21;
  ap(cl2) = ap(cl2) + A22;
  bp(cl2) = bp(cl2) + A23;

  ep(cl3) = ep(cl3) + A31;
  dp(cl3) = dp(cl3) + A32;
  ap(cl3) = ap(cl3) + A33;

  b(cl1) = b(cl1) + (B11*s(cl1) + B12*s(cl2) + B13*s(cl3))/k;
  b(cl2) = b(cl2) + (B21*s(cl1) + B22*s(cl2) + B23*s(cl3))/k;
  b(cl3) = b(cl3) + (B31*s(cl1) + B32*s(cl2) + B33*s(cl3))/k;

end

%----------------------------------
% implement the Dirichlet condition
%----------------------------------

b(ng-2) = b(ng-2) - cp(ng-2)*fL;
b(ng-1) = b(ng-1) - bp(ng-1)*fL;

%-----
% done
%-----

return;
```

Function sdq_sys: ($\longrightarrow$ Continued.) Compact assembly of the pentadiagonal linear system for one-dimensional diffusion with quadratic elements and mid-point interior nodes.

```
%==========================================
%
% CODE sdq
%
% Code for steady one-dimensional diffusion
% with quadratic elements
%
%==========================================

%-----------
% input data
%-----------

L=1.0; k=1.0; q0=-1.0; fL=0.0;

ne=4; ratio=2.0;

%----------------
% grid generation
%----------------

xe = elm_line1 (0,L,ne,ratio);

%-----------------------------
% number of unique global nodes
%-----------------------------

ng = 2*ne+1;

%-------------------------
% generate the global nodes
%-------------------------

Ic = 0;  % node counter

for i=1:ne
 Ic = Ic+1; xg(Ic) = xe(i);
 Ic = Ic+1; xg(Ic) = 0.5*(xe(i)+xe(i+1));
end

xg(ng) = xe(ne+1);

%-------------------
% specify the source
%-------------------

for i=1:ng
 s(i) = 10.0*exp(-5.0*xg(i)^2/L^2);
end

%------------------------
% compact element assembly
%------------------------
```

Code sdq: Continuing $\longrightarrow$

```
[ap,bp,cp,dp,ep,b] = sdq_sys (ne,xe,q0,fL,k,s);

%---------------
% linear solver
%---------------

f = penta (ng-1,ap,bp,cp,dp,ep,b);

f(ng) = fL;

%-----
% plot
%-----

plot(xg, f,'-o'); hold on;

plot(xe, zeros(ne+1,1),'x');

xlabel('x'); ylabel('f');

%-----
% done
%-----
```

Code sdq: ($\longrightarrow$ Continued.) Finite element code for steady one-dimensional diffusion with quadratic elements.

The graphics output generated by the code with the frivolous source function defined in (1.1.75) is shown with circles in Figure 1.3.3.

Calculations with uniformly spaced elements, *ratio*=1, and number of elements,

$$N_E = 4, \quad 8, \quad 16, \quad 32, \quad 64, \tag{1.3.35}$$

corresponding to system size

$$2\,N_E = 8, \quad 16, \quad 32, \quad 64, \quad 128, \tag{1.3.36}$$

yield the left-end value

$$f(0) = 1.9633\ 9024, \quad 1.9638\ 3352, \quad 1.9638\ 5935, \tag{1.3.37}$$
$$1.9638\ 6094, \quad 1.9638\ 61036.$$

These results are more accurate than those listed in Section 1.1.11 for linear elements. A computation with a large number of elements reveals the exact left-end value, 1.9638 6104, accurate to the eighth decimal place. The numerical error corresponding to the sequence (1.3.36), $e(0) \equiv f(0) - 1.9638\ 6104$, is

$$e(0) = -0.0004\ 7080, \quad -0.0000\ 2752, \quad -0.0000\ 0169, \tag{1.3.38}$$
$$-0.0000\ 0011, \quad -0.0000\ 0001.$$

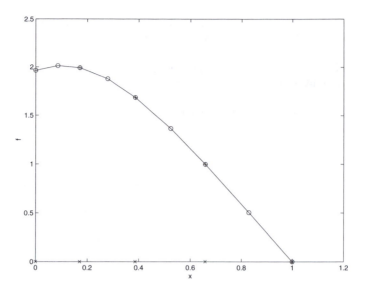

Figure 1.3.3 Numerical solution of the steady diffusion equation with four
quadratic elements, $N_E = 4$, and element stretch ratio equal to two, for
$L = 1.0$, $k = 1.0$, $q_0 = -1.0$, and $f(L) = 0.0$. The element end-points are
marked with an x on the x axis. The circles represent the results of the
FSELIB code sdq, and the coincident crosses represent the results of the
FSELIB code sdqc based on node condensation.

We observe that, as the number of elements is doubled, the numerical error
is reduced approximately by a factor of sixteen, which means that the error is
proportional to h^4, where h is the element size. Thus, the finite element method
with quadratic elements for the steady diffusion equation with a smooth source
function is fourth-order accurate in the element size.

FSELIB code sdqb (not listed in the text) solves the steady diffusion problem
with quadratic elements and an arbitrary position of the interior element node
determined by the parameter β. The element diffusion matrix is computed
using expressions (1.3.33), and the element mass matrix is computed using the
four-point Lobatto integration quadrature discussed in Section 2.3.2. The finite
element solution shows some sensitivity to β (Problem 1.3.3).

1.3.4 Node condensation

Careful inspection of the global diffusion matrix shown in (1.3.11) reveals that
the unknowns corresponding to the interior element nodes can be eliminated in
favor of the unknowns corresponding to the element end-nodes, and the size of
the linear system can be reduced from $2N_E \times 2N_E$ to $N_E \times N_E$ in a process that
is known as *node condensation*. Because the condensed system is tridiagonal,

the solution can be found efficiently using the standard Thomas algorithm. To implement node condensation, we work as follows:

- Multiply the second equation by the ratio $r_{12}^{(1)} \equiv A_{12}^{(1)}/A_{22}^{(1)}$, and subtract the outcome from the first equation.

- Multiply the second equation by the ratio $r_{32}^{(1)} \equiv A_{32}^{(1)}/A_{22}^{(1)}$, and subtract the outcome from the third equation.

- Repeat for all subsequent element blocks.

- Remove the Galerkin equations corresponding to the interior nodes.

At the end, we obtain the $N_E \times N_E$ condensed system

$$\widehat{\mathbf{D}} \cdot \widehat{\mathbf{f}} = \widehat{\mathbf{b}}, \tag{1.3.39}$$

where

$$\widehat{\mathbf{D}} = \begin{bmatrix} A_{11}^{(1)} - r_{12}^{(1)} A_{21}^{(1)} & A_{13}^{(1)} - r_{12}^{(1)} A_{23}^{(1)} & 0 \\[2mm] A_{31}^{(1)} - r_{32}^{(1)} A_{21}^{(1)} & A_{33}^{(1)} - r_{32}^{(1)} A_{23}^{(1)} + A_{11}^{(2)} - r_{12}^{(2)} A_{21}^{(2)} & A_{13}^{(2)} - r_{12}^{(2)} A_{23}^{(2)} \\[2mm] 0 & A_{31}^{(2)} - r_{32}^{(2)} A_{21}^{(2)} & \cdots \\[2mm] 0 & 0 & 0 \\[2mm] \cdots & \cdots & \cdots \\[2mm] 0 & 0 & \cdots \\[2mm] 0 & 0 & \cdots \end{bmatrix},$$

$$\tag{1.3.40}$$

is the condensed diffusion matrix, and

$$\widehat{\mathbf{f}} \equiv \begin{bmatrix} f_1 \\ f_3 \\ \vdots \\ f_{2N_E-3} \\ f_{2N_E-1} \end{bmatrix} \tag{1.3.41}$$

is the vector of function values at the end-nodes. The right-hand side of the condensed system is given by

$$\widehat{b}_1 = \frac{q_0}{k} + \left(B_{11}^{(1)} - r_{12}^{(1)} B_{21}^{(1)} \right) \frac{s_1}{k} + \left(B_{12}^{(1)} - r_{12}^{(1)} B_{22}^{(1)} \right) \frac{s_2}{k}$$
$$+ \left(B_{13}^{(1)} - r_{12}^{(1)} B_{23}^{(1)} \right) \frac{s_3}{k},$$

$$\widehat{b}_2 = \left(B_{31}^{(1)} - r_{32}^{(1)} B_{21}^{(1)} \right) \frac{s_1}{k} + \left(B_{32}^{(1)} - r_{32}^{(1)} B_{22}^{(1)} \right) \frac{s_2}{k}$$
$$+ \left(B_{33}^{(1)} - r_{32}^{(1)} B_{23}^{(1)} + B_{11}^{(2)} - r_{12}^{(2)} B_{21}^{(2)} \right) \frac{s_3}{k}$$
$$+ \left(B_{12}^{(2)} - r_{12}^{(2)} B_{22}^{(2)} \right) \frac{s_4}{k} + \left(B_{13}^{(2)} - r_{12}^{(2)} B_{23}^{(2)} \right) \frac{s_5}{k},$$

...

$$(1.3.42)$$

Once the solution of the condensed system for the element end-nodes has been found, the solution at the interior nodes can be recovered element by element, by resorting to the removed equations.

The condensed global diffusion matrix shown in (1.3.40) consists of overlapping 2×2 condensed element matrices,

$$\widehat{\mathbf{A}}^{(l)} = \begin{bmatrix} A_{11}^{(l)} - r_{12}^{(l)} A_{21}^{(l)} & A_{13}^{(l)} - r_{12}^{(l)} A_{23}^{(l)} \\ A_{31}^{(l)} - r_{32}^{(l)} A_{21}^{(l)} & A_{33}^{(l)} - r_{32}^{(l)} A_{23}^{(l)} \end{bmatrix}. \qquad (1.3.43)$$

Accordingly, condensation can be carried out on the element level, as implemented in the FSELIB function sdqc_sys, listed in the text; "sdqc" is an acronym for steady diffusion with quadratic elements and the condensed formulation. FSELIB code sdqc, listed in the text, implements the finite element method. The graphics output generated by this code with the frivolous source function defined in (1.1.75) is shown with crosses in Figure 1.3.3. The results are identical to those obtained previously in the absence of condensation.

When the interior node is located at the element mid-point, we use (1.3.28) to find $r_{12}^{(1)} = r_{32}^{(1)} = -1/2$, and

$$\widehat{\mathbf{A}}^{(l)} = \frac{1}{h_l} \begin{bmatrix} 1 & -1 \\ -1 & 1 \end{bmatrix}, \qquad (1.3.44)$$

which is precisely the diffusion matrix with linear elements shown in 1.1.62). This expression holds true for an arbitrary position of the interior node.

1.3.5   Relation to the finite difference method

If the elements are evenly spaced, $h_1 = h_2 = \ldots = h$, and the interior element nodes are located at the element mid-points, we may use expressions (1.3.30) to

```
function [at,bt,ct,b] = sdqc_sys (ne,xe,q0,fL,k,s)

%================================================================
% Compact assembly of the condensed tridiagonal linear system
% for one-dimensional steady diffusion
% with quadratic elements (sdqc)
%================================================================

%-------------
% element size
%-------------

for l=1:ne
  h(l) = xe(l+1)-xe(l);
end

%--------------------------------
% initialize the tridiagonal matrix
%--------------------------------

at = zeros(ne+1,1); bt = zeros(ne+1,1); ct = zeros(ne+1,1);

%-------------------------------
% initialize the right-hand side
%-------------------------------

b = zeros(ne+1,1); b(1) = q0/k;

%----------------------
% loop over all elements
%----------------------

for l=1:ne

  cf = 1.0/(6.0*h(l));

  A11 = 14.0*cf; A12 =-16.0*cf; A13 = 2.0*cf;
  A21 = A12;     A22 = 32.0*cf; A23 = A12;
  A31 = A13;     A32 = A23;     A33 = A11;

  cf = h(l)/30.0;

  B11 = 4.0*cf; B12 =  2.0*cf; B13 = -1.0*cf;
  B21 = B12;    B22 = 16.0*cf; B23 = B12;
  B31 = B13;    B32 = B23;     B33 = B11;

  % condense the diffusion matrix:

  r12 = A12/A22;
  A11 = A11-r12*A21; A13= A13-r12*A23;

  r32=A32/A22;
  A31 = A31-r32*A21; A33= A33-r32*A23;
```

Function sdqc_sys: Continuing $\longrightarrow$

```
at(1)   = at(1)   + A11; bt(1)   = bt(1)   + A13;
ct(1+1) = ct(1+1) + A31; at(1+1) = at(1+1) + A33;

% condense the right-hand side

cl1=2*1-1; cl2=2*1; cl3=2*1+1;

b(1)   = b(1)   +       (B11*s(cl1) + B12*s(cl2) + B13*s(cl3))/k;
b(1)   = b(1)   - r12*(B21*s(cl1) + B22*s(cl2) + B23*s(cl3))/k;
b(1+1) = b(1+1) +       (B31*s(cl1) + B32*s(cl2) + B33*s(cl3))/k;
b(1+1) = b(1+1) - r32*(B21*s(cl1) + B22*s(cl2) + B23*s(cl3))/k;

end

%-----------------------------------
% implement the Dirichlet condition
%-----------------------------------

b(ne) = b(ne)-A13*fL;

%-----
% done
%-----

return;
```

Function sdqc_sys: ($\longrightarrow$ Continued.) Compact assembly of the condensed tridiagonal linear system for one-dimensional diffusion with quadratic elements and mid-point interior nodes.

find that the right-hand side of the condensed linear system stated in (1.3.42) simplifies to

$$\widehat{b}_1 = \frac{q_0}{k} + \frac{h}{k}\frac{5}{30}(s_1 + 2s_2),$$

$$\widehat{b}_2 = \frac{h}{k}\frac{1}{3}(s_2 + s_3 + s_4), \qquad (1.3.45)$$

$$\cdots$$

The ith equation of the condensed system then takes the form

$$k\frac{f_{2i-3} - 2f_{2i-1} + f_{2i+1}}{h^2} + \frac{1}{3}(s_{2i-2} + s_{2i-1} + s_{2i}) = 0, \qquad (1.3.46)$$

where $i > 1$. For convenience, we set $2i - 1 = j$, and restate (1.3.46) as

$$k\frac{f_{j-2} - 2f_j + f_{j+2}}{h^2} + \frac{1}{3}(s_{j-1} + s_j + s_{j+1}) = 0, \qquad (1.3.47)$$

with the understanding that the index j is odd and greater than unity.

```
%=========================================
% CODE sdqc
%
% Code for steady one-dimensional diffusion
% with quadratic elements using the
% condensed formulation
%=========================================

%-----------
% input data
%-----------

L=1.0; k=1.0; q0=-1.0; fL=0.0;

ne=4; ratio=2.0;

%----------------
% grid generation
%----------------

xe = elm_line1 (0,L,ne,ratio);

%-----------------------------
% number of unique global nodes
%-----------------------------

ng = 2*ne+1;

%-------------------------
% generate the global nodes
%-------------------------

Ic = 0;  % counter

for i=1:ne
  Ic = Ic+1; xg(Ic) = xe(i);
  Ic = Ic+1; xg(Ic) = 0.5*(xe(i)+xe(i+1));
end

xg(ng) = xe(ne+1);

%-------------------
% specify the source
%-------------------

for i=1:ng
  s(i) = 10.0*exp(-5.0*xg(i)^2/L^2);
end

%------------------------
% compact element assembly
%------------------------
```

Code sdqc: Continuing $\longrightarrow$

```
[at,bt,ct,b] = sdqc_sys (ne,xe,q0,fL,k,s);

%---------------
% linear solver
%-------------

sol = thomas(ne,at,bt,ct,b);

%--------------------
% recover the solution
%--------------------

for i=1:ne
 f(2*i-1)=sol(i);
 f(2*i)=0.0;        % ignore the mid-nodes
end
f(ng)=fL;

%-----
% plot
%-----

for i=1:ne+1
 yplot(i)=f(2*i-1);
end

plot(xe, yplot,'+'); hold on;
plot(xe, zeros(ne+1,1),'x');
xlabel('x'); ylabel('f');
```

Code sdqc: ($\longrightarrow$ Continued.) Finite element code for steady one-dimensional diffusion with quadratic elements, using the condensed formulation.

To demonstrate the connection with the finite difference method, we apply the governing equation (1.1.4) at nodes $j - 1$, j, and $j + 1$, and approximate the second derivative with finite differences involving the five consecutive nodal values f_{j-2}, f_{j-1}, f_j, f_{j+1}, and f_{j+2}. Taking into consideration that these nodes are separated by the interval $h/2$, we find

$$\frac{k}{3\,h^2} \left(11\,f_{j-2} - 20\,f_{j-1} + 6\,f_j + 4\,f_{j+1} - f_{j+2}\right) + s_{j-1} = 0,$$

$$\frac{k}{3\,h^2} \left(-f_{j-2} + 16\,f_{j-1} - 30\,f_j + 16\,f_{j+1} - f_{j+2}\right) + s_j = 0, \qquad (1.3.48)$$

$$\frac{k}{3\,h^2} \left(-f_{j-2} + 4\,f_{j-1} + 6\,f_j - 20\,f_{j+1} + 11\,f_{j+2}\right) + s_{j+1} = 0,$$

The coefficients on the left-hand sides of the first and third equations are produced by the FSELIB function fdc using the method of undetermined coefficients; the corresponding coefficients in the second equation are available from differentiation tables (e.g., [43] pp. 315–318). Adding these equations and simplifying, we recover precisely (1.3.46). Thus, the finite element method with quadratic element functions and uniform elements is equivalent to the finite difference method implemented as discussed in this section with the five-point difference approximation for the interior nodes.

1.3.6   Modal expansion

The quadratic approximation of any suitable function $f(\xi)$ over the lth element takes the form

$$f(\xi) = f_1^E \, \psi_1^{(l)}(\xi) + f_2^E \, \psi_2^{(l)}(\xi) + f_3^E \, \psi_3^{(l)}(\xi), \qquad (1.3.49)$$

where f_1^E, f_2^E, and f_3^E are the element-node values, and ψ_i^E are the corresponding quadratic local interpolation functions. An alternative representation is provided by the *modal* expansion

$$f(\xi) = f_1^E \, \zeta_1^{(l)}(\xi) + c_2^E \, \zeta_2^{(l)}(\xi) + f_3^E \, \zeta_3^{(l)}(\xi), \qquad (1.3.50)$$

where

$$\zeta_1^{(l)}(\xi) = \frac{1 - \xi}{2}, \qquad \zeta_3^{(l)}(\xi) = \frac{1 + \xi}{2}, \qquad (1.3.51)$$

are *linear* end-node interpolation functions independent of the location of the interior node, and c_2^E is a constant coefficient. To satisfy the end-node interpolation conditions, $f(\xi = -1) = f_1^E$ and $f(\xi = 1) = f_3^E$ we require that the intermediate *quadratic* modal function, $\zeta_2^{(l)}(\xi)$, drops to zero at the end nodes located at $\xi = \pm 1$, and set

$$\zeta_2^{(l)}(\xi) = 1 - \xi^2. \qquad (1.3.52)$$

The Galerkin finite element method of the modal expansion is formulated as discussed previously in this section, the only difference being that the element-node interpolation functions, $\psi_i^{(l)}(\xi)$, are replaced by the modal functions $\zeta_i^{(l)}(\xi)$, and the interior node value, f_2^E, is replaced by the coefficient c_2^E. Once f_1^E, c_2^E, and f_3^E are available, the solution can be reconstructed from the three-term modal expansion. For example, the solution at the element mid-point corresponding to $\xi = 0$, is recovered as

$$f_2^E = f(\xi = 0) = \frac{f_1^E + 2\,c_2^E + f_3^E}{2}. \qquad (1.3.53)$$

In the finite element implementation, the source term over the lth-element is approximated with the quadratic modal expansion

$$s(\xi) = s_1^E \, \psi_1^{(l)}(\xi) + d_2^E \, \psi_2^{(l)}(\xi) + s_3^E \, \psi_3^{(l)}(\xi), \qquad (1.3.54)$$

where the coefficient d_2^E can be evaluated by inverting the counterpart of (1.3.53) for the source,

$$d_2^E = \frac{-s_1^E + 2\,s_2^E - s_3^E}{2}. \qquad (1.3.55)$$

The global diffusion and mass matrices are assembled from corresponding element matrices defined with respect to the modal functions, $\zeta_i^{(l)}$. Specifically, the lth-element diffusion matrix is defined as

$$A_{ij}^{(l)} = \int_{E_l} \frac{d\zeta_i^{(l)}}{dx} \frac{d\zeta_j^{(l)}}{dx} \, dx = \frac{1}{2} h_l \int_{-1}^{1} \frac{2}{h_l} \frac{d\zeta_i^{(l)}}{d\xi} \frac{2}{h_l} \frac{d\zeta_j^{(l)}}{d\xi} \, d\xi$$

$$= \frac{2}{h_l} \int_{-1}^{1} \frac{d\zeta_i^{(l)}}{d\xi} \frac{d\zeta_j^{(l)}}{d\xi} \, d\xi. \tag{1.3.56}$$

Performing the integrations, we find

$$\boldsymbol{A}^{(l)} = \frac{1}{3\,h_l} \begin{bmatrix} 3 & 0 & -3 \\ 0 & 16 & 0 \\ -3 & 0 & 3 \end{bmatrix}. \tag{1.3.57}$$

As in the case of the nodal expansion, this matrix is singular. The element mass matrix is defined as

$$B_{ij}^{(l)} = \int_{E_l} \zeta_i^{(l)} \zeta_j^{(l)} \, dx = \frac{1}{2} h_l \int_{-1}^{1} \zeta_i^{(l)} \zeta_j^{(l)} \, d\xi. \tag{1.3.58}$$

Performing the integrations, we find

$$\boldsymbol{B}^{(l)} = \frac{h_l}{30} \begin{bmatrix} 10 & 10 & 5 \\ 10 & 16 & 10 \\ 5 & 10 & 10 \end{bmatrix}. \tag{1.3.59}$$

Note that the modal element diffusion and mass matrices differ from the corresponding nodal matrices displayed in (1.3.28) and (1.3.30), with one exception: since $\zeta_2^{(l)}(\xi) = \psi_2^{(l)}(\xi) = 1 - \xi^2$, the values $A_{22}^{(l)} = 16/(3\,h_l)$ and $B_{22}^{(l)} = 8\,h_l/15$, are consistent with those shown in (1.3.28) and (1.3.30).

Because of the sparseness of the element diffusion matrix, the global diffusion matrix takes the form

$$\boldsymbol{D} = \begin{bmatrix} D_{11} & 0 & D_{13} & 0 & 0 & 0 & \dots \\ 0 & D_{22} & 0 & 0 & 0 & 0 & \dots \\ D_{31} & 0 & D_{33} & 0 & D_{35} & 0 & \dots \\ 0 & 0 & 0 & D_{44} & 0 & 0 & \dots \\ 0 & 0 & D_{53} & 0 & D_{55} & 0 & \dots \\ \dots & \dots & \dots & \dots & \dots & \dots & \dots \end{bmatrix}. \tag{1.3.60}$$

The odd-numbered unknowns representing the nodal values of the solution are decoupled from the even-number unknowns corresponding to the coefficients c_2^E of the modal functions $\zeta_2^{(l)}(\xi)$, and can be computed by solving a tridiagonal system of equations. The even-numbered unknowns can be computed immediately by dividing the corresponding entries on the right-hand side with the diagonal components. Thus, the modal expansion effectively implements node condensation. The algorithm is implemented in the FSELIB code sdqm (not listed in the text). The results are identical to those obtained previously using the nodal expansion with mid-point interior nodes.

The advantages of the modal expansion (1.3.50) will be discussed in Chapter 2 in the context of high-order and spectral element methods where an arbitrary number of interior modes, also called bubble modes, are added to the two linear end-node interpolation functions representing the so-called vertex modes displayed in (1.3.51). The bubble modes are multiplied by *a priori* unknown coefficients that are not necessarily associated with any interior nodes. The advantage of the modal approach, is that, if the modes are chosen judiciously, the resulting global diffusion or mass matrices exhibit a desirable sparsity that promotes the well-conditioning and expedites the numerical solution of the final system of linear equations.

PROBLEMS

1.3.1 *Evaluation of the diffusion and mass matrices.*

(*a*) Derive the components of the element diffusion matrix shown in (1.3.28), and then confirm that this matrix is singular. *Hint:* Observe that the sum of each row is zero.

(*b*) Derive the components of the element mass matrix shown in (1.3.30).

1.3.2 *Codes* sdq *and* sdqc.

(*a*) Execute the FSELIB code sdq with the same boundary conditions but a different source function of your choice, and discuss the results of your computation.

(*b*) Repeat (*a*) for code sdqc, and confirm that the results are identical.

1.3.3 *Arbitrary interior node.*

(*a*) Verify that the element diffusion matrix is singular for any position of the interior node corresponding to an arbitrary value of β. *Hint:* Run script edmb of FSELIB.

(*b*) Execute the code sdqb with conditions of your choice, and discuss the effect of β on the accuracy of the solution.

1.3.4 *Helmholtz equation.*

Consider the Helmholtz equation in one dimension,

$$\frac{\mathrm{d}^2 f}{\mathrm{d}x^2} + \alpha\, f = 0, \tag{1.3.61}$$

where α is a constant coefficient. The solution is to be found over the interval $[0, L]$, subject to the boundary conditions stated in (1.1.4). Modify the code `sdqb` into a code that solves the Helmholtz equation, and confirm that the position of the interior element node has no effect on the accuracy of the solution.

1.4 Unsteady diffusion in one dimension

Building on our earlier discussion, we consider the unsteady version of the heat conduction problem illustrated in Figure 1.1.1, where the rod temperature now evolves in time. Performing a heat balance over an infinitesimal section of the rod contained between x and $x + dx$, we find that the evolution of the temperature, $f(x, t)$, is governed by the unsteady heat conduction equation

$$\rho\, c_p \frac{\partial f}{\partial t} = k \frac{\partial^2 f}{\partial x^2} + s(x, t), \tag{1.4.1}$$

where ρ is the density of the rod material, and c_p is the heat capacity under constant pressure. Dividing both sides of (1.4.1) by $\rho\, c_p$, we derive the alternative form

$$\frac{\partial f}{\partial t} = \kappa \frac{\partial^2 f}{\partial x^2} + \frac{s(x, t)}{\rho\, c_p}, \tag{1.4.2}$$

where

$$\kappa \equiv \frac{k}{\rho\, c_p} \tag{1.4.3}$$

is the *thermal diffusivity* with dimensions of length squared divided by time. Our goal is to compute a numerical solution of (1.4.2) subject to a given initial condition, $f(x, t = 0) = F(x)$, and two boundary conditions, one at each end.

1.4.1 Galerkin projection

To formulate the Galerkin finite element method, we follow the steps outlined previously in this chapter for the problem of steady diffusion, the main difference being that the unknown nodal values, $f_i(t) \equiv f(x_i, t)$, are now time dependent. The ith Galerkin projection of (1.4.2) yields the counterpart of equation (1.1.23),

$$\int_0^L \phi_i \frac{\partial f}{\partial t}\, dx = -\kappa \int_0^L \frac{\partial \phi_i}{\partial x} \frac{\partial f}{\partial x}\, dx + \frac{q_0}{\rho\, c_p} \delta_{i1}$$

$$+ \frac{1}{\rho\, c_p} \int_0^L \phi_i\, s(x, t)\, dx, \tag{1.4.4}$$

where i runs over an appropriate number of nodes determined by the chosen degree of element interpolation functions and specified boundary conditions,

Dirichlet versus Neumann, as discussed in Section 1.1. Next, we substitute in (1.4.4) the finite element expansion

$$f(x) = \sum_{j=1}^{N_G} f_j(t)\,\phi_j(x), \tag{1.4.5}$$

and a corresponding expansion for the source term,

$$s(x) = \sum_{j=1}^{N_G} s_j(t)\,\phi_j(x), \tag{1.4.6}$$

where N_G is the number of unique global nodes. Rearranging, we derive a system of linear ordinary differential equations (ODEs),

$$\mathbf{M} \cdot \frac{d\mathbf{f}}{dt} + \kappa\,\mathbf{D} \cdot \mathbf{f} = \kappa\,\mathbf{b}, \tag{1.4.7}$$

where $\mathbf{M}$ is the global mass matrix, $\mathbf{D}$ is the global diffusion matrix, and $\mathbf{b}$ is a properly constructed right-hand side.

Linear elements

In the case of the piecewise linear expansion (1.1.9), subject to the boundary conditions (1.1.5) and (1.1.6), the differential system (1.4.7) is the counterpart of the algebraic system (1.1.29), involving the square, tridiagonal, $N_E \times N_E$ global mass matrix

$$
\mathbf{M} =
\begin{bmatrix}
B_{11}^{(1)} & B_{12}^{(1)} & 0 & 0 & \ldots \\
B_{21}^{(1)} & B_{22}^{(1)} + B_{11}^{(2)} & B_{12}^{(2)} & 0 & \ldots \\
0 & B_{21}^{(2)} & B_{22}^{(2)} + B_{11}^{(3)} & B_{12}^{(3)} & \ldots \\
\ldots & \ldots & \ldots & \ldots & \ldots \\
0 & 0 & \ldots & \ldots & \ldots \\
0 & 0 & \ldots & \ldots & \ldots
\end{bmatrix}
$$

$$
\begin{bmatrix}
\ldots & 0 & 0 \\
\ldots & 0 & 0 \\
\ldots & 0 & 0 \\
\ldots & & \\
\ldots & \ldots & \\
B_{21}^{(N_E-2)} & B_{22}^{(N_E-2)} + B_{11}^{(N_E-1)} & B_{12}^{(N_E-1)} \\
0 & B_{21}^{(N_E-1)} & B_{22}^{(N_E-1)} + B_{11}^{(N_E)}
\end{bmatrix}.
$$

$$\tag{1.4.8}$$

Note that $\mathbf{M}$ is identical to the bordered mass matrix $\widetilde{\mathbf{M}}$ shown in (1.1.34), except that the last column has been suppressed. The constant vector $\mathbf{b}$ on the right-hand side of (1.4.7) is defined in equations (1.1.32)–(1.1.35).

Evaluating the mass matrix, we find that the ith equation of the Galerkin finite element system (1.4.7) takes the form

$$\frac{h_{i-1}}{6}\frac{\mathrm{d}f_{i-1}}{\mathrm{d}t} + \frac{h_{i-1}+h_i}{3}\frac{\mathrm{d}f_i}{\mathrm{d}t} + \frac{h_i}{6}\frac{\mathrm{d}f_{i+1}}{\mathrm{d}t}$$

$$+\kappa \sum_{j=1}^{N_E} D_{ij}\, f_j = \kappa\, b_i, \qquad (1.4.9)$$

for $i = 2, 3, \ldots, N_E$, which can be rearranged into

$$\left(\frac{1}{3}\frac{h_{i-1}}{h_{i-1}+h_i}\right)\frac{\mathrm{d}f_{i-1}}{\mathrm{d}t} + \left(\frac{2}{3}\right)\frac{\mathrm{d}f_i}{\mathrm{d}t} + \left(\frac{1}{3}\frac{h_i}{h_{i-1}+h_i}\right)\frac{\mathrm{d}f_{i+1}}{\mathrm{d}t}$$

$$+\kappa\, \frac{2}{h_{i-1}+h_i} \sum_{j=1}^{N_E} D_{ij}\, f_j = \kappa\, \frac{2}{h_{i-1}+h_i}\, b_i. \qquad (1.4.10)$$

The first three terms on the left-hand side of (1.4.10) contribute a weighted average of the time derivative at three nodes, where the sum of the weights enclosed by the parentheses is equal to unity.

When the element size is uniform, $h_1 = h_2 = \cdots = h_{N_E} \equiv h$, the Galerkin equation (1.4.10) simplifies to

$$\frac{1}{6}\frac{\mathrm{d}f_{i-1}}{\mathrm{d}t} + \frac{2}{3}\frac{\mathrm{d}f_i}{\mathrm{d}t} + \frac{1}{6}\frac{\mathrm{d}f_{i+1}}{\mathrm{d}t} = \kappa\, \frac{f_{i-1} - 2f_i + f_{i+1}}{h^2} + \frac{\kappa}{h}\, b_i. \qquad (1.4.11)$$

The first term on the right-hand side is recognized as the central difference approximation of the second derivative, $\partial^2 f/\partial x^2$, evaluated at the ith node. The three terms on the left-hand side express a weighted average of the time derivative at the ith and adjacent nodes. A similar averaging of the source function is implicit in the constant term, b_i, on the right-hand side.

1.4.2   Integrating ODEs

Equation (1.4.7) provides us with a coupled system of first-order, linear ordinary differential equations (ODEs) in time, t, for the unknown node temperatures, $f_i(t)$. The coupling of the nodal temperatures on the left-hand side, mediated through the global mass matrix, distinguishes the finite element method from the finite difference method. In practice, the system (1.4.7) is integrated in time using a standard numerical method for solving initial-value problems involving ordinary differential equations, such as the Euler method, the Crank-Nicolson method, or a Runge-Kutta method (e.g., [43], Chapter 10).

1.4.3   Forward Euler method

In the simplest approach, the differential equation (1.4.7) is evaluated at time t, and the time derivative on the left-hand side is approximated with a first-order forward finite difference, yielding

$$\mathbf{M} \cdot \frac{\mathbf{f}(t + \Delta t) - \mathbf{f}(t)}{\Delta t} + \kappa\, \mathbf{D} \cdot \mathbf{f}(t) = \kappa\, \mathbf{b}(t), \qquad (1.4.12)$$

where Δt is a selected time step. The associated numerical error is on the order of Δt. Isolating the unknown vector $\mathbf{f}(t + \Delta t)$ on the left-hand side, we derive a tridiagonal system of equations

$$\mathbf{M} \cdot \mathbf{f}(t + \Delta t) = \mathbf{r}(t), \qquad (1.4.13)$$

where

$$\mathbf{r}(t) \equiv \mathbf{M} \cdot \mathbf{f}(t) - \Delta t\, \kappa \left[\mathbf{D} \cdot \mathbf{f}(t) - \mathbf{b}(t) \right]. \qquad (1.4.14)$$

The solution for $\mathbf{f}(t + \Delta t)$ can be found efficiently using the Thomas algorithm discussed in Section 1.1.10.

Mass lumping

Ideally, the global mass matrix $\mathbf{M}$ should be diagonal, so that the linear system (1.4.13) can be solved simply by dividing each component of the right-hand side with the corresponding diagonal component of the coefficient matrix,

$$f_i(t + \Delta t) = \frac{r_i}{M_{ii}}. \qquad (1.4.15)$$

In this ideal case, the forward Euler discretization provides us with an *explicit* numerical method, meaning that solving a linear system is not required. In later chapters, we shall see that a diagonal mass matrix is particularly desirable in two and three dimensions where the global mass matrix is sparse but not tridiagonal or pentadiagonal and the efficient Thomas algorithm may no longer be employed.

　　If we are not interested in describing the transient evolution and only want to extract the solution at steady state established after a long period of time, we can simplify the algorithm by transferring the off-diagonal elements of the mass matrix $\mathbf{M}$ in each row to the corresponding diagonal. In the finite element literature, this simplification is known as *mass lumping*. In Chapter 2, we shall see that mass lumping also arises from the inexact integration of the element mass matrix, implemented in a way that preserves the sum of the elements in each row. In the literature, the terminology *consistent formulation* implies the absence of mass lumping. Whether mass lumping is benign, tolerable, or detrimental must be considered individually for each differential equation, numerical implementation, and physical application.

1.4.4　Numerical stability

In the finite element expansion with linear elements of equal size, h, and mass lumping, the ith Galerkin projection (1.4.11) reduces to the coupled system of ordinary differential equations,

$$\frac{\mathrm{d}f_i}{\mathrm{d}t} = \kappa \frac{f_{i-1} - 2f_i + f_{i+1}}{h^2} + \frac{\kappa}{h} b_i, \qquad (1.4.16)$$

for $i = 1, 2, \ldots, N_E$, The fraction on the right-hand side is the central difference approximation of the second derivative at the ith node.

Applying the explicit Euler time discretization with time step Δt, we obtain the algebraic equation

$$\frac{f_i^{n+1} - f_i^n}{\Delta t} = \kappa \frac{f_{i-1}^n - 2f_i^n + f_{i+1}^n}{h^2} + \frac{\kappa}{h} b_i, \qquad (1.4.17)$$

where the superscript n denotes evaluation at the time level t, and the superscript $n+1$ denotes evaluation at the time level $t + \Delta t$. Rearranging, we derive the explicit formula

$$f_i^{n+1} = \alpha\, f_{i-1}^n + (1 - 2\,\alpha)\, f_i^n + \alpha f_{i+1}^n + \frac{\kappa \,\Delta t}{h} b_i, \qquad (1.4.18)$$

where

$$\alpha \equiv \frac{\Delta t\, \kappa}{h^2} \qquad (1.4.19)$$

is the dimensionless diffusion number. At steady state, $f_i^{n+1} = f_i^n$, and the solution, denoted by $\mathcal{F}$, satisfies the difference equation

$$\mathcal{F}_i = \alpha\, \mathcal{F}_{i-1} + (1 - 2\,\alpha)\, \mathcal{F}_i + \alpha\, \mathcal{F}_{i+1} + \frac{\kappa\, \Delta t}{h} b_i. \qquad (1.4.20)$$

Subtracting (1.4.20) from (1.4.18), we find that the deviation from steady state,

$$\hat{f}_i^n \equiv f_i^n - \mathcal{F}_i, \qquad (1.4.21)$$

evolves according to the homogeneous difference equation

$$\hat{f}_i^{n+1} = \alpha\, \hat{f}_{i-1}^n + (1 - 2\,\alpha)\, \hat{f}_i^n + \alpha\, \hat{f}_{i+1}^n. \qquad (1.4.22)$$

In vector notation,

$$\hat{\mathbf{f}}^{n+1} = \mathbf{P} \cdot \hat{\mathbf{f}}^n, \qquad (1.4.23)$$

where

$$\hat{\mathbf{f}}^{n+1} \equiv \begin{bmatrix} \hat{f}_p^{n+1} \\ \hat{f}_2^{n+1} \\ \vdots \\ \hat{f}_{q-1}^{n+1} \\ \hat{f}_q^{n+1} \end{bmatrix}, \qquad \hat{\mathbf{f}}^n \equiv \begin{bmatrix} \hat{f}_p^n \\ \hat{f}_2^n \\ \vdots \\ \hat{f}_{q-1}^n \\ \hat{f}_q^n \end{bmatrix}, \qquad (1.4.24)$$

are solution vectors hosting an appropriate number of $N = q - p + 1$ nodes, and $\mathbf{P}$ is an $N \times N$ tridiagonal *projection matrix*,

$$
\mathbf{P} =
\begin{bmatrix}
1 - 2\alpha & \alpha & 0 & 0 & \cdots & 0 \\
\alpha & 1 - 2\alpha & \alpha & 0 & \cdots & 0 \\
0 & \alpha & 1 - 2\alpha & \alpha & \cdots & 0 \\
\cdots & \cdots & \cdots & \cdots & \cdots & \cdots \\
\cdots & \cdots & \cdots & \cdots & \cdots & \cdots \\
\cdots & 0 & \alpha & 1 - 2\alpha & \alpha & 0 \\
0 & 0 & 0 & \alpha & 1 - 2\alpha & \alpha \\
0 & 0 & 0 & 0 & \alpha & 1 - 2\alpha
\end{bmatrix}.
$$

$$(1.4.25)$$

Each time we make a step, we effectively multiply the current solution vector, $\mathbf{f}^n$, by the projection matrix to obtain the new solution vector, $\mathbf{f}^{n+1}$. If the time step Δt is constant, $t^n = (n-1)\,\Delta t$, and

$$\hat{\mathbf{f}}^{n+1} = \mathbf{P} \cdot \hat{\mathbf{f}}^n = \mathbf{P} \cdot (\mathbf{P} \cdot \hat{\mathbf{f}}^{n-1}) = \mathbf{P}^2 \cdot \hat{\mathbf{f}}^{n-1}. \tag{1.4.26}$$

Repeating the process, we find

$$\hat{\mathbf{f}}^{n+1} = \mathbf{P}^{n+1} \cdot \hat{\mathbf{f}}^0, \tag{1.4.27}$$

where $\hat{\mathbf{f}}^0$ corresponds to the initial condition.

To determine the behavior of the solution vector, we introduce the eigenvectors of the projection matrix, $\mathbf{u}_m$, defined by the equation

$$\mathbf{P} \cdot \mathbf{u}_m = \lambda_m \, \mathbf{u}_m, \tag{1.4.28}$$

where λ_m are the corresponding eigenvalues. Next, we assume that $\mathbf{P}$ has N linearly independent eigenvectors, and express the initial deviation from the steady state as a linear combination of them,

$$\hat{\mathbf{f}}^0 = \sum_{m=1}^{N} c_m \, \mathbf{u}_m, \tag{1.4.29}$$

where c_m are appropriate coefficients. The first projection yields

$$\mathbf{P} \cdot \hat{\mathbf{f}}^0 = \sum_{m=1}^{N} c_m \, \mathbf{P} \cdot \mathbf{u}_m = \sum_{m=1}^{N} c_m \, \lambda_m \, \mathbf{u}_m. \tag{1.4.30}$$

Repeating the projections, we find

$$\hat{\mathbf{f}}^{n+1} = \mathbf{P}^{n+1} \cdot \hat{\mathbf{f}}^0 = \sum_{m=1}^{N} c_m \, \lambda_m^{n+1} \, \mathbf{u}_m, \tag{1.4.31}$$

which shows that an arbitrary initial condition will grow in time if the magnitude of at least one eigenvalue is greater than unity. Since growth is physically unacceptable in a diffusion process, it must be attributed to a numerical instability. The formal criterion for numerical stability is that the spectral radius of the projection matrix, defined as the maximum of the magnitude of the eigenvalues, is less than unity,

$$\text{Max}(|\lambda_m|) < 1. \tag{1.4.32}$$

The eigenvalues of the projection matrix (1.4.25) can be calculated exactly, and are given by

$$\lambda_m = 1 - 4\alpha \, \sin^2 \left[\frac{m\,\pi}{2(N+1)} \right], \tag{1.4.33}$$

for $m = 1, 2, \ldots, N$ ([43], p. 237). The eigenvector corresponding to the mth eigenvalue is

$$u_{m_j} = \sin \left[\frac{mj\pi}{2(N+1)} \right], \tag{1.4.34}$$

for $m, j = 1, 2, \ldots, N$. Requiring $|\lambda_m| < 1$ for any values of m and N, we derive the stability criterion

$$\alpha < \frac{1}{2}, \tag{1.4.35}$$

which can be restated as

$$\Delta t < \frac{h^2}{2\kappa}, \tag{1.4.36}$$

roughly requiring that the size of Δt is less than the size of h^2. Thus, if $h = 0.01$ in some units, Δt should be approximately less than 0.0001 in corresponding units. This stability constraint is extremely restrictive in practical applications.

Consistent formulation

The counterpart of the evolution equation (1.4.22) for the consistent formulation with elements of equal size is

$$\frac{1}{6} \hat{f}_{i-1}^{n+1} + \frac{2}{3} \hat{f}_i^{n+1} + \frac{1}{6} \hat{f}_{i+1}^{n+1} \tag{1.4.37}$$

$$= (\frac{1}{6} + \alpha) \, \hat{f}_{i-1}^n + (\frac{2}{3} - 2\,\alpha) \, \hat{f}_i^n + (\frac{1}{6} + \alpha) \, \hat{f}_{i+1}^n,$$

and the counterpart of the vector equation (1.4.23) is

$$\mathbf{Q} \cdot \hat{\mathbf{f}}^{n+1} = (\mathbf{P} + \mathbf{Q} - \mathbf{I}) \cdot \hat{\mathbf{f}}^n, \tag{1.4.38}$$

where $\mathbf{I}$ is the $N \times N$ identity matrix, and

$$
\mathbf{Q} = \frac{1}{6}
\begin{bmatrix}
4 & 1 & 0 & 0 & \cdots & \cdots & \cdots & 0 \\
1 & 4 & 1 & 0 & 0 & \cdots & \cdots & 0 \\
0 & 1 & 4 & 1 & 0 & 0 & \cdots & 0 \\
\cdots & \cdots & \cdots & \cdots & \cdots & \cdots & \cdots & \cdots \\
\cdots & \cdots & \cdots & \cdots & \cdots & \cdots & \cdots & \cdots \\
0 & \cdots & \cdots & 0 & 1 & 4 & 1 & 0 \\
0 & \cdots & \cdots & 0 & 0 & 1 & 4 & 1 \\
0 & \cdots & \cdots & 0 & 0 & 0 & 1 & 4
\end{bmatrix}.
\tag{1.4.39}
$$

Implementing mass lumping reduces $\mathbf{Q}$ to the unit matrix, yielding (1.4.23). Rearranging (1.4.38), we find

$$
\hat{\mathbf{f}}^{n+1} = \mathcal{P} \cdot \hat{\mathbf{f}}^n,
\tag{1.4.40}
$$

where

$$
\mathcal{P} \equiv \mathbf{Q}^{-1} \cdot (\mathbf{P} + \mathbf{Q} - \mathbf{I})
\tag{1.4.41}
$$

is a new projection matrix.

The eigenvalues of the matrix $\mathbf{Q}$ can be calculated exactly, and are given by

$$
\lambda_m^Q = 1 - \frac{2}{3} \sin^2 \left[\frac{m\pi}{2(N+1)} \right],
\tag{1.4.42}
$$

for $m = 1, 2, \ldots, N$ ([43], p. 237). Since the eigenvalues of the inverse of a matrix are equal to the inverses of the eigenvalues,

$$
\lambda_m^{Q^{-1}} = \frac{1}{1 - \frac{2}{3} \sin^2 \left[\frac{m\pi}{2(N+1)} \right]}.
\tag{1.4.43}
$$

The eigenvector corresponding to the mth eigenvalue is identical to the corresponding eigenvector of the matrix $\mathbf{P}$ defined in (1.4.34). Because the matrices $\mathbf{P}^{-1}$ and $\mathbf{P} + \mathbf{Q} - \mathbf{I}$ share eigenvectors, the eigenvalues of their product is equal to the product of their eigenvalues, and we may write

$$
\lambda_m^{\mathcal{P}} = \lambda_m^{Q^{-1}} \left(\lambda_m^{\mathbf{P}} + \lambda_m^{\mathbf{Q}} - 1 \right).
\tag{1.4.44}
$$

Substituting the eigenvalues, we find

$$
\lambda_m^{\mathcal{P}} = \frac{1 - \left(4\alpha + \frac{2}{3} \right) \sin^2 \left[\frac{m\pi}{2(N+1)} \right]}{1 - \frac{2}{3} \sin^2 \left[\frac{m\pi}{2(N+1)} \right]},
\tag{1.4.45}
$$

for $m = 1, 2, \ldots, N$. This formula can be verified for sample values of N and α by running the MATLAB script:

```
N=3;
al = 0.3;
P = [1-2*al al 0; al 1-2*al al; 0 al 1-2*al];
Q = [4 1 0; 1 4 1; 0 1 4]/6.0;
Pr = inv(Q)*(P+Q-eye(N));
eig(Pr)
```

and confirming that the result is consistent with the predictions of (1.4.45).

Requiring $|\lambda_i^P| < 1$ according to our earlier discussion, we derive the stability constraint

$$\alpha < \frac{1}{6}, \tag{1.4.46}$$

which can be restated as

$$\Delta t < \frac{h^2}{6\kappa}. \tag{1.4.47}$$

Comparing this result with (1.4.36), we find that the consistent formulation imposes a somewhat more stringent constraint on the time step.

1.4.5 Finite element code

FSELIB code udl.m, listed in the text, implements the forward Euler finite element method according to the following steps:

1. *Data input and parameter definition.*

2. *Element node generation.*

3. *Specification of the initial temperature distribution, and definition of the source distribution along the rod.*

4. *Compilation of the linear system.*

   This is done by calling the FSELIB function udl_sys, listed in the text, to evaluate the mass matrix and the right-hand side of (1.4.13), denoted as lhs in the code, using the definition (1.4.14).

5. *Solution of the linear system.*

   This is done using Thomas's algorithm.

6. *Graphics module.*

   A graph of the transient solution is produced every *nplot* steps, where the parameter *nplot* is set at the code input.

7. *Return to Step 4 or terminate the time stepping.*

```
%=============================================
% CODE udl
%
% Finite element code for unsteady diffusion
% with linear elements using Euler's method
%=============================================

%-----------
% input data
%-----------

L=1.0;
k=1.0; rho=1.0; cp=1.0;
q0=-1.0; fL=0.0;

Dt=0.0017;

nsteps=2000; nplot=100;

ne=10; ratio=1.0;

%--------
% prepare
%--------

kappa = k/(rho*cp);

%----------------
% grid generation
%----------------

xe = elm_line1 (0,L,ne,ratio);

%----------------------------
% initial condition and source
%----------------------------

for i=1:ne+1
  f(1,i) = fL;
  s(i) = 10.0*exp(-5.0*xe(i)^2/L^2);
end

%-----
% plot
%-----

plot(xe, f,'x'); hold on; xlabel('x'); ylabel('f');

%--------------------
% begin time stepping
%--------------------

icount = 0; % step counter
```

Code udl: Continuing $\longrightarrow$

```
for irun=1:nsteps

%--------------------------------
% generate the tridiagonal system
%--------------------------------

[at,bt,ct,rhs] = udl_sys (ne,xe,q0,f,fL,k,kappa,s,Dt);

%--------------
% linear solver
%--------------

sol = thomas (ne,at,bt,ct,rhs);

f = [sol fL];     % include the value at the right end

%---------
% plotting
%---------

if(icount==nplot)
 plot(xe, f,'-o'); icount = 0;
end

icount = icount+1;

end

%---------------------
% end of time stepping
%---------------------

%-----
% done
%-----
```

Code udl: (⟶ Continued.) Finite element code for one-dimensional unsteady diffusion with linear elements using the forward Euler method.

Figure 1.4.1 shows the graphics display of the finite element code for ten evenly spaced elements of size $h = 0.1$, $\kappa = 1$, and time steps $\Delta t = 0.0016$ and 0.0017, corresponding to diffusion numbers $\alpha = 0.16$ and 0.17. A graph is produced every 100 time steps. Our earlier analysis suggested that the threshold for numerical stability is

$$\alpha = \frac{1}{6} \simeq 0.1667. \qquad (1.4.48)$$

Consistent with the theoretical predictions regarding numerical stability, the first numerical solution smoothly tends to the steady-steady solution, whereas the second solution develops an unphysical numerical oscillation dominated by a saw-tooth wave in spite of the smallness of the time step.

```
function [at,bt,ct,rhs] = udl_sys (ne,xe,q0,f,fL,k,kappa,s,Dt)

%=======================================================
% Compact assembly of the tridiagonal linear system
% for one-dimensional unsteady diffusion
% with linear elements (udl)
%=======================================================

%-------------
% element size
%-------------

for l=1:ne
  h(l) = xe(l+1)-xe(l);
end

%----------------------------------
% initialize the tridiagonal matrix
%----------------------------------

at = zeros(ne,1);
bt = zeros(ne,1);
ct = zeros(ne,1);
```

Function udl_sys: Continuing $\longrightarrow$

(a) (b)

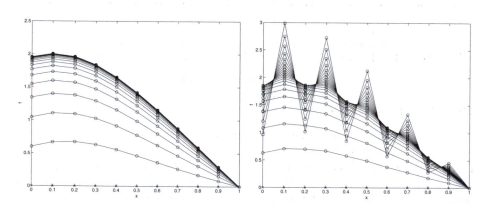

Figure 1.4.1 Graphics output of the finite element code udl for the source
function shown in (1.1.75), with a time step corresponding to diffusion
numbers $\alpha = 0.16$ and 0.17. Theoretical analysis shows that the threshold
for numerical stability is $\alpha = \frac{1}{6} = 0.1667$.

```
%--------------------------------
% initialize the right-hand side
%--------------------------------

rhs = zeros(ne,1); rhs(1) = Dt*q0/k;

%-------------------------------
% loop over the first ne-1 elements
%-------------------------------

for l=1:ne-1

    A11 = 1/h(l); A12 =-A11;       % diffusion matrix
    A21 = A12;     A22 = A11;

    B11 = h(l)/3.0; B12 = 0.5*B11;  % mass matrix
    B21 = B12;        B22 = B11;

    at(l)   = at(l) + B11;          % tridiagonal matrix components
    bt(l)   = bt(l) + B12;
    ct(l+1) = ct(l+1) + B21;
    at(l+1) = at(l+1) + B22;

    rhs(l)   = rhs(l)   + B11*f(l) + B12*f(l+1);
    rhs(l+1) = rhs(l+1) + B21*f(l) + B22*f(l+1);

    rhs(l)   = rhs(l)   - Dt*kappa*(A11*f(l) + A12*f(l+1));
    rhs(l+1) = rhs(l+1) - Dt*kappa*(A21*f(l) + A22*f(l+1));

    rhs(l)   = rhs(l)   + Dt*kappa*(B11*s(l) + B12*s(l+1))/k;
    rhs(l+1) = rhs(l+1) + Dt*kappa*(B12*s(l) + B22*s(l+1))/k;

end

%----------------------
% last element is special
%----------------------

A11 = 1.0/h(ne); A12 =-A11;
B11 = h(ne)/3.0; B12 = 0.5*B11;
at(ne) = at(ne) + B11;

rhs(ne) = rhs(ne) + (B11*f(ne) + B12*fL);
rhs(ne) = rhs(ne) - Dt*kappa*(A11*f(ne) + A12*fL);
rhs(ne) = rhs(ne) + Dt*kappa*(B11*s(ne) + B12*s(ne+1))/k;

%-----
% done
%-----

return;
```

Function udl_sys: ($\longrightarrow$ Continued.) Assembly of the linear system for one-dimensional unsteady diffusion with linear elements using the forward Euler method.

1.4.6 Crank-Nicolson integration

To circumvent the restriction on the time step, Δt, we integrate the differential equations using an *implicit* method. In the Crank-Nicolson discretization, we select a time step, Δt, evaluate the differential equation (1.4.7) at time $t + \frac{1}{2}\Delta t$, approximate the time derivative on the left-hand side with a centered finite difference and the rest of the terms with averages to obtain

$$\mathbf{M} \cdot \frac{\mathbf{f}^{n+1} - \mathbf{f}^n}{\Delta t} + \frac{\kappa}{2} \mathbf{D} \cdot \left[\mathbf{f}^n + \mathbf{f}^{n+1} \right] = \frac{\kappa}{2} \left[\mathbf{b}^n + \mathbf{b}^{n+1} \right]. \qquad (1.4.49)$$

The associated error is on the order of Δt^2. Rearranging, we derive the linear system

$$\mathbf{C} \cdot \mathbf{f}^{n+1} = \mathbf{r}, \qquad (1.4.50)$$

where the tridiagonal coefficient matrix on the left-hand side is given by

$$\mathbf{C} = \mathbf{M} + \frac{1}{2} \kappa \, \Delta t \, \mathbf{D}, \qquad (1.4.51)$$

and the right-hand side is given by

$$\mathbf{r} = \left[\mathbf{M} - \frac{1}{2} \kappa \, \Delta t \, \mathbf{D} \right] \cdot \mathbf{f}^n + \frac{1}{2} \kappa \, \Delta t \left[\mathbf{b}^n + \mathbf{b}^{n+1} \right]. \qquad (1.4.52)$$

The numerical procedure involves solving the linear system (1.4.50) from a specified initial state for a sequence of steps with a constant or variable time step, Δt. The solution at every step can be found efficiently using the Thomas algorithm.

A stability analysis conducted as discussed previously in this section for the Euler method shows that the spectral radius of the associated projection matrix is less than unity for any time step Δt, and the numerical method is unconditionally stable (e.g., [43]).

PROBLEMS

1.4.1 *Eigenvalues and eigenvectors of the projection matrix.*

Confirm the eigenvalues and eigenvectors of the matrix $\mathbf{P}$ given in (1.4.33) and (1.4.34).

1.4.2 *Code with mass lumping.*

Modify the function udl_sys to implement mass lumping, run the modified code, and discuss the onset of numerical instability with reference to the theoretical predictions.

1.4.3 *Code for the Crank-Nicolson method.*

Modify the function `udl_sys` to implement the Crank-Nicolson discretization, run the code `udl`, and confirm that the method is stable independent of the size of the time step.

1.4.4 *Consistency analysis and hyperdiffusivity.*

A consistency analysis is carried out working backwards from an algebraic difference equation to the so-called modified differential equation (MDE). If the MDE reduces to the governing partial differential equation as the time step and element size independently tend to zero, then the algebraic difference equation is consistent.

To carry out the consistency analysis of (1.4.22), we express all discrete variables in terms of the value at the ith node at time level n, writing, for example,

$$f_i^{n+1} = f_i^n + \left(\frac{\partial f}{\partial t}\right)_i^n \Delta t + \frac{1}{2}\left(\frac{\partial^2 f}{\partial t^2}\right)_i^n \Delta t^2 + \dots$$

$$f_{i+1}^n = f_i^n + \left(\frac{\partial f}{\partial x}\right)_i^n \Delta x + \frac{1}{2}\left(\frac{\partial^2 f}{\partial x^2}\right)_i^n \Delta x^2 + \dots \tag{1.4.53}$$

Substituting these expressions in (1.4.22) and simplifying, we find

$$\left(\frac{\partial f}{\partial t}\right)_i^n \Delta t + \frac{1}{2}\left(\frac{\partial^2 f}{\partial t^2}\right)_i^n \Delta t^2 + \dots$$
$$= \alpha \left(\frac{\partial^2 f}{\partial x^2}\right)_i^n \Delta x^2 + \frac{\alpha}{12}\left(\frac{\partial^4 f}{\partial x^4}\right)_i^n \Delta x^4 + \dots, \tag{1.4.54}$$

which can be rearranged into

$$\left(\frac{\partial f}{\partial t}\right)_i^n = \kappa \left(\frac{\partial^2 f}{\partial x^2}\right)_i^n - \frac{1}{2}\left(\frac{\partial^2 f}{\partial t^2}\right)_i^n \Delta t + \frac{\kappa}{12}\Delta x^2 \left(\frac{\partial^4 f}{\partial x^4}\right)_i^n + \dots. \tag{1.4.55}$$

Because as $\Delta t \to 0$ and $\Delta x \to 0$ independently the modified differential equation (1.4.57) reduces to the unsteady diffusion equation, the numerical method is consistent.

Now, differentiating the unsteady heat conduction equation with respect to t, we find that the solution satisfies the high-order equation

$$\frac{\partial^2 f}{\partial t^2} = \kappa^2 \frac{\partial^4 f}{\partial x^4}. \tag{1.4.56}$$

Using this equation to eliminate the second derivative with respect to t on the right-hand side of (1.4.55), and rearranging, we obtain

$$\left(\frac{\partial f}{\partial t}\right)_i^n = \kappa \left(\frac{\partial^2 f}{\partial x^2}\right)_i^n - \mathcal{D}\left(\frac{\partial^4 f}{\partial x^4}\right)_i^n + \dots. \tag{1.4.57}$$

where the coefficient

$$\mathcal{D} \equiv \frac{\kappa}{2} \, \Delta x^2 \, (\frac{1}{6} - \alpha) = \frac{\kappa}{2} \, (\frac{1}{6} \, \Delta x^2 - \kappa \, \Delta t) \qquad (1.4.58)$$

is called the hyperdiffusivity. Derive the hyperdiffusivity of the consistent formulation expressed by (1.4.38).

1.5   One-dimensional convection

The problem of one-dimensional convection discussed in this section is complementary to the problem of one-dimensional unsteady diffusion discussed in Section 1.4. Our present goal is to compute a time-dependent function, $f(x, t)$, that satisfies the convection equation

$$\frac{\partial f}{\partial t} + u \, \frac{\partial f}{\partial x} = 0, \qquad (1.5.1)$$

where $u(x, f, t)$ is a prescribed convection velocity. The solution is to be found over a specified interval of the x axis, $a \leq x \leq b$, subject to a given initial condition $f(x, t = 0) = F(x)$, and a suitable boundary condition.

If the convection velocity, u, is independent of the convected field, f, that is $u(x, t)$ is a function of x and t explicitly but not implicitly through $f(x, t)$, the convection equation is quasi-linear, and we say that the field $f(x, t)$ is *advected* with the specified position- and time-dependent advection velocity. In the special case where u is a constant, the convection equation is linear.

The implementation of the Galerkin finite element method follows the steps outlined previously in this section for the diffusion equation. In the case of quasi-linear convection, the Galerkin projection takes the form

$$\int_a^b \phi_i \, \frac{\partial f}{\partial t} \, \mathrm{d}x + \int_a^b u(x, t) \, \phi_i \, \frac{\partial f}{\partial x} \, \mathrm{d}x = 0, \qquad (1.5.2)$$

where the index i runs over an appropriate number of nodes. Substituting in place of $f(x)$ the finite element expansion involving the N_G global interpolation functions,

$$f(x, t) = \sum_{j=1}^{N_G} f_j(t) \, \phi_j(x), \qquad (1.5.3)$$

we obtain

$$\sum_{j=1}^{N_G} \left(\int_a^b \phi_i \, \phi_j \, \mathrm{d}x \right) \frac{\mathrm{d} f_j}{\mathrm{d}t} + \sum_{j=1}^{N_G} \left(\int_a^b u(x, t) \, \phi_i \, \frac{\partial \phi_j}{\partial x} \, \mathrm{d}x \right) f_j = 0, \qquad (1.5.4)$$

which can be recast into the form

$$\sum_{j=1}^{N_G} M_{ij} \, \frac{\mathrm{d} f_j}{\mathrm{d}t} + \sum_{j=1}^{N_G} N_{ij} \, f_j = 0, \qquad (1.5.5)$$

where M_{ij} is the global mass matrix, and

$$N_{ij} \equiv \int_a^b u(x) \, \phi_i \, \frac{\partial \phi_j}{\partial x} \, dx \qquad (1.5.6)$$

is the *global advection matrix*. Equation (1.5.5) provides us with a system of linear ordinary differential equations,

$$\mathbf{M} \cdot \frac{d\mathbf{f}}{dt} + \mathbf{N} \cdot \mathbf{f} = \mathbf{0}. \qquad (1.5.7)$$

To evaluate the global advection matrix, we express the convection velocity in the isoparametric form

$$u(x, t) = \sum_{k=1}^{N_G} u_k(t) \, \phi_k(x). \qquad (1.5.8)$$

Substituting this expansion in (1.5.6), we find

$$N_{ij} = \sum_{k=1}^{N_G} u_k \int_a^b \phi_k \, \phi_i \, \frac{\partial \phi_j}{\partial x} \, dx \equiv \sum_{k=1}^{N_G} u_k \, \Pi_{kij}, \qquad (1.5.9)$$

where

$$\Pi_{kij} \equiv \int_a^b \phi_k \, \phi_i \, \frac{\partial \phi_j}{\partial x} \, dx \qquad (1.5.10)$$

is a three-index global advection tensor. In terms of $\mathbf{\Pi}$, equation (1.5.7) takes the form

$$\mathbf{M} \cdot \frac{d\mathbf{f}}{dt} + \mathbf{u} \cdot \mathbf{\Pi} \cdot \mathbf{f} = \mathbf{0}. \qquad (1.5.11)$$

In practice, the tensor $\mathbf{\Pi}$ is assembled from the individual element advection matrices.

1.5.1 Linear elements

In the case of linear elements, the first derivative of the jth global interpolation function on the right-hand side of (1.5.6) is equal to $1/h_{j-1}$ over the $(j-1)$ element, $-1/h_j$ over the jth element, and 0 over all other elements. Thus, the only nonzero components of the matrix N_{ij} are the tridiagonal elements $N_{i-1,j}$, $N_{i,j}$, and $N_{i+1,j}$, given by

$$N_{ii} = \frac{1}{h_{i-1}} \int_{x_{i-1}}^{x_i} (u_{i-1} \, \phi_{i-1} + u_i \, \phi_i) \, \phi_i \, dx$$

$$- \frac{1}{h_i} \int_{x_i}^{x_{i+1}} (u_i \, \phi_i + u_{i+1} \, \phi_{i+1}) \, \phi_i \, dx,$$

$$N_{i-1,i} = -\frac{1}{h_{i-1}} \int_{x_{i-1}}^{x_i} (u_{i-1} \, \phi_{i-1} + u_i \, \phi_i) \, \phi_i \, dx, \qquad (1.5.12)$$

$$N_{i+1,i} = \frac{1}{h_i} \int_{x_i}^{x_{i+1}} (u_i \, \phi_i + u_{i+1} \, \phi_{i+1}) \, \phi_i \, dx.$$

Accordingly, the global advection matrix simplifies to

$$N_{ij} = (\delta_{i,j} - \delta_{i-1,j}) \frac{1}{h_{i-1}} \int_{x_{i-1}}^{x_i} (u_{i-1}\,\phi_{i-1} + u_i\,\phi_i)\,\phi_i\,dx$$

$$+(\delta_{i+1,j} - \delta_{i,j}) \frac{1}{h_i} \int_{x_i}^{x_{i+1}} (u_i\,\phi_i + u_{i+1}\,\phi_{i+1})\,\phi_i\,dx. \qquad (1.5.13)$$

Rearranging, we find

$$N_{ij} = u_{i-1}\,(\delta_{i,j} - \delta_{i-1,j}) \frac{1}{h_{i-1}} \int_{x_{i-1}}^{x_i} \phi_{i-1}\,\phi_i\,dx$$

$$+u_i \left[(\delta_{i,j} - \delta_{i-1,j}) \frac{1}{h_{i-1}} \int_{x_{i-1}}^{x_i} \phi_i^2\,dx + (\delta_{i+1,j} - \delta_{i,j}) \frac{1}{h_i} \int_{x_i}^{x_{i+1}} \phi_i^2\,dx \right]$$

$$+u_{i+1}\,(\delta_{i+1,j} - \delta_{i,j}) \frac{1}{h_i} \int_{x_i}^{x_{i+1}} \phi_{i+1}\,\phi_i\,dx$$

$$= u_{i-1}\,(\delta_{i,j} - \delta_{i-1,j}) \frac{1}{6} + u_i\,(\delta_{i+1,j} - \delta_{i-1,j}) \frac{1}{3} + u_{i+1}\,(\delta_{i+1,j} - \delta_{i,j}) \frac{1}{6}.$$

$$(1.5.14)$$

The contribution of the advection term to the ith Galerkin projection is

$$\sum_{j=1}^{N_G} N_{ij}\,f_j = \frac{1}{6} \left[u_{i-1}\,(f_i - f_{i-1}) + 4\,u_i\,\frac{f_{i+1} - f_{i-1}}{2} + u_{i+1}\,(f_{i+1} - f_i) \right].$$

$$(1.5.15)$$

Note that the right-hand side involves a particular combination of backward, central, and forward differences.

Linear advection

If the convection velocity, u, is constant and equal to U, the global advection matrix reduces to

$$\mathbf{N} \equiv \frac{U}{2} \begin{bmatrix} -1 & 1 & 0 & 0 & \ldots & 0 & 0 & 0 \\ -1 & 0 & 1 & 0 & \ldots & 0 & 0 & 0 \\ 0 & -1 & 0 & 1 & \ldots & 0 & 0 & 0 \\ \ldots & \ldots & \ldots & \ldots & \ldots & \ldots & \ldots & \ldots \\ 0 & 0 & \ldots & \ldots & -1 & 0 & 1 & 0 \\ 0 & 0 & \ldots & \ldots & 0 & -1 & 0 & 1 \\ 0 & 0 & \ldots & \ldots & 0 & 0 & -1 & 1 \end{bmatrix}, \qquad (1.5.16)$$

independent of the element size. Apart from the first and last diagonal elements, the global advection matrix shown in (1.5.16) is skew-symmetric, that is, it is equal to the negative of its transpose. A skew-symmetric matrix has zeros along the diagonal.

The global advection matrix shown in (1.5.16) is composed of overlapping diagonal blocks of corresponding element matrices. The lth-element advection matrix is given by

$$\mathbf{C}^{(l)} = \frac{U}{2} \begin{bmatrix} -1 & 1 \\ -1 & 1 \end{bmatrix}. \tag{1.5.17}$$

Cancellation of the alternating 1 and -1 diagonal entries during the assembly process yields zeros along the diagonal, as shown on the right-hand side of (1.5.16).

Using the mass matrix displayed in (1.4.8), we find that the ith equation of the GFEM system (1.5.7) takes the form

$$\frac{h_{i-1}}{6} \frac{\mathrm{d}f_{i-1}}{\mathrm{d}t} + \frac{h_{i-1} + h_i}{3} \frac{\mathrm{d}f_i}{\mathrm{d}t} + \frac{h_i}{6} \frac{\mathrm{d}f_{i+1}}{\mathrm{d}t} + \frac{U}{2}(f_{i+1} - f_{i-1}) = 0, \tag{1.5.18}$$

which can be rearranged into

$$(\frac{1}{3} \frac{h_{i-1}}{h_{i-1} + h_i}) \frac{\mathrm{d}f_{i-1}}{\mathrm{d}t} + (\frac{2}{3}) \frac{\mathrm{d}f_i}{\mathrm{d}t} + (\frac{1}{3} \frac{h_i}{h_{i-1} + h_i}) \frac{\mathrm{d}f_{i+1}}{\mathrm{d}t}$$
$$+ U \frac{f_{i+1} - f_{i-1}}{h_{i-1} + h_i} = 0. \tag{1.5.19}$$

When the element size is uniform, $h_1 = h_2 = \cdots = h_{N_E} \equiv h$, we obtain the simplified form

$$\frac{1}{6} \frac{\mathrm{d}f_{i-1}}{\mathrm{d}t} + \frac{2}{3} \frac{\mathrm{d}f_i}{\mathrm{d}t} + \frac{1}{6} \frac{\mathrm{d}f_{i+1}}{\mathrm{d}t} + \frac{U}{2h} (f_{i+1} - f_{i-1}) = 0. \tag{1.5.20}$$

The first three terms on the left-hand side of (1.5.19) contribute a weighted average of the time derivative at three nodes. It is reassuring to observe that the sum of the weights enclosed by the parentheses is equal to unity, ensuring mass conservation. The last term on the left-hand side of (1.5.19) is the central difference approximation of the derivative $\partial f / \partial x$ at the ith node, computed over an interval of length $h_{i-1} + h_i$. Overall, (1.5.19) expresses a filtered or smeared finite-difference approximation.

When the convection velocity is not constant but varies in x, equation (1.5.18) can still be used with U on the left-hand side replaced by the nodal velocity u_i, as an alternative to the primary form (1.5.15).

Numerical dispersion due to spatial discretization

Consider a solution in the form of a periodic wave with period λ, and assume that the solution of the discrete system (1.5.20) is given by the real or imaginary part of the function

$$f(x, t) = A \, \exp[Ik(x - ct)], \tag{1.5.21}$$

where I is the imaginary unit, $I^2 = -1$, $A = A_R + IA_I$ is the complex wave amplitude, the subscripts R and I denote the real and imaginary part, $k = 2\pi/\lambda$ is the wavenumber, and $c = c_R + Ic_I$ is the complex phase velocity. The complex angular frequency of the oscillation is given by $\omega \equiv kc$.

We shall assume that the length of the solution domain, $L = hN_E$, is a multiple of the wavelength λ, that is,

$$\frac{L}{\lambda} = n, \tag{1.5.22}$$

where $n = 1, 2, \ldots, N_E$, and N_E is the number of available elements. Substituting (1.5.21) in (1.5.20), writing $x_i = (i-1)h$ and simplifying, we find

$$-Ikc \left[\frac{1}{6} \exp(Ikh) + \frac{2}{3} + \frac{1}{6} \exp(-Ikh) \right]$$
$$+ \frac{U}{2h} \left[\exp(Ikh) - \exp(-Ikh) \right] = 0, \tag{1.5.23}$$

or

$$-Ikc \left[\frac{2}{3} + \frac{1}{3} \cos(kh) \right] + I \frac{U}{h} \sin(kh) = 0. \tag{1.5.24}$$

Solving for the phase velocity, we obtain

$$\frac{c}{U} = \frac{\sin(kh)}{kh} \frac{3}{2 + \cos(kh)}. \tag{1.5.25}$$

The corresponding angular frequency, $\omega \equiv kc$, reduced by the time scale U/L, is given by

$$\frac{\omega L}{U} = \frac{kcL}{U} = 2\pi n \frac{c}{U} = \frac{2\pi n}{kh} \sin(kh) \frac{3}{2 + \cos(kh)}. \tag{1.5.26}$$

Now, $kh = 2\pi h/\lambda$, and since $h = L/N_E$, $h/\lambda = n/N_E$, and

$$kh = \frac{2\pi n}{N_E}, \tag{1.5.27}$$

for $n = 1, 2, \ldots, N_E$. Thus,

$$\frac{\omega L}{U} = N_E \sin(kh) \frac{3}{2 + \cos(kh)}. \tag{1.5.28}$$

When the ratio n/N_E is small, we may perform a Taylor series expansion to find

$$\frac{\omega L}{U} \simeq 2\pi n. \tag{1.5.29}$$

In the lumped mass approximation, the fractions $3/[2 + \cos(kh)]$ on the right-hand sides of (1.5.25)–(1.5.28) are replaced by unity (Problem 1.5.2).

The exact solution of the linear convection equation states that the initial distribution travels with the specified velocity U, which means that $c = U$, and $\omega L/U = 2\pi n$, consistent with the asymptotic result (1.5.29). Expression (1.5.28) reveals that the spatial discretization modifies the exact phase velocity by an amount that depends on the reduced wave number, kh, thereby introducing numerical dispersion.

The following FSELIB script `dispersion` produces a graph of $\omega L/(UN_E)$ against, kh, for the consistent and mass-lumped approximation,

```
ne = 32;
pi2 = 2*pi;

x     = zeros(ne,1);
s     = zeros(ne,1);
slump = zeros(ne,1);

for n=1:ne
   x(n) = n/ne;
   kh   = pi2*x(n);
   s(n) = sin(kh) * 3/(2+cos(kh));
   slump(n) = sin(kh);
end

plot(x, s,'o'); hold on; plot(x, s);            % consistent

plot(x, slump,'x'); hold on; plot(x, slump);  % lumped

xe=[0 0.5]; ye=[0 0.5*pi2]; plot(xe,ye,'--'); % exact

axis([0 1 -3 3]);

xlabel('n/N_E'); ylabel('s/N_E');
```

Results for thirty-two elements, $N_E = 32$, shown in Figure 1.5.1, suggest that the spatial discretization significantly affects the wave speed of moderately or poorly resolved waves, and thus introduces significant numerical dispersion.

1.5.2 Quadratic elements

The implementation of the Galerkin finite element method with quadratic elements follows the general guidelines described previously in this section for linear elements. In the case of quadratic elements with interior nodes situated at the element mid-points, the element advection matrix is given by

$$\mathbf{C}^{(l)} = \frac{U}{6} \begin{bmatrix} -3 & 4 & -1 \\ -4 & 0 & -4 \\ 1 & -4 & 3 \end{bmatrix}. \tag{1.5.30}$$

The global advection matrix is composed of overlapping diagonal blocks of the element advection matrix given in (1.5.30) (Problem 1.5.1).

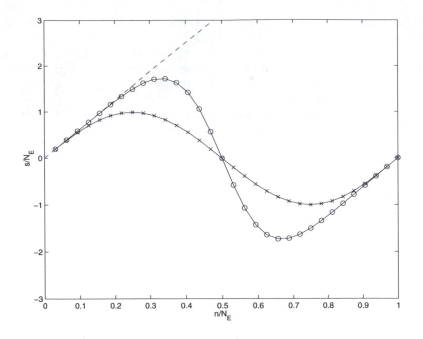

Figure 1.5.1 Graph of the reduced angular frequency $\omega L/(UN_E)$ plotted
against the wavenumber kh for $N_E = 32$ elements, generated by the
FSELIB script `dispersion`. The circles represent the predictions of
(1.5.28), the $\times$ symbols correspond to the mass lumped approximation,
and the dashed line represents the exact solution recovered for small values of the wavenumber, kh.

1.5.3   Integrating ODEs

Equation (1.5.7) provides us with a coupled system of first-order linear ordinary
differential equations for the unknown nodal values, $f_i(t)$. Implementing the
simplest possible time discretization expressed by the forward Euler method,
we write

$$\mathbf{M} \cdot \frac{\mathbf{f}(t + \Delta t) - \mathbf{f}(t)}{\Delta t} + \mathbf{N} \cdot \mathbf{f}(t) = \mathbf{0}, \tag{1.5.31}$$

where Δt is the time step. Rearranging, we derive a tridiagonal system of
equations for the unknown vector $\mathbf{f}(t + \Delta t)$,

$$\mathbf{M} \cdot \mathbf{f}(t + \Delta t) = \left[\mathbf{M} - \Delta t\, \mathbf{N} \right] \cdot \mathbf{f}(t). \tag{1.5.32}$$

Implementing mass lumping to diagonalize the coefficient matrix on the left-
hand side, we recover the explicit forward-time, centered-space (FTCS) finite-
difference equation. In the case of linear elements with uniform size h, the ith

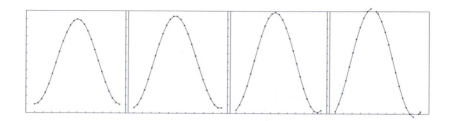

Figure 1.5.2 Evolution of a sinusoidal wave using the FTCS method computed
with $N_E = 24$ elements, and a time step corresponding to $c = 0.05$. The
depicted profiles correspond to times $t = 0, L/U, 2L/U, 3L/U$.

equation takes the form

$$f_i(t + \Delta t) = f_i(t) - \mathcal{C} \left[f_{i+1}(t) - f_{i-1}(t) \right], \tag{1.5.33}$$

where

$$\mathcal{C} \equiv \frac{\Delta t \, U}{h} \tag{1.5.34}$$

is the convection number.

Unfortunately, a stability analysis of the FTCS discretization shows that
the method is unstable for any value of $\mathcal{C}$, that is, no matter how much we
reduce the size of the time step (e.g., [43], pp. 556–557). Figure 1.5.2 shows the
evolution of a sinusoidal wave computed using the FTCS method with $N_E = 24$
elements over a period L, corresponding to $n = 1$, and a time step corresponding
to $\mathcal{C} = 0.05$. Profiles are depicted at times $t = 0, L/U, 2L/U, 3L/U$. Two
important features are evident in these illustrations: the unphysical growth of
the wave amplitude, and a reduction in the phase speed with respect to U,
which is consistent with our earlier discussion on dispersion.

More generally, the non-diagonal structure of the advection matrix arising
from the Galerkin finite element method introduces serious or even incurable
numerical instabilities in the case of explicit time integration, and significant
numerical diffusivity in the case of implicit time integration. Moreover, the
seemingly innocuous discretization (1.5.15) is the source of a detrimental *alias-
ing instability* (e.g., [23], p. 59). These difficulties underscore the sensitivity of
the finite element method in problems dominated by convection. In practice,
time integration is performed by a variety of explicit or semi-implicit numerical
methods, such as the Runge-Kutta method (e.g., [23, 43]).

1.5.4 Nonlinear convection

A prototypical nonlinear convection equation used to investigate the performance of numerical methods is the inviscid Burgers' equation

$$\frac{\partial f}{\partial t} + f \frac{\partial f}{\partial x} = 0. \tag{1.5.35}$$

Consider an initial condition where the function $f(x)$ is a sinusoidal wave. Because the convection velocity multiplying the spatial derivative, $\partial f/\partial x$, is equal to f, the larger the magnitude of the solution, the faster the magnitude of the local convection velocity. Accordingly, the crests of a sinusoidal wave with zero mean will travel to the right, the troughs will travel to the left, and the nodes will remain stationary. The evolution leads to wave steepening followed by the eventual formation of a step discontinuity in the form of a shock wave.

Applying (1.5.11) with $\mathbf{f}$ in place of $\mathbf{u}$, we derive the nonlinear GFEM equations

$$\mathbf{M} \cdot \frac{d\mathbf{f}}{dt} + \mathbf{f} \cdot \mathbf{\Pi} \cdot \mathbf{f} = \mathbf{0}. \tag{1.5.36}$$

The quadratic nonlinearity requires careful attention to prevent the onset of numerical instability and successfully obtain the structure of the solution at steady state.

PROBLEMS

1.5.1 *Quadratic elements.*

Display the structure of the global advection matrix for linear convection with the quadratic elements discussed in Section 1.5.2.

1.5.2 *Dispersion with mass lumping.*

Show that, in the lumped mass approximation, the second fraction on the right-hand sides of (1.5.25)–(1.5.28) is replaced by unity.

1.6 One-dimensional convection–diffusion

In this section, we synthesize the complementary cases of diffusion and convection discussed previously in this chapter into the unsteady convection–diffusion equation

$$\frac{\partial f}{\partial t} + u \frac{\partial f}{\partial x} = \kappa \frac{\partial^2 f}{\partial x^2} + \frac{s(x,t)}{\rho\, c_p}, \tag{1.6.1}$$

where $u(x, f, t)$ is a prescribed convection velocity, and $\kappa = k/(\rho\, c_p)$ is the medium diffusivity. The solution of (1.6.1) is to be found subject to the initial condition, $f(x, t = 0) = F(x)$, and a suitable number of boundary conditions.

1.6.1   Steady linear convection–diffusion

To illustrate the development of the Galerkin finite element method, we first discuss the generalization of the steady diffusion problem introduced in Section 1.1, where the steady-state heat conduction equation (1.1.4) is replaced by the linear convection–diffusion equation

$$\rho\, c_p\, U\, \frac{\mathrm{d}f}{\mathrm{d}x} = k\, \frac{\mathrm{d}^2 f}{\mathrm{d}x^2} + s(x), \qquad (1.6.2)$$

where U is a constant advection velocity. The importance of convection relative to diffusion is expressed by the dimensionless Péclet number

$$Pe \equiv \frac{\rho\, c_p\, L U}{k} = \frac{L U}{\kappa}, \qquad (1.6.3)$$

where L is a characteristic length, presently identified with the size of the solution domain.

Applying the Galerkin finite element method with linear elements, we derive a $N_E \times N_E$ linear algebraic system that is similar to that shown in (1.1.29),

$$\mathbf{Q} \cdot \mathbf{f} = \mathbf{c} + \frac{1}{k}\, \widetilde{\mathbf{M}} \cdot \mathbf{s}. \qquad (1.6.4)$$

When the elements are evenly spaced with uniform size h, the $N_E \times N_E$ coefficient matrix on the right-hand side of (1.6.4) is given by

$$\mathbf{Q} = \frac{1}{h} \begin{bmatrix} 1 - \frac{1}{2}Pe_c & -1 + \frac{1}{2}Pe_c & 0 & 0 & \cdots \\ -1 - \frac{1}{2}Pe_c & 2 & -1 + \frac{1}{2}Pe_c & 0 & \cdots \\ 0 & -1 - \frac{1}{2}Pe_c & 2 & -1 + \frac{1}{2}Pe_c & 0 \\ \cdots & \cdots & \cdots & \cdots & \cdots \\ 0 & 0 & \cdots & \cdots & \cdots \\ 0 & 0 & \cdots & \cdots & \cdots \\ 0 & 0 & \cdots & \cdots & \cdots \end{bmatrix}$$

$$\begin{bmatrix} \cdots & \cdots & 0 & 0 & 0 \\ \cdots & \cdots & 0 & 0 & 0 \\ \cdots & \cdots & 0 & 0 & 0 \\ \cdots & \cdots & \cdots & \cdots & \cdots \\ 0 & -1 - \frac{1}{2}Pe_c & 2 & -1 + \frac{1}{2}Pe_c & 0 \\ \cdots & 0 & -1 - \frac{1}{2}Pe_c & 2 & -1 + \frac{1}{2}Pe_c \\ \cdots & 0 & 0 & -1 - \frac{1}{2}Pe_c & 2 + \frac{1}{2}Pe_c \end{bmatrix},$$

$$(1.6.5)$$

where

$$Pe_c \equiv \frac{\rho\, c_p\, h U}{k} = \frac{h U}{\kappa} \qquad (1.6.6)$$

is the *cell Péclet number*. In the absence of convection, $Pe_c = 0$, we recover the diffusion matrix shown in (1.1.38), $\mathbf{Q} = \mathbf{D}$. The N_E-dimensional vector, $\mathbf{c}$, on the right-hand side of (1.6.4) is given by

$$
\mathbf{c} \equiv \begin{bmatrix} q_0/k \\ 0 \\ \vdots \\ 0 \\ (1 + \frac{1}{2} Pe_c)\, f_L/h_{N_E} \end{bmatrix}, \tag{1.6.7}
$$

and the bordered mass matrix, $\widetilde{\mathbf{M}}$, is given in (1.1.39).

As Pe_c becomes large, the coefficient matrix $\mathbf{Q}$ tends to the skew-symmetric form $\frac{1}{\kappa} \mathbf{N}$, where $\mathbf{N}$ is the global advection matrix shown in (1.5.16). In this limit, the odd- and even-numbered unknowns are coupled only by means of the first equation represented by the first row on the right-hand side of (1.6.5). From a practical standpoint, because the coefficient matrix ceases to be diagonally dominant, numerical methods for solving the linear system either fail or become unreliable.

Finite element code

The assembly of the linear system for linear elements is implemented in the FSELIB function scdl_sys, listed in the text. This function can accommodate the Neumann or the Robin boundary condition at the right end of the solution domain, as discussed in Problem 1.1.5. FSELIB code scdl, listed in the text, implements the finite element code.

Results with the Robin boundary condition and parameter values $L = 1.0$, $k=1.0$, $h_T = 1.0$, $f_\infty=0.0$, $f_L = 0.0$, $\rho = 1.0$, $c_p = 1.0$, $N_E = 16$ elements of equal size, and $U = 0$ (highest curve), 20, 40, 60, 80, and 100 (lowest curve) are shown in Figure 1.6.1. As the convection velocity U becomes larger, and thus the Péclet number is raised, the temperature distribution becomes increasingly steep, and numerical oscillations appear. In practical applications, Pe can be on the order of 10^3 or even higher. This difficulty underlines the notion that the Galerkin finite element method performs best for conduction- or diffusion-dominated transport.

The poor performance of the Galerkin finite element method for convection-dominated transport is remedied by the Petrov-Galerkin method, where the differential equation to be solved is weighted by a linear combination of the global interpolation functions and their derivatives, with position-dependent coefficients (e.g., [51, 66]). This modification effectively biases the difference equations resulting from the Petrov-Galerkin projection in a desired direction, and thereby promotes the overall stability of the numerical method. In this sense, the Petrov-Galerkin method is the counterpart of upwind differencing in finite different implementations.

```
function [at,bt,ct,b] ...
   ...
     = scdl_sys (ne,xe,q0,ht,finf,fL,k,U,rho,cp,s)

%==========================================================
% Compact assembly of the tridiagonal linear system for
% one-dimensional steady convection-diffusion
% with linear elements (scdl)
%==========================================================

%-------------
% element size
%-------------

for l=1:ne
  h(l) = xe(l+1)-xe(l);
end

%----------
% initialize
%----------

at = zeros(ne,1);
bt = zeros(ne,1);
ct = zeros(ne,1);
b  = zeros(ne,1);

%-------------------
% boundary conditions
%-------------------

% b(1) = q0/k;                      % Neumann
  b(1) = -ht*finf/k; at(1) = ht/k;  % Robin

cf = 0.5*U*rho*cp/k;

C11 = -cf; C12 = cf; % advection matrix
C21 = -cf; C22 = cf;

%----------------------
% loop over ne-1 elements
%----------------------
```

Function scdl_sys: Continuing $\longrightarrow$

1.6.2 Nonlinear convection–diffusion

A prototypical convection–diffusion equation incorporating the effect of convection nonlinearity is the viscous Burgers' equation

$$\frac{\partial f}{\partial t} + f\,\frac{\partial f}{\partial x} = \kappa\,\frac{\partial^2 f}{\partial x^2}, \tag{1.6.8}$$

which is a generalization of the inviscid form shown in (1.5.35). Remarkably, when the solution domain extends over the whole x axis, the solution can be

```
for l=1:ne-1

   A11 = 1/h(l); A12=-A11; A22=A11;
   B11 = h(l)/3.0; B12=0.5*B11; B22=B11;

   at(l)   = at(l)   + A11 + C11;
   bt(l)   = bt(l)   + A12 + C12;
   ct(l+1) = ct(l+1) + A12 + C21;
   at(l+1) = at(l+1) + A22 + C22;

   b(l)   = b(l)   + (B11*s(l) + B12*s(l+1))/k;
   b(l+1) = b(l+1) + (B12*s(l) + B22*s(l+1))/k;

end

%-------------
% last element
%-------------

A11 = 1/h(ne);   A12=-A11;
B11 = h(ne)/3.0; B12=0.5*B11;

at(ne) = at(ne) + A11+C11;
b(ne) = b(ne) + (B11*s(ne) + B12*s(ne+1))/k - (A12+C12)*fL;

%-----
% done
%-----

return;
```

Function scdl_sys: ($\longrightarrow$ Continued.)   Assembly of the linear system for steady one-dimensional convection–diffusion with linear elements.

found exactly for any initial condition by means of an integral transformation that reduces the viscous Burgers' equation to the unsteady diffusion equation (e.g., [43]).

The Galerkin finite element method results in a system of quadratic ordinary differential equations (ODEs) for the time-dependent nodal solution vector, $\mathbf{f}(t)$,

$$\mathbf{M} \cdot \frac{d\mathbf{f}}{dt} + \mathbf{f} \cdot \mathbf{\Pi} \cdot \mathbf{f} = -\mathbf{D} \cdot \mathbf{f} + \mathbf{b}, \qquad (1.6.9)$$

which is a generalization of (1.6.4) and (1.5.36), where $\mathbf{\Pi}$ is the three-index global advection tensor defined in (1.5.10). At steady state, we obtain a quadratic system of nonlinear equations

$$\mathbf{f} \cdot \mathbf{\Pi} \cdot \mathbf{f} = -\mathbf{D} \cdot \mathbf{f} + \mathbf{b}, \qquad (1.6.10)$$

that must be solved by iterative methods. In practice, the solution is found using Newton or modified Newton iterations such as Broyden's algorithm (e.g., [43]).

```
%=========================================================
% CODE scdl
%
% Code for steady one-dimensional convection-diffusion
% with linear elements (scdl)
%=========================================================

%-----------
% input data
%-----------

L=1.0; k=1.0; q0 = -1.00; ht=1.0; finf=0.0; fL=0.0;
rho=1.0; cp=1.0; U=100.0; ne=16; ratio=1.0;

%----------------
% grid generation
%----------------

xe = elm_line1 (0,L,ne,ratio);

%-------------------
% specify the source
%-------------------

for i=1:ne+1
 s(i) = 10.0*exp(-5.0*xe(i)^2/L^2);
end

%----------------
% element assembly
%----------------

[at,bt,ct,b] = scdl_sys (ne,xe,q0,ht,finf,fL,k,U,rho,cp,s);

%--------------
% linear solver
%--------------

f = thomas (ne,at,bt,ct,b);

f(ne+1) = fL;

%-----
% plot
%-----

plot(xe, f,'-o'); xlabel('x'); ylabel('f');

%-----
% done
%-----
```

Code scdl: Finite element code for steady one-dimensional convection–diffusion with linear elements.

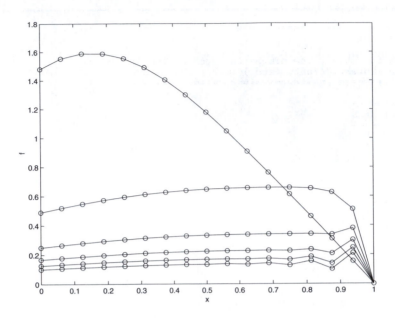

Figure 1.6.1 Finite element solution of the steady convection–diffusion equation with linear elements generated by the FSELIB code sdcl, for $Pe = 0$ (highest curve), 20, 40, 60, 80, and 100. Artificial oscillations appear as the Péclet number is raised yielding a convection-dominated transport.

PROBLEMS

1.6.1 *Convection dominated transport.*

(a) Run the FSELIB code scdl with the Robin boundary condition at the left, parameter values L=1.0, k=1.0, h_T =1.0, f_∞=0.0, f_L =0.0, ρ =1.0, c_p = 1.0, $U = 100$, and number of evenly-spaced elements $N_E = 2, 4, 8, 16, 32$, and 64, and discuss the behavior of the numerical solution.

(b) Repeat (a) with sixteen elements, and investigate the effect of element clustering near the right end of the computation domain where sharp gradients may arise.

1.6.2 *Neumann boundary condition.*

Repeat Problem 1.6.1(a) and (b) with the Neumann boundary condition at the left end.

1.7   Beam bending

The finite element method enjoys important applications in the field of structural mechanics where it is used to describe the various modes of deformation and the dynamical response of beams, plates, and shells subject to a permanent or transient load.

The simplest member of a structure is a slender beam that bends under an imposed transverse load, while exhibiting insignificant twisting and elongation or compression. A straight beam with a uniform cross-section is called prismatic. Beams made of wood, steel, and concrete are familiar from everyday experience. In engineering design, beam deformation is assessed by solving ordinary differential equations by numerical methods that are similar to those discussed previously in this chapter for the convection–diffusion equation. However, because beam deflection is governed by a fourth-order differential equation, a new set of element interpolation functions must be introduced to ensure continuity of the derivatives of the finite element expansion.

1.7.1   Euler-Bernoulli beam

Figure 1.7.1(a, b) illustrates a straight beam before and after deformation. The deflection is due to a distributed vertical load with force density, w, defined as the force per unit length along the x axis, and reckoned to be positive when directed downward against the y axis. For simplicity, we shall assume that the cross-section of the beam is symmetric with respect to the xy plane. Figure 1.7.1(c) illustrates the *axial stress* developing due to the deformation, defined as the force per unit cross-sectional area and denoted by σ, and the *transverse shear force*, defined as the integrated transverse shear stress over the entire beam cross-section and denoted by Q.

To develop the equations governing the beam deflection, we perform a vertical force balance over an infinitesimal section of the beam with length $\mathrm{d}x$, regarded as a free body. Under the sign conventions defined in Figure 1.7.1(c), we find

$$Q(x) + w(x)\,\mathrm{d}x = Q(x + \mathrm{d}x). \tag{1.7.1}$$

Rearranging, we derive the differential relation

$$w = \frac{\mathrm{d}Q}{\mathrm{d}x}. \tag{1.7.2}$$

The axial stress, σ, varies across the beam cross-section due to the stretching of axial fibers near the upper surface and the compression of axial fibers near the lower surface. The *neutral surface* of the beam is a material surface normal to the xy plane, located at $y = y_N$, that undergoes neither stretching nor compression, and is thus unstressed in the deformed configuration, $\sigma = 0$. The variation of the axial stress over the beam cross-section gives rise to a

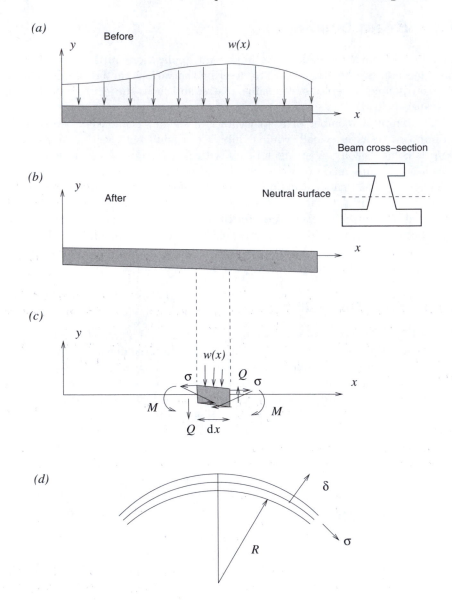

Figure 1.7.1 Schematic illustration of a straight beam bending (*a*) before, and
(*b*) after deformation. (*c*) Distribution of the axial stress, σ, over the cross-
section, and depiction of the transverse shear force, Q. (*d*) An exaggerated
local view of a deformed beam used to derive a constitutive equation for
the bending moment; δ is the distance from the neutral surface.

macroscopic bending moment about that z axis computed with respect to the neutral surface, given by

$$M = \iint \sigma\, \delta\, \mathrm{d}A, \qquad (1.7.3)$$

where $\delta = y - y_N$ is the distance of a point from the neutral surface, as shown in Figure 1.7.1(d), and the integration is performed over the beam cross-section.

Performing a torque balance over an infinitesimal section of the beam with length $\mathrm{d}x$, we find

$$Q(x)\, \mathrm{d}x + M(x) = M(x + \mathrm{d}x). \qquad (1.7.4)$$

Note that the neglected contribution of the vertical load, $w(x)$, is quadratic in $\mathrm{d}x$. Rearranging, we derive the differential relation

$$Q = \frac{\mathrm{d}M}{\mathrm{d}x}. \qquad (1.7.5)$$

Substituting this expression in the force balance (1.7.2), we derive the equilibrium equation

$$\frac{\mathrm{d}^2 M}{\mathrm{d}x^2} = w(x). \qquad (1.7.6)$$

Next, we relate the bending moment to the curvature of the neutral surface in the xy plane, denoted by κ, using the constitutive equation

$$M = EI\kappa, \qquad (1.7.7)$$

where E is the modulus of elasticity of the beam material, and I is the principal moment of inertia of the beam cross-section with respect to the z axis that is normal to the xy plane,

$$I \equiv \iint \delta^2\, \mathrm{d}A, \qquad (1.7.8)$$

To derive (1.7.7), we consider the simple configuration depicted in Figure 1.7.1(d), where a straight beam has been locally deformed into a circular arc with centerline radius, R, and introduce the fundamental assumption of the Euler-Bernoulli theory: *a material plane that is normal to the beam before deformation remains normal to the beam after deformation.* The arc radius across the beam is $r = R + \delta$, and the axial stretch of fibers along the beam is $\epsilon = (R + \delta)/R$. Adopting a linear constitutive equation for the axial stress expressed by Hooke's law, we write $\sigma = E\,(\epsilon - 1) = E\,\delta/R$. Substituting this expression in (1.7.3), and setting $\kappa = 1/R$, we derive (1.7.7). Note that $\sigma = 0$ at the neutral surface, $\delta = 0$, as required.

For small vertical deflection, v, the curvature can be approximated with the negative of the second derivative,

$$\kappa = -\frac{\mathrm{d}^2 v}{\mathrm{d}x^2}. \qquad (1.7.9)$$

Note that, for the bending configuration illustrated in Figure 1.7.1, the vertical deflection is negative, $v < 0$. Substituting (1.7.9) in the preceding expressions for the bending moment and transverse shear force, we find

$$M = -EI \, \frac{\mathrm{d}^2 v}{\mathrm{d}x^2},$$

$$Q = -\frac{\mathrm{d}}{\mathrm{d}x}\left(EI \, \frac{\mathrm{d}^2 v}{\mathrm{d}x^2} \right).$$

(1.7.10)

Substituting further the expression for the bending moment in (1.7.6), we derive the governing differential equation

$$\frac{\mathrm{d}^2}{\mathrm{d}x^2}\left(EI \, \frac{\mathrm{d}^2 v}{\mathrm{d}x^2} \right) = -w(x).$$

(1.7.11)

In the case of a homogeneous prismatic beam, the product EI is constant. Rearranging, we obtain

$$\frac{\mathrm{d}^4 v}{\mathrm{d}x^4} = -\frac{1}{EI} \, w(x).$$

(1.7.12)

To solve this fourth-order differential equation by the Galerkin finite element method, we must use at least a cubic polynomial expansion over the individual elements, and require a C^1 solution, meaning, that the solution and its first derivative must be continuous across the element nodes. In contrast, previously in this chapter we discussed heat conduction governed by a second-order differential equation, and were able to produce a solution using linear elements while requiring C^0 continuity of the global expansion.

Boundary conditions

The beam bending equation is to be solved subject to boundary conditions that reflect the physical circumstances of the problem under consideration. Since beam bending is governed by a fourth-order differential equation, we require two scalar boundary conditions at each end, adding up to a total of four boundary conditions. Possible choices include the following:

- When the end of a beam is *fixed*, *clamped*, or *built-in*, the deflection and its derivative are required to be zero,

$$v = 0, \qquad \frac{\partial v}{\partial x} = 0.$$

(1.7.13)

  An aircraft wing can be regarded as a beam that is built into the aircraft on one end.

- When the end of a beam is *simply supported*, the deflection and the bending moment are required to be zero,

$$v = 0, \qquad M = 0.$$

(1.7.14)

  Physically, a beam is simply supported when it rests on a ledge.

- When the end of a beam is free, the transverse shear force and the bending moment are required to be zero,

$$Q = 0, \qquad M = 0. \qquad (1.7.15)$$

An aircraft wing can be regarded as a beam that is free at the far end.

A beam that is clamped on one end and free at the other end is called a cantilever.

Variational formulation

In Section 1.2, we discussed the variational formulation of the steady diffusion equation in the presence of a source. An analogous formulation is possible for the beam bending equation forced by a vertical load. The variational formulation of the Euler-Bernoulli beam states that computing the solution of (1.7.11) over the interval $[a, b]$ for a beam that is simply supported at both ends, is equivalent to maximizing the functional

$$\mathcal{F} < w(x) > \equiv -\frac{1}{2} \int_a^b EI \left(\frac{d^2 w}{dx^2}\right)^2 dx + \int_a^b w(x)\, \omega(x)\, dx, \qquad (1.7.16)$$

over the set of all C^1 functions that satisfy the homogeneous boundary-conditions $w(a) = 0$, $w''(a) = 0$, $w(b) = 0$, and $w''(b) = 0$. The proof is carried out as discussed in Section 1.2 for the steady diffusion equation. Similar variational formulations are possible for other types of boundary conditions.

Timoshenko's theory

The Euler-Bernoulli theory is based on the fundamental assumption that a material plane that is normal to the beam before deformation remains normal to the beam after deformation. In Timoshenko's theory, this assumption is relaxed by introducing the displacement of point particles along the x and y axes, denoted by v_x and v_y, and stipulating

$$v_x = u(x) - y \left[\frac{\partial v}{\partial x} - \gamma(x)\right],$$

$$(1.7.17)$$

$$v_y = v(x) - y \left[\frac{\partial v}{\partial x} - \gamma(x)\right],$$

where $u(x)$ and $v(x)$ are the x and y displacements of point particles in the neutral surface, and $\gamma(x)$ is the shear rotation (e.g., [41]). The Euler-Bernoulli theory arises by setting $u(x) = 0$ and $\gamma(x) = 0$, computing the axial strain $\epsilon_{xx} \equiv \partial v_x/\partial x = -y\, \partial^2 v/\partial x^2$, and then using Hooke's law to express the axial stress as $\sigma = E\, \epsilon_{xx} = -E\, y\, \partial^2 v/\partial x^2$. Closure in Timoshenko's theory is achieved by relating $\gamma(x)$ to the transverse shear force, $Q(x)$, using an appropriate constitutive equation involving the shear modulus of elasticity, G, and an effective shear area, A_s, as $\gamma = Q/(GA_s)$. Finite element methods for the Timoshenko formulation are generalizations of those for the Euler-Bernoulli formulation discussed in the remainder of this section.

1.7.2   Cubic Hermitian elements

To implement the finite element method, we divide the beam in N_E elements, where the lth element is subtended between the first end-node, $X_1^{(l)}$, and the second end-node, $X_2^{(l)}$. To implement the cubic expansion, we introduce a dimensionless parameter, η, ranging in the interval $[0, 1]$, and express the x position over the element as

$$x = X_1^{(l)} + h_l\, \eta, \qquad (1.7.18)$$

where $h_l \equiv X_2^{(l)} - X_1^{(l)}$ is the element length. When $\eta = 0$, $x = X_1^{(l)}$, and when $\eta = 1$, $x = X_2^{(l)}$.

Next, we introduce the element-node vertical deflections and their derivatives expressing the slope,

$$v_1^{(l)}, \qquad v_1'^{(l)} \equiv \left(\frac{\mathrm{d}v}{\mathrm{d}x}\right)_1^{(l)}, \qquad v_2^{(l)}, \qquad v_2'^{(l)} \equiv \left(\frac{\mathrm{d}v}{\mathrm{d}x}\right)_2^{(l)}. \qquad (1.7.19)$$

The requirement of a C^1 solution demands that these values are shared by neighboring elements at corresponding positions. The cubic expansion over the lth element takes the form

$$v(\eta) = v_1^{(l)}\, \psi_1^{(l)}(\eta) + v_1'^{(l)}\, \psi_2^{(l)}(\eta) + v_2^{(l)}\, \psi_3^{(l)}(\eta) + v_2'^{(l)}\, \psi_4^{(l)}(\eta), \qquad (1.7.20)$$

where

$$\psi_1^{(l)}(\eta) = 2\eta^3 - 3\eta^2 + 1,$$

$$\psi_2^{(l)}(\eta) = h_l\, \eta\, (\eta^2 - 2\eta + 1),$$

$$\psi_3^{(l)}(\eta) = \eta^2\, (-2\eta + 3), \qquad (1.7.21)$$

$$\psi_4^{(l)}(\eta) = h_l\, \eta^2\, (\eta - 1),$$

are cubic *Hermitian* local interpolation functions illustrated in Figure 1.7.2. Each one of these functions satisfies a set of four boundary conditions. The first function satisfies

$$\psi_1^{(l)}(\eta = 0) = 1, \qquad \left(\frac{\mathrm{d}\psi_1^{(l)}}{\mathrm{d}\eta}\right)_{\eta=0} = 0,$$

$$\psi_1^{(l)}(\eta = 1) = 0, \qquad \left(\frac{\mathrm{d}\psi_1^{(l)}}{\mathrm{d}\eta}\right)_{\eta=1} = 0, \qquad (1.7.22)$$

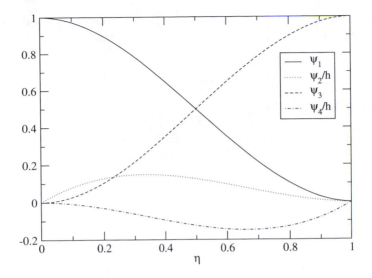

Figure 1.7.2 Graphs of the four Hermitian element interpolation functions over a cubic element of length h.

the second function satisfies

$$\psi_2^{(l)}(\eta = 0) = 0, \qquad \left(\frac{\mathrm{d}\psi_2^{(l)}}{\mathrm{d}\eta}\right)_{\eta=0} = h_l,$$

$$\psi_2^{(l)}(\eta = 1) = 0, \qquad \left(\frac{\mathrm{d}\psi_2^{(l)}}{\mathrm{d}\eta}\right)_{\eta=1} = 0, \qquad (1.7.23)$$

the third function satisfies

$$\psi_3^{(l)}(\eta = 0) = 0, \qquad \left(\frac{\mathrm{d}\psi_3^{(l)}}{\mathrm{d}\eta}\right)_{\eta=0} = 0,$$

$$\psi_3^{(l)}(\eta = 1) = 1, \qquad \left(\frac{\mathrm{d}\psi_3^{(l)}}{\mathrm{d}\eta}\right)_{\eta=1} = 0, \qquad (1.7.24)$$

and the fourth function satisfies

$$\psi_4^{(l)}(\eta = 0) = 0, \qquad \left(\frac{\mathrm{d}\psi_4^{(l)}}{\mathrm{d}\eta}\right)_{\eta=0} = 0,$$

$$\psi_4^{(l)}(\eta = 1) = 0, \qquad \left(\frac{\mathrm{d}\psi_4^{(l)}}{\mathrm{d}\eta}\right)_{\eta=1} = h_l. \qquad (1.7.25)$$

An element that supports these interpolation functions is called a cubic Hermitian element.

Evaluating (1.7.20) at $\eta = 0$ and 1, invoking the boundary conditions satisfied by the element interpolation functions, and writing

$$\frac{dv}{dx} = \frac{dv}{d\eta}\frac{d\eta}{dx} = \frac{dv}{d\eta}\frac{1}{h_l}, \tag{1.7.26}$$

we confirm the required properties

$$v(\eta = 0) = v_1^{(l)}, \qquad \left(\frac{dv}{dx}\right)_{\eta=0} = v_1'^{(l)},$$

$$\tag{1.7.27}$$

$$v(\eta = 1) = v_2^{(l)}, \qquad \left(\frac{dv}{dx}\right)_{\eta=1} = v_2'^{(l)}.$$

Global interpolation functions

The end-nodes of the N_E elements comprise a set of $N_G = N_E + 1$ unique global nodes. The ith global node hosts an odd-numbered deflection global interpolation function, $\phi_{2i-1}(x)$, and an even-numbered deflection-slope global interpolation function, $\phi_{2i}(x)$, where $i = 1, 2, \ldots, N_G$. These global interpolation functions consist of the union of corresponding element interpolation functions, as discussed earlier in this section.

Specifically, since the ith cubic Hermitian element is subtended between the global nodes numbered i and $i + 1$, the odd-numbered deflection global interpolation function, $\phi_{2i-1}(x)$, is comprised of the union of $\psi_1^{(i)}(\eta)$ and $\psi_3^{(i-1)}(\eta)$, as illustrated in Figure 1.7.3(a). Similarly, the even-numbered deflection-slope global interpolation function, $\phi_{2i}(x)$, is comprised of the union of $\psi_2^{(i)}(\eta)$ and $\psi_4^{(i-1)}(\eta)$, as illustrated in Figure 1.7.3(b).

In terms of the global interpolation functions, the finite element expansion over the entire length of the beam is described by

$$v(x) = \sum_{j=1}^{N_G} v_j\,\phi_{2j-1}(x) + \sum_{j=1}^{N_G} v_j'\,\phi_{2j}(x), \tag{1.7.28}$$

where v_j is the deflection, and v_j' is its slope at the position of the jth global node.

1.7.3 Galerkin projection

To carry out the Galerkin projection, we multiply the governing equation (1.7.12) by each of the global interpolation functions, and integrate the product over the solution domain. For a beam extending from $x = 0$ to L, we write

$$\int_0^L \phi_i\,\frac{d^4v}{dx^4}\,dx = -\frac{1}{EI}\int_0^L \phi_i\,w(x)\,dx. \tag{1.7.29}$$

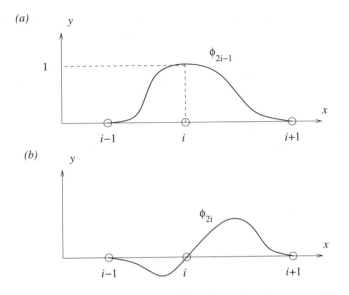

Figure 1.7.3 Schematic illustration of (*a*) the deflection, and (*b*) deflection-slope global interpolation function associated with the *i*th global node.

Integrating by parts on the left-hand side to reduce the order of the fourth derivative, we find

$$
\int_0^L \phi_i \frac{\mathrm{d}^4 v}{\mathrm{d}x^4}\,\mathrm{d}x = \int_0^L \phi_i \frac{\mathrm{d}}{\mathrm{d}x}\left(\frac{\mathrm{d}^3 v}{\mathrm{d}x^3}\right)\mathrm{d}x
$$

$$
= \left[\phi_i \frac{\mathrm{d}^3 v}{\mathrm{d}x^3}\right]_0^L - \int_0^L \frac{\mathrm{d}\phi_i}{\mathrm{d}x}\frac{\mathrm{d}^3 v}{\mathrm{d}x^3}\,\mathrm{d}x, \qquad (1.7.30)
$$

where the square brackets signify the change of the enclosed expression between the upper and lower limits of integration. Repeating the integration by parts for the last integral on the right-hand side, we find

$$
\int_0^L \frac{\mathrm{d}\phi_i}{\mathrm{d}x}\frac{\mathrm{d}^3 v}{\mathrm{d}x^3}\,\mathrm{d}x = \int_0^L \frac{\mathrm{d}\phi_i}{\mathrm{d}x}\frac{\mathrm{d}}{\mathrm{d}x}\left(\frac{\mathrm{d}^2 v}{\mathrm{d}x^2}\right)\mathrm{d}x
$$

$$
= \left[\frac{\mathrm{d}\phi_i}{\mathrm{d}x}\frac{\mathrm{d}^2 v}{\mathrm{d}x^2}\right]_0^L - \int_0^L \frac{\mathrm{d}^2\phi_i}{\mathrm{d}x^2}\frac{\mathrm{d}^2 v}{\mathrm{d}x^2}\,\mathrm{d}x. \qquad (1.7.31)
$$

Combining the preceding expressions and substituting the result in (1.7.29), we obtain

$$
\int_0^L \frac{\mathrm{d}^2\phi_i}{\mathrm{d}x^2}\frac{\mathrm{d}^2 v}{\mathrm{d}x^2}\,\mathrm{d}x + \left[\phi_i \frac{\mathrm{d}^3 v}{\mathrm{d}x^3} - \frac{\mathrm{d}\phi_i}{\mathrm{d}x}\frac{\mathrm{d}^2 v}{\mathrm{d}x^2}\right]_0^L = -\frac{1}{EI}\int_0^L \phi_i\, w(x)\,\mathrm{d}x.
$$

$$
(1.7.32)
$$

In terms of the transverse shear force, Q, and bending moment, M, given in (1.7.10),

$$EI \int_0^L \frac{d^2 v}{dx^2} \frac{d^2 \phi_i}{dx^2} \, dx = \left[\phi_i \, Q - \frac{d\phi_i}{dx} M \right]_0^L - \int_0^L \phi_i \, w(x) \, dx, \qquad (1.7.33)$$

for $i = 1, 2, \ldots, 2N_G$.

Substituting the finite element expansion (1.7.28) in (1.7.33), and compiling the resulting equations, we formulate the linear system

$$EI \, \mathbf{K} \cdot \mathbf{u} = \mathbf{b}, \qquad (1.7.34)$$

where $\mathbf{K}$ is the global stiffness matrix with components

$$K_{ik} = \int_0^L \frac{d^2 \phi_i}{dx^2} \frac{d^2 \phi_k}{dx^2} \, dx, \qquad (1.7.35)$$

for $i, k = 1, 2, \ldots, 2N_G$, and

$$\mathbf{u} \equiv \begin{bmatrix} v_1 \\ v_1' \\ v_2 \\ v_2' \\ \vdots \\ v_{N_G} \\ v_{N_G}' \end{bmatrix} \qquad (1.7.36)$$

is the vector of nodal deflections and slopes. The first two components of the vector $\mathbf{b}$ on the right-hand side are given by

$$b_1 = \left[\phi_1 \, Q - \frac{d\phi_1}{dx} M \right]_0^L - \int_0^L \phi_1 \, w(x) \, dx$$

$$= -Q(0) - \int_0^L w(x) \, \phi_1 \, dx,$$

$$\qquad (1.7.37)$$

$$b_2 = \left[\phi_2 \, Q - \frac{d\phi_2}{dx} M \right]_0^L - \int_0^L w(x) \, \phi_2 \, dx$$

$$= M(0) - \int_0^L \phi_2 \, w(x) \, dx.$$

The intermediate components are given by

$$b_i = - \int_0^L \phi_i \, w(x) \, dx, \qquad (1.7.38)$$

for $i = 3, 4, \ldots, 2N_G - 2$, and the last two components are given by

$$b_{2N_G-1} = \left[\phi_{2N_G-1} \, Q - \frac{\mathrm{d}\phi_{2N_G-1}}{\mathrm{d}x} \, M \right]_0^L - \int_0^L \phi_{2N_G-1} \, w(x) \, \mathrm{d}x$$

$$= Q(L) - \int_0^L \phi_{2N_G-1} \, w(x) \, \mathrm{d}x,$$

(1.7.39)

$$b_{2N_G} = \left[\phi_{2N_G} \, Q - \frac{\mathrm{d}\phi_{2N_G}}{\mathrm{d}x} \, M \right]_0^L - \int_0^L \phi_{2N_G} \, w(x) \, \mathrm{d}x$$

$$= -M(L) - \int_0^L \phi_{2N_G} \, w(x) \, \mathrm{d}x.$$

Implementing the boundary conditions by specifying selected values of the generalized deflection vector $\mathbf{u}$, end-node transverse shear forces $Q(0)$, $Q(L)$, and bending moments, $M(0)$, $M(L)$, we derive the final system of equations for the remaining unknowns.

If the load density function and its first derivative are continuous over the entire length of the beam, we may introduce the familiar Hermite expansion

$$w(x) = \sum_{j=1}^{N_G} w_j \, \phi_{2j-1}(x) + \sum_{i=j}^{N_G} w'_j \, \phi_{2j}(x),$$

(1.7.40)

where w_j and w'_j are the nodal values and slopes. Accordingly, we may write

$$\int_0^L \phi_i \, w(x) \, \mathrm{d}x = \sum_{j=1}^{N_G} w_j \int_0^L \phi_i \, \phi_{2j-1} \, \mathrm{d}x + \sum_{i=j}^{N_G} w'_j \int_0^L \phi_{2j} \, \phi_i \, \mathrm{d}x, \quad (1.7.41)$$

and identify the integrals on the right-hand side with the components of the generalized global mass matrix,

$$M_{ik} = \int_0^L \phi_i \, \phi_k \, \mathrm{d}x,$$

(1.7.42)

where $i, k = 1, 2, \ldots, 2N_G$.

1.7.4   Element stiffness and mass matrices

In practice, the global stiffness matrix, $\mathbf{K}$, is assembled from 4×4 element stiffness matrices with components

$$H_{ij}^{(l)} = \int_{E_l} \frac{\mathrm{d}^2 \psi_i^{(l)}}{\mathrm{d}x^2} \frac{\mathrm{d}^2 \psi_j^{(l)}}{\mathrm{d}x^2} \, \mathrm{d}x,$$

(1.7.43)

for $i, j = 1, 2, 3, 4$. Substituting the Hermitian element interpolation functions, and carrying out the integrations, we find

$$\mathbf{H}^{(l)} = \frac{1}{h_l^3} \begin{bmatrix} 12 & 6\,h_l & -12 & 6\,h_l \\ 6\,h_l & 4\,h_l^2 & -6\,h_l & 2\,h_l^2 \\ -12 & -6\,h_l & 12 & -6\,h_l \\ 6\,h_l & 2\,h_l^2 & -6\,h_l & 4\,h_l^2 \end{bmatrix}. \tag{1.7.44}$$

The assembled stiffness matrix, $\mathbf{K}$, takes the hexadiagonal form

$$\begin{bmatrix} H_{11}^{(1)} & H_{12}^{(1)} & H_{13}^{(l)} & H_{14}^{(1)} & 0 & 0 & \cdots & 0 \\ H_{21}^{(1)} & H_{22}^{(1)} & H_{23}^{(l)} & H_{24}^{(1)} & 0 & 0 & \cdots & 0 \\ H_{31}^{(1)} & H_{32}^{(1)} & H_{33}^{(1)} + H_{11}^{(2)} & H_{34}^{(1)} + H_{12}^{(2)} & H_{13}^{(2)} & H_{14}^{(2)} & 0 & \cdots \\ H_{41}^{(1)} & H_{42}^{(1)} & H_{43}^{(1)} + H_{21}^{(2)} & H_{44}^{(1)} + H_{22}^{(2)} & H_{23}^{(2)} & H_{24}^{(2)} & 0 & \cdots \\ 0 & 0 & H_{31}^{(2)} & H_{32}^{(2)} & H_{33}^{(2)} + H_{11}^{(3)} & H_{34}^{(2)} + H_{12}^{(3)} & \cdots & \cdots \\ 0 & 0 & H_{41}^{(2)} & H_{42}^{(2)} & H_{43}^{(2)} + H_{21}^{(3)} & H_{44}^{(2)} + H_{22}^{(3)} & \cdots & \cdots \\ \cdots & \cdots & \cdots & \cdots & \cdots & \cdots & \cdots & \cdots \\ 0 & 0 & 0 & 0 & 0 & 0 & \cdots & \cdots \\ 0 & 0 & 0 & 0 & 0 & 0 & \cdots & \cdots \\ 0 & 0 & 0 & 0 & 0 & 0 & \cdots & \cdots \end{bmatrix},$$

$$\tag{1.7.45}$$

with the understanding that the individual equations of the linear system are arranged in the order of the Galerkin projection with ϕ_1, ϕ_2, ϕ_3, ϕ_4, ... "Hexadiagonal" derives from the Greek words $\epsilon\xi\iota$, which means "six," and $\delta\iota\alpha\gamma\omega\nu\iota o\varsigma$, which means "diagonal." The assembly is carried out according to Algorithm 1.7.1 implemented in the FSELIB function **beam_sys**, listed in the text.

Similarly, the generalized global mass matrix can be assembled from 4×4 element mass matrices with components

$$B_{ij}^{(l)} = \int_{E_l} \psi_i^{(l)} \, \psi_j^{(l)} \mathrm{d}x, \tag{1.7.46}$$

for $i, j = 1, 2, 3, 4$. Substituting the Hermitian element interpolation functions and carrying out the integrations, we find

$$\mathbf{B}^{(l)} = \frac{h_l}{420} \begin{bmatrix} 156 & 22\,h_l & 54 & -13\,h_l \\ 22\,h_l & 4\,h_l^2 & 13\,h_l & -3\,h_l^2 \\ 54 & 13\,h_l & 156 & -22\,h_l \\ -13\,h_l & -3\,h_l^2 & -22\,h_l & 4\,h_l^2 \end{bmatrix}. \tag{1.7.47}$$

The assembled global mass matrix takes a hexadiagonal form similar to that shown in (1.7.45). The assembly process requires a simple modification of Algorithm 1.7.1.

Specify the number of elements, N_E

$N_G = N_E + 1$ *Number of unique global nodes*

Do $i = 1, 2, \ldots, 2N_G$ *Initialize to zero*
     Do $j = 1, 2, \ldots, 2N_G$
         $K_{ij} = 0$
     End Do
End Do

Do $l = 1, 2, \ldots, N_E$ *Run over the elements*
     Compute the element stiffness matrix $\mathbf{H}^{(l)}$ using (1.7.44)
     $i_1 = 2l - 2$

     Do $i = 1, 2, 3, 4$
         Do $j = 1, 2, 3, 4$
             $K_{i_1+i,i_1+j} = K_{i_1+i,i_1+j} + H_{ij}^{(l)}$
         End Do
     End Do
End Do

Algorithm 1.7.1 Assembly of the global stiffness matrix for beam bending.

One-element cantilever beam

As an example, we consider the bending of a cantilever beam due to a tip load of magnitude F, as shown in Figure 1.7.4(a). The beam load density can be described in terms of the one-dimensional Dirac delta function, $\delta(x - x_0)$, as

$$w(x) = F\,\delta(x - L). \tag{1.7.48}$$

Note that the one-dimensional delta function has units of inverse length; accordingly, the load density has the expected units of force over length.

     Requiring the free-end boundary conditions

$$Q(L) = 0, \qquad M(L) = 0, \tag{1.7.49}$$

and using the distinctive properties of the delta function, we find that the right-hand side of the preliminary linear system simplifies to

```
function [stiff] = beam_sys (ne,xe)

%=====================================================
% Assembly of the stiffness matrix for beam bending
% with cubic Hermitian elements
%
% ne: number of elements,      xe: element end-points
% stiff: global stiffness matrix
%=====================================================

%----------------------------------------
% element size and number of global nodes
%----------------------------------------

for l=1:ne
  h(l) = xe(l+1)-xe(l);
end

ng = ne+1;    % number of unique global nodes

%-----------
% initialize
%-----------

stiff = zeros(2*ng,2*ng);

%-----------------------
% loop over the elements
%-----------------------

for l=1:ne
    esm(1,1) = 12.0; esm(1,2) = 6.0*h(l); esm(1,3) = -12.0;
    esm(1,4) = 6.0*h(l);
    esm(2,1) = esm(1,2); esm(2,2) = 4.0*h(l)^2; esm(2,3) = -6.0*h(l);
    esm(2,4) = 2.0*h(l)^2;
    esm(3,1) = esm(1,3); esm(3,2) = esm(2,3); esm(3,3) = 12.0;
    esm(3,4) = -6.0*h(l);
    esm(4,1) = esm(1,4); esm(4,2) = esm(2,4); esm(4,3) = esm(3,4);
    esm(4,4) = 4.0*h(l)^2;
    esm = esm/h(l)^3;   i1 = 2*(l-1);
    for i=1:4
      for j=1:4
        stiff(i1+i,i1+j) = stiff(i1+i,i1+j) + esm(i,j);
      end
    end
end

%-----
% done
%-----

return;
```

Function beam_sys: Assembly of the global stiffness matrix for beam bending.

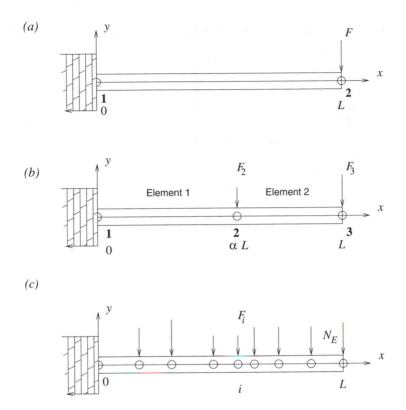

(a)

(b)

(c)

Figure 1.7.4 (*a*) Illustration of a beam subjected to a tip load; (*b*) two-element and (*c*) many-element discretization of a beam subject to a nodal load. The bold-faced numbers in (*a*) and (*b*) are the labels of the global nodes.

$$b_1 = -Q(0), \qquad b_2 = M(0),$$

$$b_3 = -\int_0^L \phi_3 \, w(x) \, dx = -F \int_0^L \phi_3 \, \delta(x - L) \, dx = -F \, \phi_3(x = L) = -F,$$

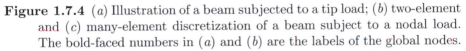

$$b_4 = 0. \tag{1.7.50}$$

The Galerkin linear system takes the form

$$\frac{EI}{L^3} \begin{bmatrix} 12 & 6L & -12 & 6L \\ 6L & 4L^2 & -6L & 2L^2 \\ -12 & -6L & 12 & -6L \\ 6L & 2L^2 & -6L & 4L^2 \end{bmatrix} \cdot \begin{bmatrix} v_1 \\ v_1' \\ v_2 \\ v_2' \end{bmatrix} = \begin{bmatrix} -Q(0) \\ M(0) \\ -F \\ 0 \end{bmatrix}. \tag{1.7.51}$$

Specifying on physical grounds the built-in or clamped boundary conditions

$$v_1 = 0, \qquad v_1' = 0, \tag{1.7.52}$$

we obtain a system of four linear equations for the four unknowns:

$$v_2, \qquad v_2', \qquad Q(0), \qquad M(0). \tag{1.7.53}$$

Because boundary conditions on the deflection and its slope are specified at the left end, only the projections with ϕ_3 and ϕ_4 expressed by the last two equations in (1.7.51) are necessary. In practice, it is expedient to sacrifice some computational cost and implement these boundary conditions after the complete preliminary system has been compiled. Replacing the redundant finite element equations with the specified boundary conditions reduces the preliminary system (1.7.51) to the final system

$$\frac{EI}{L^3} \begin{bmatrix} 1 & 0 & 0 & 0 \\ 0 & 1 & 0 & 0 \\ 0 & 0 & 12 & -6\,L \\ 0 & 0 & -6\,L & 4\,L^2 \end{bmatrix} \cdot \begin{bmatrix} v_1 \\ v_1' \\ v_2 \\ v_2' \end{bmatrix} = \begin{bmatrix} 0 \\ 0 \\ -F \\ 0 \end{bmatrix}, \tag{1.7.54}$$

which effectively encapsulates the 2×2 system

$$\frac{EI}{L^3} \begin{bmatrix} 12 & -6\,L \\ -6\,L & 4\,L^2 \end{bmatrix} \cdot \begin{bmatrix} v_2 \\ v_2' \end{bmatrix} = \begin{bmatrix} -F \\ 0 \end{bmatrix}. \tag{1.7.55}$$

The solution is readily found to be

$$v_2 = -\frac{FL^3}{3\,EI}, \qquad v_2' = -\frac{FL^3}{2\,EI}, \tag{1.7.56}$$

which, in fact, is the exact solution of the problem under consideration.

The practice of implementing the boundary conditions after the complete preliminary finite element system has been compiled will be discussed further in this section and then in Section 3.1.7 with reference to the finite element implementation in two dimensions.

1.7.5   Code for cantilever beam with nodal loads

Consider a cantilever beam discretized in N_E elements, as illustrated in Figure 1.7.4(c), subject to concentrated nodal loads described by the load density function

$$w(x) = \sum_{i=1}^{N_G} F_i\, \delta(x - x_i), \tag{1.7.57}$$

where $N_G = N_E + 1$ is the number of global nodes defining the element endnodes, F_i is the ith nodal force reckoned to be positive when directed downward, x_i is the position of the ith global node, and $\delta(x)$ is Dirac's one-dimensional delta function.

Requiring the free-end boundary conditions, $Q(L) = 0$ and $M(L) = 0$, and using the distinctive properties of the delta function, we find that the first component of the right-hand side of the preliminary linear system simplifies to

$$b_1 = -Q(0) - \int_0^L \phi_1 \, w(x) \, dx = -Q(0) - F_1 \int_0^L \phi_1 \, \delta(x - x_1^G) \, dx$$
$$= -Q(0) - F_1 \, \phi_1(x_2^G) = Q(0) - F_1, \qquad (1.7.58)$$

and the second component simplifies to

$$b_2 = M(0) - \int_0^L \phi_2 \, w(x) \, dx = M(0). \qquad (1.7.59)$$

Subsequent components simplify to

$$
\begin{array}{lll}
b_3 = -F_2, & b_4 = 0, & \cdots \\
b_{2i-1} = -F_i, & b_{2i} = 0, & \cdots \qquad (1.7.60) \\
b_{2N_G-1} = -F_{N_G} & b_{2N_G} = 0.
\end{array}
$$

Code `beam`, listed in the text, implements the finite element method. First, the preliminary linear system is assembled, with the right-hand side computed using the FSELIB function `beam_sys` based on the preceding expressions. Second, the left-end boundary conditions, $v_1 = 0$ and $v_1' = 0$, are implemented in three steps:

1. All elements in the first and second columns and rows of the global stiffness matrix, **K**, are set to 0.

2. The first two diagonal elements, K_{11} and K_{22}, are set to unity.

3. The first and second entries of the right-hand side, b_1 and b_2, are replaced by zeros.

In the case of one element, $N_E = 1$, the algorithm amounts to replacing the preliminary system (1.7.51) with the final system (1.7.54). This way of implementing the boundary conditions is attractive in that it preserves the symmetry of the stiffness matrix.

For simplicity, the linear system is solved using an internal function embedded in MATLAB, implemented in symbolic form by dividing a row vector by a matrix in the line *sol=b/stiff'*, which seemingly defies the rules of matrix calculus. Note that, because the solution, *sol*, is also a row vector, the transpose of the coefficient matrix indicated by the prime must be entered in the denominator according to MATLAB convention, though in this case *stiff'=stiff*. Methods for solving systems of linear equations in finite element applications are discussed in Appendix C.

The graphics display of the solution under uniform nodal loading and parameter values listed in the code is shown in Figure 1.7.5.

```
%=======================================
% CODE beam
%
% Code for cantilever beam bending with
% Hermitian cubic elements, subject
% to nodal forcing
%=======================================

%-----------
% input data
%-----------

L=1.0; E=1.0; I=1.0; ne=16; ratio=1.0;

%----------------
% grid generation
%----------------

xe = elm_line1 (0,L,ne,ratio);

ng = ne+1;  % number of unique global nodes

%----------------------
% specify the nodal load
%----------------------

for i=1:ng
 F(i) = 1.0/ne;
end

%----------------------------
% compile the stiffness matrix
%----------------------------

stiff = beam_sys (ne,xe); stiff = E*I*stiff;

%--------------------------
% preliminary right-hand side
%--------------------------

for i=1:ng
 b(2*i-1) = - F(i); b(2*i) = 0.0;
end

%-------------------------------------------
% implement the left-end boundary conditions
%-------------------------------------------

for i=1:2*ng
 stiff(1,i) = 0.0; stiff(2,i) = 0.0;
 stiff(i,1) = 0.0; stiff(i,2) = 0.0;
end
```

Code beam: Continuing $\longrightarrow$

```
stiff(1,1) = 1.0; stiff(2,2) = 1.0;

b(1) = 0.0; b(2) = 0.0;

%--------------
% linear solver
%--------------

sol = b/stiff';

%-----------------------
% extract the deflections
%-----------------------

for i=1:ng
 v(i) = sol(2*i-1);
end

%----------------
% prepare to plot
%----------------

xlabel('x'); ylabel('y');
axis([0 1.1*L -0.2*L 0.10*L]);

%---------------
% plot the nodes
%---------------

plot (xe, zeros(ng),'-o'); hold on;

%---------------------
% plot the nodal forces
%---------------------

for i=1:ng
 plotx(1) = xe(i); ploty(1) = 0.0;
 plotx(2) = xe(i); ploty(2) = -F(i);
 plot(plotx, ploty);
 plttx(1) = xe(i); pltty(1) = -F(i);
 plot(plttx, pltty,'v');
end

%--------------------
% plot the deflections
%--------------------

plot(xe, v,'-o');

%-----
% done
%-----
```

Code beam: ($\longrightarrow$ Continued.) Code for solving the equations of beam bending subject to a discrete nodal load.

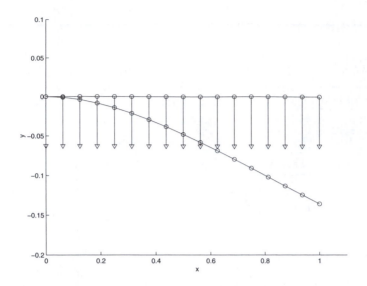

Figure 1.7.5 Beam deflection under the influence of a uniform transverse
nodal load indicated by the vertical arrows. The figure shows the graphics
output of the FSELIB code **beam**.

PROBLEMS

1.7.1 *Element stiffness matrix.*

Confirm that the determinant of the element stiffness matrix displayed in (1.7.44)
is zero, which indicates that the matrix is singular.

1.7.2 *Two-element cantilever beam.*

Consider the two-element discretization of a cantilever beam subject to nodal
load, as shown in Figure 1.7.4(b). The length of the first element is αL, where
α is a dimensionless coefficient taking values in the interval $(0, 1)$.

Write out the 6×6 linear system arising from the finite element formulation in
terms of E, I, L, α, F_2, F_3, $Q(0)$, and $M(0)$, before, and after the implementation of the boundary conditions (1.7.52).

1.7.3 *Code* **beam**.

Consider a cantilever beam where the load is zero at all nodes, except at the
last node located at the free end. Modify accordingly the FSELIB code **beam**,
run the modified code for several discretization levels, compare and discuss the
results of your computation.

1.8 Beam buckling

We continue the study of beams by considering deformation under the combined action of a transverse load, as discussed in Section 1.7, and a compressive axial force, P, as illustrated in Figure 1.8.1. Performing a force balance in the direction of the y axis, we recover the differential relation (1.7.2) for the transverse shear force. Performing an analogous torque balance over the infinitesimal section with length $\mathrm{d}x$, as shown in Figure 1.8.1(c), we find

$$Q(x)\,\mathrm{d}x + M(x) + P\,\mathrm{d}v = M(x + \mathrm{d}x), \tag{1.8.1}$$

where v is the vertical deflection. Rearranging, we find

$$Q = -P\frac{\mathrm{d}v}{\mathrm{d}x} + \frac{\mathrm{d}M}{\mathrm{d}x}. \tag{1.8.2}$$

Substituting this expression in the force balance (1.7.2), we derive the equilibrium equation

$$\frac{\mathrm{d}^2 M}{\mathrm{d}x^2} - P\frac{\mathrm{d}^2 v}{\mathrm{d}x^2} = w(x). \tag{1.8.3}$$

Finally, we express the bending moment in terms of the curvature using (1.7.10), and derive the governing equation. For a homogeneous prismatic beam, we find

$$EI\frac{\mathrm{d}^4 v}{\mathrm{d}x^4} + P\frac{\mathrm{d}^2 v}{\mathrm{d}x^2} = -w(x). \tag{1.8.4}$$

When $P = 0$, we recover the governing equation (1.7.12).

The Galerkin finite element formulation produces the following generalization of (1.7.33),

$$EI\int_0^L \frac{\mathrm{d}^2\phi_i}{\mathrm{d}x^2}\frac{\mathrm{d}^2 v}{\mathrm{d}x^2}\,\mathrm{d}x - P\int_0^L \frac{\mathrm{d}\phi_i}{\mathrm{d}x}\frac{\mathrm{d}v}{\mathrm{d}x}\,\mathrm{d}x \tag{1.8.5}$$

$$= \left[\phi_i\,Q - \frac{\mathrm{d}\phi_i}{\mathrm{d}x}\,M - \phi_i\frac{\mathrm{d}v}{\mathrm{d}x}\,P\right]_0^L - \int_0^L \phi_i\,w(x)\,\mathrm{d}x,$$

for $i = 1, 2, \ldots, 2N_G$. Substituting the finite element expansion (1.7.28) in (1.8.5), and compiling the resulting equations, we derive the linear system

$$(EI\,\mathbf{K} - P\,\mathbf{D}) \cdot \mathbf{u} = \mathbf{b}, \tag{1.8.6}$$

where $\mathbf{K}$ is the global stiffness matrix defined in (1.7.35), $\mathbf{D}$ is the generalized global diffusion matrix, also called the global *geometrical stiffness matrix* with components

$$D_{ij} = \int_0^L \frac{\mathrm{d}\phi_i}{\mathrm{d}x}\frac{\mathrm{d}\phi_j}{\mathrm{d}x}\,\mathrm{d}x, \tag{1.8.7}$$

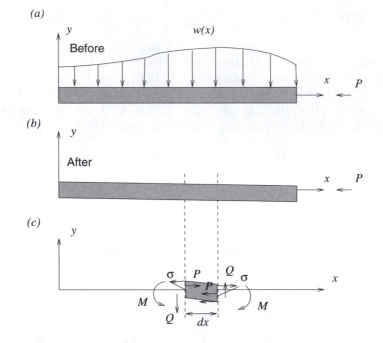

Figure 1.8.1 Schematic illustration of combined beam bending and buckling under the influence of a transverse load and an axial compressive force, P, (*a*) before, and (*b*) after deformation. (*c*) Distribution of the axial stress, σ, over a beam cross-section, and depiction of the transverse shear force, Q.

for $i, j = 1, 2, \ldots, 2N_G$, and the vector $\mathbf{u}$ is defined in (1.7.36). The first two components of the vector $\mathbf{b}$ on the right-hand side are given by

$$
b_1 = \left[Q\,\phi_1 - M\,\frac{\mathrm{d}\phi_1}{\mathrm{d}x} - P\,\phi_1\,\frac{\mathrm{d}v}{\mathrm{d}x} \right]_0^L - \int_0^L \phi_1\,w(x)\,\mathrm{d}x
$$

$$
= -Q(0) + P\,v_1' - \int_0^L \phi_1\,w(x)\,\mathrm{d}x,
$$

$$(1.8.8)$$

$$
b_2 = \left[Q\,\phi_2 - M\,\frac{\mathrm{d}\phi_2}{\mathrm{d}x} - P\,\phi_2\,\frac{\mathrm{d}v}{\mathrm{d}x} \right]_0^L - \int_0^L \phi_2\,w(x)\,\mathrm{d}x
$$

$$
= M(0) - \int_0^L \phi_2\,w(x)\,\mathrm{d}x.
$$

The intermediate components are given by

$$b_i = -\int_0^L \phi_i\, w(x)\mathrm{d}x, \qquad (1.8.9)$$

for $i = 1, 2, \ldots, 2N_G - 2$, and the last two components are given by

$$b_{2N_G-1} = \left[\phi_{2N_G-1}\, Q - \frac{\mathrm{d}\phi_{2N_G-1}}{\mathrm{d}x}\, M - \phi_{2N_G-1}\frac{\mathrm{d}v}{\mathrm{d}x}\, P\right]_0^L - \int_0^L \phi_{2N_G-1}\, w(x)\,\mathrm{d}x$$

$$= Q(L) - P\, v'_{N_G} - \int_0^L \phi_{2N_G-1}\, w(x)\,\mathrm{d}x,$$

$$\qquad (1.8.10)$$

$$b_{2N_G} = \left[Q\,\phi_{2N_G} - \frac{\mathrm{d}\phi_{2N_G}}{\mathrm{d}x}\, M - \phi_{2N_G}\frac{\mathrm{d}v}{\mathrm{d}x}\, P\right]_0^L - \int_0^L \phi_{2N_G}\, w(x)\,\mathrm{d}x$$

$$= -M(L) - \int_0^L \phi_{2N_G}\, w(x)\mathrm{d}x.$$

In practice, the global stiffness and geometrical stiffness matrices are assembled from corresponding 4×4 element matrices. The element stiffness matrix was given in (1.7.44). A detailed calculation shows that the lth-element geometrical stiffness matrix corresponding to **D** is

$$\mathbf{G}^{(l)} = \frac{1}{30\, h_l}\begin{bmatrix} 36 & 3\, h_l & -36 & 3\, h_l \\ 3\, h_l & 4\, h_l^2 & -3\, h_l & -h_l^2 \\ -36 & -3\, h_l & 36 & -3\, h_l \\ 3\, h_l & -h_l^2 & -3\, h_l & 4\, h_l^2 \end{bmatrix}. \qquad (1.8.11)$$

Implementing the boundary conditions by specifying selected values of the generalized deflection vector **u**, end-node transverse shear forces $Q(0)$, $Q(L)$, and bending moments, $M(0)$, $M(L)$, provides us with the final system of equations for the remaining unknowns.

1.8.1 Buckling under a compressive tip force

In the absence of a transverse load, $w = 0$, the right-hand side of (1.8.4) is zero, yielding the homogeneous differential equation

$$EI\frac{\mathrm{d}^4v}{\mathrm{d}x^4} + P\frac{\mathrm{d}^2v}{\mathrm{d}x^2} = 0, \qquad (1.8.12)$$

which admits the trivial solution, $v = 0$. Nontrivial solutions are possible for certain values of the compressive force, P, corresponding to different buckling modes.

To illustrate how these modes arise from the finite element formulation, we consider the one-element discretization resulting in four Galerkin equations encapsulated in the linear system

$$\left(\frac{EI}{L^3} \begin{bmatrix} 12 & 6\,L & -12 & 6\,L \\ 6\,L & 4\,L^2 & -6\,L & 2\,L^2 \\ -12 & -6\,L & 12 & -6\,L \\ 6\,L & 2\,L^2 & -6\,L & 4\,L^2 \end{bmatrix} \right. \tag{1.8.13}$$

$$\left. -\frac{P}{30\,L} \begin{bmatrix} 36 & 3\,L & -36 & 3\,L \\ 3\,L & 4\,L^2 & -3\,L & -L^2 \\ -36 & -3\,L & 36 & -3\,L \\ 3\,L & -L^2 & -3\,L & 4\,L^2 \end{bmatrix} \right) \cdot \begin{bmatrix} v_1 \\ v_1' \\ v_2 \\ v_2' \end{bmatrix} = \begin{bmatrix} -Q(0) + P\,v_1' \\ M(0) \\ Q(L) - P\,v_2' \\ -M(L) \end{bmatrix}.$$

For a beam that is simply supported at both ends, we require

$$v_1 = 0, \qquad M(0) = 0, \qquad v_2 = 0, \qquad M(L) = 0, \tag{1.8.14}$$

and obtain a system of linear equations for the four unknowns:

$$v_1', \qquad Q(0), \qquad v_2', \qquad Q(L).$$

Physically, the beam is a toothpick pinched by two fingers at both ends.

Implementing the boundary conditions (1.8.14), we find that the second and fourth equations in (1.8.13) reduce to the homogeneous system

$$\begin{bmatrix} 4 & 2 \\ 2 & 4 \end{bmatrix} \cdot \begin{bmatrix} v_1' \\ v_2' \end{bmatrix} = \frac{\hat{P}}{30} \begin{bmatrix} 4 & -1 \\ -1 & 4 \end{bmatrix} \cdot \begin{bmatrix} v_1' \\ v_2' \end{bmatrix}, \tag{1.8.15}$$

where

$$\hat{P} \equiv \frac{PL^2}{EI} \tag{1.8.16}$$

is the dimensionless compressive force. Equation (1.8.15) has the standard form of a generalized eigenvalue problem expressed by the algebraic equation

$$\mathbf{A} \cdot \mathbf{x} = \lambda\,\mathbf{B} \cdot \mathbf{x}, \tag{1.8.17}$$

where $\mathbf{A}$ and $\mathbf{B}$ are two matrices with the same dimensions, λ is an eigenvalue, and $\mathbf{x}$ is the corresponding eigenvector. A non-trivial solution of the 2×2 system (1.8.15) exists only for two eigenvalues, $\hat{P}$. Once these have been found along with the corresponding eigensolutions, they can be substituted in the first and third equations in (1.8.13) to yield the transverse shear forces, $Q(0)$ and $Q(L)$.

The eigenvalues can be found expeditiously using the MATLAB function eigen in the following short session:

```
>> A = [4 2; 2 4];
>> B = [4 -1; -1 4]/30;
>> eig(A,B)
```

```
ans =

    12
    60
```

The exact solution predicts that the eigenvalues form an infinite sequence,

$$\hat{P}_n = n^2 \, \pi^2, \tag{1.8.18}$$

where n is an integer; the numerical values displayed above are approximations to the first two modes, $n = 1, 2$, corresponding to $\hat{P}_1 = 9.9$ and $\hat{P}_2 = 39.5$.

PROBLEMS

1.8.1 *Element stiffness matrix.*

Confirm that the determinant of the geometric element stiffness matrix displayed in (1.8.11) is zero, which indicates that the matrix is singular.

1.8.2 *Buckling of a simply supported beam.*

Derive the analytical expression for the eigenvalues and eigensolutions of (1.8.12) for a beam that is simply supported at both ends.

1.8.3 *Buckling of a beam with different types of support.*

A beam that is simply supported at both ends is said to have a pin-pin support. Other types of support are the fixed-pin support, the fixed-fixed support, and the fixed-free support. Solving (1.8.12) by elementary analytical methods, we find that the eigenvalues are given by

$$\hat{P}_n = \left(\frac{n \, \pi}{\alpha} \right)^2, \tag{1.8.19}$$

where the coefficient α depends on the boundary conditions. For the pin-pin support, $\alpha = 1$, as discussed in the text.

(*a*) Show that, for a beam with a fixed-pin support, $\alpha = 0.7$; derive and solve the one-element eigenvalue problem.

(*b*) Show that, for a beam with a fixed-fixed support, $\alpha = 0.5$; derive and solve the one-element eigenvalue problem.

(*c*) Show that, for a beam with a fixed-free support, $\alpha = 2$; derive and solve the one-element eigenvalue problem.

1.8.4 *Buckling of a heavy vertical beam.*

Consider the buckling of a heavy vertical beam, as illustrated in Figure 1.8.2, before and after deformation.

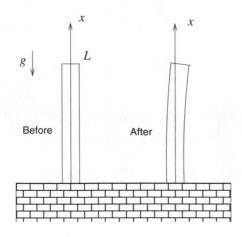

Figure 1.8.2 A heavy vertical beam buckling under the influence of its own
      weight.

(*a*) Show that the beam deflection is governed by the following modified version
of (1.8.12),

$$\frac{d^4v}{dx^4} + \frac{\rho g A}{EI} \frac{d}{dx}\left[(L - x) \frac{dv}{dx}\right] = 0, \tag{1.8.20}$$

where A is the beam cross-sectional area, ρ is the beam density, and g is the
acceleration of gravity.

(*b*) Integrating equation (1.8.20) once, we derive a second-order equation for
the linearized slope angle, $\theta \equiv dv/dx$,

$$\frac{d^2\theta}{dx^2} + \frac{\rho g A}{EI} (L - x)\, \theta = c, \tag{1.8.21}$$

where c is a constant. Requiring the free-end condition $Q(L) = 0$, we find $c = 0$.
In terms of the new spatial variable,

$$\hat{x} \equiv \left(\frac{\rho g A}{EI}\right)^{1/3} (L - x), \tag{1.8.22}$$

we obtain Stokes's equation

$$\frac{d^2\theta}{d\hat{x}^2} + \hat{x}\, \theta = 0. \tag{1.8.23}$$

Show that the general solution of (1.8.23) is

$$\theta = \sqrt{\hat{x}} \left(c_1\, J_{1/3}(X) + c_2\, J_{-1/3}(X) \right), \tag{1.8.24}$$

where c_1 and c_2 are constants,

$$X \equiv \frac{2}{3} \hat{x}^{3/2}, \tag{1.8.25}$$

and $J_{\pm 1/3}$ are Bessel functions of fractional order (e.g., [1], pp. 446–447).

(*c*) Derive the specific form of the solution subject to the second free-end condition, $M(x = L) = 0$, and the fixed-end condition, $\theta(x = 0) = 0$. Specifically, impose these boundary conditions to derive a homogeneous system of two equations for the two constants, c_1 and c_2. For a non-trivial solution to exist, the determinant of the coefficient matrix must be zero. This observation provides us with a secular equation for the computation of the critical beam height for buckling.

(*d*) Derive the one-element Galerkin finite element system corresponding to (1.8.23), and solve the generalized eigenvalue problem.

High-order and spectral elements in one dimension

2

In Chapter 1, we introduced the Galerkin finite element method in one dimension with linear, quadratic, and cubic Hermitian elements. High-order finite element methods employ high-order polynomial expansions over the individual elements. The polynomial expansions themselves are defined by an appropriate number of element *interpolation* nodes that are generally distinct from the two shared *geometrical* element end-nodes.

For example, in Chapter 1, we discussed quadratic elements defined by three interpolation nodes, including the two shared geometrical end-nodes, and one interior interpolation node, yielding a total of $N_G = 2 N_E + 1$ global interpolation nodes, where N_E is the number of elements. Cubic elements are defined by four interpolation nodes, including the two shared geometrical end-nodes, and two interior interpolation nodes, yielding a total of $N_G = 3 N_E + 1$ global interpolation nodes. More generally, pth-order elements are defined by $p + 1$ interpolation nodes, including the two shared geometrical end-nodes, and $p - 1$ interior interpolation nodes, yielding a total of $N_G = p N_E + 1$ global interpolation nodes.

Improving the accuracy of the finite element solution by decreasing the element size, h, while holding the polynomial order fixed, is classified as an *h*-refinement. Typically, the error decreases like a power of h, where the exponent is determined by the polynomial order and smoothness of the solution. On the other hand, improving the accuracy by raising the polynomial order, p, for a fixed element discretization, is classified as a *p*-refinement, typically associated with spectral convergence. The terminology "spectral" means that the numerical error decreases faster than any power of $1/p$, where p is the order of the polynomial expansion. Combinations of the two refinement strategies yield the *hp*-refinement.

To implement the Galerkin finite element method for a high-order expansion, we integrate the product of the governing equation with each of the global interpolation functions associated with the global nodes, and thus derive a system of algebraic or differential equations. The system is subsequently modified by implementing the stipulated Neumann, Dirichlet, or other boundary conditions to obtain its final form, and then solved by standard numerical methods.

In practice, we would like to use low-order expansions in regions where the solution is expected to vary smoothly, and high-order expansions in regions where the solution is expected to exhibit rapid spatial variations. Moreover, we would like to achieve the best possible accuracy for a given number of interpolation nodes. For linear and homogeneous differential equations, the node distribution has no effect on the numerical solution, and only determines the structure and standing of the global matrices quantified by the condition number. For inhomogeneous and nonlinear equations, the node distribution may play in important role in the accuracy and convergence of the solution, as will be discussed in Section 2.1. Theoretical analysis of the interpolation error shows that, given the number of interpolation nodes to be distributed over an element, the highest interpolation accuracy is achieved when the interior nodes are distributed at positions corresponding to the zeros of certain families of orthogonal polynomials. When this is done, we obtain a spectral element expansion and associated spectral element method.

In this chapter, we discuss the theoretical foundation and implementation of high-order and spectral element methods in one dimension. The theoretical development relies heavily on the theory of function interpolation and orthogonal polynomials reviewed in Appendices A and B. These two appendices should be briefly reviewed, and preferably studied in detail, before continuing with this chapter. Readers who are not presently interested in the spectral element method can proceed without penalty to the discussion on the finite element method in two dimensions, beginning in Chapter 3. The spectral element method in two and three dimensions is discussed in Chapters 4 and 6.

2.1   Nodal bases

Assume that we have decided to approximate the solution over the lth element with a $p^{(l)}$-degree polynomial, defined by $p^{(l)} + 1$ element interpolation nodes, including the two geometrical end-nodes and $p^{(l)} - 1$ interior interpolation nodes. To simplify the notation, heretoforth we denote $p^{(l)}$ by m, with the understanding that m may vary from an element to another.

To facilitate the computations, we map the lth element onto the standard interval $[-1, 1]$ of the dimensionless ξ axis, as shown in Figure 1.3.2. The mapping is mediated by the function

$$x(\xi) = \frac{1}{2}\left(X_2^{(l)} + X_1^{(l)}\right) + \frac{1}{2}\left(X_2^{(l)} - X_1^{(l)}\right)\xi, \qquad (2.1.1)$$

where $X_1^{(l)}$ is the first element end-node, and $X_2^{(l)}$ is the second element end-node on the x axis. As ξ increases from -1 to 1, the field point x moves from the first element end-node to the second element end-node.

2.1.1   Lagrange interpolation

The element interpolation functions can be conveniently identified with the mth-degree Lagrange interpolating polynomials discussed in Appendix A,

$$\psi_i(\xi) = \frac{(\xi - \xi_1) \dots (\xi - \xi_{i-1})(\xi - \xi_{i+1}) \dots (\xi - \xi_{m+1})}{(\xi_i - \xi_1) \dots (\xi_i - \xi_{i-1})(\xi_i - \xi_{i+1}) \dots (\xi_i - \xi_{m+1})}, \tag{2.1.2}$$

for $i = 1, 2, \dots, m+1$. Cursory inspection reveals that these polynomials satisfy the cardinal interpolation property

$$\psi_i(\xi_j) = \begin{cases} 1 & \text{if } i = j \\ 0 & \text{if } i \neq j \end{cases}. \tag{2.1.3}$$

Stated differently, $\psi_i(\xi_j)$ is the identity matrix, $\psi_i(\xi_j) = \delta_{ij}$.

An equivalent representation is

$$\psi_i(\xi) = \frac{\Phi_{m+1}(\xi)}{(\xi - \xi_i)\, \Phi'_{m+1}(\xi_i)}, \tag{2.1.4}$$

where

$$\Phi_{m+1}(\xi) \equiv (\xi - \xi_1)(\xi - \xi_2) \dots (\xi - \xi_m)(\xi - \xi_{m+1}) \tag{2.1.5}$$

is the $(m + 1)$-degree Lagrange generating polynomial defined in terms of *all* element interpolation nodes, as discussed in Appendix A.

A function of interest, $f(\xi)$, defined over the lth element can be approximated with the mth-degree interpolating polynomial, $P_m(\xi)$, which can be constructed explicitly in terms of specified or *a priori* unknown nodal values, $f(\xi_i)$, and the Lagrange interpolating polynomials, as

$$f(\xi) \simeq P_m(\xi) = \sum_{i=1}^{m+1} f(\xi_i)\, \psi_i(\xi). \tag{2.1.6}$$

The aforementioned cardinal property of the Lagrange polynomials, $\psi_i(\xi_j) = \delta_{ij}$, ensures that

$$P_m(\xi_j) = \sum_{i=1}^{m+1} f(\xi_i)\, \psi_i(\xi_j) = \sum_{i=1}^{m+1} f(\xi_i)\, \delta_{ij} = f(\xi_j), \tag{2.1.7}$$

as required.

Evenly spaced nodes

Figure 2.1.1 shows graphs of the node interpolation functions over the canonical interval of the ξ axis, [-1, 1], for data sets of five and eleven evenly spaced points. As the number of interpolation points is raised, oscillations occur near the ends of the interpolation domain, described as the *Runge effect*. Unless a low-order polynomial approximation is used, an even distribution of interpolation nodes may be detrimental to the accuracy of the interpolation, and consequently to the reliability of the finite element solution. This potential failure can be circumvented by judiciously deploying the interpolation nodes over each element, guided by the zeros of orthogonal polynomials. For the moment, we bypass this important issue and discuss the computation of the element mass and diffusion matrices for an arbitrary node distribution.

(a)

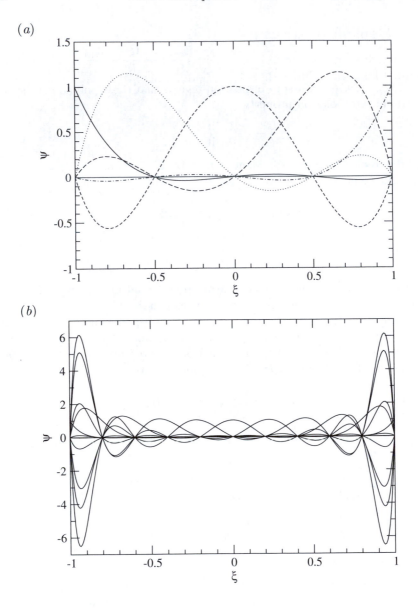

(b)

Figure 2.1.1 Graphs of the (a) five node interpolation functions corresponding
to polynomial order $m = 4$, and (b) eleven node interpolation functions
corresponding to $m = 10$. In both cases, the nodes are evenly spaced
between $\xi_1 = -1$ and $\xi_{m+1} = 1$. Runge oscillations occur near the two
ends of the interpolation domain as the polynomial order is raised.

2.1.2   Element matrices

Expressing the physical variable x in terms of the canonical variable ξ using relation (2.1.1), and noting that $\mathrm{d}x = \frac{h_l}{2}\mathrm{d}\xi$, where $h_l \equiv X_2^{(l)} - X_1^{(l)}$ is the element size, we find that the lth-element diffusion matrix is given by

$$A_{ij}^{(l)} \equiv \int_{X_1^{(l)}}^{X_2^{(l)}} \frac{\mathrm{d}\psi_i(x)}{\mathrm{d}x} \frac{\mathrm{d}\psi_j(x)}{\mathrm{d}x}\, \mathrm{d}x = \frac{2}{h_l} \int_{-1}^{1} \frac{\mathrm{d}\psi_i(\xi)}{\mathrm{d}\xi} \frac{\mathrm{d}\psi_j(\xi)}{\mathrm{d}\xi}\, \mathrm{d}\xi, \qquad (2.1.8)$$

for $i, j = 1, 2, \ldots, m+1$. For the physical reason discussed at the end of Section 1.1.8, this matrix is singular irrespective of the number and location of the element interpolation nodes. Working in a similar fashion, we find that the element mass matrix is given by

$$B_{ij}^{(l)} \equiv \int_{X_1^{(l)}}^{X_2^{(l)}} \psi_i(x)\, \psi_j(x)\, \mathrm{d}x = \frac{h_l}{2} \int_{-1}^{1} \psi_i(\xi)\, \psi_j(\xi)\, \mathrm{d}\xi, \qquad (2.1.9)$$

for $i, j = 1, 2, \ldots, m+1$. The integrals with respect to ξ on the right-hand sides can be calculated by a combination of analytical and numerical methods, as will be discussed later in this chapter for specific node distributions.

Consider the computation of the derivatives of the element interpolation functions. Differentiating expression (2.1.4) using the rules of quotient differentiation, we find

$$\frac{\mathrm{d}\psi_i(\xi)}{\mathrm{d}\xi} = \frac{\Phi'_{m+1}(\xi)(\xi - \xi_i) - \Phi_{m+1}(\xi)}{(\xi - \xi_i)^2\, \Phi'_{m+1}(\xi_i)}. \qquad (2.1.10)$$

For future reference, we evaluate these derivatives at the interpolation nodes. Observing that $\Phi_{m+1}(\xi_j) = 0$, we derive the *node differentiation matrix*,

$$d_{ij} \equiv \left(\frac{\mathrm{d}\psi_i}{\mathrm{d}\xi}\right)_{\xi = \xi_j} = \begin{cases} \dfrac{\Phi'_{m+1}(\xi_j)}{(\xi_j - \xi_i)\, \Phi'(\xi_j)} & \text{if } \ i \neq j \\[2ex] \dfrac{\Phi''_{m+1}(\xi_i)}{2\, \Phi'_{m+1}(\xi_i)} & \text{if } \ i = j \end{cases}. \qquad (2.1.11)$$

The expression for $i = j$ arises by expanding $\Phi_{m+1}(\xi)$ and $\Phi'_{m+1}(\xi)$ in the numerator of (2.1.10) in Taylor series about the point ξ_i, and then taking the limit $\xi \to \xi_i$, or else by using the l'Hôpital rule (Problem 2.1.1). Further expressions for the derivatives of the element interpolation functions and node differentiation matrix are given in Section A.2.3 of Appendix A.

2.1.3 C^0 continuity and shared element nodes

The Galerkin finite element implementation for the steady heat conduction problem discussed in Section 1.1 culminates in the linear system $\mathbf{D} \cdot \mathbf{f} = \mathbf{b}$, where $\mathbf{f}$ is the vector of unknown nodal temperatures, $\mathbf{D}$ is the global diffusion matrix, and $\mathbf{b}$ is a properly constructed right-hand side.

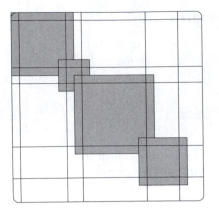

Figure 2.1.2 Structure of the global diffusion or mass matrix consisting of overlapping diagonal blocks hosting the element diffusion matrices.

The importance of using a C^0 (continuous) global finite element expansion implemented by shared element end-nodes is underscored by two key observations. First, the Galerkin finite element equations were derived on the assumption of continuous global interpolation functions. Second, if the elements did not share end-nodes, the global diffusion matrix $\mathbf{D}$ would be composed of non-overlapping diagonal element blocks, and the Galerkin equations corresponding to groups of element nodes would be decoupled. Because the element diffusion matrix is singular, the solution of the sub-blocks could be found only up to an arbitrary constant, and the finite element formulation would catastrophically end up in an ill-posed problem.

To ensure continuity of the solution represented by the union of the element expansions, we place the first interpolation node at the first geometrical element end-node, and the last interpolation node at the second geometrical element end-node,

$$\xi_1 = -1, \qquad \xi_{m+1} = 1, \tag{2.1.12}$$

and require that the nodal values are shared by neighboring elements. The structure of the global diffusion matrix consisting of partially overlapping element diffusion matrices is illustrated in Figure 2.1.2. Note that overlap occurs only on one diagonal matrix element corresponding the first or second element end-node. Algorithm 2.1.1 establishes the correspondence between the element and global nodes by means of a connectivity matrix, and counts the number of unique global interpolation nodes, N_G.

2.1.4   Change of nodal base

Consider two different element nodal sets, each consisting of $m + 1$ nodes, denoted as I and II, corresponding to the same polynomial order, m, subject to

---

$Ic = 2$ *Initialize the global node counter*

Do $l = 1, 2, \ldots, N_E$ *Run over the elements*
  $Ic = Ic - 1$ *Account for the common end-node*
  Do $i = 1, 2, \ldots, N_p(l) + 1$ *Run over the element nodes*
    $c(l, i) = Ic$ *Connectivity matrix*
    $Ic = Ic + 1$
  End Do
End Do

$N_G = Ic - 1$ *Total number of unique global nodes*

Algorithm 2.1.1 Correspondence between element and global nodes, and counting of the total number of unique global nodes, N_G.

the restrictions imposed by (2.1.12). The cardinal interpolation functions of the first set, $\psi_i^{\mathrm{I}}(\xi)$, are related to the cardinal interpolation functions of the second set, $\psi_i^{\mathrm{II}}(\xi)$, by the Lagrange interpolation formula

$$\psi_i^{\mathrm{I}}(\xi) = \sum_{j=1}^{m+1} \psi_i^{\mathrm{I}}(\xi_j^{\mathrm{II}}) \, \psi_j^{\mathrm{II}}(\xi), \qquad (2.1.13)$$

for $i = 1, 2, \ldots, m + 1$. Accordingly, the corresponding nodal interpolation function vectors are related by the linear transformation

$$\boldsymbol{\psi}^{\mathrm{I}}(\xi) = \mathbf{V}_{\mathrm{I,II}} \cdot \boldsymbol{\psi}^{\mathrm{II}}(\xi), \qquad (2.1.14)$$

where

$$\mathbf{V}_{\mathrm{I,II}} = \begin{bmatrix} \psi_1^{\mathrm{I}}(\xi_1^{\mathrm{II}}) & \psi_1^{\mathrm{I}}(\xi_2^{\mathrm{II}}) & \cdots & \psi_1^{\mathrm{I}}(\xi_m^{\mathrm{II}}) & \psi_1^{\mathrm{I}}(\xi_{m+1}^{\mathrm{II}}) \\ \psi_2^{\mathrm{I}}(\xi_1^{\mathrm{II}}) & \psi_2^{\mathrm{I}}(\xi_2^{\mathrm{II}}) & \cdots & \psi_2^{\mathrm{I}}(\xi_m^{\mathrm{II}}) & \psi_2^{\mathrm{I}}(\xi_{m+1}^{\mathrm{II}}) \\ \cdots & \cdots & \cdots & \cdots & \cdots \\ \psi_m^{\mathrm{I}}(\xi_1^{\mathrm{II}}) & \psi_m^{\mathrm{I}}(\xi_2^{\mathrm{II}}) & \cdots & \psi_m^{\mathrm{I}}(\xi_m^{\mathrm{II}}) & \psi_m^{\mathrm{I}}(\xi_{m+1}^{\mathrm{II}}) \\ \psi_{m+1}^{\mathrm{I}}(\xi_1^{\mathrm{II}}) & \psi_{m+1}^{\mathrm{I}}(\xi_2^{\mathrm{II}}) & \cdots & \psi_{m+1}^{\mathrm{I}}(\xi_m^{\mathrm{II}}) & \psi_{m+1}^{\mathrm{I}}(\xi_{m+1}^{\mathrm{II}}) \end{bmatrix}$$

$$(2.1.15)$$

is a generalized Vandermonde matrix. Because the end-nodes are common in the two distributions, the first and last columns of $\mathbf{V}_{\mathrm{I,II}}$ are filled with zeros, with the exception of the first and last diagonal elements that are equal to unity, that is, $\psi_i^{\mathrm{I}}(\xi_1^{\mathrm{II}}) = \delta_{i,1}$ and $\psi_i^{\mathrm{I}}(\xi_{m+1}^{\mathrm{II}}) = \delta_{i,m+1}$.

As an example, we identify $\psi_i^{\mathrm{I}}(\xi)$ with the quadratic element interpolation functions corresponding to the mid-point interior node shown in (1.3.26), repeated here for convenience,

$$\psi_1^{\mathrm{I}}(\xi) = \frac{1}{2}\,\xi\,(\xi-1), \quad \psi_2^{\mathrm{I}}(\xi) = 1 - \xi^2, \quad \psi_3^{\mathrm{I}}(\xi) = \frac{1}{2}\,\xi\,(\xi+1), \qquad (2.1.16)$$

and $\psi_i^{\mathrm{II}}(\xi)$ with the quadratic element interpolation functions corresponding to an arbitrary interior node located at $\xi_2^{\mathrm{II}} = \beta$; in both cases, $\xi_1 = -1$ and $\xi_3 = 1$. The associated 3×3 generalized Vandermonde matrix takes the form

$$\mathbf{V}_{\mathrm{I,II}} = \begin{bmatrix} 1 & \frac{1}{2}\,\beta\,(\beta-1) & 0 \\ 0 & 1-\beta^2 & 0 \\ 0 & \frac{1}{2}\,\beta\,(\beta+1) & 1 \end{bmatrix}, \qquad (2.1.17)$$

whose determinant is equal to $1-\beta^2$. Note that, when $\beta = 0$, this Vandermonde matrix reduces to the identity matrix. The second set of interpolation functions can be deduced from the first set by solving the linear system (2.1.14). The resulting expressions are identical to those displayed in (1.3.32), repeated here for convenience,

$$\psi_1^{\mathrm{II}}(\xi) = \frac{1}{2\,(1+\beta)}\,(\xi - \beta)\,(\xi - 1),$$

$$\psi_2^{\mathrm{II}}(\xi) = \frac{1 - \xi^2}{1 - \beta^2}, \qquad (2.1.18)$$

$$\psi_3^{\mathrm{II}}(\xi) = \frac{1}{2\,(1-\beta)}\,(\xi - \beta)\,(\xi + 1).$$

Conversely, we may write

$$\boldsymbol{\psi}^{\mathrm{II}}(\xi) = \mathbf{V}_{\mathrm{II,I}} \cdot \boldsymbol{\psi}^{\mathrm{I}}(\xi), \qquad (2.1.19)$$

where

$$\mathbf{V}_{\mathrm{II,I}} = \begin{bmatrix} \psi_1^{\mathrm{II}}(\xi_1^{\mathrm{I}}) & \psi_1^{\mathrm{II}}(\xi_2^{\mathrm{I}}) & \cdots & \psi_1^{\mathrm{II}}(\xi_m^{\mathrm{I}}) & \psi_1^{\mathrm{II}}(\xi_{m+1}^{\mathrm{I}}) \\ \psi_2^{\mathrm{II}}(\xi_1^{\mathrm{I}}) & \psi_2^{\mathrm{II}}(\xi_2^{\mathrm{I}}) & \cdots & \psi_2^{\mathrm{II}}(\xi_m^{\mathrm{I}}) & \psi_2^{\mathrm{II}}(\xi_{m+1}^{\mathrm{I}}) \\ \cdots & \cdots & \cdots & \cdots & \cdots \\ \psi_m^{\mathrm{II}}(\xi_1^{\mathrm{I}}) & \psi_m^{\mathrm{II}}(\xi_2^{\mathrm{I}}) & \cdots & \psi_m^{\mathrm{II}}(\xi_m^{\mathrm{I}}) & \psi_m^{\mathrm{II}}(\xi_{m+1}^{\mathrm{I}}) \\ \psi_{m+1}^{\mathrm{II}}(\xi_1^{\mathrm{I}}) & \psi_{m+1}^{\mathrm{II}}(\xi_2^{\mathrm{I}}) & \cdots & \psi_{m+1}^{\mathrm{II}}(\xi_m^{\mathrm{I}}) & \psi_{m+1}^{\mathrm{II}}(\xi_{m+1}^{\mathrm{I}}) \end{bmatrix}$$

$$(2.1.20)$$

is another generalized Vandermonde matrix. Comparing (2.1.14) with (2.1.19), we find

$$\mathbf{V}_{\mathrm{II,I}} = \mathbf{V}_{\mathrm{I,II}}^{-1}. \qquad (2.1.21)$$

For example, in the case of the quadratic elements discussed earlier in this section, $\xi_1^{\mathrm{I}} = -1$, $\xi_2^{\mathrm{I}} = 0$, and $\xi_3^{\mathrm{I}} = 1$, and

$$\mathbf{V}_{\mathrm{II,I}} = \begin{bmatrix} 1 & \frac{\beta}{2\,(\beta+1)} & 0 \\ 0 & \frac{1}{1-\beta^2} & 0 \\ 0 & \frac{\beta}{2\,(\beta-1)} & 1 \end{bmatrix}, \qquad (2.1.22)$$

which is the inverse of the matrix $\mathbf{V}_{\mathrm{I,II}}$ shown in (2.1.17).

Relation between the element matrices

Using (2.1.14), we find that the element diffusion and mass matrices of the first set are related to those of the second set by

$$\mathbf{A}^\mathrm{I} = \mathbf{V}_\mathrm{I,II} \cdot \mathbf{A}^\mathrm{II} \cdot \mathbf{V}_\mathrm{I,II}^T, \qquad \mathbf{B}^\mathrm{I} = \mathbf{V}_\mathrm{I,II} \cdot \mathbf{B}^\mathrm{II} \cdot \mathbf{V}_\mathrm{I,II}^T, \qquad (2.1.23)$$

or

$$\mathbf{A}^\mathrm{I} = \mathbf{V}_\mathrm{II,I}^{-1} \cdot \mathbf{A}^\mathrm{II} \cdot \mathbf{V}_\mathrm{II,I}^{-1\,T}, \qquad \mathbf{B}^\mathrm{I} = \mathbf{V}_\mathrm{II,I}^{-1} \cdot \mathbf{B}^\mathrm{II} \cdot \mathbf{V}_\mathrm{II,I}^{-1\,T}, \qquad (2.1.24)$$

where the superscript T denotes the matrix transpose. Conversely, the element matrices of the second set are related to those of the first set by

$$\mathbf{A}^\mathrm{II} = \mathbf{V}_\mathrm{II,I} \cdot \mathbf{A}^\mathrm{I} \cdot \mathbf{V}_\mathrm{II,I}^T, \qquad \mathbf{B}^\mathrm{II} = \mathbf{V}_\mathrm{II,I} \cdot \mathbf{B}^\mathrm{I} \cdot \mathbf{V}_\mathrm{II,I}^T, \qquad (2.1.25)$$

or

$$\mathbf{A}^\mathrm{II} = \mathbf{V}_\mathrm{I,II}^{-1} \cdot \mathbf{A}^\mathrm{I} \cdot \mathbf{V}_\mathrm{I,II}^{-1\,T}, \qquad \mathbf{B}^\mathrm{II} = \mathbf{V}_\mathrm{I,II}^{-1} \cdot \mathbf{B}^\mathrm{I} \cdot \mathbf{V}_\mathrm{I,II}^{-1\,T}. \qquad (2.1.26)$$

These relations can be used to assess the sensitivity of the numerical solution to the choice of a nodal base.

In the case of the Laplace or Helmholtz equation, the finite element formulation with the second nodal set produces the element-level linear system

$$(\mathbf{A}^\mathrm{II} - \alpha\,\mathbf{B}^\mathrm{II}) \cdot \mathbf{f}^\mathrm{II} = \mathbf{0}, \qquad (2.1.27)$$

where α is a constant, with $\alpha = 0$ corresponding to Laplace's equation. Using (2.1.23), we derive the equivalent form

$$(\mathbf{A}^\mathrm{I} - \alpha\,\mathbf{B}^\mathrm{I}) \cdot \mathbf{f}^\mathrm{I} = \mathbf{0}, \qquad (2.1.28)$$

where

$$\mathbf{V}_\mathrm{I,II}^T \cdot \mathbf{f}^\mathrm{I} = \mathbf{f}^\mathrm{II}. \qquad (2.1.29)$$

Because this equation produces the II nodal values from the I nodal values by Lagrange interpolation, the choice of a nodal base is immaterial.

On the other hand, in the case of the Poisson equation, the finite element formulation with the second nodal set culminates in the linear system

$$\mathbf{A}^\mathrm{II} \cdot \mathbf{f}^\mathrm{II} = \mathbf{B}^\mathrm{II} \cdot \mathbf{s}^\mathrm{II}, \qquad (2.1.30)$$

where the vector $\mathbf{s}^\mathrm{II}$ contains the II nodal values of the source. Using (2.1.23), we derive the equivalent form

$$\mathbf{A}^\mathrm{I} \cdot \mathbf{f}^\mathrm{I} = \mathbf{B}^\mathrm{I} \cdot \mathbf{s}^\mathrm{I}, \qquad (2.1.31)$$

where

$$\mathbf{V}_\mathrm{I,II}^T \cdot \mathbf{f}^\mathrm{I} = \mathbf{f}^\mathrm{II}, \qquad \mathbf{V}_\mathrm{I,II}^T \cdot \mathbf{s}^\mathrm{I} = \mathbf{s}^\mathrm{II}. \qquad (2.1.32)$$

These equations produce the II solution and source vectors from the corresponding I vectors by Lagrange interpolation. Because the interpolated values of the source are generally different from the exact values, finite element solutions with different nodal sets will not necessarily be the same.

PROBLEMS

2.1.1 *Node differentiation matrix.*

Working as indicated in the text, prove the expression for $i = j$ shown in (2.1.11).

2.1.2 *Lebesgue constant for evenly spaced nodes.*

The Lebesgue function is defined as

$$\mathcal{L}_m(\xi) \equiv \sum_{i=1}^{m+1} |\psi_i(\xi)|, \qquad (2.1.1)$$

and the associated Lebesgue constant is defined as

$$\Lambda_m \equiv \text{Max}\Big[\mathcal{L}(\xi)\Big], \qquad (2.1.2)$$

where Max[•] signifies the maximum value of the enclosed function over the interpolation interval $[-1, 1]$.

Prepare a table of the Lebesgue constant against the polynomial order, m, for evenly spaced nodes, and discuss its dependence on m.

2.2   Spectral interpolation

In the spectral element method, the element interpolation nodes are placed at the zeros of an appropriate family of orthogonal polynomials over the canonical ξ interval, subject to the mandatory constraint that the first node is assigned to $\xi_1 = -1$, and the last node is assigned to $\xi_{m+1} = 1$, where m is the order of the polynomial expansion defined by $m + 1$ nodes.

2.2.1   Lobatto nodal base

The theory of polynomial interpolation discussed in Appendix A in conjunction with the theory of orthogonal polynomials discussed in Appendix B dictates that the intermediate $m - 1$ nodes should be distributed at the zeros of the $(m - 1)$-degree Lobatto polynomial, $Lo_{m-1}(\xi)$, discussed in Section B.6 of Appendix B. The first few Lobatto polynomials are listed in Table 2.2.1 along

Lobatto polynomials

$Lo_0(\xi) = 1$

$Lo_1(\xi) = 3\,\xi$

$Lo_2(\xi) = \frac{3}{2}\left(5\,\xi^2 - 1\right)$

$Lo_3(\xi) = \frac{5}{2}\left(7\,\xi^2 - 3\right)\xi$

$Lo_4(\xi) = \frac{15}{8}\left(21\,\xi^4 - 14\,\xi^2 + 1\right)$

$Lo_5(\xi) = \frac{1}{8}\left(693\,\xi^4 - 630\,\xi^2 + 105\right)\xi$

$Lo_6(\xi) = \frac{1}{16}\left(3003\,\xi^6 - 3465\,\xi^4 + 945\,\xi^2 - 35\right)$

$\dots$

$Lo_i(\xi) = \frac{1}{2^{i+1}(i+1)!}\frac{\mathrm{d}^{i+2}}{\mathrm{d}\xi^{i+2}}\left(\xi^2 - 1\right)^{i+1} = \frac{i\,(2(i+1))!}{2^{i+1}(i+1)!}\,\xi^i + \dots$

Table 2.2.1 The first few members of the triangular family of Lobatto polynomials defined in the interval $-1 \le \xi \le 1$. In the generating formula shown in the last entry, $i! = 1 \cdot 2 \cdot 3 \dots \cdot i$, is the factorial. The Lobatto polynomials are related to the Legendre polynomials by $Lo_i(\xi) = L'_{i+1}(\xi)$.

with a generating formula, and their graphs are displayed in Figure 2.2.1. By construction, the Lobatto polynomials satisfy the orthogonality property

$$\int_{-1}^{1} Lo_i(\xi)\,Lo_j(\xi)\,(1 - \xi^2)\,\mathrm{d}\xi = \frac{2\,(i+1)(i+2)}{2i+3}\,\delta_{ij}, \qquad (2.2.1)$$

where δ_{ij} is Kronecker's delta. Further properties are listed in Section B.6 of Appendix B.

The main reason for choosing the zeros of the Lobatto polynomials is that the corresponding node interpolation functions are guaranteed to vary in the range $[-1, 1]$, independent of the order of the polynomial approximation, m, that is,

$$|\psi_i(\xi)| \le 1, \qquad (2.2.2)$$

with the equality holding only when $\xi = \xi_i$, as shown in Section A.5 of Appendix A. Because of this property, the Runge oscillations are suppressed, and the rate of convergence of the interpolation with respect to m is spectral, that is, faster than any power of $1/m$.

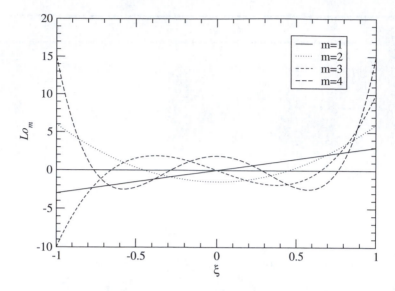

Figure 2.2.1 Graphs of the first few Lobatto polynomials, $Lo_m(\xi)$, in their native domain of definition.

Numerical values of the zeros of the first few Lobatto polynomials are shown in the second column of Table 2.2.2. The third column lists the corresponding integration weights used in the Lobatto integration quadrature to be discussed in Section 2.3 (see also Section B.3 of Appendix B). FSELIB function `lobatto`, listed in the text, assigns numerical values to the roots and corresponding weights from a lookup table.

To compute the roots of $Lo_4(\xi)$ using MATLAB, we define the polynomial coefficient vector by issuing the command:

```
>> c = [21, 0, -14, 0, 1]
```

where ">>" is the standard MATLAB prompt, and then invoke the MATLAB root finder function by typing:

```
>> roots(c)
```

MATLAB responds by printing the values:

```
ans =
     0.7651
    -0.7651
     0.2852
    -0.2852
```

which are consistent with those listed in Table 2.2.2 for $i = 4$.

	Zeros of Lo_i (Lobatto quadrature base points)	Integration weights
$i = 1$	$t_1 = 0$	$\widehat{w}_1 = \frac{4}{3}$
$i = 2$	$t_1 = -1/\sqrt{5}$ $t_2 = -t_1$	$\widehat{w}_1 = \frac{5}{6}$ $\widehat{w}_2 = \widehat{w}_1$
$i = 3$	$t_1 = -\sqrt{3/7}$ $t_2 = 0.0$ $t_3 = -t_1$	$\widehat{w}_1 = 49/60$ $\widehat{w}_2 = 32/45$ $\widehat{w}_3 = \widehat{w}_1$
$i = 4$	$t_1 = -0.76505532392946$ $t_2 = -0.28523151648064$ $t_3 = -t_2$ $t_4 = -t_1$	$\widehat{w}_1 = 0.37847495629785$ $\widehat{w}_2 = 0.55485837703549$ $\widehat{w}_3 = \widehat{w}_2$ $\widehat{w}_4 = \widehat{w}_1$
$i = 5$	$t_1 = -0.83022389627857$ $t_2 = -0.46884879347071$ $t_3 = 0.0$ $t_4 = -t_2$ $t_5 = -t_1$	$\widehat{w}_1 = 0.27682604736157$ $\widehat{w}_2 = 0.43174538120986$ $\widehat{w}_3 = 0.48761904761905$ $\widehat{w}_4 = \widehat{w}_2$ $\widehat{w}_5 = \widehat{w}_1$
$i = 6$	$t_1 = -0.87174014850961$ $t_2 = -0.59170018143314$ $t_3 = -0.20929921790248$ $t_4 = -t_3$ $t_5 = -t_2$ $t_6 = -t_1$	$\widehat{w}_1 = 0.21070422714350$ $\widehat{w}_2 = 0.34112269248350$ $\widehat{w}_3 = 0.41245879465870$ $\widehat{w}_4 = \widehat{w}_3$ $\widehat{w}_5 = \widehat{w}_2$ $\widehat{w}_6 = \widehat{w}_1$

Table 2.2.2 Zeros of the first few Lobatto polynomials and corresponding weights for the $(i + 1)$-point Lobatto integration quadrature. The hat over the weights indicates that the node labels will be shifted by one count when the Lobatto quadrature is applied.

```
function [Z, W] = lobatto(i)

%=================================================
% Zeros of the ith-degree Lobatto polynomial,
% and corresponding weights for the Lobatto
% integration quadrature
%
% This table contains values for i = 1,2,..., 6
% The default value is i=6
%=================================================

%------
% check
%------

if(i>6)
   disp('-->'); disp(' lobatto: Chosen lobatto order');
   disp('   is not available; Will take i=6');
   i=6;
end

%-------
if(i==1)
%-------

Z(1) = 0.0; W(1) = 4.0/3.0;

%-----------
else if(i==2)
%-----------

Z(1) = -1.0D0/sqrt(5.0); Z(2) = -Z(1);
W(1) = 5.0/6.0; W(2) = W(1);

%------------
else if(i==3)
%------------

Z(1) = -sqrt(3.0/7.0); Z(2) = 0.0; Z(3) = -Z(1);
W(1) = 49.0/90.0; W(2) = 32.0/45.0; W(3) = W(1);

%------------
else if(i==4)
%-----------

Z(1) = -0.76505532392946; Z(2) = -0.28523151648064;
Z(3) = -Z(2); Z(4) = -Z(1);
W(1) = 0.37847495629785; W(2) = 0.55485837703549;
W(3) = W(2); W(4) = W(1);

%------------
else if(i==5)
%------------
```

Function lobatto: Continuing $\longrightarrow$

```
Z(1) = -0.83022389627857; Z(2) = -0.46884879347071;
Z(3) = 0.0; Z(4) = -Z(2); Z(5) = -Z(1);
W(1) = 0.27682604736157; W(2) = 0.43174538120986;
W(3) = 0.48761904761905;
W(4) = W(2); W(5) = W(1);

%------------
else if(i==6)
%-----------

Z(1) = - 0.87174014850961; Z(2) = - 0.59170018143314;
Z(3) = - 0.20929921790248;
Z(4) = -Z(3); Z(5) = -Z(2); Z(6) = -Z(1);
W(1) = 0.21070422714350; W(2) = 0.34112269248350;
W(3) = 0.41245879465870;
W(4) = W(3); W(5) = W(2); W(6) = W(1);

%--
end
%--

%-----
% done
%-----

return;
```

Function lobatto: ($\longrightarrow$ Continued.) Evaluation of the zeros of the Lobatto polynomials, and corresponding integration weights in the Lobatto integration quadrature.

With the Lobatto interpolation nodal base, the $m + 1$ element interpolation nodes are distributed at

$$\begin{aligned} \xi_1 &= -1, \\ \xi_2 &= t_1, \quad \xi_2 = t_2, \quad \cdots \quad \xi_m = t_{m-1}, \\ \xi_{m+1} &= 1, \end{aligned} \qquad (2.2.3)$$

where $t_1, t_2, \ldots, t_{m-1}$ are the zeros of the $(m-1)$-degree Lobatto polynomial. The $m+1$ interpolation nodes are thus the roots of the $(m+1)$-degree *completed Lobatto polynomial*, defined as

$$Lo^c_{m+1}(t) \equiv (1 - t^2)\, Lo_{m-1}(t), \qquad (2.2.4)$$

where the superscript "c" stands for "completed."

For example, having chosen $m = 1$, we place the interpolation nodes at

$$\xi_1 = -1, \qquad \xi_2 = 1, \qquad (2.2.5)$$

and obtain the linear expansion discussed in Section 1.1. Stepping up to $m = 2$, we note that $t_1 = 0$ is the single zero of $Lo_1(\xi)$, and place the interpolation nodes at

$$\xi_1 = -1.0, \qquad \xi_2 = 0, \qquad \xi_3 = 1.0, \tag{2.2.6}$$

to obtain the quadratic expansion discussed in Section 1.3, where the interior element node lies at the element mid-point. Stepping up once more to $m = 3$, we note that $t_1 = -1/\sqrt{5}$ and $t_2 = 1/\sqrt{5}$ are the two zeros of $Lo_2(\xi)$, and deploy the interpolation nodes at

$$\xi_1 = -1.0, \qquad \xi_2 = -\frac{1}{\sqrt{5}}, \qquad \xi_3 = \frac{1}{\sqrt{5}}, \qquad \xi_4 = 1.0, \tag{2.2.7}$$

to obtain a cubic expansion.

To demonstrate the advantages of the Lobatto node distribution, we approximate the Runge function

$$f(\xi) = \frac{1}{1 + 25\,\xi^2} \tag{2.2.8}$$

in the interval $[-1, 1]$ with an mth-degree interpolating polynomial that passes through $m+1$ data points, ξ_i, where $i = 1, 2, \ldots, m+1$, $\xi_1 = -1$, and $\xi_{m+1} = 1$. Figure 2.2.2(a) shows graphs of several interpolating polynomials of increasing order, for evenly distributed interpolation nodes. The seemingly innocuous Runge function is plotted with the heavy solid line. As the number of data points, and thus the polynomial order, is increased, the interpolation worsens near the edges of the interpolation domain due to the Runge effect discussed in Section 2.1. Physically, because the interpolated polynomial is reconstructed from evenly spaced data that pay equal attention to the middle as well as to the ends of the interpolation domain, insufficient information on the interpolated function is provided beyond the boundaries of the interpolation domain, $\xi < -1$ and $\xi > 1$, and this results in oscillations. Consequently, as the polynomial order is increased, a finite element solution based on evenly spaced element interpolation nodes will either become notably sensitive to the numerical error or fail.

Figure 2.2.2(b) shows graphs of the fourth- and sixth-degree interpolating polynomials based on the Lobatto nodes. In this case, as the number of data points, and thus the polynomial order, is increased, the interpolating polynomial converges uniformly throughout the interpolation domain. Comparison of the dashed with the dot-dashed line corresponding, respectively, to Lobatto or evenly spaced nodes with $m = 6$, demonstrates the advantages of the Lobatto nodes.

A further assessment of the interpolation accuracy is made in Figure 2.2.3 for the interpolated function

$$f(\xi) = \frac{1}{1 + 4\,\xi^2} \tag{2.2.9}$$

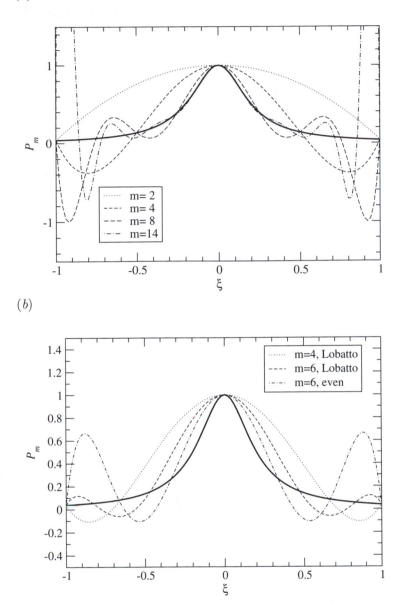

Figure 2.2.2 Effect of node distribution on the polynomial interpolation of the Runge function $f(\xi) = 1/(1+25\xi^2)$, for (a) evenly spaced data points with $\xi_1 = -1$ and $\xi_{m+1} = 1$, and (b) the Lobatto node distribution.

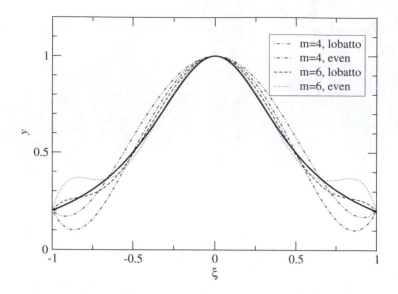

Figure 2.2.3 Comparison of the interpolating polynomials for the scaled
Runge function $f(\xi) = 1/(1 + 4\xi^2)$ with evenly spaced or Lobatto nodes.

in the interval $[-1, 1]$, which essentially consists of a smaller portion of the
Runge function described by (2.2.8). In this case, the sixth-degree interpolation
polynomial based on the Lobatto nodes faithfully approximates the interpolated
function, whereas the interpolation polynomial based on evenly spaced nodes
exhibits significant deviations near the ends of the interpolation domain.

2.2.2 Discretization code

FSELIB function `discr_s`, listed in the text, defines the element end-points,
generates the Lobatto interpolation nodes, and produces the global nodes and
connectivity matrix according to Algorithm 2.1.1. The driver script `discr_s_dr`,
listed in the text, calls `discr_s_dr` and displays the generated node distribution. Figure 2.2.4 shows results for a two-element arrangement, $N_E = 2$, and
polynomial expansion order $m = 6$ for the first element, and $m = 4$ for the
second element.

2.2.3 Legendre polynomials

The ith-degree Lobatto polynomial, Lo_i, derives from the $(i+1)$-degree Legendre polynomial, $L_{i+1}(\xi)$, as

$$Lo_i(\xi) = L'_{i+1}(\xi), \tag{2.2.10}$$

where a prime denotes a derivative with respect to ξ, as discussed in Section
B.6 of Appendix B.

```
function [xe,xen,xg,c,ng] = discr_s (x1,x2,ne,ratio,np)

%=========================================================
% Discretize the interval (x1, x2) into "ne" elements,
% and generate the np-order Lobatto
% interpolation nodes
%
% xe:  element end-nodes
% xen: element interpolation nodes
% xg:  unique global nodes
% c:   connectivity matrix
% ng:  number of unique global nodes
% np:  polynomial order over each element
%=========================================================

%-----------------------------------
% define the element end-nodes (xe)
%-----------------------------------

xe = elm_line1 (x1,x2,ne,ratio);

%------------------------------------------------
% define the element interpolation nodes (xen)
%------------------------------------------------

for l=1:ne

   mx = 0.5*(xe(l+1)+xe(l)); dx = 0.5*(xe(l+1)-xe(l));
   xen(l,1) = xe(l);
   if(np(l)>1)
     [zlob, wlob] = lobatto(np(l)-1);
     for je=1:np(l)-1
       xen(l,je+1) = mx + zlob(je)*dx;
     end
   end
   xen(l,np(l)+1) = xe(l+1);

end

%--------------------------------------------------------------
% define the connectivity matrix and the global nodes (xg)
%--------------------------------------------------------------

Ic = 2;   % node counter

for l=1:ne
   Ic = Ic-1;
   for je=1:np(l)+1
     c(l,je) = Ic; xg(Ic) = xen(l,je); Ic = Ic+1;
   end
end

ng = Ic-1;   % number of global nodes
```

Function discr_s: Continuing $\longrightarrow$

```
%-----
% done
%-----

return;
```

Function discr_s: Discretization of the solution domain into N_E elements with Lobatto element nodes. This function also counts the unique global nodes and generates the connectivity matrix.

```
%===============================================================
% Code: discr_s_dr
%
% Driver script for discretizing the interval (0, L) into "ne"
% elements, and defining the Lobatto interpolation nodes
%===============================================================

%-----------
% input data
%-----------

L=1.0;     % discretization length
ne=2; ratio=0.5;  % number of elements, stretch ratio
np(1)=6; np(2)=4;  % element polynomial orders

%-----------
% discretize
%-----------

[xe,xen,xg,c,ng] = discr_s(0,L,ne,ratio,np);

%-----
% plot
%-----

yg= zeros(ng,1); plot (xg,yg,'-d'); hold on;  % global nodes
ye= zeros(ne+1,1); plot (xe,ye,'o')    % end nodes
axis ([ 0 L -0.10*L 0.10*L])

%-----
% done
%-----
```

Code discr_s_dr: Driver code for discretizing the solution domain into N_E elements and defining the Lobatto interpolation nodes.

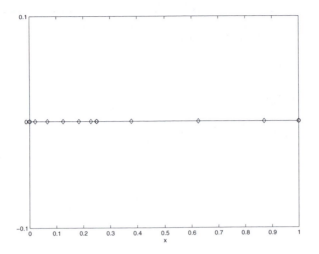

Figure 2.2.4 Spectral node distribution on two elements generated by the FSELIB code `discr_s`. The circles mark the position of the element endnodes, and the diamonds mark the position of the Lobatto interpolation nodes.

By construction, the ith-degree Legendre polynomial satisfies the second-order differential equation

$$\left[(\xi^2 - 1)\, L_i'(\xi) \right]' = \left[(\xi^2 - 1)\, Lo_{i-1}(\xi) \right]' = i\, (i+1)\, L_i(\xi). \qquad (2.2.11)$$

An arbitrary pair of Legendre polynomials of order i and j satisfy the orthogonality property

$$\int_{-1}^{1} L_i(\xi)\, L_j(\xi)\, \mathrm{d}\xi = \frac{2}{2i+1}\, \delta_{ij}, \qquad (2.2.12)$$

where δ_{ij} is Kronecker's delta. The first few Legendre polynomials are shown in Table 2.2.3 along with a generating formula, and their graphs are presented in Figure 2.2.5.

The Legendre polynomials can be computed from the recursion relation

$$L_{i+1}(\xi) = \frac{2i+1}{i+1}\, \xi\, L_i(\xi) - \frac{i}{i+1}\, L_{i-1}(\xi). \qquad (2.2.13)$$

The derivatives of the Legendre polynomials, and thus the Lobatto polynomials themselves, can be computed from the recursion relation

$$(1 - \xi^2)\, L_{i+1}'(\xi) = (1 - \xi^2)\, Lo_i(\xi)$$
$$= (i+1) \left(L_i(\xi) - \xi\, L_{i+1}(\xi) \right). \qquad (2.2.14)$$

Legendre polynomials

$L_0(\xi) = 1$

$L_1(\xi) = \xi$

$L_2(\xi) = \frac{1}{2}\left(3\,\xi^2 - 1\right)$

$L_3(\xi) = \frac{1}{2}\left(5\,\xi^2 - 3\right)\xi$

$L_4(\xi) = \frac{1}{8}\left(35\,\xi^4 - 30\,\xi^2 + 3\right)$

$L_5(\xi) = \frac{1}{8}\left(63\,\xi^4 - 70\,\xi^2 + 15\right)\xi$

$L_6(\xi) = \frac{1}{16}\left(231\,\xi^6 - 315\,\xi^4 + 105\,\xi^2 - 5\right)$

$L_7(\xi) = \frac{1}{16}\left(429\,\xi^6 - 693\,\xi^4 + 315\,\xi^2 - 35\right)\xi$

$\cdots$

$L_i(\xi) = \frac{1}{2^i i!}\frac{\mathrm{d}^i}{\mathrm{d}\xi^i}(\xi^2 - 1)^i = \frac{(2i)!}{2^i i!}\,\xi^i + \cdots$

Table 2.2.3 The first few members of the triangular family of Legendre polynomials defined in the interval $-1 \le \xi \le 1$. In the generating formula shown in the last entry, $i! = 1 \cdot 2 \cdot 3 \ldots \cdot i$, is the factorial. The Lobatto polynomials are the derivatives of the Legendre polynomials, $Lo_i(\xi) = L'_{i+1}(\xi)$.

Replacing i with $i+1$ in (2.2.13) and rearranging, we find

$$\xi\,L_{i+1}(\xi) = \frac{i+2}{2i+3}\,L_{i+2}(\xi) + \frac{i+1}{2i+3}\,L_i(\xi). \tag{2.2.15}$$

Substituting this expression in the right-hand side of (2.2.14) and simplifying, we obtain

$$(1 - \xi^2)\,L'_{i+1}(\xi) = (1 - \xi^2)\,Lo_i(\xi) \tag{2.2.16}$$
$$= \frac{(i+1)(i+2)}{2i+3}\,\Big(L_i(\xi) - L_{i+2}(\xi)\Big).$$

These expressions find applications in the computation of the element diffusion and mass matrices discussed in the next section.

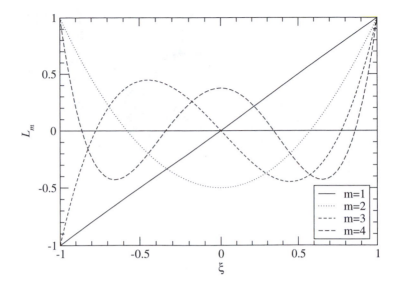

Figure 2.2.5 Graphs of the first few Legendre polynomials, $L_m(\xi)$, in their native domain of definition. Note that the Legendre polynomials vary in the range $[-1,1]$, whereas the Lobatto polynomials illustrated in Figure 2.2.1 vary over a wider range.

Finally, we use (2.2.11) and (2.2.16) and find that the completed Lobatto polynomials can be expressed in terms of the Legendre polynomials, L_i, as

$$Lo_{i+1}^c(\xi) \equiv (1 - \xi^2)\, Lo_{i-1}(\xi) = -i\,(i+1) \int_{-1}^{\xi} L_i(\xi')\, \mathrm{d}\xi'$$

$$= \frac{i\,(i+1)}{2\,i+1}\left(L_{i-1}(\xi) - L_{i+1}(\xi)\right). \tag{2.2.17}$$

2.2.4 Chebyshev second-kind nodal base

An alternative to the Lobatto node distribution is the Chebyshev second-kind distribution based on the Chebyshev polynomials of the second kind, $\mathcal{T}_m(t)$, discussed in Appendix B. With this choice, the element interpolation nodes are distributed at the zeros of the $(m+1)$-degree *completed Chebyshev polynomial of the second kind*, defined as

$$T_{m+1}^c(t) \equiv (1 - t^2)\, \mathcal{T}_{m-1}(t). \tag{2.2.18}$$

The nodal positions simply arise by projecting $m+1$ evenly-spaced points along the upper or lower half of the unit semi-circle onto the horizontal ξ axis passing through the circle center, yielding

$$\xi_i = \cos\left[\frac{(i-1)\pi}{m}\right], \tag{2.2.19}$$

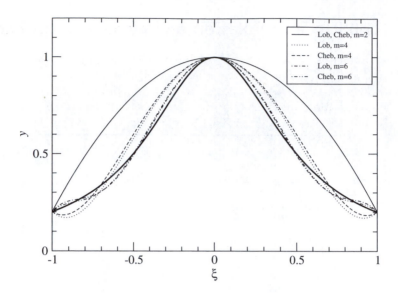

Figure 2.2.6 Comparison of the interpolating polynomials of the function
$f(\xi) = 1/(1+4\xi^2)$, plotted with the bold line, for the Lobatto and second-
kind Chebyshev nodal base.

for $i = 1, 2, \ldots, m+1$. Figure 2.2.6 compares interpolating polynomials of several degrees for the narrow Runge function $f(\xi) = 1/(1+4\xi^2)$, computed using the Lobatto or second-kind Chebyshev nodal set. The quadratic polynomials for $m = 2$ are identical. Although higher-order polynomials differ, the differences are small. The second-kind Chebyshev nodal basis is favored in spectral methods for viscous flow (e.g., [40]).

PROBLEMS

2.2.1 *Polynomial root finder.*

Use the MATLAB function **roots** to compute the zeros of the Lobatto polynomial $Lo_5(\xi)$, and compare the results with those listed in Table 2.2.1.

2.2.2 *Lobatto interpolation base.*

(*a*) Prepare graphs of the *m*th-degree interpolating polynomials for the function

$$f(\xi) = \frac{1}{1+\xi^2}, \qquad (2.2.1)$$

in the interval [-1, 1], based on the Lobatto nodal base with *m*=2, 3, 4, 5, and discuss the behavior of the interpolation error.

(*b*) Repeat (*a*) for the second-kind Chebyshev nodal base.

2.3 Lobatto interpolation and element matrices

In the remainder of this chapter, we confine our attention to the Lobatto nodal base introduced in Section 2.2. The quadratic element interpolation functions for polynomial order $m = 2$ are given in (1.3.26). Graphs of the four cubic element interpolation functions corresponding to $m = 3$, and of the five quartic interpolation functions corresponding to $m = 4$, are displayed in Figure 2.3.1. Note that each interpolation function achieves the maximum value of unity at the corresponding interpolation node, in agreement with inequality (2.2.2). More generally, it can be shown that the $2m$-degree polynomial

$$G_{2m}(\xi) = \sum_{i=1}^{m+1} \psi_i^2(\xi) \tag{2.3.1}$$

reaches the maximum value of unity only at the spectral nodes themselves, as discussed in Section A.5 of Appendix A. The existence of this upper bound guarantees that the magnitude of the global interpolation functions will not increase as the interpolation order is raised, and thereby guarantees a fast-converging interpolation.

For the Lobatto nodal base, the generating function defined in (2.1.5) takes the form

$$\Phi_{m+1}(\xi) = \frac{1}{c_{m-1}} (\xi^2 - 1) \, Lo_{m-1}(\xi), \tag{2.3.2}$$

where c_{m-1} is the coefficient of the highest power of the $(m-1)$-degree Lobatto polynomial; for example, $c_3 = 3$. Substituting (2.3.2) into the right-hand side of (2.1.4), we obtain the lth-element interpolation functions

$$\psi_i(\xi) = \frac{1}{[(\xi^2 - 1) \, Lo_{m-1}(\xi)]'_{\xi=\xi_i}} \frac{(\xi^2 - 1) \, Lo_{m-1}(\xi)}{\xi - \xi_i}, \tag{2.3.3}$$

which are simply the Lagrange interpolating polynomials corresponding to the Lobatto nodes, complemented by the end-nodes, $\xi = \pm 1$. A more explicit expression arises by using property (2.2.11) to simplify the denominator of the first fraction on the right-hand side of (2.3.3), obtaining

$$\psi_i(\xi) = \frac{1}{m \, (m + 1) \, L_m(\xi_i)} \frac{(\xi^2 - 1) \, Lo_{m-1}(\xi)}{\xi - \xi_i}. \tag{2.3.4}$$

The cardinal interpolation property $\psi_i(\xi_i) = 1$ requires

$$\lim_{\xi \to \xi_i} \left[\frac{(\xi^2 - 1) \, Lo_{m-1}(\xi)}{\xi - \xi_i} \right] = m \, (m + 1) \, L_m(\xi_i). \tag{2.3.5}$$

As an example, we recall the mandatory choice $\xi_1 = -1$ and $\xi_{m+1} = 1$, and use (2.3.4) to compute the first and last interpolation functions

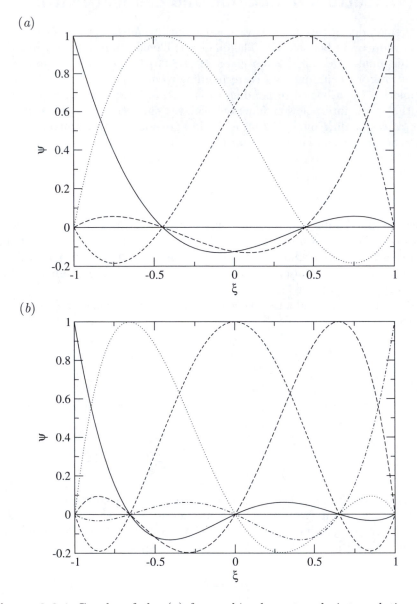

Figure 2.3.1 Graphs of the (*a*) four cubic element-node interpolation functions corresponding to $m = 3$, and (*b*) five quartic interpolation functions corresponding to $m = 4$, for the Lobatto nodal base.

$$\psi_1(\xi) = \frac{1}{m\,(m+1)\,L_m(\xi = -1)}\,(\xi - 1)\,Lo_{m-1}(\xi),$$

$$\psi_{m+1}(\xi) = \frac{1}{m\,(m+1)\,L_m(\xi = 1)}\,(\xi + 1)\,Lo_{m-1}(\xi). \tag{2.3.6}$$

Choosing $m = 1$, and recalling that the first-degree Legendre polynomial is the linear function, $L_1(\xi) = \xi$, and the zeroth-degree Lobatto polynomial is equal to unity, $Lo_0(\xi) = 1$, we find

$$\psi_1(\xi) = \frac{1}{2\,L_1(\xi = -1)}\,(\xi - 1)\,Lo_0(\xi) = \frac{1}{2}\,(1 - \xi),$$

$$\psi_2(\xi) = \frac{1}{2\,Lo_0(\xi = 1)}\,(1 + \xi)\,Lo_0(\xi) = \frac{1}{2}\,(1 + \xi), \tag{2.3.7}$$

which are precisely the cardinal interpolation functions of the linear elements associated with the first and second element nodes.

2.3.1 Lobatto mass matrix

For convenience, we recast the expression for the element mass matrix given in (2.1.9) into the form

$$B_{ij}^{(l)} = \frac{h_l}{2}\,\Theta_{ij}, \tag{2.3.8}$$

for $i, j = 1, 2, \ldots, m + 1$, where

$$\Theta_{ij} \equiv \int_{-1}^{1} \psi_i(\xi)\,\psi_j(\xi)\,d\xi \tag{2.3.9}$$

is the dimensionless mass matrix. Using expression (2.3.4) for the node interpolation functions, we find

$$\Theta_{ij} \equiv \frac{1}{m^2\,(m+1)^2\,L_m(\xi_j)\,L_m(\xi_i)}\int_{-1}^{1} \mathcal{P}_{ij}(\xi)\,d\xi, \tag{2.3.10}$$

where

$$\mathcal{P}_{ij}(\xi) \equiv \frac{(1 - \xi)^2(1 + \xi)^2}{(\xi - \xi_i)(\xi - \xi_j)}\,Lo_{m-1}^2(\xi)$$

$$= \frac{(\xi^2 - 1)^2}{(\xi - \xi_i)(\xi - \xi_j)}\,Lo_{m-1}^2(\xi) \tag{2.3.11}$$

is a $2m$-degree polynomial. As ξ tends to -1, the numerator of the fraction in the last expression of (2.3.11) tends to zero quadratically, and the fraction tends to zero provided that $i \neq 1$ and $j \neq 1$. When $i = 1$ and $j = 1$, the fraction tends to the value of four. Similarly, as ξ tends to 1, the fraction tends to zero provided that $i \neq m + 1$ and $j \neq m + 1$. When $i = m + 1$ and $j = m + 1$, the fraction tends to the value of four.

Accordingly, the matrices $\mathcal{P}_{ij}(\xi = \pm 1)$ are null, except for the first element $\mathcal{P}_{11}(\xi = -1)$, and last element, $\mathcal{P}_{m+1,m+1}(\xi = 1)$. In compact notation,

$$\mathcal{P}_{ij}(\xi = -1) = 4\,\delta_{i1}\,\delta_{j1}\, Lo^2_{m-1}(\xi = -1),$$

(2.3.12)

$$\mathcal{P}_{ij}(\xi = 1) = 4\,\delta_{i,m+1}\,\delta_{j,m+1}\, Lo^2_{m-1}(\xi = 1),$$

where δ_{ij} is Kronecker's delta.

2.3.2   Lobatto integration quadrature

In practice, the integral on the right-hand side of (2.3.10) is computed by numerical methods. The $(k + 1)$-point Lobatto integration quadrature, discussed in Section B.3 of Appendix B, provides us with the approximation

$$\int_{-1}^{1} f(\xi)\,d\xi \simeq \sum_{p=1}^{k+1} f(\xi = z_p)\, w_p,$$

(2.3.13)

for some smooth function $f(\xi)$, where $k \geq 1$ is the selected order of the quadrature, z_p are the quadrature base points, and w_p are the associated integration weights, defined as follows:

- $z_1 = -1$.

- z_p, for $p = 2, 3, \ldots, k$ are the zeros of the $(k-1)$-degree Lobatto polynomial, Lo_{k-1}, denoted by t_i. Thus, $z_p = t_{p-1}$, or

$$z_2 = t_1, \quad z_3 = t_2, \quad \ldots, \quad z_k = t_{k-1}.$$

(2.3.14)

- $z_{k+1} = 1$.

- The integration weights are given by

$$w_1 = \frac{2}{k(k+1)},$$

$$w_p = \frac{2}{k(k+1)} \frac{1}{L_k^2(z_p)}, \qquad p = 2, 3, \ldots, k.$$

(2.3.15)

$$w_{k+1} = \frac{2}{k(k+1)},$$

where L_k is a Legendre polynomial. Numerical values for the weights at the interior base points are given in Table 2.2.1, subject to the index shifting convention, $w_p = \widehat{w}_{p-1}$, for $p = 2, 3, \ldots, k$.

In Section B.3 of Appendix B, it is shown that the numerical result obtained by applying the Lobatto quadrature is exact if the function $f(\xi)$ is a $(2k-1)$-degree polynomial. Thus, if $k = 5$, the quadrature integrates exactly ninth-degree polynomials, even though it only employs six base points.

FSELIB function int_lob, listed in the text, implements the quadrature for an arbitrary integrand, $f(\xi)$, defined inside the function. The integration base points and weights are read from the FSELIB function lobatto. Executing int_lob with $k = 6$ to integrate the function $f(\xi) = \xi^{10}$ produces the exact answer, in agreement with the theoretical predictions.

2.3.3 Computation of the Lobatto mass matrix

Applying the Lobatto quadrature to compute the integral on the right-hand side of (2.3.10), and taking into consideration (2.3.12), we find

$$
\int_{-1}^{1} \mathcal{P}_{ij}(\xi) \, \mathrm{d}\xi \simeq \sum_{p=1}^{k+1} \mathcal{P}_{ij}(\xi = z_p) \, w_p
$$

$$
= \delta_{ij} \, \delta_{i1} \, 4 \, Lo_{m-1}^2(\xi = -1) \, w_1 + \sum_{p=2}^{k} \mathcal{P}_{ij}(\xi = z_p) \, w_p \qquad (2.3.16)
$$

$$
+ \delta_{ij} \, \delta_{i(m+1)} \, 4 \, Lo_{m-1}^2(\xi = 1) \, w_{k+1}.
$$

Since the integrand on the left-hand side is a $2m$-degree polynomial in ξ, the integration will be exact if $2m \leq 2k - 1$, which is true if

$$
k \geq m + 1, \qquad (2.3.17)
$$

that is, if the order of the Lobatto quadrature is higher than the size of the nodal interpolation set by at least one unit.

FSELIB code emm_lob, listed in the text, computes the dimensionless mass matrix Θ introduced in (2.3.8), based on the Lobatto quadrature. Table 2.3.3 displays numerical values for several polynomial orders, m. Note that the sum of all elements of Θ is equal to two for any value of m. The top and tail-end of the FSELIB function emm that evaluates this matrix from a table is listed in the text; the complete function is included in FEMLIB. This function will be used in the next section for building a spectral element code.

Karniadakis & Sherwin ([30], p. 44) present graphs of the condition number of the mass matrix for the Lobatto spectral nodes and for the evenly-spaced nodes, and demonstrate that the mass matrix for the evenly-spaced nodes is significantly worse conditioned for polynomial orders higher than five. A high condition number carries an increased risk of numerical instability and failure due to the growth of the computer round-off error.

```
function integral = int_lob(k)

%========================================
% int_lob
%
% Numerical integration of a function
% using the Lobatto quadrature
%========================================

%----------------------------------------------------
% make up the integration base points and weights
%----------------------------------------------------

z(1) = -1.0; z(k+1) = 1.0;
w(1) = 2/(k*(k+1)); w(k+1) = w(1);

if(k>1)

  [zL, wL] = lobatto(k-1);

  for i=2:k
   z(i) = zL(i-1);
   w(i) = wL(i-1);
   end

end

%----------------------------
% generate the function values
%----------------------------

for i=1:k+1
   xi = z(i);
   f(i) = xi^2;     % example, can also be generated by a function
end

%-----------------------
% perform the quadrature
%-----------------------

integral = 0.0;

for i=1:k+1
   integral = integral + f(i)*w(i);
end

%-----
% done
%-----

return
```

Function int_lob: Lobatto integration quadrature expressed by the integra-
    tion rule (2.3.13).

```
%========================================
%  Code: emm_lob
%
%  Evaluation of the dimensionless
%  element mass matrix using
%  the lobatto quadrature
%
%  m: order of polynomial expansion
%  k: order of Lobatto quadrature
%
%  setting m=k implements mass lumping
%========================================

%----------------------------
% lobatto interpolation nodes
%----------------------------

m = input('Enter the order of the polynomial expansion, m: ');

%----------------------------
% generate the spectral nodes
%----------------------------

xi(1) = -1.0;

if(m>1)
   [tL, wL] =  lobatto(m-1);
   for i=2:m
    xi(i) = tL(i-1);
   end
end

xi(m+1) = 1.0;

%------------------------------------------
% generate the lobatto integration points
% and corresponding weights
%------------------------------------------

k = input('Enter the order of the lobatto quadrature, k: ');

z(1) = -1.0;
w(1) = 2.0/(k*(k+1));

if(k>1)
   [zL, wL] = lobatto(k-1);
   for i=2:k
     z(i) = zL(i-1);
     w(i) = wL(i-1);
   end
end

z(k+1) = 1.0;
w(k+1) = w(1);
```

Code emm_lob: Continuing $\longrightarrow$

```
%----------------------------
% compute the mass matrix
% by the Lobatto quadrature
%----------------------------

for ind=1:m+1    % loop over interpolation nodes, closes at echidna

   for jnd=1:m+1    % loop over interpolation nodes

   % Generate the integrand function values
   % consisting of the product of the interpolation functions
   % for nodes ind and jnd

        for j=1:k+1

         f(j) = 1.0;

         for i=1:m+1
          if(i ~= ind)
            f(j) = f(j)*(z(j)-xi(i))/(xi(ind)-xi(i));
          end
         end

         for i=1:m+1
          if(i ~= jnd)
            f(j) = f(j)*(z(j)-xi(i))/(xi(jnd)-xi(i));
          end
         end

        end

   % perform the (k+1)-point Lobatto quadrature
   % for the computed function values

        msmat = 0.0;

        for i=1:k+1
           msmat = msmat + f(i)*w(i);
        end

        mm(ind,jnd) = msmat;

   end

end    % echidna

%-----
% done
%-----
```

Code emm_lob: ($\longrightarrow$ Continued.) Evaluation of the dimensionless mass matrix Θ defined in (2.3.9).

Exact → Lumped

$m = 1$:
$$\frac{1}{3}\begin{bmatrix} 2 & 1 \\ 1 & 2 \end{bmatrix} \rightarrow \begin{bmatrix} 1 & 0 \\ 0 & 1 \end{bmatrix}$$

$m = 2$:
$$\frac{1}{15}\begin{bmatrix} 4 & 2 & -1 \\ 2 & 16 & 2 \\ -1 & 2 & 4 \end{bmatrix} \rightarrow \frac{1}{3}\begin{bmatrix} 1 & 0 & 0 \\ 0 & 4 & 0 \\ 0 & 0 & 1 \end{bmatrix}$$

$m = 3$:

$$\begin{bmatrix} 0.14285714 & 0.05323971 & -0.05323971 & 0.02380952 \\ 0.05323971 & 0.71428572 & 0.11904762 & -0.05323971 \\ -0.05323971 & 0.11904762 & 0.71428572 & 0.05323971 \\ 0.02380952 & -0.05323971 & 0.05323971 & 0.14285714 \end{bmatrix} \rightarrow$$

$$\rightarrow \begin{bmatrix} 0.16666667 & 0.0 & 0.0 & 0.0 \\ 0.0 & 0.83333333 & 0.0 & 0.0 \\ 0.0 & 0.0 & 0.83333333 & 0.0 \\ 0.0 & 0.0 & 0.0 & 0.16666667 \end{bmatrix}$$

Table 2.3.3 Continuing ⟶

Inexact integration and mass lumping

The inexact choice, $k = m$, has the practical advantage that $\mathcal{P}_{ij}(\xi = z_p) \neq 0$ only if $i = j = p$, whereupon (2.3.16) simplifies to

$$\int_{-1}^{1} \mathcal{P}_{ij}(\xi)\, d\xi \simeq \delta_{ij}\left[\delta_{i1}\, 4\, Lo_{m-1}^2(\xi = -1)\, w_1\right.$$

$$\left. +\delta_{ip}\, \alpha_p\, w_p + \delta_{i(m+1)}\, 4\, Lo_{m-1}^2(\xi = 1)\, w_{m+1}\right], \qquad (2.3.18)$$

for $p = 2, 3, \ldots, m$. Using (2.3.5), we find

$$\alpha_p \equiv \mathcal{P}_{ij}(\xi = z_p) = m^2(m+1)^2\, L_m^2(z_p), \qquad (2.3.19)$$

for $p = 2, 3, \ldots, m$. We may now define

$$\alpha_1 \equiv 4\, Lo_{m-1}^2(\xi = -1), \qquad \alpha_{m+1} \equiv 4\, Lo_{m-1}^2(\xi = 1), \qquad (2.3.20)$$

$m = 4$:

$$\begin{bmatrix} 0.08888888 & 0.02592591 & -0.02962963 \\ 0.02592591 & 0.48395066 & 0.06913582 \\ -0.02962963 & 0.06913582 & 0.63209873 \\ 0.02592593 & -0.06049384 & 0.06913582 \\ -0.01111111 & 0.02592593 & -0.02962963 \end{bmatrix}$$

$$\begin{bmatrix} 0.02592593 & -0.01111111 \\ -0.06049384 & 0.02592593 \\ 0.06913582 & -0.02962963 \\ 0.48395066 & 0.02592591 \\ 0.02592591 & 0.08888888 \end{bmatrix} \rightarrow$$

$$\rightarrow \begin{bmatrix} 0.1 & 0.0 & 0.0 & 0.0 & 0.0 \\ 0.0 & 0.54444444 & 0.0 & 0.0 & 0.0 \\ 0.0 & 0.0 & 0.71111111 & 0.0 & 0.0 \\ 0.0 & 0.0 & 0.0 & 0.54444444 & 0.0 \\ 0.0 & 0.0 & 0.0 & 0.0 & 0.1 \end{bmatrix}$$

Table 2.3.3 ($\longrightarrow$ Continued.) The dimensionless mass matrix Θ defined in (2.3.10), for polynomial orders $m = 1, 2, 3, 4$. The arrows point to the diagonalized mass-lumped approximation. Complete high-accuracy tables are included in the FSELIB function emm.

and recast (2.3.18) into the compact form

$$\int_{-1}^{1} \mathcal{P}_{ij}(\xi) \, \mathrm{d}\xi \simeq \delta_{ij} \sum_{p=1}^{m+1} \delta_{ip} \, \alpha_p \, w_p. \tag{2.3.21}$$

We have found that the inexact integration effectively diagonalizes the mass matrix by implementing mass lumping for any polynomial order, m.

Diagonalization of the mass matrix is especially desirable in time-dependent problems where the finite element method culminates in a system of differential equations for the nodal values, as will be discussed in Section 2.5. In these cases, forward time discretization leads to an explicit method that circumvents the need for solving a system of linear equations at each time step. However, because of numerical stability considerations, the explicit method is of interest primarily in convection-dominated transport.

```
function elm_mm = emm(m)

%=================================================
% Element mass matrix for an mth-order polynomial
% expansion, and m = 1-6
%=================================================

%-----
% trap
%-----

if(m>5)
  disp('');
  disp(' elmm: Chosen number of points is not available');
  break;
end

%-------
if(m==1)
%-------

elm_mm = [2/3 1/3; 1/3 2/3];

%-----------
elseif(m==2)
%-----------

elm_mm = [
 0.266666666667   0.133333333333 -0.066666666667;
 0.133333333333   1.066666666667  0.133333333333;
-0.066666666667   0.133333333333  0.266666666667];

%-----------
elseif(m==3)
%-----------

elm_mm =[
 0.14285714   0.05323971 -0.05323971   0.02380952;
 0.05323971   0.71428571  0.11904761 -0.05323971;
-0.05323971   0.11904761  0.71428571  0.05323971;
 0.02380952 -0.05323971  0.05323971  0.14285714];

......

%--
end
%--

return;
```

Function emm: Evaluation of the dimensionless element mass matrix Θ. The six dots near the bottom indicate unprinted code. Higher accuracy data is implemented in the function included in FSELIB.

Substituting (2.3.21) in (2.3.10), we find that the diagonal entries of the inexact dimensionless mass matrix, indicated by a prime, are given by

$$\Theta'_{11} = \frac{8\, Lo^2_{m-1}(\xi = -1)}{m^3\,(m+1)^3\, L^2_m(\xi = -1)},$$

$$\Theta'_{ii} = w_{i-1}, \qquad \text{for } i = 2, \ldots, m, \tag{2.3.22}$$

$$\Theta'_{(m+1)(m+1)} = \frac{8\, Lo^2_{m-1}(\xi = 1)}{m^3\,(m+1)^3\, L^2_m(\xi = 1)}.$$

For example, when $m = 1$, corresponding to linear interpolation, we obtain the diagonal elements

$$\Theta'_{11} = \frac{8\, Lo^2_0(\xi = -1)}{1 \cdot 8\, L^2_1(\xi = -1)} = \frac{8 \cdot (1)^2}{1 \cdot 8 \cdot (-1)^2} = 1,$$

$$\tag{2.3.23}$$

$$\Theta'_{22} = \frac{8\, Lo^2_0(\xi = 1)}{1 \cdot 8\, L^2_1(\xi = 1)} = \frac{8 \cdot (1)^2}{1 \cdot 8 \cdot (1)^2} = 1.$$

The inexact element mass matrix is thus given by

$$\mathbf{\Theta}' = \frac{h_l}{2} \begin{bmatrix} 1 & 0 \\ 0 & 1 \end{bmatrix}. \tag{2.3.24}$$

A comparison with the exact value displayed in (1.1.62) confirms that the inexact integration effectively lumps the off-diagonal elements of each row to the corresponding diagonal. Stated differently, the diagonal element of the inexact form is the sum of the elements in the corresponding row of the exact form.

As a further example, we consider the quadratic interpolation corresponding to $m = 2$, and compute the diagonal elements

$$\Theta'_{11} = \frac{8\, Lo^2_1(\xi = -1)}{8 \cdot 27\, L^2_2(\xi = -1)} = \frac{8 \cdot (-3)^2}{8 \cdot 27 \cdot 1} = \frac{1}{3},$$

$$\Theta'_{22} = w_1 = \frac{2}{2 \cdot 3}\, \frac{1}{L^2_2(\xi = 0)} = \frac{2}{2 \cdot 3}\, 4 = \frac{4}{3}, \tag{2.3.25}$$

$$\Theta'_{33} = \frac{8\, Lo_1(\xi = 1)^2}{8 \cdot 27\, L^2_2(\xi = 1)} = \frac{8 \cdot 3^2}{8 \cdot 27 \cdot 1} = \frac{1}{3}.$$

The inexact element mass matrix is thus given by

$$\mathbf{\Theta}' = \frac{h_l}{30} \begin{bmatrix} 5 & 0 & 0 \\ 0 & 20 & 0 \\ 0 & 0 & 5 \end{bmatrix}. \tag{2.3.26}$$

A comparison with the exact result displayed in (1.3.30) confirms again that the inexact integration lumps the off-diagonal elements in each row to the diagonal entry.

Table 2.3.3 displays numerical values of the lumped dimensionless matrix Θ' for several polynomial orders, m. FSELIB function `emm_lump` (not listed in the text) evaluates this matrix in the spirit of the function `emm`.

2.3.4 Computation of the Lobatto diffusion matrix

Next, we consider the computation of the element diffusion matrix given in (2.1.8). To isolate the effect of element size, we write

$$A_{ij}^{(l)} = \frac{2}{h_l}\,\Psi_{ij}, \tag{2.3.27}$$

for $i, j = 1, 2, \ldots, m+1$, where $\boldsymbol{\Psi}$ is the dimensionless diffusion matrix,

$$\Psi_{ij} \equiv \int_{-1}^{1} \frac{\mathrm{d}\psi_i}{\mathrm{d}\xi} \frac{\mathrm{d}\psi_j}{\mathrm{d}\xi}\,\mathrm{d}\xi. \tag{2.3.28}$$

Since the integrand is a $(2m-2)$-degree polynomial in ξ, the integral can be computed exactly using the Gauss-Lobatto integration quadrature (2.3.13) with $k = m$, yielding

$$\Psi_{ij} = \left(\frac{\mathrm{d}\psi_i}{\mathrm{d}\xi}\frac{\mathrm{d}\psi_j}{\mathrm{d}\xi}\right)_{\xi=-1} \times w_1$$
$$+ \sum_{p=2}^{m}\left[\left(\frac{\mathrm{d}\psi_i}{\mathrm{d}\xi}\frac{\mathrm{d}\psi_j}{\mathrm{d}\xi}\right)_{\xi=z_p} \times w_p\right]$$
$$+ \left(\frac{\mathrm{d}\psi_i}{\mathrm{d}\xi}\frac{\mathrm{d}\psi_j}{\mathrm{d}\xi}\right)_{\xi=1} \times w_{m+1}. \tag{2.3.29}$$

In terms of the differentiation matrix, d_{ij}, defined in (2.1.11), the integration quadrature reads

$$\Psi_{ij} = \sum_{p=1}^{m+1} d_{ip}\,d_{jp}\,w_p, \tag{2.3.30}$$

without any approximation.

Evaluation of the node differentiation matrix

To evaluate the differentiation matrix, d_{ip}, we resort to (2.1.11) and use property (2.2.11) to find

$$d_{ip} = \frac{L_m(\xi_p)}{L_m(\xi_i)}\,\frac{1}{\xi_p - \xi_i}, \tag{2.3.31}$$

for $i \neq p$, and

$$d_{ii} = \frac{L'_m(\xi_i)}{2\,L_m(\xi_i)} = \frac{Lo_{m-1}(\xi_i)}{2\,L_m(\xi_i)}, \qquad (2.3.32)$$

for $i = p$. Because ξ_i is a root of $Lo_{m-1}(\xi)$, for $i = 2, 3, \ldots, m$,

$$d_{ii} = 0, \qquad (2.3.33)$$

for $i = 2, 3, \ldots, m$. The first and last diagonal values are found to be

$$d_{11} = \frac{Lo_{m-1}(\xi_i = -1)}{2\,L_m(\xi_i = -1)} = -\frac{1}{4}\,m\,(m+1),$$

$$(2.3.34)$$

$$d_{m+1,m+1} = \frac{Lo_{m-1}(\xi_i = 1)}{2\,L_m(\xi_i = 1)} = \frac{1}{4}\,m\,(m+1).$$

Expressions (2.3.31)–(2.3.33) provide us with a complete definition of the node differentiation matrix, which can be used to evaluate the dimensionless diffusion matrix.

As an example, we consider the case $m = 1$, involving two interpolation nodes located at $\xi_1 = -1$ and $\xi_2 = 1$. Recalling that $L_1(\xi) = \xi$, we obtain the values

$$d_{1,1} = -\frac{1}{2}, \qquad d_{1,2} = -\frac{1}{2},$$

$$(2.3.35)$$

$$d_{2,1} = \frac{1}{2}, \qquad d_{2,2} = \frac{1}{2},$$

which are consistent with the values obtained by directly differentiating (2.3.7) (Problem 2.3.4).

When $m = 2$, the interpolation nodes are located at $\xi_1 = -1$, $\xi_2 = 0$, and $\xi_3 = 1$. Recalling that $L_2(\xi) = \frac{1}{2}\,(3\,\xi^2 - 1)$, we obtain the values

$$d_{1,1} = -\frac{3}{2}, \qquad d_{1,2} = -\frac{1}{2}, \qquad d_{1,3} = \frac{1}{2},$$

$$d_{2,1} = 2, \qquad d_{2,2} = 0, \qquad d_{1,2} = -2, \qquad (2.3.36)$$

$$d_{3,1} = -\frac{1}{2}, \qquad d_{3,2} = \frac{1}{2}, \qquad d_{3,3} = \frac{3}{2},$$

which are also consistent with those obtained by directly differentiating (1.3.26) (Problem 2.3.4).

Alternatively, the node differentiation matrix can be evaluated using the general expressions (A.2.22) and (A.2.23) given in Appendix A for an arbitrary distribution of the element interpolation nodes,

$$d_{ij} = \frac{1}{\xi_j - \xi_i} \frac{(\xi_j - \xi_1) \dots (\xi_j - \xi_{j-1})(\xi_j - \xi_{j+1}) \dots (\xi_j - \xi_{m+1})}{(\xi_i - \xi_1) \dots (\xi_i - \xi_{i-1})(\xi_i - \xi_{i+1}) \dots (\xi_i - \xi_{m+1})}, \tag{2.3.37}$$

for $i \neq j$, and

$$d_{ii} = \frac{1}{\xi_i - \xi_1} + \dots + \frac{1}{\xi_i - \xi_{i-1}} + \frac{1}{\xi_i - \xi_{i+1}} + \dots + \frac{1}{\xi_i - \xi_{m+1}}, \tag{2.3.38}$$

for the diagonal components.

FSELIB code edm_lob, listed in the text, computes the dimensionless diffusion matrix Ψ based on the Lobatto quadrature shown in (2.3.30). The algorithm uses formulas (2.3.37) and (2.3.38) to evaluate the node differentiation matrix. The top and tail-end of the FSELIB function edm that evaluates this matrix from tabulated values is listed in the text. The complete function containing high-accuracy data is included in FEMLIB. This function will be used in the next section for building a spectral element code.

PROBLEMS

2.3.1 *Quadratic element interpolation functions.*

Apply (2.3.4) for $m = 2$ to recover the quadratic element interpolation functions shown in (1.3.26).

2.3.2 *Lobatto quadrature.*

Confirm by an example of your choice that the numerical result obtained by applying the $(k+1)$-point Lobatto quadrature is exact if the integrated function, $f(\xi)$, is a $(2k - 1)$-degree polynomial.

2.3.3 *Evaluation of the element mass matrix.*

Run the code emm_lob to evaluate the dimensionless mass matrix, Θ, for $m = 1, 2, 3,$ and 4, and verify that the results are consistent with those shown in Table 2.3.3, for both the consistent and the mass-lumped formulation.

2.3.4 *Evaluation of the node differentiation matrix.*

(a) Derive the components of the node differentiation matrix shown in (2.3.35), by directly differentiating the element interpolation functions.

(b) Repeat (a) for (2.3.36).

2.3.5 *Evaluation of the element diffusion matrix.*

(a) Run the code edm_lob to evaluate the dimensionless diffusion matrix, Ψ, for $m = 1, 2, 3,$ and 4, and verify that the results are consistent with those shown in Table 2.3.4.

```
%================================================
%  Code: edm_lob
%
%  Evaluation of the dimensionless
%  element diffusion matrix
%  using the node differentiation matrix
%  and the lobatto integration quadrature
%
%  The Lobatto base points and integration
%  weights are read from function "lobatto"
%
%  The node differentiation matrix is computed
%  using formulas (2.3.37) and (2.3.38)
%
%================================================

%---------------------------
% lobatto interpolation nodes
%---------------------------

m = input('Enter the order of the polynomial expansion, m: ');

%-------------------------------
% generate the lobatto nodes and
% integration weights
%-------------------------------

xi(1) = -1.0;
w(1)  = 2.0/(m*(m+1));

if(m>1)

  [tL, wL] = lobatto(m-1);

  for i=2:m
   xi(i) = tL(i-1);
   w(i)  = wL(i-1);
   end

end

xi(m+1) = 1.0;
w(m+1)  = w(1);

%------------------------------------------
% compute the node differentiation matrix
%------------------------------------------
```

Code edm_lob: Continuing ⟶

(*b*) The determinant of a matrix can be evaluated using the MATLAB function
det, and the inverse of a matrix can be evaluated using the MATLAB function
inv. Repeat (*a*), also computing the determinant and inverse of Ψ using these
MATLAB functions, and discuss the results of your computation.

```
for i=1:m+1
  for j=1:m+1

      %---
      if(i ~= j)
      %---

        ndm(i,j) = 1.0/(xi(j)-xi(i));
        for l=1:m+1
          if(l ~= j) ndm(i,j) = ndm(i,j)*(xi(j)-xi(l)); end
          if(l ~= i) ndm(i,j) = ndm(i,j)/(xi(i)-xi(l)); end
        end

      %---
      else
      %---

        ndm(i,i) = 0.0;

        for l=1:m+1
          if(i ~= l) ndm(i,i) = ndm(i,i)+1.0/(xi(i)-xi(l)); end
        end

      %---
      end
      %---
  end
end

%-----------------------------
% compute the diffusion matrix
% by the Lobatto quadrature
%-----------------------------

for i=1:m+1     % loop over interpolation nodes

    for j=1:m+1          % loop over interpolation nodes

      dm(i,j) = 0.0;

      for p=1:m+1
        dm(i,j) = dm(i,j) + ndm(i,p)*ndm(j,p)*w(p);
      end

    end

end

%-----
% done
%-----
```

Code edm_lob: ($\longrightarrow$ Continued.) Evaluation of the dimensionless diffusion matrix $\mathbf{\Psi}$ defined in (2.3.27), using the Lobatto quadrature.

$m = 1:$

$$\frac{1}{2} \begin{bmatrix} 1 & -1 \\ -1 & 1 \end{bmatrix}$$

$m = 2:$

$$\frac{1}{12} \begin{bmatrix} 14 & -16 & 2 \\ -16 & 32 & -16 \\ 2 & -16 & 14 \end{bmatrix}$$

$m = 3:$

$$\begin{bmatrix} 2.1666666 & -2.4392091 & 0.3558758 & -0.0833333 \\ -2.4392091 & 4.1666666 & -2.0833333 & 0.3558758 \\ 0.3558758 & -2.0833333 & 4.1666666 & -2.4392091 \\ -0.0833333 & 0.3558758 & -2.4392091 & 2.1666666 \end{bmatrix}$$

$m = 4:$

$$\begin{bmatrix} 3.500000 & -3.912885 & 0.533333 & -0.170448 & 0.050000 \\ -3.912885 & 6.351852 & -2.903703 & 0.635185 & -0.170448 \\ 0.533333 & -2.903703 & 4.740740 & -2.903703 & 0.533333 \\ -0.170448 & 0.635185 & -2.903703 & 6.351852 & -3.912885 \\ 0.050000 & -0.170448 & 0.533333 & -3.912885 & 3.500000 \end{bmatrix}$$

Table 2.3.4 The dimensionless diffusion matrix $\mathbf{\Psi}$ defined in (2.3.27), for polynomial orders m=1, 2, 3, and 4. Complete high-accuracy tables are included in the FSELIB function `edm`.

```
function elm_dm = edm(m)

%===========================================================
% Element diffusion matrix for an mth-order polynomial
% expansion, and m = 1-6
%===========================================================

%-----
% trap
%-----

if(m>5)
  disp('');
  disp(' edm: Chosen number of points is not available');
  break;
end

%-------
if(m==1)
%-------

elm_dm = [0.5 -0.5; -0.5 0.5];

%-----------
elseif(m==2)
%-----------

elm_dm = [
  1.16666667 -1.33333333  0.16666667;
 -1.33333333  2.66666667 -1.33333333;
  0.16666667 -1.33333333  1.16666667];

%-----------
elseif(m==3)
%-----------

elm_dm =[
  2.1666666 -2.4392091  0.3558758 -0.0833333;
 -2.4392091  4.1666666 -2.0833333  0.3558758;
  0.3558758 -2.0833333  4.1666666 -2.4392091;
 -0.0833333  0.3558758 -2.4392091  2.1666666];

......

%--
end
%--

return;
```

Function edm: FSELIB function for evaluating the dimensionless element diffusion matrix **Ψ**. The six dots toward the end of the listing denote unprinted code. Higher accuracy data is implemented in the function included in FSELIB.

2.4 Spectral code for steady diffusion

Algorithm 2.4.1 assembles the linear system $\mathbf{D} \cdot \mathbf{f} = \mathbf{b}$ for the steady diffusion problem discussed in Section 1.1. The unknown vector, $\mathbf{f}$, contains the values of the solution at all nodes, expect at the very last node where the Dirichlet boundary condition is specified,

$$
\mathbf{f} \equiv \begin{bmatrix} f_1 \\ f_2 \\ \vdots \\ f_{N_G-2} \\ f_{N_G-1} \end{bmatrix}, \tag{2.4.1}
$$

where N_G is the number of global interpolation nodes. The $(N_G-1) \times (N_G-1)$ global diffusion matrix, $\mathbf{D}$, consists of partially overlapping element diffusion matrices, as illustrated in Figure 2.1.2. Algorithms 1.1.1 for linear elements and Algorithm 1.2.1 for quadratic elements are special implementations of Algorithm 2.4.1 arising, respectively, when $N_p(l) = 1$ or 2, for all elements, $l = 1, 2, \ldots, N_E$, where $N_p(l)$ is the element polynomial order.

FSELIB function sds_sys, listed in the text, implements Algorithm 2.4.1; "sds" is an acronym for **s**teady **d**iffusion with **s**pectral elements. Code sds, listed in the text, implements the spectral element code in five modules:

1. *Data input and parameter definition:*

   For simplicity, the data are hard-coded in the main program.

2. *Element and interpolation node definition:*

   This is accomplished by the function discr_s discussed earlier in this section.

3. *System assembly:*

   This is accomplished by the function sds_sys. For programming simplicity, this function produces the full $N_G \times N_G$ global diffusion matrix, denoted as gdm in the code, and associated right-hand side, denoted as b in the code.

4. *Linear solver module:*

   For simplicity, we use the linear solver embedded in MATLAB. For this purpose, we use a MATLAB function to remove the extraneous last column and row of the global diffusion matrix, as well as the last entry of the right-hand side, so as to reduce the dimension of the linear system to the proper size, $(N_G-1) \times (N_G-1)$. Methods for solving systems of linear equations in finite element applications are discussed in Appendix C.

5. *Graphics module:*

   This is a standard xy plot.

Do $i = 1, 2, \ldots, N_G - 1$ *Initialize to zero*
     $b_i = 0.0$
     Do $j = 1, 2, \ldots, N_G - 1$
         $D_{ij} = 0.0$
     End Do
End Do

$b_1 = q_0/k$ *Neumann boundary condition at the left end*

Do $l = 1, 2, \ldots, N_E$ *Run over the elements*

     Compute the element matrices $\mathbf{A}^{(l)}$ and $\mathbf{B}^{(l)}$

     Do $i = 1, 2, \ldots, N_p(l) + 1$ *Run over the spectral element nodes*
         $i_1 = c(l, i)$
         Do $j = 1, 2, \ldots, N_p(l) + 1$ *Run over the spectral element nodes*
             $i_2 = c(l, j)$
             $D_{i_1,i_2} = D_{i_1,i_2} + A_{ij}^{(l)}$
             $b_{i_1} = b_{i_1} + (B_{ij}^{(l)} \times s_{i_2})/k$
         End Do
     End Do

End Do

$m = N_p(N_E)$ *Process the last node*

Do $i = 1, 2, \ldots, m$ *Run over the last element nodes*
     $i_1 = c(N_e, i)$
     $b_{i_1} = b_{i_1} - B_{i,m+1}^{(N_E)} \times f_L$
End Do

Algorithm 2.4.1 Assembly of the $(N_G - 1) \times (N_G - 1)$ linear system $\mathbf{D} \cdot \mathbf{f} = \mathbf{b}$ for the spectral element expansion. $N_p(l)$ is the specified polynomial order over the lth element.

```
function [gdm,b] = sds_sys (ne,xe,np,ng,c,q0,fL,k,s)

%============================================================
%
% Assembly of the linear system for one-dimensional
% steady diffusion with spectral elements (sds)
%
% gdm: global diffusion matrix
% b:   right-hand side
% c:   connectivity matrix
%
%============================================================

%-------------
% element size
%-------------

for l=1:ne
  h(l) = xe(l+1)-xe(l);
end

%-----------
% initialize
%-----------

gdm = zeros(ng,ng); b = zeros(1,ng);

b(1) = q0/k;

%-----------------------------------------------
% loop over the elements to compose gdm and b
%-----------------------------------------------

for l=1:ne

   m = np(l);

   elm_mm = 0.5*h(l)*emm(m);    % element mass matrix
   elm_dm = 2.0*edm(m)/h(l);    % element diffusion matrix

   for i=1:m+1

     i1 = c(l,i);

     for j=1:m+1
       j1 = c(l,j);
       gdm(i1,j1) = gdm(i1,j1) + elm_dm(i,j);
       b(i1) = b(i1) + elm_mm(i,j)*s(j1);
     end

   end

end

end
```

Function sds_sys: Continuing $\longrightarrow$

```
%-------------------------------------
% implement the Dirichlet condition
% at the last node
%-------------------------------------

m = np(ne);

for i=1:m
  i1 = c(ne,i);
  b(i1) = b(i1) - elm_dm(i,m+1)*fL
end

%-----
% done
%-----

return;
```

Function sds_sys: ($\longrightarrow$ Continued.) Assembly of the linear system for one-dimensional diffusion with spectral elements.

The graphics output generated by **sds** for the source function displayed in (1.1.75) is shown in Figure 2.4.1. In this calculation, three elements are used, $N_E = 3$, with polynomial expansion order $N_p(1) = 2$ for the first element, $N_p(2) = 2$ for the second element, and $N_p(3) = 4$ for the third element.

To assess whether implementing the high-order method is worthy of the additional analytical and programming effort, we compare the spectral element solution with the solution of the entry-level finite element method that employs linear elements, as discussed in Section 1.1. Table 2.4.1 lists the properly reduced value of the solution at the left end of the solution domain, $f(x = 0)$, computed with (a) the consistent spectral element method, (b) the mass-lumped spectral element method, and (c) the linear element method, which is nearly identical to the central-differencing finite difference method. In all cases, the elements are evenly spaced.

The results listed on the left part of the table show that the effect of mass lumping is insignificant for all but the crudest discretization. Comparison of the numerical solutions for a fixed number of global nodes, N_G, clearly demonstrates the superiority of the spectral element method over the linear element method. For example, the spectral element solution with sixteen global nodes, $N_G = 16$, is accurate to the fifth significant figure, whereas the finite element solution with one hundred twenty-eight global nodes, $N_G = 128$, is accurate only to the fourth significant figure.

```
%===========================================
% CODE: sds
%
% Spectral element code for steady diffusion
%===========================================

%-----------
% input data
%-----------

L=1.0; k=1.0; q0 = -1.0; fL=0.0;

ne=3; ratio=5.0;                    % number of elements, stretch ratio
np(1)=2; np(2)=2; np(3)=4;   % element polynomial order

%-----------------------
% element node generation
%-----------------------

[xe,xen,xg,c,ng] = discr_s (0,L,ne,ratio,np);

%--------------------
% specify the source
%--------------------

for i=1:ng
  s(i) = 10.0*exp(-5.0*xg(i)^2/L^2);
end

%-----------------
% element assembly
%-----------------

[gdm,b] = sds_sys (ne,xe,np,ng,c,q0,fL,k,s);

%--------------
% linear solver
%--------------

gdm(:,ng) = [];   % remove the last (ng) column
gdm(ng,:) = [];   % remove the last (ng) row
b(:,ng) = [];     % remove the last (ng) element
f = b/gdm';       % solve the linear system
f = [f fL];       % include the value at the right end

%-----
% plot
%-----

plot(xg, f,'x-'); hold on;
ye= zeros(ne+1,1); plot(xe,ye,'o');
xlabel('x'); ylabel('f');
```

Code sds: Spectral element code for steady one-dimensional diffusion.

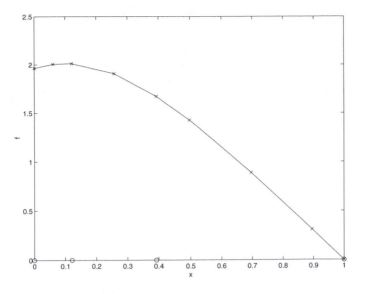

Figure 2.4.1 Graphics output of the FSELIB spectral element code `sds` for steady diffusion in the presence of the source displayed in (1.1.75). In this computation, three elements are used, $N_E = 3$, with polynomial order expansions $N_p(1) = 2$, $N_p(2) = 2$, and $N_p(3) = 4$. The crosses on the graph denote the computed nodal values at the interpolation nodes, and the circles on the x axis mark the location of the element end-nodes.

| Spectral: | | | | | Linear uniform: | |
N_G	N_E	N_p	$f(0)$ consistent	$f(0)$ lumped	$N_E = N_G$	$f(0)$
3	2	1	1.8024	2.2163	2	1.8024
5	2	2	1.9512	1.9512	4	1.9140
7	2	3	1.9643	$\leftarrow$	6	1.9412
9	2	4	1.9638	$\leftarrow$	8	1.9510
11	2	5	1.9639	$\leftarrow$	10	1.9556
16	3	5	1.9639	$\leftarrow$	15	1.9602
					32	1.9632
					64	1.9637
					128	1.9638

Table 2.4.1 Accuracy and convergence of the spectral element method compared to the finite element method with linear elements.

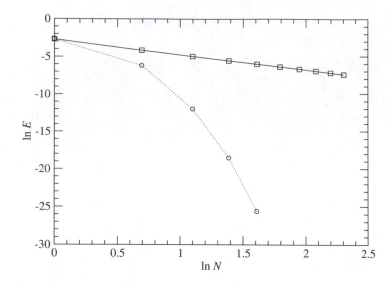

Figure 2.4.2 Error in the left-end value for a model problem, plotted against
the available degrees of freedom, N, for linear elements (solid line), and
Lobatto nodes (circles). The slope of the solid line corresponding to the
linear elements tends to the value of -2, and the numerical error behaves
like $1/N^2$. The slope of the dotted line tracing the results of the spectral
code continues to decrease, indicating that the error decreases at a rate
that is faster than any power of $1/N$.

Spectral accuracy

The performance of the spectral element method can be further quantified by
considering steady diffusion in the interval $[0, L]$, where the source term is given
by $s(x) = -s_0 \exp(x/L)$, with s_0 being a constant. The boundary conditions,
$q_0 = -L s_0$ and $f_L = f_0 e$, are designed so that the exact solution is the
exponential function, $f(x) = f_0 \exp(x/L)$, where $f_0 = s_o L^2/k$.

Figure 2.4.2 shows a graph of the relative error in the left-end value, defined
as $E \equiv |f(0)/f_0 - 1.0|$, plotted against the available degrees of freedom, N, on
a log-log scale. In the case of the finite element method, N is the number of
evenly spaced linear elements, N_E; in the case of the spectral element method,
N is the expansion order, m, for the one-element solution, $N_E = 1$. As N is
increased, the slope of the solid line corresponding to the linear elements tends
to the value of -2, indicating that the numerical error behaves like $1/N^2$. On
the other hand, the slope of the dotted line tracing the results of the spectral
element code continues to decrease, indicating that the error decreases at a rate
that is faster than any power of $1/N$, and is thus classified as spectral. This
behavior is typical of a p-type convergence.

As a second case study, we consider the solution of the one-dimensional Helmholtz equation

$$\frac{d^2 f}{dx^2} + \left(\frac{\pi}{L}\right)^2 f = 0, \tag{2.4.2}$$

over the interval $[0, L]$, subject to the boundary conditions $(df/dx)_{x=0} = f'_0$ and $f(x = L) = f_L$, where f_L and f'_0 are given constants. The exact solution for $f_L = 0$ is a sinusoid, $f = A \sin(\pi x/L)$, where $A = f'_0 L/\pi$. Figure 2.4.3 shows a graph of the error in the left-end value, $E \equiv |f(0)/A|$, plotted against the available degrees of freedom, N, on a log-log scale. In the case of the finite element method, N is the number of global nodes with evenly spaced quadratic elements, $N_G = 2N_E + 1$. In the case of the spectral element method, N is the number of global nodes for the two-element solution, $N_E = 2$. As N is increased, the slope of the solid line corresponding to the quadratic elements tends to the value of -4, indicating that the numerical error behaves like $1/N^4$. On the other hand, the slope of the dotted line tracing the results of the spectral element code continues to decrease, revealing a spectral convergence. It should be emphasized that, for the reasons discussed in Section 2.1.3, the results shown in Figure 2.4.3 are indifferent to the precise location of the interior nodes in the quadratic expansion as well as in the high-order element expansion. Thus, evenly spaced element nodes would produce exactly the same answer.

As a third and more stringent test, we consider steady diffusion in the interval $[0, L]$ in the presence of the source function

$$s(x) = 200 \, s_0 \, \frac{1 - 75 \, \hat{x}^2}{(1 + 25 \, \hat{x}^2)^3}, \tag{2.4.3}$$

where s_0 is a constant, and $\hat{x} = 2 \, (x/L) - 1$ is a dimensionless parameter varying between -1 and 1. The boundary conditions, $q_0 = -100 \, L \, s_0/26^2$ and $f_L = 26 \, A$, are designed so that the exact solution is the Runge function,

$$f(x) = \frac{A}{1 + 25 \, \hat{x}^2}, \tag{2.4.4}$$

where $A = s_0 L^2/k$. Figure 2.4.4 shows numerical solutions with two elements of equal size, $N_E = 2$, and an increasing number of (a) evenly spaced, and (b) spectral nodes. The exact solution is represented by the smooth line in both graphs. The numerical solution for evenly spaced nodes was generated by the FSELIB code sde (not listed in the text). The computations with the evenly spaced nodes are haunted by oscillations and fail to converge as the polynomial order is raised from four (circles) to eight (crosses). The computations with the spectral nodes behave much better, exhibiting a uniform improvement as the polynomial order is raised from three (circles) to six (triangles).

2.4.1 Node condensation

In Section 1.3, we demonstrated that the unknowns corresponding to the interior nodes of quadratic elements can be eliminated in favor of those correspond-

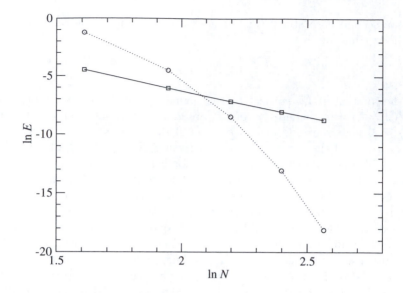

Figure 2.4.3 Error in the left-end value for a model problem, plotted against
    the available degrees of freedom, N, for quadratic elements (solid line),
    and Lobatto nodes (circles). The slope of the solid line corresponding
    to the quadratic elements is nearly equal to -4, and the numerical error
    behaves like $1/N^4$. The slope of the dotted line tracing the results of
    the spectral element code continues to decrease, indicating that the error
    decreases at a rate that is faster than any power of $1/N$.

ing to the element end-nodes by the process of *node condensation*. As a result,
the size of the final linear system is reduced nearly by a factor of two. Because
the condensed system is tridiagonal, the solution can be found efficiently using
the standard Thomas algorithm.

Node condensation in the spectral element method can be implemented at
the element level by a slight modification of the method of Gauss-Jordan reduction used for solving systems of linear equations, as discussed in Section C.1.6
of Appendix C. Consider the contribution of the lth element to the Galerkin
equations, expressed by the constituent linear system

$$\sum_{j=1}^{m+1} A_{ij}^{(l)} f_j^{(l)} = b_i^{(l)}, \qquad (2.4.5)$$

where $i = 1, 2, \ldots, m + 1$, m is the order of the element expansion, $f_j^{(l)}$ are the
unknown nodal values of the solution, and $A_{ij}^{(l)}$ is the element diffusion matrix
represented by the shaded blocks in Figure 2.1.2. The right-hand side of (2.4.5)
is given by

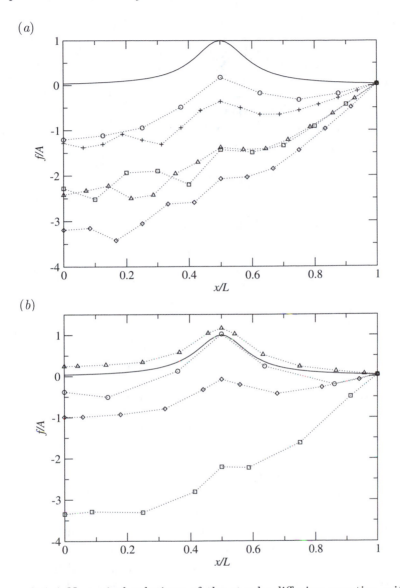

(a)

(b)

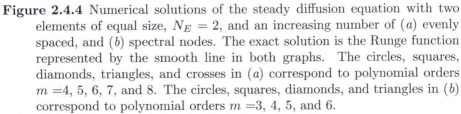

Figure 2.4.4 Numerical solutions of the steady diffusion equation with two elements of equal size, $N_E = 2$, and an increasing number of (a) evenly spaced, and (b) spectral nodes. The exact solution is the Runge function represented by the smooth line in both graphs. The circles, squares, diamonds, triangles, and crosses in (a) correspond to polynomial orders $m = 4$, 5, 6, 7, and 8. The circles, squares, diamonds, and triangles in (b) correspond to polynomial orders $m = 3$, 4, 5, and 6.

$$b_i^{(l)} \equiv \frac{1}{k} \sum_{j=1}^{m+1} B_{ij}^{(l)} \, s_j^{(l)}, \qquad (2.4.6)$$

where $s_j^{(l)}$ are the known nodal values of the source. To implement node condensation, we transform the dense system (2.4.5) into the bordered diagonal system

$$\sum_{j=1}^{m+1} \mathcal{C}_{ij}^{(l)} \, f_j^{(l)} = \beta_i^{(l)}, \qquad (2.4.7)$$

where the new coefficient matrix has the form

$$\mathcal{C}^{(l)} = \begin{bmatrix} \mathcal{C}_{11}^{(l)} & 0 & 0 & \cdots & 0 & 0 & \mathcal{C}_{1,m+1}^{(l)} \\ \mathcal{C}_{21}^{(l)} & \mathcal{C}_{22}^{(l)} & 0 & \cdots & 0 & 0 & \mathcal{C}_{2,m+1}^{(l)} \\ \mathcal{C}_{31}^{(l)} & 0 & \mathcal{C}_{33}^{(l)} & \cdots & 0 & 0 & \mathcal{C}_{3,m+1}^{(l)} \\ \cdots & \cdots & \cdots & \cdots & \cdots & \cdots & \cdots \\ \mathcal{C}_{m-1,1}^{(l)} & 0 & 0 & \cdots & \mathcal{C}_{m-1,m-1}^{(l)} & 0 & \mathcal{C}_{m-1,m+1}^{(l)} \\ \mathcal{C}_{m,1}^{(l)} & 0 & 0 & \cdots & 0 & \mathcal{C}_{m,m}^{(l)} & \mathcal{C}_{m,m+1}^{(l)} \\ \mathcal{C}_{m+1,1}^{(l)} & 0 & 0 & \cdots & 0 & 0 & \mathcal{C}_{m+1,m+1}^{(l)} \end{bmatrix}. \qquad (2.4.8)$$

Note that all elements, except those along the diagonal and down the first and last columns, are zero. The redeeming feature of this reduction is that, because the interior nodes are decoupled from the end-nodes, they can be removed from the Galerkin system, yielding a condensed system where the end-nodes only appear.

To reduce the element equations into the form shown in (2.4.7), we multiply the second equation by the ratio $r_{12} \equiv A_{12}^{(l)}/A_{22}^{(l)}$, and subtract the outcome from the first equation to eliminate $f_2^{(l)}$. Repeating this calculation, we eliminate $f_2^{(l)}$ from the third, fourth, fifth, and all subsequent equations. In the next cycle, we eliminate $f_3^{(l)}$ from the first, second, fourth, fifth, and all subsequent equations. The process continues until we have eliminated $f_m^{(l)}$ from the first, second, third, all the way up to the $m-1$ equation, and then from the $m+1$ equation. The condensed global diffusion matrix consists of overlapping 2×2 condensed element matrices,

$$\widehat{\mathbf{A}}^{(l)} = \begin{bmatrix} \mathcal{C}_{11}^{(l)} & \mathcal{C}_{1,m+1}^{(l)} \\ \mathcal{C}_{m+1,1}^{(l)} & \mathcal{C}_{m+1,m+1}^{(l)} \end{bmatrix} = \frac{1}{h_l} \begin{bmatrix} 1 & -1 \\ -1 & 1 \end{bmatrix}, \qquad (2.4.9)$$

which is precisely the diffusion matrix with linear elements shown in (1.1.62). FSELIB function sdsc_sys, listed in the text, implements the algorithm to assemble the condensed $N_E \times N_E$ system of equations; "sdsc" is an acronym for steady diffusion with spectral elements with the condensed formulation.

```
function [gdm,b] = sdsc_sys (ne,xe,np,c,q0,fL,k,s)

%================================================================
%
% Assembly of the condensed linear system for one-dimensional
% steady diffusion with spectral elements (sds)
%
% gdm: global diffusion matrix
% b:   right-hand side
% c:   connectivity matrix
%
%================================================================

%-------------
% element size
%-------------

for l=1:ne
  h(l) = xe(l+1)-xe(l);
end

%----------
% initialize
%----------

gdm = zeros(ne+1,ne+1); b = zeros(1,ne+1);

b(1) = q0/k;

%----------------------
% loop over the elements
%----------------------

for l=1:ne      % closes at echidna

  m = np(l);    % element polynomial expansion order

  elm_dm = 2.0*edm(m)/h(l);          % element diffusion matrix
  elm_mm = 0.5*h(l)*emm(m);          % element mass matrix

  % Compute the element-equations right-hand side

  for i=1:m+1

    belm(i) = 0.0;
    for j=1:m+1
     j1 = c(l,j)
     belm(i) = belm(i) + elm_mm(i,j)*s(j1);
    end
    belm(i) = belm(i)/k;

  end
```

Function sdsc_sys: Continuing $\longrightarrow$

```
%-----------
   if(m > 1)        % Gauss-Jordan reduction; closes at squirrel
%-----------

   for i=2:m       % loop over interior nodes
       for j=1:m+1   % loop over all element nodes
%------
        if(i ~= j)    % skip the self node

           rji = elm_dm(j,i)/elm_dm(i,i);
           for p=1:m+1
            elm_dm(j,p) = elm_dm(j,p) - rji*elm_dm(i,p);
           end
           belm(j) = belm(j) - rji*belm(i);

        end
%------

    end
    end

%-----------
    end     % squirrel
%-----------

    % Assign element equations to global equations
    % The first and last element equations are assigned
    % to the 1 and 1+1 global equations of the condensed system

    gdm(1,  1)   = gdm(1,1)       + elm_dm(1,  1);
    gdm(1,  1+1) = gdm(1, 1+1)    + elm_dm(1,  m+1);
    gdm(1+1,1)   = gdm(1+1,1)     + elm_dm(m+1,1);
    gdm(1+1,1+1) = gdm(1+1, 1+1)  + elm_dm(m+1,m+1);

    b(1)   = b(1)   + belm(1);
    b(1+1) = b(1+1) + belm(m+1);

end     % echidna

%-----------
% last node
%-----------

 b(ne) = b(ne) - elm_dm(1,m+1)*fL

%-----
% done
%-----

return;
```

Function sdsc_sys: ($\longrightarrow$ Continued.) Assembly of the condensed linear system for one-dimensional diffusion with spectral elements.

```
%===========================================================
%
% CODE: sdsc
%
% Steady one-dimensional diffusion with spectral elements
% using the condensed formulation where only the end-nodes
% appear in the final system of equations
%===========================================================

%-----------
% input data
%-----------

L=1.0; k=1.0; q0=-1.0; fL=0.0;

ne=3; ratio=1.0;                % number of elements, stretch ratio

np(1)=3; np(2)=3; np(3)=2;   % element expansion order

%------------------------
% element node generation
%------------------------

[xe,xen,xg,c,ng] = discr_s (0,L,ne,ratio,np);

%-------------------
% specify the source
%-------------------

for i=1:ng
  s(i) = 10.0*exp(-5.0*xg(i)^2/L^2);
end

%-----------------
% element assembly
%-----------------

[gdm,b] = sdsc_sys (ne,xe,np,c,q0,fL,k,s);
```

Code sdsc: Continuing $\longrightarrow$

Code `sdsc`, listed in the text, implements the spectral element code. For simplicity, the condensed system is solved using the MATLAB linear solver instead of Thomas's algorithm. The results are identical with those obtained with the non-condensed code `sds`.

PROBLEM

2.4.1 *Codes* `sds` *and* `sdsc`.

(*a*) Execute code `sds` with the same boundary conditions but a different source

```
%---------------
% linear solver
%---------------

gdm(:,ne+1) = [];   % remove the last (ne+1) column

gdm(ne+1,:) = [];   % remove the last (ne+1) row

b(:,ne+1) = [];     % remove the last (ne+1) node

f = b/gdm';         % solve the linear system

f = [f fL];         % add the value at the right end

%-----
% plot
%-----

plot(xe, f,'x-'); hold on;

ye= zeros(ne+1,1); plot(xe,ye,'o');

xlabel('x'); ylabel('f');

%-----
% done
%-----
```

Code sdsc: ($\longrightarrow$ Continued.)   Spectral element code for steady one-dimensional diffusion with the condensed formulation.

function using a number of elements and element nodes of your choice, and discuss the results of your computations.

(*a*) Repeat (*b*) for the condensed code sdsc, and verify that the results are identical to those produced by sds.

(*c*) Repeat (*a*), but also prepare the counterpart of Table 2.4.1 and discuss the accuracy of the spectral element solution as compared to the finite element solution with linear elements.

2.5   Spectral code for unsteady diffusion

To illustrate the implementation of the spectral element method for time-dependent problems, we consider unsteady heat conduction in a rod, as discussed in Section 1.4. The distribution of the temperature along the rod, $f(x,t)$, is governed by equation (1.4.2), repeated here for convenience,

$$\frac{\partial f}{\partial t} = \kappa \frac{\partial^2 f}{\partial x^2} + \frac{s(x,t)}{\rho \, c_p}. \tag{2.5.1}$$

The solution is to be found subject to the Neumann boundary condition (1.1.5) and the Dirichlet boundary condition (1.1.6), repeated here for convenience,

$$q_0 \equiv -k \left(\frac{\partial f}{\partial x} \right)_{x=0}, \tag{2.5.2}$$

and

$$f(x = L) \equiv f_L. \tag{2.5.3}$$

For simplicity, we assume that the left-end flux, q_0, the right-end temperature, f_L, and the source term, $s(x)$, are all constant, independent of time. Since the temperature is prescribed at the right end of the rod, the Galerkin projection produces a system of $N_G - 1$ ordinary differential equations for the temperatures at all but the last node,

$$\mathbf{M} \cdot \frac{d\mathbf{f}}{dt} + \kappa \, \mathbf{D} \cdot \mathbf{f} = \kappa \, \mathbf{b}, \tag{2.5.4}$$

where N_G is the number of global nodes. FSELIB function gdmm, listed in the text, generates the $N_G \times N_G$ global diffusion matrix, $\mathbf{D}$, and the associated global mass matrix, $\mathbf{M}$, for an arbitrary spectral element discretization.

2.5.1 Crank-Nicolson discretization

The Crank-Nicolson discretization of the differential equation (2.5.4) produces a linear system of equations for the vector $\mathbf{f}^{n+1}$ containing the unknown nodal values at the $n+1$ time level,

$$\mathbf{C} \cdot \mathbf{f}^{n+1} = \mathbf{r}, \tag{2.5.5}$$

as discussed in Section 1.4. The coefficient matrix on the left-hand side is given by

$$\mathbf{C} = \mathbf{M} + \frac{1}{2} \kappa \, \Delta t \, \mathbf{D}, \tag{2.5.6}$$

and the right-hand side is given by

$$\mathbf{r} = \left[\mathbf{M} - \frac{1}{2} \kappa \, \Delta t \, \mathbf{D} \right] \cdot \mathbf{f}^n + \kappa \, \Delta t \, \mathbf{b}, \tag{2.5.7}$$

where Δt is the time step, and the vector $\mathbf{f}^n$ contains the known nodal values at the nth time level.

FSELIB code uds_cn.m, listed in the text, implements the spectral element method according to the following steps:

```
function [gdm,gmm] = gdmm(ne,xe,np,ng,c)

%=================================================
%
% Assembly of the global diffusion matrix (gdm)
% and gobal mass matrix (gmm)
%
% ne:      number of elements
% np(l):   polynomial order over the lth element
% ng:      number of distinct global nodes
% c:       connectivity matrix
%
%=================================================

%--------------
% element sizes
%--------------

for l=1:ne
  h(l) = xe(l+1)-xe(l);
end

%-----------
% initialize
%-----------

gdm = zeros(ng,ng);
gmm = zeros(ng,ng);

%----------------------
% loop over the elements
%----------------------

for l=1:ne

    m = np(l);
    elm_dm = 2.0*edm(m)/h(l);          % element diffusion matrix
    elm_mm = 0.5*h(l)*emm(m);          % element mass matrix

    for ip=1:m+1
       i1 = c(l,ip);
       for jp=1:m+1
        i2 = c(l,jp);
        gdm(i1,i2) = gdm(i1,i2) + elm_dm(ip,jp);
        gmm(i1,i2) = gmm(i1,i2) + elm_mm(ip,jp);
       end
    end

end

return;
```

Function gdmm: Assembly of the global diffusion and mass matrices for one-dimensional unsteady diffusion with the spectral element method.

```
%=================================================
% CODE uds_cn
%
% Spectral-element code for unsteady diffusion
% with the Crank-Nicolson method
%
% SYMBOLS:
%
% ne:   number of elements
% np(l):  polynomial order over the lth element
% ng:   number of distinct global nodes
% c:    connectivity matrix
%=================================================

%-----------
% input data
%-----------

L=1.0;                      % length of the rod
rho = 1.0; cp=1.0; k=1.0;   % physical properties
q0 = -1.0; fL=0.0;          % boundary conditions
ne=3; ratio=5.0;            % number of elements, stretch ratio
np(1)=4; np(2)=2; np(3)=4;  % polynomial element order
dt = 0.01; nsteps = 100;    % time stepping
nplot=10;                   % will display after nplot steps

%--------
% prepare
%--------

kappa = k/(rho*cp);         % thermal diffusivity

%------------------
% initial condition
%------------------

for l=1:ng
   f(1,l) = fL;
end

%------------------------
% element node generation
%------------------------

[xe,xen,xg,c,ng] = discr_s (0,L,ne,ratio,np);

%------------------------
% source (time independent)
%------------------------

for l=1:ng
   s(i) = 10.0*exp(-5.0*xg(i)^2/L^2);
end
```

Code uds_cn: Continuing $\longrightarrow$

```
%-----
% plot
%-----

plot(xg, f,'-x'); hold on;        %  spectral nodes
xlabel('x'); ylabel('f');
ye= zeros(ne+1,1); plot(xe,ye,'o');   % element end-nodes

%----------------------------------
% global diffusion and mass matrices
%----------------------------------

[gdm,gmm] = gdmm(ne,xe,np,ng,c);

%---------------------------------
% generate the vector b on the rhs
%---------------------------------

b = zeros(1,ng-1);

b(1) = q0/k;

for ip=1:ng-1
  for jp=1:ng
    b(ip) = b(ip) + gmm(ip,jp)*s(jp);
  end
end

m = np(ne);

for i=1:m
  i1 = c(ne,i);
  b(i1) = b(i1) - gdm(i1,ng)*fL;
end

%--------------------------------------------
% generate the matrix on the left-hand side
%--------------------------------------------

lhs = zeros(ng-1,ng-1);

for ip=1:ng-1
  for jp=1:ng-1
    lhs(ip,jp) = gmm(ip,jp) + 0.5*dt*kappa*gdm(ip,jp);
  end
end

%--------------------
% begin time stepping
%--------------------

icount = 0;    % step counter
```

Code uds_cn: $\longrightarrow$ Continuing $\longrightarrow$

```
for irun=1:nsteps

%----------------------------
% generate the right-hand side
%----------------------------

rhs = zeros(1,ng-1);

for ip=1:ng-1

  rhs(ip) = dt*kappa*b(ip);

  for jp=1:ng
    rhs(ip) = rhs(ip) + (gmm(ip,jp) ...
      ...
       - 0.5*dt*kappa*gdm(ip,jp))*f(jp);
  end

end

%--------------
% linear solver
%--------------

sol = rhs/lhs';      % solve the (ng-1)x(ng-1) linear system

f = [sol fL];        % add the value at the right end

%---------
% plotting
%---------

if(icount==nplot)
 plot(xg, f,'-o'); icount = 0;
end

icount = icount+1;
end

%--------------------
% end of time stepping
%--------------------

%-----
% done
%-----
```

Code uds_cn: ($\longrightarrow$ Continued.) Spectral-element code for one-dimensional unsteady diffusion with the Crank-Nicolson method.

1. *Data input and parameter definition.*

2. *Element and node definition:*

   This is accomplished by the function `discr_s`, as discussed in Section 2.2.

3. *Specification of the initial temperature distribution, and definition of the source distribution along the rod.*

4. *Computation of the global diffusion and mass matrices:*

   This is accomplished by the function `gdmm` listed previously in this section.

5. *Compilation of the linear system:*

   This step involves the computation of the coefficient matrix $\mathbf{C}$ defined in (2.5.6), denoted as "lhs" in the code, and the computation of the vector $\mathbf{b}$ on the right-hand side of (2.5.7). Because vector $\mathbf{b}$ is time-independent, it is evaluated only once, outside the time-stepping loop. More generally, this vector will be updated inside the time-stepping loop.

6. *Initiation of time stepping.*

7. *Computation of the right-hand side of the linear system (2.5.5).*

8. *Linear solver:*

   For simplicity, we use the linear solver embedded in MATLAB.

9. *Graphics module:*

   A graph of the transient solution is produced every $Nplot$ steps, where the parameter $Nplot$ is set at the code input.

10. *Return to Step 8 or terminate the time stepping.*

Figure 2.5.1(a) shows the graphics output generated by the code. The right-end Dirichlet condition specifies $f_L = 0$, the initial condition specifies $f(x, t = 0) = 0$, and other parameters are defined in the code. The element end-nodes are marked with circles, and the interpolation nodes are marked with x along the x axis. As time progresses, the transient solution smoothly tends to the steady-state solution discussed earlier in this chapter, represented by the x symbols on the uppermost graph.

2.5.2   Forward Euler discretization

To demonstrate the importance of numerical stability, we consider time discretization according to the forward Euler method. Working in the familiar way, we derive the linear system

$$\mathbf{M} \cdot \mathbf{f}^{n+1} = \mathbf{r}, \tag{2.5.8}$$

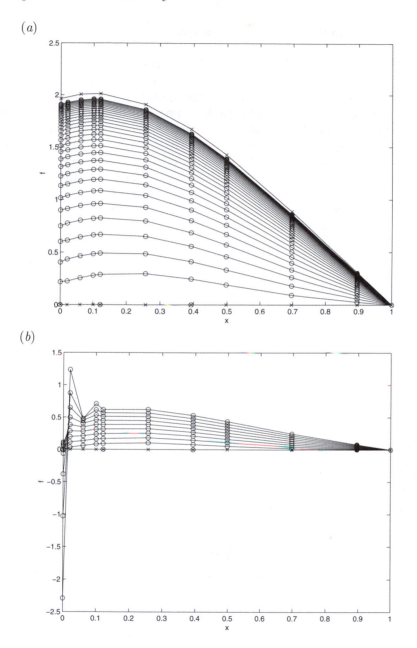

Figure 2.5.1 Graphics output of the FSELIB spectral element code (*a*) `uds_cn` implementing the Crank-Nicolson method, and (*b*) `uds_fe` implementing the forward Euler method, for the source function shown in (1.1.75).

where the vector on the right-hand side is given by

$$\mathbf{r} = \left[\mathbf{M} - \kappa\,\Delta t\,\mathbf{D}\right] \cdot \mathbf{f}^n + \kappa\,\Delta t\,\mathbf{b}. \tag{2.5.9}$$

FSELIB code `uds_fe.m` (not listed in the text) implements the spectral element method. Figure 2.5.1(b) shows the graphics output generated by the code for conditions that are identical to those listed in the text for the Crank-Nicolson code `uds_cn`, except that the time step is $\Delta t = 0.000167$. The developing numerical instability in spite of the smallness of the time step underscores the significance of the implicit time discretization underlying the Crank-Nicolson method.

In the calculation shown in Figure 2.5.1(b), the solution domain is divided into ten intervals separated by the spectral nodes, and the size of the smallest interval is $h \simeq 0.02$. The stability criterion (1.4.47) imposes the approximate restriction

$$\Delta t < \frac{h^2}{2\kappa} \simeq 0.000066, \tag{2.5.10}$$

which is consistent with the results of the numerical simulation.

PROBLEMS

2.5.1 *Significance of mass lumping.*

Modify the FSELIB code `uds_cn` to implement mass lumping, and discuss the significance of mass lumping by comparison with the consistent formulation.

2.5.2 *Forward Euler method.*

Run the FSELIB code `uds_fe` for several time step sizes, Δt, and discuss the stability of the solution with regard to the theoretical stability criterion shown in (1.4.47).

2.5.3 *Node condensation for unsteady diffusion.*

Discuss whether it is possible to implement node condensation in the case of unsteady diffusion.

2.6 Modal expansion

The mth-order polynomial expansion over the lth element can be expressed in terms of a new set of $m + 1$ basis functions, $\zeta_i(\xi)$, with associated expansion coefficients, c_i, in the modal form

$$f(\xi) = c_1^{(l)}\,\zeta_1(\xi) + \sum_{i=2}^{m} c_i^{(l)}\,\zeta_i(\xi) + c_{m+1}^{(l)}\,\zeta_{m+1}(\xi), \tag{2.6.1}$$

where

$$\zeta_1(\xi) = \frac{1-\xi}{2}, \qquad \zeta_{m+1}(\xi) = \frac{1+\xi}{2}, \qquad (2.6.2)$$

are the first- and last-node *linear* interpolation functions. The interior or bubble modes, $\zeta_i(\xi)$, are at most mth-degree polynomials anchored to zero at the end-nodes,

$$\zeta_i(\xi = \pm 1) = 0, \qquad (2.6.3)$$

for $i = 2, 3, \ldots, m$. Evaluating (2.6.1) at $\xi = \pm 1$ and using (2.6.3), we find

$$c_1^{(l)} = f_1^{(l)}, \qquad c_{m+1}^{(l)} = f_{m+1}^{(l)}. \qquad (2.6.4)$$

To ensure the C^0 continuity of the solution, the end-node values of $f_1^{(l)}$ and $f_{m+1}^{(l)}$ must be shared by neighboring elements at the corresponding positions.

To guarantee the satisfaction of the bubble-mode conditions (2.6.3), we set

$$\zeta_i(\xi) = (1 - \xi^2)\, Q_{i-2}(\xi), \qquad (2.6.5)$$

for $i = 2, 3, \ldots, m$, where the functions $Q_{i-2}(\xi)$ are at most $(m-2)$-degree polynomials. For example, when $m = 2$, corresponding to the quadratic approximation, we obtain a single interior node, $\zeta_2 = (1 - \xi^2)\, Q_0(\xi)$. Setting $Q_0(\xi) = c$, and choosing for convenience the value of the constant $c = 1$, we recover the quadratic modal expansion (1.3.51).

Relation to the nodal expansion

The modal functions can be expressed in terms of a set of $m + 1$ nodal interpolation functions constructed with reference to an arbitrary set of element nodes, ξ_j, subject to the restrictions $\xi_1 = -1$ and $\xi_{m+1} = 1$. An example is the set of the Lobatto nodal interpolation functions displayed in (2.3.4).

Viewing the nodal interpolation functions as Lagrange polynomials, as discussed in Appendix A, we write

$$\zeta_i(\xi) = \sum_{j=1}^{m+1} \zeta_i(\xi_j)\, \psi_j(\xi). \qquad (2.6.6)$$

For example, when $i = 1$, we find

$$\zeta_1(\xi) = \frac{1}{2} \sum_{j=1}^{m+1} (1 - \xi_j)\, \psi_j(\xi), \qquad (2.6.7)$$

and when $i = m + 1$, we find

$$\zeta_{m+1}(\xi) = \frac{1}{2} \sum_{j=1}^{m+1} (1 + \xi_j)\, \psi_j(\xi). \qquad (2.6.8)$$

The modal function vector may then be related to the nodal function vector by the linear transformation

$$\zeta = \mathbf{V} \cdot \psi, \tag{2.6.9}$$

where

$$\mathbf{V} = \begin{bmatrix} 1 & \zeta_1(\xi_2) & \cdots & \zeta_1(\xi_m) & 0 \\ 0 & \zeta_2(\xi_2) & \cdots & \zeta_2(\xi_m) & 0 \\ \cdots & \cdots & \cdots & \cdots & \cdots \\ 0 & \zeta_m(\xi_2) & \cdots & \zeta_m(\xi_m) & 0 \\ 0 & \zeta_{m+1}(\xi_2) & \cdots & \zeta_{m+1}(\xi_m) & 1 \end{bmatrix}, \tag{2.6.10}$$

is a generalized Vandermonde matrix. Note that the first and last columns of $\mathbf{V}$ are filled with zeros, with the exception of the first and last diagonal elements that are equal to unity.

For $m = 1$, corresponding to the linear expansion, we find that $\mathbf{V}$ is the 2×2 identity matrix. For $m = 2$ with $\xi_1 = -1$, $\xi_2 = 0$, and $\xi_3 = 1$, we find

$$\mathbf{V} = \begin{bmatrix} 1 & \frac{1}{2} & 0 \\ 0 & 1 & 0 \\ 0 & \frac{1}{2} & 1 \end{bmatrix}. \tag{2.6.11}$$

The element diffusion matrix associated with the modal expansion is defined as

$$\mathcal{A}_{ij}^{(l)} = \frac{2}{h_l} \int_{-1}^{1} \frac{d\zeta_i}{d\xi} \frac{d\zeta_j}{d\xi} \, d\xi. \tag{2.6.12}$$

Using (2.6.9), we find

$$\mathcal{A}^{(l)} = \mathbf{V} \cdot \mathbf{A}^{(l)} \cdot \mathbf{V}^T, \tag{2.6.13}$$

where $\mathbf{A}^{(l)}$ is the element diffusion matrix corresponding to a nodal expansion, and the superscript T indicates the matrix transpose. Conversely,

$$\mathbf{A}^{(l)} = \mathbf{V}^{-1} \cdot \mathcal{A}^{(l)} \cdot \mathbf{V}^{-1^T}. \tag{2.6.14}$$

The element mass matrix associated with the modal expansion is defined as

$$\mathcal{B}_{ij}^{(l)} = \frac{h_l}{2} \int_{-1}^{1} \zeta_i \, \zeta_j \, d\xi. \tag{2.6.15}$$

Using (2.6.9), we find

$$\mathcal{B}^{(l)} = \mathbf{V} \cdot \mathbf{B}^{(l)} \cdot \mathbf{V}^T, \tag{2.6.16}$$

where $\mathbf{B}^{(l)}$ is the element mass matrix of a nodal expansion. Conversely,

$$\mathbf{B}^{(l)} = \mathbf{V}^{-1} \cdot \mathcal{B}^{(l)} \cdot \mathbf{V}^{-1^T}. \tag{2.6.17}$$

These relations allow us to compute the element nodal diffusion and mass matrices from the corresponding modal matrices, and *vice versa*, in terms of the Vandermonde matrix based on the modal functions.

2.6.1 Implementation

The Galerkin finite element method for the modal expansion is implemented as discussed previously in this chapter for the nodal expansion, the only difference being that the element-node interpolation functions, $\psi_i(\xi)$, are replaced by the non-cardinal modal functions, $\zeta_i(\xi)$, and the interior nodal values, $f_i^{(l)}$, are replaced by the expansion coefficients, $c_i^{(l)}$.

To implement the method for the steady diffusion problem discussed in Section 1.1, we approximate the source term over the lth element with an mth-degree polynomial, $\mathcal{S}_m$, expressed in the familiar modal form

$$\mathcal{S}_m(\xi) = \sum_{i=1}^{m+1} d_i^{(l)}\, \zeta_i(\xi), \tag{2.6.18}$$

and compute the $m + 1$ modal coefficients, $d_i^{(l)}$, for $i = 1, 2, \ldots, m + 1$, by requiring the interpolation conditions $\mathcal{S}(\xi_j) = s(\xi_j)$, for $i = 1, 2, \ldots, m + 1$, where ξ_j is a set of element nodes. Enforcing these conditions yields the linear system

$$\mathbf{V}^T \cdot \mathbf{d}^{(l)} = \mathbf{s}^{(l)}, \tag{2.6.19}$$

where $\mathbf{V}$ is the Vandermonde matrix defined in (2.6.10). Once the finite element solution has been completed, the unknown function can be evaluated at the nodes using the counterpart of (2.6.19), yielding

$$\mathbf{f}^{(l)} = \mathbf{V}^T \cdot \mathbf{c}^{(l)}. \tag{2.6.20}$$

The lth-element equation block of the Galerkin finite element system with the modal expansion takes the form

$$\mathcal{A}^{(l)} \cdot \mathbf{c}^{(l)} = \mathcal{B}^{(l)} \cdot \mathbf{d}^{(l)}. \tag{2.6.21}$$

Recasting this equation in terms of the element matrices for the nodal expansion, we obtain

$$\mathbf{V} \cdot \mathbf{A}^{(l)} \cdot \mathbf{V}^T \cdot \mathbf{c}^{(l)} = \mathbf{V} \cdot \mathbf{B}^{(l)} \cdot \mathbf{V}^T \cdot \mathbf{d}^{(l)} = \mathbf{V} \cdot \mathbf{B}^{(l)} \cdot \mathbf{s}^{(l)}. \tag{2.6.22}$$

Combining this equation with (2.6.20), we find

$$\mathbf{A}^{(l)} \cdot \mathbf{f}^{(l)} = \mathbf{B}^{(l)} \cdot \mathbf{s}^{(l)}, \tag{2.6.23}$$

which shows that, in fact, the modal and nodal expansions are equivalent.

2.6.2 Lobatto modal expansion

A *hierarchical* modal expansion arises by specifying that the functions $Q_{i-2}(\xi)$ introduced in (2.6.5) are $(i - 2)$-degree polynomials. For reasons that will become evident in hindsight, we further identify the polynomial $Q_{i-2}(\xi)$ with the $(i-2)$-degree Lobatto polynomial $Lo_{i-2}(\xi)$ for $i = 2, 3, \ldots, m$, yielding the interior modes

$$\zeta_i(\xi) = (1 - \xi^2) \, Lo_{i-2}(\xi) \equiv Lo_i^c(\xi), \tag{2.6.24}$$

for $i = 2, 3, \ldots, m$, where Lo_i^c is a completed Lobatto polynomial [38]. Using relations (2.2.17), we find the alternative representations [61, 30]

$$\zeta_i(\xi) = -(i-1)\, i \int_{-1}^{\xi} L_{i-1}(\xi') \, \mathrm{d}\xi'$$

$$= -\frac{(i-1)\, i}{2i - 1} \left(L_i(\xi) - L_{i-2}(\xi) \right). \tag{2.6.25}$$

Explicitly, the first few modes are given by

$$\zeta_2(\xi) = 1 - \xi^2,$$
$$\zeta_3(\xi) = 3\,(1 - \xi^2)\, \xi,$$
$$\zeta_4(\xi) = \frac{3}{2}\,(1 - \xi^2)\,(5\,\xi^2 - 1), \tag{2.6.26}$$
$$\ldots$$

arising for $m \geq 2$.

Element mass matrix

Substituting in (2.6.15) the modal functions for the interior modes, $i, j = 2, 3, \ldots, m$, we find

$$\mathcal{B}_{ij}^{(l)} = \frac{h_l}{2} \int_{-1}^{1} (1 - \xi^2)^2 \, Lo_{i-2}(\xi) \, Lo_{j-2}(\xi) \, \mathrm{d}\xi. \tag{2.6.27}$$

Using property (2.2.16), we obtain

$$\mathcal{B}_{ij}^{(l)} = \frac{h_l}{2} \frac{(i-1)(j-1)\,ij}{(2i-1)(2j-1)} \int_{-1}^{1} \left(L_{i-2} - L_i \right)\left(L_{j-2} - L_j \right) \mathrm{d}\xi. \tag{2.6.28}$$

Because of the orthogonality condition (2.2.12), $\mathcal{B}_{ij}^{(l)}$ is non-zero only when $j = i$ or $j = i \pm 2$. The diagonal components are given by

$$\mathcal{B}_{ii}^{(l)} = \frac{h_l}{2} \frac{(i-1)^2 i^2}{(2i-1)^2} \int_{-1}^{1} \left(L_{i-2}^2 + L_i^2 \right) \mathrm{d}\xi$$

$$= h_l \, \frac{2\,(i-1)^2\, i^2}{(2i-3)(2i-1)(2i+1)}, \tag{2.6.29}$$

and the off-diagonal components are given by

$$\mathcal{B}^{(l)}_{i(i+2)} = -\frac{h_l}{2} \frac{(i-1)(i+1)i(i+2)}{(2i-1)(2i+3)} \int_{-1}^{1} L_i^2 \, d\xi$$

$$= -h_l \frac{(i-1)i(i+1)(i+2)}{(2i-1)(2i+1))(2i+3)}, \tag{2.6.30}$$

$$\mathcal{B}^{(l)}_{i(i-2)} = -\frac{h_l}{2} \frac{(i-1)(i-3)i(i-2)}{(2i-1)(2i-5)} \int_{-1}^{1} L_{i-2}^2 \, d\xi$$

$$= -h_l \frac{(i-3)(i-2)(i-1)i}{(2i-5)(2i-3)(2i-1)}, \tag{2.6.31}$$

for $i = 2, 3, \ldots, m$. Invoking the orthogonality of the quadratic and higher-degree Lobatto polynomials against the linear polynomials associated with the first- and last-node linear interpolation functions, we find that the $(m+1) \times (m+1)$ element-mass matrix takes the nearly tridiagonal banded form

$$\mathcal{B}^{(l)} = \begin{bmatrix} \mathcal{B}^{(l)}_{11} & \mathcal{B}^{(l)}_{12} & \mathcal{B}^{(l)}_{13} & 0 & 0 & 0 & 0 & 0 & \cdots \\ \mathcal{B}^{(l)}_{21} & \mathcal{B}^{(l)}_{22} & 0 & \mathcal{B}^{(l)}_{24} & 0 & 0 & 0 & 0 & \cdots \\ \mathcal{B}^{(l)}_{31} & 0 & \mathcal{B}^{(l)}_{33} & 0 & \mathcal{B}^{(l)}_{35} & 0 & 0 & 0 & \cdots \\ 0 & \mathcal{B}^{(l)}_{42} & 0 & \mathcal{B}^{(l)}_{44} & 0 & \mathcal{B}^{(l)}_{46} & 0 & 0 & \cdots \\ \cdots & \cdots & \cdots & \cdots & \cdots & \cdots & \cdots & \cdots & \cdots \\ 0 & 0 & 0 & 0 & 0 & 0 & 0 & 0 & \cdots \\ \mathcal{B}^{(l)}_{(m+1)1} & \mathcal{B}^{(l)}_{(m+1)2} & \mathcal{B}^{(l)}_{(m+1)3} & 0 & 0 & 0 & 0 & 0 & \cdots \end{bmatrix}$$

$$\begin{bmatrix} 0 & 0 & \cdots & 0 & 0 & 0 & \mathcal{B}^{(l)}_{1(m+1)} \\ 0 & 0 & \cdots & 0 & 0 & 0 & \mathcal{B}^{(l)}_{2(m+1)} \\ 0 & 0 & \cdots & 0 & 0 & 0 & \mathcal{B}^{(l)}_{3(m+1)} \\ 0 & 0 & \cdots & 0 & 0 & 0 & 0 \\ \cdots & \cdots & \cdots & \cdots & \cdots & \cdots & \cdots \\ 0 & 0 & \cdots & \mathcal{B}^{(l)}_{m(m-2)} & 0 & \mathcal{B}^{(l)}_{mm} & 0 \\ 0 & 0 & 0 & \cdots & 0 & 0 & \mathcal{B}^{(l)}_{(m+1)(m+1)} \end{bmatrix}. \tag{2.6.32}$$

The interior elements are computed using expressions (2.6.29)–(2.6.31), and the bordering elements are given by

$$\mathcal{B}^{(l)}_{11} = \mathcal{B}^{(l)}_{(m+1)(m+1)} = \frac{h_l}{3},$$

$$\mathcal{B}^{(l)}_{1(m+1)} = \mathcal{B}^{(l)}_{(m+1)1} = \frac{h_l}{6},$$

$$\mathcal{B}_{12}^{(l)} = \mathcal{B}_{21}^{(l)} = \frac{h_l}{2} \int_{-1}^{1} \frac{1-\xi}{2} \left(1 - \xi^2\right) \mathrm{d}\xi = \frac{h_l}{3},$$

$$\mathcal{B}_{13}^{(l)} = \mathcal{B}_{31}^{(l)} = \frac{h_l}{2} \int_{-1}^{1} \frac{1-\xi}{2} \left(1 - \xi^2\right) 3\,\xi\,\mathrm{d}\xi = -\frac{h_l}{5}, \qquad (2.6.33)$$

$$\mathcal{B}_{2(m+1)}^{(l)} = \mathcal{B}_{(m+1)2}^{(l)} = \frac{h_l}{2} \int_{-1}^{1} \frac{1+\xi}{2} \left(1 - \xi^2\right) \mathrm{d}\xi = \frac{h_l}{3},$$

$$\mathcal{B}_{3(m+1)}^{(l)} = \mathcal{B}_{(m+1)3}^{(l)} = \frac{h_l}{2} \int_{-1}^{1} \frac{1+\xi}{2} \left(1 - \xi^2\right) 3\,\xi\,\mathrm{d}\xi = \frac{h_l}{5}.$$

To evaluate the integral $\int_{-1}^{1} \frac{1-\xi}{2} \left(1 - \xi^2\right) \mathrm{d}\xi$ using MATLAB, we may issue the commands:

```
>> syms x
>> int(0.5*(1-x)*(1-x^2),x,-1,1)
```

where >> is the MATLAB command line prompt. The first command declares x as a symbolic object, and the second command computes the indefinite integral with respect to x, and then it evaluates it between the specified limits of integration, -1 and 1.

For example, when $m = 2$, corresponding to the quadratic expansion, we obtain the dense element mass matrix

$$\boldsymbol{\mathcal{B}}^{(l)} = \frac{h_l}{30} \begin{bmatrix} 10 & 10 & 5 \\ 10 & 16 & 10 \\ 5 & 10 & 10 \end{bmatrix}, \qquad (2.6.34)$$

also displayed in (1.3.59). When $m = 3$, corresponding to the cubic expansion, we find

$$\boldsymbol{\mathcal{B}}^{(l)} = \begin{bmatrix} \mathcal{B}_{11}^{(l)} & \mathcal{B}_{12}^{(l)} & \mathcal{B}_{13}^{(l)} & \mathcal{B}_{14}^{(l)} \\ \mathcal{B}_{21}^{(l)} & \mathcal{B}_{22}^{(l)} & 0 & \mathcal{B}_{24}^{(l)} \\ \mathcal{B}_{31}^{(l)} & 0 & \mathcal{B}_{33}^{(l)} & \mathcal{B}_{34}^{(l)} \\ \mathcal{B}_{41}^{(l)} & \mathcal{B}_{42}^{(l)} & \mathcal{B}_{43}^{(l)} & \mathcal{B}_{44}^{(l)} \end{bmatrix}, \qquad (2.6.35)$$

where the filled elements are non-zero. When $m = 4$, corresponding to the quartic expansion, we find

$$\boldsymbol{\mathcal{B}}^{(l)} = \begin{bmatrix} \mathcal{B}_{11}^{(l)} & \mathcal{B}_{12}^{(l)} & \mathcal{B}_{13}^{(l)} & 0 & \mathcal{B}_{15}^{(l)} \\ \mathcal{B}_{21}^{(l)} & \mathcal{B}_{22}^{(l)} & 0 & \mathcal{B}_{24}^{(l)} & \mathcal{B}_{25}^{(l)} \\ \mathcal{B}_{31}^{(l)} & 0 & \mathcal{B}_{33}^{(l)} & 0 & \mathcal{B}_{35}^{(l)} \\ 0 & \mathcal{B}_{42}^{(l)} & 0 & \mathcal{B}_{44}^{(l)} & 0 \\ \mathcal{B}_{51}^{(l)} & \mathcal{B}_{52}^{(l)} & \mathcal{B}_{53}^{(l)} & 0 & \mathcal{B}_{55}^{(l)} \end{bmatrix}, \qquad (2.6.36)$$

where the filled elements are non-zero.

Element diffusion matrix

Using (1.1.62), we readily evaluate the four corner entries of the element diffusion matrix defined in (2.6.13),

$$A_{11}^{(l)} = A_{m+1,m+1}^{(l)} = \frac{1}{h_l}, \qquad A_{1,m+1}^{(l)} = A_{m+1,1}^{(l)} = -\frac{1}{h_l}. \qquad (2.6.37)$$

According to property (2.2.11),

$$\frac{\mathrm{d}\zeta_i}{\mathrm{d}\xi} = -(i-1)\, i\, L_{i-1}(\xi), \qquad (2.6.38)$$

for $i = 2, 3, \ldots, m$, where L_{i-1} is a Legendre polynomial. Using this relation, we find

$$A_{ij}^{(l)} = \frac{2}{h_l}\,(i-1)(j-1)\,i\,j \int_{-1}^{1} L_{i-1}(\xi)\, L_{j-1}(\xi)\, \mathrm{d}\xi, \qquad (2.6.39)$$

for $i, j = 2, 3, \ldots, m$. Invoking the orthogonality condition (2.2.12), we obtain the final expression

$$A_{ij}^{(l)} = \frac{1}{h_l}\, \frac{4\,(i-1)^2\, i^2}{2i-1}\, \delta_{ij}, \qquad (2.6.40)$$

for $i, j = 2, 3, \ldots, m$. Moreover,

$$
\begin{aligned}
A_{1j}^{(l)} = A_{j1}^{(l)} &= \frac{2}{h_l} \int_{-1}^{1} \frac{\mathrm{d}\zeta_1}{\mathrm{d}\xi}\, \frac{\mathrm{d}\zeta_j}{\mathrm{d}\xi}\, \mathrm{d}\xi \\
&= \frac{j\,(j-1)}{h_l} \int_{-1}^{1} L_{j-1}(\xi)\, \mathrm{d}\xi = 0, \qquad (2.6.41)
\end{aligned}
$$

for $j = 2, 3, \ldots, m$, due to the orthogonality of the Legendre polynomials. Working in a similar fashion, we find $A_{m+1,j}^{(l)} = A_{j,m+1}^{(l)} = 0$, for $j = 2, 3, \ldots, m$.

Combining these results, we find that the $(m+1) \times (m+1)$ element diffusion matrix has the attractive nearly diagonal form

$$
\mathcal{A}^{(l)} = \frac{1}{h_l}
\begin{bmatrix}
1 & 0 & 0 & \ldots & 0 & \ldots & 0 & -1 \\
0 & \frac{16}{3} & 0 & \ldots & 0 & \ldots & 0 & 0 \\
0 & 0 & 0 & \ldots & \ldots & \ldots & 0 & 0 \\
0 & 0 & 0 & \ldots & \frac{4\,(i-1)^2\, i^2}{2i-1} & \ldots & 0 & 0 \\
\ldots & \ldots & \ldots & \ldots & \ldots & \ldots & \ldots & \ldots \\
0 & 0 & 0 & \ldots & \ldots & \ldots & 0 & 0 \\
0 & 0 & 0 & \ldots & 0 & 0 & \frac{4\,(m-1)^2\, m^2}{2m-1} & 0 \\
-1 & 0 & 0 & \ldots & 0 & 0 & 0 & 1
\end{bmatrix}.
$$

$$(2.6.42)$$

For example, when $m = 3$, we obtain the 4×4 matrix

$$\mathcal{A}^{(l)} = \frac{1}{h_l} \begin{bmatrix} 1 & 0 & 0 & -1 \\ 0 & \frac{16}{3} & 0 & 0 \\ 0 & 0 & \frac{144}{7} & 0 \\ -1 & 0 & 0 & 1 \end{bmatrix}. \tag{2.6.43}$$

Applying the Laplace expansion of the determinant with respect to the first or last column or row, we find that the element diffusion matrix is singular. If we had used instead the rescaled modes,

$$\widehat{\zeta}_i = -\left(\frac{2}{2i-1}\right)^{1/2} i\,(i-1)\,\zeta_i, \tag{2.6.44}$$

for $i = 2, 3, \ldots, m$, we would have found that the interior diagonal elements of the matrix on the right-hand side of (2.6.42) are constant and equal to two [61].

We see now that the choice of Lobatto polynomials has the significant advantage of diagonalizing the portion of the element diffusion matrix corresponding to the interior modes, but not to the end-modes. Thus, the modal expansion effectively implements node condensation.

Spectral modal expansion

The modal spectral element method arises by identifying the element interpolation nodes in (2.6.19) with the Lobatto spectral nodes discussed earlier in this chapter. To compute the coefficient vector of the source, we may set $d_1^{(l)} = s_1^{(l)}$ and $d_{m+1}^{(l)} = s_{m+1}^{(l)}$, and rearrange the jth interpolation condition, $\mathcal{S}(\xi_j) = s(\xi_j)$, for $j = 2, 3 \ldots, m$, to derive a linear system for the coefficients of the bubble modes,

$$\sum_{i=2}^{m} d_i^{(l)}\,\zeta_i(t_j) = s(t_j) - s_1^{(l)}\,\zeta_1(t_j) - s_{m+1}^{(l)}\,\zeta_{m+1}(t_j), \tag{2.6.45}$$

where t_j, for $j = 1, 2, \ldots, m-1$, are the zeros of the $(m-1)$-degree Lobatto polynomial. FSELIB code sdms (not listed in the text) implements the spectral element method, assisted by an assembly algorithm implemented in the FSELIB function sdms_sys (not listed in the text). The results are identical to those obtained previously using the Lobatto spectral nodal expansion.

PROBLEMS

2.6.1 *Modal expansion.*
Evaluate the exact expansion coefficients, c_i, for the following functions defined over the interval $-1 \leq \xi \leq 1$, and expansion orders, m: (a) $f(\xi) = 1 + 4\xi$, $m = 1$, (b) $f(\xi) = 1 + 4\xi + \xi^2$, $m = 2$, (c) $f(\xi) = 1 + 4\xi + \xi^2 + \xi^3$, $m = 3$.

2.6.2 *Element mass matrix.*

Evaluate the element mass matrix (2.6.35).

2.6.3 *Element diffusion matrix.*

Write out the 5×5 element diffusion matrix for $m = 4$, and use the MATLAB function `det` to compute its determinant.

2.6.4 *Legendre modal expansion.*

Discuss the structure of the element mass matrix for the Legendre modal expansion,

$$\zeta_i(\xi) = (1 - \xi^2)\, L_{i-2}(\xi), \tag{2.6.46}$$

for $i = 2, 3, \ldots, m$, where $L_{i-2}(\xi)$ are the Legendre polynomials discussed in Section B.5 of Appendix B.

The finite element method in two dimensions

3

Having discussed the fundamental concepts underlying the formulation and implementation of the Galerkin finite element method in one spatial dimension, we proceed to extend the methodology to two dimensions. The generalized framework pursued in this chapter will allow us to develop algorithms for solving the Laplace equation, the Poisson equation, the unsteady heat conduction equation, the convection equation, the convection–diffusion equation, the equations of linear elasticity in solid mechanics, and the equations of viscous flow in hydrodynamics. The discourse will demonstrate one of the most powerful features of the finite element method, which is the ability to readily accommodate solution domains with arbitrary and even complex geometry.

The implementation of the Galerkin finite element method in two dimensions follows the general blueprint of the one-dimensional case discussed in Chapter 1, involving the following basic modules:

1. *Formulation of the Galerkin projection using the global interpolation functions.*

   This is followed by the application of Gauss's divergence theorem to reduce the order of the highest derivatives.

2. *Domain discretization into finite elements defined by geometrical nodes, assignment of the element interpolation nodes, and computation of the node and element connectivity matrices, as required.*

   To ensure the C^0 continuity of the finite element expansion, adjacent elements must share at least some interpolation nodes. Unless a physical discontinuity or a singularity is anticipated, the solution must be common at the shared element nodes.

3. *Derivation of a system of algebraic or differential equations by substituting the finite element expansion in the Galerkin projection.*

4. *Computation of the element diffusion, mass, and advection matrices, as required.*

   These computations are done mostly numerically with the aid of integration quadratures pertinent to the selected element shape.

5. *Assembly of the global linear system in terms of the element matrices.*

   The algorithm relies on a connectivity matrix that associates element interpolation nodes to unique global nodes.

6. *Implementation of the boundary conditions.*

   This is done most conveniently by modifying the coefficient matrix and right-hand side of the linear system arising from the projection of the global interpolation functions corresponding to *all* global nodes, possibly with the addition of contour integrals implementing the Neumann boundary condition, as required.

7. *Solution of the final algebraic system.*

   For simplicity, we shall use the linear solver embedded in MATLAB. In practice, because the dimension of the linear system can be large, custom-made algorithms are employed, as discussed in Appendix C.

8. *Time stepping.*

   For time-dependent problems, the algorithm is applied for a sequence of time steps to yield the evolution of the solution from a specified initial state.

   In this chapter, the formulation and implementation of these basic procedures will be demonstrated by discussing selected modular cases of the unsteady convection–diffusion equation. Once the general methodology has been established, high-order and spectral element methods will arise as natural extensions, as discussed in Chapter 4. Applications in solid, structural, and fluid mechanics will be discussed in Chapter 5.

3.1   Convection–diffusion in two dimensions

Consider unsteady heat transport in the xy plane, in the presence of a distributed source due, for example, to a chemical reaction, as illustrated in Figure 3.1.1(a). The evolution of the temperature field, $f(x, y, t)$, is governed by the convection–diffusion equation

$$\rho\, c_p \left(\frac{\partial f}{\partial t} + u_x\, \frac{\partial f}{\partial x} + u_y\, \frac{\partial f}{\partial y} \right) = k \left(\frac{\partial^2 f}{\partial x^2} + \frac{\partial^2 f}{\partial y^2} \right) + s, \qquad (3.1.1)$$

where ρ, c_p, and k are the medium density, heat capacity, and thermal conductivity, $u_x(x, y, f, t)$ and $u_y(x, y, f, t)$ are the x and y components of the velocity, and $s(x, y, f, t)$ is the distributed source. Note that (3.1.1) is the two-dimensional counterpart of the one-dimensional rod equation (1.6.1). For simplicity, we assume that the convection velocity and source depend on x, y and t explicitly, but not implicitly through f. To signify this, we write $u_x(x, y, t)$, $u_y(x, y, t)$ and $s(x, y, t)$.

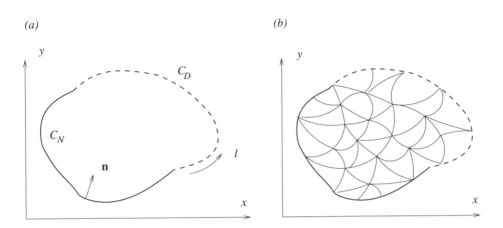

Figure 3.1.1 (*a*) Illustration of heat conduction in a plate with arbitrary geometry in the xy plane, showing the Dirichlet contour, C_D, and the Neumann contour, C_N. (*b*) Finite element discretization of the solution domain into triangular elements with curved sides.

In vector notation, the transport equation (3.1.1) assumes the compact form

$$\frac{\partial f}{\partial t} + \mathbf{u} \cdot \nabla f = \kappa \, \nabla^2 f + \frac{s}{\rho \, c_p}, \qquad (3.1.2)$$

where $\kappa \equiv k/(\rho \, c_p)$ is the thermal diffusivity,

$$\nabla f = \left(\frac{\partial f}{\partial x}, \ \frac{\partial f}{\partial y} \right) \qquad (3.1.3)$$

is the two-dimensional gradient, and

$$\nabla^2 f \equiv \nabla \cdot \nabla f = \frac{\partial^2 f}{\partial x^2} + \frac{\partial^2 f}{\partial y^2} \qquad (3.1.4)$$

is the two-dimensional Laplacian, equal to the divergence of the gradient. Thus,

$$\mathbf{u} \cdot \nabla f = u_x \, \frac{\partial f}{\partial x} + u_y \, \frac{\partial f}{\partial y}. \qquad (3.1.5)$$

Equation (3.1.2) is to be solved in a domain, D, that is enclosed by the contour C, subject to two complementary boundary conditions: the *Neumann* boundary condition that specifies the inward or outward normal derivative of the unknown function, and the *Dirichlet* boundary condition that specifies the boundary distribution of the unknown function. In the present problem, the boundary conditions prescribe:

- The heat flux along the Neumann portion of C, denoted by C_N and drawn with the solid line in Figure 3.1.1(a),

$$-k\,\mathbf{n}\cdot\nabla f \equiv -k\,\frac{\partial f}{\partial l_n} = q(l), \qquad (3.1.6)$$

where $\mathbf{n} = (n_x, n_y)$ is the unit vector normal to C pointing *into* the solution domain, $q(l)$ is a given function of arc length, l, along C, as shown in Figure 3.1.1(a), and

$$\mathbf{n}\cdot\nabla f = n_x\,\frac{\partial f}{\partial x} + n_y\,\frac{\partial f}{\partial y}. \qquad (3.1.7)$$

The notation $\partial/\partial l_n$ designates the derivative with respect to the inward distance normal to the boundary.

If $q(l) > 0$, in which case $\mathbf{n}\cdot\nabla f < 0$, heat enters the solution domain, whereas if $q(l) < 0$, in which case $\mathbf{n}\cdot\nabla f > 0$, heat escapes from the solution domain across C_N.

- The temperature distribution along the Dirichlet portion of the boundary, denoted by C_D and drawn with the broken line in Figure 3.1.1(a),

$$f = g(l), \qquad (3.1.8)$$

where $C_D = C - C_N$ is the complement of C_N, and $g(l)$ is a specified function.

Other boundary conditions, including the mixed boundary condition, also called a convection or Robin boundary condition, can be handled by similar methods.

To develop the Galerkin finite element method, we follow the general steps outlined in the introduction of this chapter, beginning with the Galerkin projection.

3.1.1   Galerkin projection

In the first step, we carry out the Galerkin projection of the governing differential equation (3.1.1), using as weighting functions the global interpolation functions, $\phi_i(x, y)$. These are defined according to the selected element types and location of the associated element interpolation nodes, as will be discussed later in this section in detail. For the moment, we assume that these functions are available.

Multiplying first the right-hand side of (3.1.1) by $\phi_i(x, y)$, integrating the product over the solution domain, D, and manipulating the integrand to form a divergence, we find

$$\iint_D \phi_i \left[k \nabla^2 f + s \right] dx\, dy = \iint_D \left[k\, \phi_i \left(\nabla \cdot \nabla f \right) + \phi_i\, s \right] dx\, dy$$

$$= \iint_D \left[k \left(\nabla \cdot (\phi_i \nabla f) - \nabla \phi_i \cdot \nabla f \right) + \phi_i\, s \right] dx\, dy \qquad (3.1.9)$$

$$= k \iint_D \nabla \cdot (\phi_i \nabla f)\, dx\, dy + \iint_D \left[-k \nabla \phi_i \cdot \nabla f + \phi_i\, s \right] dx\, dy.$$

Next, we apply the Gauss divergence theorem under the assumption that ϕ_i is continuous throughout D, and find

$$\iint_D \phi_i \left[k \nabla^2 f + s \right] dx\, dy \qquad (3.1.10)$$

$$= -k \oint_C \phi_i\, \mathbf{n} \cdot \nabla f\, dl + \iint_D \left[-k \nabla \phi_i \cdot \nabla f + \phi_i\, s \right] dx\, dy.$$

Substituting the boundary flux definition in the first integral on the right-hand side, $q \equiv -k\, \mathbf{n} \cdot \nabla f$, and rearranging, we find

$$\iint_D \phi_i \left[k \nabla^2 f + s \right] dx\, dy = -k \iint_D \nabla \phi_i \cdot \nabla f\, dx\, dy + Q_i + S_i, \qquad (3.1.11)$$

where

$$Q_i \equiv \oint_C \phi_i\, q\, dl, \qquad S_i \equiv \iint_D \phi_i\, s\, dx\, dy, \qquad (3.1.12)$$

are boundary and domain integrals involving the boundary flux and distributed source, weighted by the ith global interpolation function. Note that, if the ith node associated with ϕ_i is an interior node, ϕ_i is zero along the whole of the contour C, and $Q_i = 0$.

Next, we multiply the left-hand side of (3.1.1) by ϕ_i, integrate the product over the solution domain, D, set the resulting expression equal to the right-hand side of (3.1.11), and divide by $\rho\, c_p$ to obtain the Galerkin equation

$$\iint_D \phi_i \frac{\partial f}{\partial t}\, dx\, dy + \iint_D \phi_i\, \mathbf{u} \cdot \nabla f\, dx\, dy$$

$$= \frac{1}{\rho\, c_p} \left(-k \iint_D \nabla \phi_i \cdot \nabla f\, dx\, dy + Q_i + S_i \right), \qquad (3.1.13)$$

which provides us with the foundation of the Galerkin finite element method (GFEM).

3.1.2 Domain discretization and interpolation

To implement the finite element method, we discretize the solution domain, D, into a collection of N_E elements, as illustrated in Figure 3.1.1(*b*). In practice, the elements have triangular or rectangular shapes defined by a small group of geometrical element nodes. The collection of all geometrical element nodes comprises a set of N_{GN} *unique* geometrical global nodes. If an element node coincides with a neighboring element node, the two nodes are mapped to the same global node through a connectivity matrix, as will be discussed later in this section. Accordingly, the element nodes are assigned *global labels* ranging from 1 to N_{GN}.

If all elements are defined by the same number of nodes, q, then, because of node sharing, the number of unique global nodes is less than $q N_E$, where N_E is the number of elements. For example, if all elements are triangles with straight edges defined by the three vertices, $q = 3$, then the number of unique global nodes is less than $3 N_E$.

Interpolation nodes

To represent the *a priori* unknown solution in numerical form, we introduce interpolation element nodes, not all of which necessarily coincide with the geo-metrical element nodes. The collection of all interpolation element nodes comprises a set of N_G *unique* global interpolation nodes.

In the *isoparametric interpolation*, the set of interpolation element nodes coincides with the set of geometrical element nodes, and the number of global interpolation nodes is equal to the number of global geometrical nodes, $N_G = N_{GN}$. In this chapter, we discuss the isoparametric interpolation, and in Chapter 4 we shall discuss the super-parametric interpolation in the context of the spectral element method where $N_G > N_{GN}$.

Boundary flags and the connectivity matrix

In the second stage of the implementation, we introduce boundary flags and a connectivity matrix with two main goals:

- *Map the local element nodes to the unique global nodes.*

- *Designate whether a certain element or global node is a boundary node and store the boundary conditions.*

The connectivity matrix, $c(i, j)$, is defined such that $c(i, j)$ is the global label of the jth node of the ith element, where $i = 1, 2, \ldots, N_E$. Thus, $c(i, j)$ takes values in the range: $1, 2, \ldots, N_G$, where N_G is the number of unique global nodes.

Assume now that the ith element hosts m interpolation nodes. To specify the boundary conditions, we introduce the element-node flag, $efl(i, j)$, defined such that

$$efl(i, j) = \begin{cases} 0 & \text{if the } j\text{th node is not a boundary node} \\ \neq 0 & \text{if the } j\text{th node is a boundary node} \end{cases}, \quad (3.1.14)$$

for $i = 1, 2, \ldots, N_E$, and $j = 1, 2, \ldots, m$. If the solution domain is enclosed by a single contiguous contour, we may set $efl(i, j) = 1$ for the nodes located on that contour. If the solution domain is enclosed by two distinct contours, we may set $efl(i, j) = 1$ for the nodes located on the first contour, and $efl(i, j) = 2$ for the nodes located on the second contour. Multiply connected domains enclosed by a higher number of distinct boundary contours can be accommodated in a similar fashion.

Correspondingly, we introduce the global node flag $gfl(i, j)$, defined such that

$$gfl(i, 1) = \left\{ \begin{array}{ll} 0 & \text{if the } i\text{th global node is not a boundary node} \\ \neq 0 & \text{if the } i\text{th global node is a boundary node} \end{array} \right. ,$$

(3.1.15)

where $i = 1, 2, \ldots, N_G$ and $j = 1, 2$. In addition, we set $gfl(i, 1) = 1$ if the ith global node is a boundary node where the Dirichlet condition is specified, and $gfl(i, 1) = 2$ if the ith global node is a boundary node where the Neumann condition is specified. Other types of boundary conditions can be flagged by similar designations. When $gfl(i, 1) \neq 0$, the second entry, $gfl(i, 2)$, contains the specified boundary data. In some finite element codes, these values are stored in a separate array.

In practice, the connectivity matrix and the boundary flags are generated in the process of triangulation, as will be discussed later in the next section.

Isoparametric interpolation

In the third step of the implementation, we express the requisite solution in terms of the *a priori* unknown values at the global interpolation nodes, f_j, and associated *global cardinal interpolation functions*, ϕ_j, in the isoparametric form

$$f(x, y, t) = \sum_{j=1}^{N_G} f_j(t)\, \phi_j(x, y).$$

(3.1.16)

By definition, the global interpolation function, $\phi_j(x, y)$, takes the value of unity at the jth global interpolation node and the value of zero at all other global interpolation nodes, as illustrated in Figure 3.1.2. Specific expressions will be given later in this chapter for triangular elements defined by three vertex nodes, and in Chapter 4 for triangular elements defined by six nodes, including three vertex nodes and three edge nodes.

3.1.3 GFEM equations

Inserting expansion (3.1.16) and a similar expansion for the source term s in (3.1.13), we derive the Galerkin finite element equations

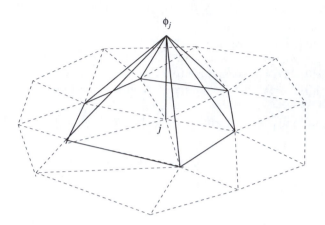

Figure 3.1.2 Illustration of a tent-like global interpolation function associated with the jth global node. By definition, $\phi_j(x, y)$ takes the value of unity at the jth global interpolation node and the value of zero at all other global interpolation nodes.

$$\sum_{i=j}^{N_G} M_{ij} \frac{\mathrm{d}f_j}{\mathrm{d}t} + \sum_{i=j}^{N_G} N_{ij} f_j$$

$$= \frac{1}{\rho \, c_p} \oint_C \phi_i \, q \, \mathrm{d}l - \kappa \sum_{i=j}^{N_G} D_{ij} \, f_j + \frac{1}{\rho \, c_p} \sum_{i=j}^{N_G} M_{ij} \, s_j, \qquad (3.1.17)$$

where

$$D_{ij} \equiv \iint_D \nabla \phi_i \cdot \nabla \phi_j \, \mathrm{d}x \, \mathrm{d}y \qquad (3.1.18)$$

is the global diffusion matrix,

$$M_{ij} \equiv \iint_D \phi_i \, \phi_j \, \mathrm{d}x \, \mathrm{d}y \qquad (3.1.19)$$

is the global mass matrix, and

$$N_{ij} \equiv \iint_D \phi_i \, \mathbf{u} \cdot \nabla \phi_j \, \mathrm{d}x \, \mathrm{d}y \qquad (3.1.20)$$

is the global advection matrix.

Applying (3.1.17) for the global interpolation functions associated with the interpolation nodes where the Dirichlet boundary condition is *not* prescribed, we obtain a system of ordinary differential equations for the unknown nodal values,

$$\mathbf{M} \cdot \frac{d\mathbf{f}}{dt} + \mathbf{N} \cdot \mathbf{f} = \kappa \left(-\mathbf{D} \cdot \mathbf{f} + \mathbf{b} \right), \qquad (3.1.21)$$

where the vector $\mathbf{b}$ on the right-hand side incorporates the given source term and the boundary conditions,

$$b_i \equiv \frac{1}{k} \left(\oint_C \phi_i\, q\, dl + \sum_{j=1}^{N_G} M_{ij}\, s_j \right). \qquad (3.1.22)$$

The first integral on the right-hand side of (3.1.22) is nonzero only if the ith node is a boundary node where the Neumann condition is specified.

Element matrices

Following the discussion of Section 1.1, we assemble the domain integrals in (3.1.19), (3.1.18), and (3.1.20), in terms of corresponding element integrals. The lth-element diffusivity matrix is given by

$$A_{ij}^{(l)} \equiv \iint_{E_l} \nabla \psi_i^{(l)} \cdot \nabla \psi_j^{(l)}\, dx\, dy, \qquad (3.1.23)$$

the corresponding element mass matrix is given by

$$B_{ij}^{(l)} \equiv \iint_{E_l} \psi_i^{(l)}\, \psi_j^{(l)}\, dx\, dy, \qquad (3.1.24)$$

and the corresponding element advection matrix is given by

$$C_{ij}^{(l)} \equiv \iint_{E_l} \psi_i^{(l)}\, \mathbf{u} \cdot \nabla \psi_j^{(l)}\, dx\, dy, \qquad (3.1.25)$$

where the integration is performed over the element area, E_l. The assembly algorithm hinges on the observation that the element integrals are nonzero only if nodes i and j lie inside or around the contour of the lth element. Algorithm 3.1.1 assembles the $N_G \times N_G$ global diffusion matrix with the assistance of the connectivity matrix, $c(i, j)$, that relates element nodes to global nodes. The mass and advection matrices are assembled in a similar fashion.

3.1.4   Implementation of the Dirichlet boundary condition

When the Dirichlet boundary condition is specified at a boundary node, that is, the node lies on the Dirichlet contour, C_D, where the unknown function is specified, it is not necessary to carry out the corresponding Galerkin projection. In practice, it is expedient to ignore this exception and implement it after the whole preliminary $N_G \times N_G$ system has been compiled, at a minimal computational cost.

Do $i = 1, 2, \ldots, N_G$ *Initialize to zero*
     Do $j = 1, 2, \ldots, N_G$
         $D_{ij} = 0.0$
     End Do
End Do

Do $l = 1, 2, \ldots, N_E$ *Run over the elements*
     Compute the element diffusion matrix $\mathbf{A}^{(l)}$
     Do $i = 1, 2, \ldots, N(l)$ *Run over the element interpolation nodes*
       $i_1 = c(l, i)$
       Do $j = 1, 2, \ldots, N(l)$ *Run over the element interpolation nodes*
         $i_2 = c(l, j)$
         $D_{i_1, i_2} = D_{i_1, i_2} + A_{ij}^{(l)}$
       End Do
     End Do

End Do

Algorithm 3.1.1 Assembly of the $N_G \times N_G$ global diffusion matrix, $\mathbf{D}$, in terms of the element diffusion matrices.

For example, in the case of steady heat conduction, the system of ordinary differential equations (3.1.21) reduces to the linear algebraic system

$$\mathbf{D} \cdot \mathbf{f} = \mathbf{b}. \tag{3.1.26}$$

Assume that the boundary conditions specify that $f = f_m$ at the mth global node, where f_m is given. In the finite element implementation, after the Galerkin system (3.1.26) has been compiled for all nodes, the mth constituent equation is discarded and replaced by the aforementioned Dirichlet boundary condition, in four steps:

1. Replace all entries on the right-hand side, b_i, with $b_i - D_{im} f_m$, for $i = 1, 2, \ldots, N_G$.

2. Set all elements in the mth column and mth row of $\mathbf{D}$ equal to zero.

3. Set the diagonal element, D_{mm}, equal to unity.

4. Replace the mth entry of the right-hand side, b_m, with f_m.

Although the algorithm involves some redundancy, the preserved symmetry of the coefficient matrix after the implementation of the boundary conditions is a significant advantage.

Do $m = 1, 2, \ldots, N_G$ *Run over all global nodes*

    If($gfl(m, 1)$=1) then   *Dirichlet boundary node*

       Do $i = 1, 2, \ldots, N_G$ *Run over all global nodes*
         $b_i = b_i - D_{im} \times gfl(m, 2)$
         $D_{im} = 0$
         $D_{mi} = 0$
       End Do

       $D_{mm} = 1.0$
       $b_m = gfl(m, 2)$

    End If

End Do

Algorithm 3.1.2 Implementation of the Dirichlet boundary condition for steady two-dimensional conduction. The entry $gfl(m, 2)$ contains the prescribed boundary value of the solution at the mth global node.

The method is implemented in Algorithm 3.1.2, subject to the convention that the entry $gfl(m, 2)$ contains the prescribed boundary value of the solution specified by the Dirichlet boundary condition at the position of the mth global node.

3.1.5   Split nodes

When a boundary node lies at a corner or junction where different conditions are applied on either side, or when a discontinuity is expected across a shared element node, the node develops a split personality. One way to handle this dichotomy is to map the unfortunate node to two ghost global nodes, one corresponding to the left, and the other to the right side of the boundary. By doing so, we effectively introduce an artificial crack extending up to the nearest interior node, as illustrated in Figure 3.1.3. Because of the implicit presence of the crack, the ghost nodes support two distinct global interpolation functions. Appropriate boundary conditions may then be enforced at the two ghost nodes, as will be discussed in Section 3.8.

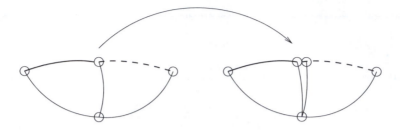

Figure 3.1.3 Illustration of node splitting to accommodate a discontinuity or
a corner.

3.1.6   Variational formulation

In Section 1.2, we demonstrated the intimate relationship between Galerkin's
method and Ritz's implementation of Rayleigh's variational formulation for the
problem of steady one-dimensional diffusion. Moreover, we showed that the
finite element method is a specific implementation of the method of weighted
residuals.

In the case of steady two-dimensional diffusion governed by the Poisson
equation,

$$k \, \nabla^2 f + s(x, y) = 0, \tag{3.1.27}$$

subject to the boundary conditions (3.1.6) and (3.1.8), the variational formulation states that computing the solution, f, is equivalent to maximizing the
functional

$$\mathcal{F} < w(x, y) >= -\frac{k}{2} \iint_D \left[\left(\frac{\partial w}{\partial x} \right)^2 + \left(\frac{\partial w}{\partial y} \right)^2 \right] \mathrm{d}x \, \mathrm{d}y$$

$$+ \iint_D s \, w \, \mathrm{d}x \, \mathrm{d}y + \int_{C_N} q(l) \, w \, \mathrm{d}l, \tag{3.1.28}$$

over all functions, $w(x, y)$, that satisfy the Dirichlet boundary condition (3.1.8),
where C_N in the last integral of (3.1.28) is the Neumann contour. The satisfaction of the Neumann boundary condition is enforced by the last term on the
right-hand side of (3.1.28).

Manipulating the first integral on the right-hand side of (3.1.28) by use of
the divergence theorem, we find

$$\iint_D \left[\left(\frac{\partial w}{\partial x} \right)^2 + \left(\frac{\partial w}{\partial y} \right)^2 \right] \mathrm{d}x \, \mathrm{d}y$$

$$= \iint_D \left[\frac{\partial}{\partial x} \left(w \, \frac{\partial w}{\partial x} \right) - w \, \frac{\partial^2 w}{\partial x^2} + \frac{\partial}{\partial y} \left(w \, \frac{\partial w}{\partial y} \right) - w \, \frac{\partial^2 w}{\partial y^2} \right] \mathrm{d}x \, \mathrm{d}y \tag{3.1.29}$$

$$= \iint_D \left[\nabla \cdot (w \, \nabla w) - w \, \nabla^2 w \right] \mathrm{d}x \, \mathrm{d}y = - \oint_C w \, \mathbf{n} \cdot \nabla w \, \mathrm{d}l - \iint_D w \, \nabla^2 w \, \mathrm{d}x \, \mathrm{d}y,$$

where $\mathbf{n}$ is the unit vector normal to C, pointing *into* the solution domain, D. Substituting this expression in (3.1.28), we derive the alternative form

$$
\mathcal{F} < \omega(x,y) >= \frac{k}{2}\Big[\iint_D \omega\, \nabla^2\omega\, dx\, dy + \oint_C \omega\, \mathbf{n}\cdot\nabla\omega\, dl \Big]
$$
$$
+ \iint_D s(x,y)\,\omega(x,y)\, dx\, dy + \int_{C_N} q(l)\,\omega\, dl, \qquad (3.1.30)
$$

which can be simplified to

$$
\mathcal{F} < \omega(x,y) >= \frac{k}{2} \iint_D \omega\, \nabla^2\omega\, dx\, dy + \iint_D s(x,y)\,\omega(x,y)\, dx\, dy
$$
$$
+ \frac{k}{2} \int_{C_D} g(l)\, \mathbf{n}\cdot\nabla\omega\, dl + \frac{1}{2}\int_{C_N} q(l)\,\omega\, dl. \qquad (3.1.31)
$$

If only the Dirichlet boundary condition is specified, the last terms on the right-hand sides of (3.1.28) and (3.1.30) do not appear.

To prove the equivalence, we perturb the solution $f(x,y)$ to $f(x,y)+\epsilon v(x,y)$, where the disturbance function, $v(x,y)$, also called a variation, is required to satisfy the homogeneous Dirichlet boundary condition $v = 0$ over the Dirichlet contour C_D, and ϵ is a dimensionless number whose magnitude is much less than unity. Next, we consider the difference

$$
\delta\mathcal{F} \equiv \mathcal{F} < f(x,y) + \epsilon\, v(x,y) > -\mathcal{F} < f(x,y) >, \qquad (3.1.32)
$$

and use expression (3.1.28) to find

$$
\delta\mathcal{F} = \epsilon\Big(-k \iint_D \nabla f\cdot\nabla v\, dx\, dy + \iint_D s(x,y)\, v\, dx\, dy
$$
$$
+ \int_{C_N} q(l)\, v\, dl \Big) + O(\epsilon^2). \qquad (3.1.33)
$$

Applying the divergence theorem, we write

$$
\iint_D \nabla f\cdot\nabla v\, dx\, dy = \iint_D \Big[\nabla\cdot(v\,\nabla f) - v\,\nabla^2 f\Big]\, dx\, dy
$$
$$
= -\oint_C v\,\mathbf{n}\cdot\nabla f\, dl - \iint_D v\,\nabla^2 f\, dx\, dy, \qquad (3.1.34)
$$

and obtain

$$
\delta\mathcal{F} = \epsilon \iint_D \big(k\,\nabla^2 f + s\big)\, v\, dx\, dy + \epsilon\, k \oint_C v\,\mathbf{n}\cdot\nabla f\, dl + \epsilon \int_{C_N} q(l)\, v\, dl + O(\epsilon^2).
$$
$$
(3.1.35)
$$

Finally, we recall that $v = 0$ on C_D, and invoke the definition of the boundary flux, $q \equiv -k\,\mathbf{n}\cdot\nabla f$, to find

$$
\delta\mathcal{F} = \epsilon \iint_D \big(k\,\nabla^2 f + s\big)\, v\, dx\, dy + O(\epsilon^2). \qquad (3.1.36)
$$

If the function $f(x, y)$ satisfies (3.1.27) subject to the aforementioned boundary conditions, the first variation vanishes, and the functional is stationary.

Alternatively, we may use (3.1.30) to compute the variation,

$$\delta \mathcal{F} = \epsilon \frac{k}{2} \left[\iint_D (v \, \nabla^2 f + f \, \nabla^2 v) \, \mathrm{d}x \, \mathrm{d}y + \oint_C (v \, \mathbf{n} \cdot \nabla f + f \, \mathbf{n} \cdot \nabla v) \, \mathrm{d}l \right]$$
$$+ \epsilon \left[\iint_D s(x, y) \, v \, \mathrm{d}x \, \mathrm{d}y + \int_{C_N} q(l) \, v \, \mathrm{d}l \right] + O(\epsilon^2). \qquad (3.1.37)$$

Using Green's second identity,

$$\iint_D f \, \nabla^2 v \, \mathrm{d}x \, \mathrm{d}y = \iint_D f \, \nabla^2 v \, \mathrm{d}x \, \mathrm{d}y - \oint_C (f \, \mathbf{n} \cdot \nabla v - v \, \mathbf{n} \cdot \nabla f) \, \mathrm{d}l, \quad (3.1.38)$$

we simplify the first integral on the right-hand side to obtain

$$\delta \mathcal{F} = \epsilon \, k \left(\iint_D v \, \nabla^2 f \, \mathrm{d}x \, \mathrm{d}y + \oint_C v \, \mathbf{n} \cdot \nabla f \, \mathrm{d}l \right)$$
$$+ \epsilon \iint_D s(x, y) \, v \, \mathrm{d}x \, \mathrm{d}y + \epsilon \int_{C_N} q(l) \, v \, \mathrm{d}l + O(\epsilon^2), \qquad (3.1.39)$$

which reduces to (3.1.36).

Ritz's implementation of the variational formulation and the Galerkin implementation of the method of weighted residuals are similar to those discussed in Section 1.2 in one dimension with Dirichlet/Neumann boundary conditions.

PROBLEM

3.1.1 *Variational formulation.*

(*a*) Show that (3.1.28) is the two-dimensional counterpart of the one-dimensional form (1.2.21).

(*b*) Consider the linear equation

$$k \, \nabla^2 f + g(x, y) \, f + s(x, y) = 0, \qquad (3.1.40)$$

where $g(x, y)$ is a specified function, subject to the boundary conditions (3.1.6) and (3.1.8). Show that the pertinent Rayleigh functional is given by (3.1.28), with the additional term

$$\frac{1}{2} \iint_D g(x, y) \, \omega^2 \, \mathrm{d}x \, \mathrm{d}y \qquad (3.1.41)$$

included on the right-hand side.

3.2   3-node triangles

Having outlined the foundation of the Galerkin finite element method in two dimensions, we proceed to discuss the particulars of the implementation and build a complete finite element code.

In the simplest implementation, the solution domain in the xy plane is discretized into triangular elements with straight edges defined by three geometrical *vertex nodes*,

$$\mathbf{x}_i^E = (x_i^E, y_i^E), \tag{3.2.1}$$

where $i = 1, 2, 3$, as illustrated on the left of Figure 3.2.1(a). Note that the vertices are numbered in the counterclockwise fashion around the element contour. The superscript E emphasizes that these are local or "element" nodes, which can be mapped to the unique global nodes by means of the connectivity matrix.

To describe the element in parametric form, we map it from the physical xy plane to a right isosceles triangle in the $\xi\eta$ parametric plane, as shown in Figure 3.2.1(a). The first element node is mapped to the origin, $\xi = 0, \eta = 0$, the second to the point $\xi = 1, \eta = 0$ on the ξ axis, and the third to the point $\xi = 0, \eta = 1$ on the η axis. The mapping from the physical to the parametric plane is mediated by the function

$$\mathbf{x} = \mathbf{x}_1^E \, \psi_1(\xi, \eta) + \mathbf{x}_2^E \, \psi_2(\xi, \eta) + \mathbf{x}_3^E \, \psi_3(\xi, \eta), \tag{3.2.2}$$

where $\psi_i(\xi, \eta)$ are element interpolation functions satisfying the familiar cardinal property: $\psi_i = 1$ at the ith element node, and $\psi_i = 0$ at the other two element nodes,

$$\psi_i(\xi_j, \eta_j) = \delta_{ij}, \tag{3.2.3}$$

for $i, j = 1, 2, 3$, where δ_{ij} is Kronecker's delta, and

$$(\xi_1, \eta_1) = (0, 0), \quad (\xi_2, \eta_2) = (1, 0), \quad (\xi_3, \eta_3) = (0, 1), \tag{3.2.4}$$

are the coordinates of the vertices in the $\xi\eta$ plane. To derive the node interpolation functions, we write

$$\psi_i(\xi, \eta) = a_i + b_i \, \xi + c_i \, \eta, \tag{3.2.5}$$

and compute the coefficients a_i, b_i, and c_i to satisfy the aforementioned cardinal interpolation conditions. For example, for node 1, we require

$$\begin{aligned}
\psi_1(\xi_1, \eta_1) &= a_1 + b_1 \, \xi + c_1 \, \eta = a_1 = 1, \\
\psi_1(\xi_2, \eta_2) &= a_1 + b_1 \times 1.0 + c_1 \times 0.0 = a_1 + b_1 = 0, \\
\psi_1(\xi_3, \eta_3) &= a_1 + b_1 \times 0.0 + c_1 \times 1.0 = a_1 + c_1 = 0,
\end{aligned} \tag{3.2.6}$$

(*a*)

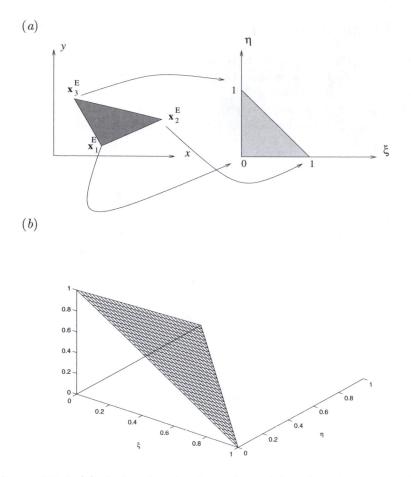

(*b*)

Figure 3.2.1 (*a*) A 3-node triangle with straight edges in the xy plane is mapped to a right isosceles triangle in the parametric $\xi\eta$ plane. (*b*) Graph of the element interpolation function, ψ_1, associated with the first vertex node.

which yields $a_1 = 1$, $b_1 = -1$, and $c_1 = -1$. Substituting these values in (3.2.5) and repeating the calculation for the other two nodes, we find

$$\psi_1(\xi, \eta) = \zeta, \qquad \psi_2(\xi, \eta) = \xi, \qquad \psi_3(\xi, \eta) = \eta, \qquad (3.2.7)$$

where

$$\zeta = 1 - \xi - \eta. \qquad (3.2.8)$$

The trio of variables (ξ, η, ζ) comprise the triangle *barycentric coordinates*. Physically, $\zeta = A_1/A$, $\xi = A_2/A$, and $\eta = A_3/A$, where A_1, A_2, and A_3 are

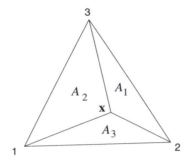

Figure 3.2.2 The barycentric coordinates are defined with respect to the areas of the triangles defined by an arbitrary field point, **x**, and two vertices.

the areas of the sub-triangles defined by the field point, **x**, and a pair ofa pair of vertices, as shown in Figure 3.2.2.

Figure 3.2.1(*b*) shows a graph of the first interpolation function, ψ_1, over the area of the triangle in the $\xi\eta$ plane, produced by the FSELIB function `psi_t3` (not listed in the text). The other two interpolation functions have similar shapes.

Substituting the interpolation functions (3.2.7) in (3.2.2), we obtain a mapping function that is a *complete* linear function in ξ and η, consisting of a constant term, a term that is linear in ξ, and a term that is linear in η,

$$\mathbf{x} = \mathbf{x}_1^E + (\mathbf{x}_2^E - \mathbf{x}_1^E)\,\xi + (\mathbf{x}_3^E - \mathbf{x}_1^E)\,\eta. \tag{3.2.9}$$

Explicitly, the x and y components of the mapping are given by

$$x = x_1^E + (x_2^E - x_1^E)\,\xi + (x_3^E - x_1^E)\,\eta,$$

$$y = y_1^E + (y_2^E - y_1^E)\,\xi + (y_3^E - y_1^E)\,\eta. \tag{3.2.10}$$

For a specified location (x, y), we can find the corresponding coordinates (ξ, η) over the parametric triangle by solving the system of two linear equations (3.2.10).

Integral of a function over the triangle

The Jacobian matrix of the mapping from the xy to the $\xi\eta$ plane is defined as

$$\mathbf{J} \equiv \begin{bmatrix} \frac{\partial x}{\partial \xi} & \frac{\partial x}{\partial \eta} \\[2mm] \frac{\partial y}{\partial \xi} & \frac{\partial y}{\partial \eta} \end{bmatrix}. \tag{3.2.11}$$

The determinant of the Jacobian matrix is the *surface metric coefficient*,

$$h_S \equiv \mathrm{Det}[\mathbf{J}] = \frac{\partial x}{\partial \xi}\frac{\partial y}{\partial \eta} - \frac{\partial x}{\partial \eta}\frac{\partial y}{\partial \xi} = \left|\frac{\partial \mathbf{x}}{\partial \xi} \times \frac{\partial \mathbf{x}}{\partial \eta}\right|. \tag{3.2.12}$$

The last expression in (3.2.12) is the magnitude of the outer product of the vectors $\partial \mathbf{x}/\partial \xi$ and $\partial \mathbf{x}/\partial \eta$ (see Appendix D). Since these vectors lie in the xy plane, their outer product points along the z axis that is normal to the xy plane.

Using the linear mapping functions shown in (3.2.10), we find

$$\frac{\partial \mathbf{x}}{\partial \xi} = \mathbf{x}_2^E - \mathbf{x}_1^E, \qquad \frac{\partial \mathbf{x}}{\partial \eta} = \mathbf{x}_3^E - \mathbf{x}_1^E, \tag{3.2.13}$$

and calculate

$$\mathbf{J} = \begin{bmatrix} x_2^E - x_1^E & x_3^E - x_1^E \\ y_2^E - y_1^E & y_3^E - y_1^E \end{bmatrix}, \tag{3.2.14}$$

and then

$$h_S = |(\mathbf{x}_2^E - \mathbf{x}_1^E) \times (\mathbf{x}_3^E - \mathbf{x}_1^E)|$$

$$= (x_2^E - x_1^E)(y_3^E - y_1^E) - (x_3^E - x_1^E)(y_2^E - y_1^E). \tag{3.2.15}$$

Invoking the geometrical interpretation of the outer vector product, we find that h_S is equal to the area of a rectangle with two sides coinciding with the vectors $\mathbf{x}_2^E - \mathbf{x}_1^E$ and $\mathbf{x}_3^E - \mathbf{x}_1^E$, which is equal to twice the area of the triangle in the physical xy plane. The surface metric coefficient for a 3-node triangle is then

$$h_S = 2\,A, \tag{3.2.16}$$

and A is the area of the triangle in the physical xy plane.

Using elementary calculus, we find that the integral of a function $f(x, y)$ over the physical triangle in the xy plane can be expressed as an integral over the parametric triangle in the $\xi\eta$ plane, as

$$\iint f(x, y)\, \mathrm{d}x\, \mathrm{d}y = \iint f(\xi, \eta)\, h_S\, \mathrm{d}\xi\, \mathrm{d}\eta = 2\,A \iint f(\xi, \eta)\, \mathrm{d}\xi\, \mathrm{d}\eta. \tag{3.2.17}$$

The integrals with respect to ξ and η can be computed analytically in simple cases, or numerically in more involved cases, as will be discussed later in this section.

3.2.1 Element matrices

In the *isoparametric interpolation*, a function of interest defined over an element is expressed in a form that is analogous to that shown in (3.2.2),

$$f(x, y, t) = \sum_{i=1}^{3} f_i^E(t)\, \psi_i(\xi, \eta), \tag{3.2.18}$$

with the understanding that the point $\mathbf{x} = (x, y)$ is mapped to (ξ, η) and *vice versa* through (3.2.2).

Using the preceding integration formulas, we find that the *l*th element diffusion matrix is given by

$$A_{ij}^{(l)} \equiv \iint_{E_l} \nabla \psi_i \cdot \nabla \psi_j \, dx \, dy = 2 \, A \iint \nabla \psi_i \cdot \nabla \psi_j \, d\xi \, d\eta, \qquad (3.2.19)$$

for $i, j = 1, 2, 3$, where

$$\nabla \psi_i = \left(\frac{\partial \psi_i}{\partial x}, \frac{\partial \psi_i}{\partial y} \right). \qquad (3.2.20)$$

The element mass matrix is given by

$$B_{ij}^{(l)} = \iint_{E_l} \psi_i \, \psi_j \, dx \, dy = 2 \, A \iint \psi_i \, \psi_j \, d\xi \, d\eta, \qquad (3.2.21)$$

and the element advection matrix is given by

$$C_{ij}^{(l)} = \iint_{E_l} \psi_i \, \mathbf{u} \cdot \nabla \psi_j \, dx \, dy = 2 \, A \iint \psi_i \, \mathbf{u} \cdot \nabla \psi_j \, d\xi \, d\eta, \qquad (3.2.22)$$

where $\mathbf{u} = (u_x, u_y)$ is the advection velocity.

Computation of the gradient

To compute the element diffusion and advection matrices, we require the gradients of the element interpolation functions, $\nabla \psi_i$, for $i = 1, 2, 3$. These can be found readily using the relations

$$\frac{\partial \mathbf{x}}{\partial \xi} \cdot \nabla \psi_i = \frac{\partial \psi_i}{\partial \xi}, \qquad \frac{\partial \mathbf{x}}{\partial \eta} \cdot \nabla \psi_i = \frac{\partial \psi_i}{\partial \eta}, \qquad (3.2.23)$$

which state that the projection of the gradient vector in the direction of the ξ-line vector, $\partial \mathbf{x}/\partial \xi$, is the partial derivative of $\mathbf{x}$ with respect to ξ, and the projection of the gradient vector in the direction of the η-line vector, $\partial \mathbf{x}/\partial \eta$, is the partial derivative of $\mathbf{x}$ with respect to η. Explicitly, equations (3.2.23) read

$$\frac{\partial x}{\partial \xi} \frac{\partial \psi_i}{\partial x} + \frac{\partial y}{\partial \xi} \frac{\partial \psi_i}{\partial y} = \frac{\partial \psi_i}{\partial \xi},$$

$$(3.2.24)$$

$$\frac{\partial x}{\partial \eta} \frac{\partial \psi_i}{\partial x} + \frac{\partial y}{\partial \eta} \frac{\partial \psi_i}{\partial y} = \frac{\partial \psi_i}{\partial \eta},$$

which can be compiled in the linear system,

$$\mathbf{J}^T \cdot \nabla \psi_i = \begin{bmatrix} \frac{\partial \psi_i}{\partial \xi} \\[2mm] \frac{\partial \psi_i}{\partial \eta} \end{bmatrix}, \qquad (3.2.25)$$

where $\mathbf{J}^T$ is the transpose of the Jacobian matrix defined in (3.2.11),

$$
\mathbf{J}^T \equiv \begin{bmatrix} \frac{\partial x}{\partial \xi} & \frac{\partial y}{\partial \xi} \\[2mm] \frac{\partial x}{\partial \eta} & \frac{\partial y}{\partial \eta} \end{bmatrix}. \tag{3.2.26}
$$

The determinant of $\mathbf{J}^T$ is equal to the determinant of $\mathbf{J}$, which is equal to the surface metric coefficient, $h_S = 2\,A$.

Using (3.2.13), we find that equations (3.2.23) take the specific form,

$$
(\mathbf{x}_2^E - \mathbf{x}_1^E) \cdot \nabla \psi_i = \frac{\partial \psi_i}{\partial \xi}, \qquad (\mathbf{x}_3^E - \mathbf{x}_1^E) \cdot \nabla \psi_i = \frac{\partial \psi_i}{\partial \eta}. \tag{3.2.27}
$$

Accordingly,

$$
\mathbf{J}^T = \begin{bmatrix} x_2^E - x_1^E & y_2^E - y_1^E \\ x_3^E - x_1^E & y_3^E - y_1^E \end{bmatrix}, \tag{3.2.28}
$$

which is consistent with (3.2.14). Substituting the specific expressions for the cardinal interpolation functions given in (3.2.7), we derive the linear systems

$$
\mathbf{J}^T \cdot \nabla \psi_1 = -\begin{bmatrix} 1 \\ 1 \end{bmatrix}, \qquad \mathbf{J}^T \cdot \nabla \psi_2 = \begin{bmatrix} 1 \\ 0 \end{bmatrix},
$$

$$
\mathbf{J}^T \cdot \nabla \psi_3 = \begin{bmatrix} 0 \\ 1 \end{bmatrix}, \tag{3.2.29}
$$

whose solutions are

$$
\nabla \psi_1 = \frac{1}{2A} \begin{bmatrix} -(y_3^E - y_2^E) \\ x_3^E - x_2^E \end{bmatrix},
$$

$$
\nabla \psi_2 = \frac{1}{2A} \begin{bmatrix} -(y_1^E - y_3^E) \\ x_1^E - x_3^E \end{bmatrix}, \tag{3.2.30}
$$

$$
\nabla \psi_3 = \frac{1}{2A} \begin{bmatrix} -(y_2^E - y_1^E) \\ x_2^E - x_1^E \end{bmatrix}.
$$

Note that $\nabla \psi_1$ is perpendicular to the side 23, $\nabla \psi_2$ is perpendicular to the side 31, and $\nabla \psi_3$ is perpendicular to the side 12. Expressions (3.2.30) are used to evaluate the element diffusion and advection matrices.

3.2.2   Computation of the element diffusion matrix

Substituting (3.2.30) in (3.2.19), we find that the element diffusion matrix is given by

$$
\mathbf{A}^{(l)} = \frac{1}{4A} \begin{bmatrix} |\mathbf{d}_{32}|^2 & \mathbf{d}_{32} \cdot \mathbf{d}_{13} & \mathbf{d}_{32} \cdot \mathbf{d}_{21} \\ \mathbf{d}_{32} \cdot \mathbf{d}_{13} & |\mathbf{d}_{13}|^2 & \mathbf{d}_{13} \cdot \mathbf{d}_{21} \\ \mathbf{d}_{32} \cdot \mathbf{d}_{21} & \mathbf{d}_{13} \cdot \mathbf{d}_{21} & |\mathbf{d}_{21}|^2 \end{bmatrix}, \tag{3.2.31}
$$

```
function edm = edm3 (x1,y1,x2,y2,x3,y3)

%================================================
% Evaluation of the element diffusion matrix
% for a 3-node triangle
%================================================

d32x = x3-x2; d32y = y3-y2;
d13x = x1-x3; d13y = y1-y3;
d21x = x2-x1; d21y = y2-y1;

A = 0.5*(d13x*d21y - d31y*d21x);   % element area

fc = 0.25*A;

edm(1,1) = fc*(d32x*d32x + d32y*d32y);
edm(1,2) = fc*(d32x*d13x + d32y*d13y);
edm(1,3) = fc*(d32x*d21x + d32y*d21y);

edm(2,1) = fc*(d13x*d32x + d13y*d32y);
edm(2,2) = fc*(d13x*d13x + d13y*d13y);
edm(2,3) = fc*(d13x*d21x + d13y*d21y);

edm(3,1) = fc*(d21x*d32x + d21y*d32y);
edm(3,2) = fc*(d21x*d13x + d21y*d13y);
edm(3,3) = fc*(d21x*d21x + d21y*d21y);

%-----
% done
%-----

return;
```

Function edm3: Evaluation of the element diffusion matrix from the coordinates of the vertices of a 3-node triangle.

where

$$\mathbf{d}_{32} = \mathbf{x}_3^E - \mathbf{x}_2^E, \qquad \mathbf{d}_{13} = \mathbf{x}_1^E - \mathbf{x}_3^E, \qquad \mathbf{d}_{21} = \mathbf{x}_2^E - \mathbf{x}_1^E. \qquad (3.2.32)$$

Note that, if the triangle in the xy plane is orthogonal with the right angle occurring at the first element node labeled 1, then $\mathbf{d}_{31} \cdot \mathbf{d}_{12} = 0$, and two off-diagonal elements of $\mathbf{A}^{(l)}$ are zero. Similar simplifications arise if the right angle occurs at one of the other two nodes.

FSELIB function edm3, listed in the text, evaluates the element diffusion matrix from the coordinates of the nodes, based on the preceding expressions. It can be confirmed by direct evaluation that the determinant of this matrix is zero, which means that the matrix is singular (Problem 3.2.2). The physical reason was discussed at the end of Section 1.1.8.

3.2.3   Computation of the element mass matrix

Substituting (3.2.7) in (3.2.21), we derive an expression for the element mass matrix,

$$\mathbf{B}^{(l)} = 2\,A \iint \begin{bmatrix} \zeta^2 & \zeta\xi & \zeta\eta \\ \xi\zeta & \xi^2 & \xi\eta \\ \eta\zeta & \eta\xi & \eta^2 \end{bmatrix} d\xi\,d\eta. \tag{3.2.33}$$

To compute the double integral, we use the integration formula

$$\iint \zeta^p\,\xi^q\,\eta^r\,d\xi\,d\eta = \frac{p!\,q!\,r!}{(p+q+r+2)!}, \tag{3.2.34}$$

to be proved later in this section, where p, q, and r are non-negative integers, and an exclamation mark denotes the factorial, $m! = 1 \cdot 2 \ldots \cdot m$. A straightforward calculation yields

$$\mathbf{B}^{(l)} = \frac{A}{12} \begin{bmatrix} 2 & 1 & 1 \\ 1 & 2 & 1 \\ 1 & 1 & 2 \end{bmatrix}. \tag{3.2.35}$$

Note that the element mass matrix depends only on the element area, A, and is independent of the element shape. The sum of all entries of the element mass matrix is equal to the area of the triangle, A.

The mass-lumped diagonal element mass matrix, designated by a hat, is given by

$$\widehat{\mathbf{B}}^{(l)} = \frac{A}{3} \begin{bmatrix} 1 & 0 & 0 \\ 0 & 1 & 0 \\ 0 & 0 & 1 \end{bmatrix}. \tag{3.2.36}$$

The trace of the lumped element mass matrix is also equal to the area of the triangle, A.

It is instructive to note that the lumped form arises from the inexact integration of the right-hand side of (3.2.33), using the trapezoidal rule. In this approximation, the entries of the matrix inside the integral are assigned constant values, equal to the mean of the three values at the three vertices. This results in the value of $1/3$ for the diagonal elements, and the value of zero for the off-diagonal elements, as shown in (3.2.36).

Proof of the integration formula

To prove the integration formula (3.2.34), we map the standard triangle in the $\xi\eta$ plane to the standard square in the $\xi'\eta'$ parametric plane using Duffy's transformation, as shown in Figure 3.2.3. The mapping is mediated by the functions

$$\xi = \frac{1+\xi'}{2}\,\frac{1-\eta'}{2}, \qquad \eta = \frac{1+\eta'}{2}, \tag{3.2.37}$$

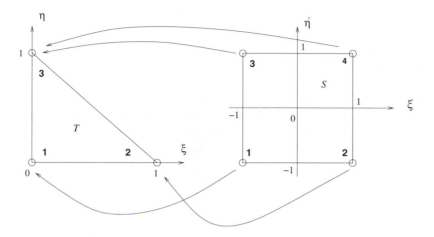

Figure 3.2.3 Mapping of the standard triangle to the standard square using Duffy's transformation. The bold-faced numbers are the element node labels.

yielding

$$\zeta \equiv 1 - \xi - \eta = \frac{1 - \xi'}{2} \frac{1 - \eta'}{2}, \tag{3.2.38}$$

where $-1 < \xi' < 1$ and $-1 < \eta' < 1$. For future reference, we note that

$$1 - \eta = 1 - \frac{1 + \eta'}{2} = \frac{1 - \eta'}{2}. \tag{3.2.39}$$

The inverse mapping functions are given by

$$\xi' = \frac{2\xi}{1 - \eta} - 1, \qquad \eta' = 2\eta - 1. \tag{3.2.40}$$

The integral of a function $f(\xi, \eta)$ over the area of the triangle in the $\xi\eta$ plane can be expressed as an integral over the standard square, as

$$\iint f(\xi, \eta) \, d\xi \, d\eta = \int_{-1}^{1} \int_{-1}^{1} f(\xi, \eta) \, h'_S \, d\xi' \, d\eta'. \tag{3.2.41}$$

The metric coefficient of the transformation, h'_S, is equal to the determinant of the Jacobian matrix of the transformation,

$$h'_S = \text{Det}\left(\begin{bmatrix} \frac{\partial \xi}{\partial \xi'} & \frac{\partial \eta}{\partial \xi'} \\ \frac{\partial \xi}{\partial \eta'} & \frac{\partial \eta}{\partial \eta'} \end{bmatrix} \right) = \text{Det}\left(\begin{bmatrix} \frac{1}{4}(1 - \eta') & 0 \\ -\frac{1}{4}(1 + \xi') & \frac{1}{2} \end{bmatrix} \right) = \frac{1}{8}(1 - \eta').$$

$$\tag{3.2.42}$$

Thus,

$$\iint f(\xi, \eta) \, \mathrm{d}\xi \, \mathrm{d}\eta = \frac{1}{8} \int_{-1}^{1} \int_{-1}^{1} f(\xi, \eta) \, (1 - \eta') \, \mathrm{d}\xi' \, \mathrm{d}\eta'. \tag{3.2.43}$$

The double integral on the right-hand side over the standard square can be computed accurately by applying a Gaussian quadrature twice, once with respect to ξ', and the second time with respect to η'.

Applying the integration formula (3.2.43) for the function $f(\xi, \eta) = \zeta^p \, \xi^q \, \eta^r$, we find

$$\iint \zeta^p \, \xi^q \, \eta^r \, \mathrm{d}\xi \, \mathrm{d}\eta = \frac{1}{8} \int_{-1}^{1} \int_{-1}^{1} \left(\frac{1 - \xi'}{2} \right)^p \left(\frac{1 - \eta'}{2} \right)^p \left(\frac{1 + \xi'}{2} \right)^q$$

$$\times \left(\frac{1 - \eta'}{2} \right)^q \left(\frac{1 + \eta'}{2} \right)^r (1 - \eta') \, \mathrm{d}\xi' \, \mathrm{d}\eta', \tag{3.2.44}$$

or

$$\iint \zeta^p \, \xi^q \, \eta^r \, \mathrm{d}\xi \, \mathrm{d}\eta = \frac{1}{4} \left[\int_{-1}^{1} \left(\frac{1 - \xi'}{2} \right)^p \left(\frac{1 + \xi'}{2} \right)^q \mathrm{d}\xi' \right]$$

$$\times \left[\int_{-1}^{1} \left(\frac{1 - \eta'}{2} \right)^{p+q+1} \left(\frac{1 + \eta'}{2} \right)^r \mathrm{d}\eta' \right]. \tag{3.2.45}$$

The integral with respect to ξ' enclosed by the first square brackets on the right-hand side can be evaluated by introducing the new variable $\omega \equiv \frac{1}{2} (1 - \xi')$, and writing

$$\int_{-1}^{1} \left(\frac{1 - \xi'}{2} \right)^p \left(\frac{1 + \xi'}{2} \right)^q \mathrm{d}\xi' = 2 \int_{0}^{1} \omega^p (1 - \omega)^q \, \mathrm{d}\omega$$

$$\equiv 2 \, B(p + 1, q + 1) = 2 \, \frac{p! \, q!}{(p + q + 1)!}, \tag{3.2.46}$$

where $B(k, l)$ is the beta function (e.g., [1]). The integral with respect to η' can be evaluated in a similar fashion, yielding

$$\int_{-1}^{1} \left(\frac{1 - \eta'}{2} \right)^{p+q+1} \left(\frac{1 + \eta'}{2} \right)^r \mathrm{d}\eta' = 2 \, \frac{(p + q + 1)! \, r!}{(p + q + r + 1)!}. \tag{3.2.47}$$

Putting these results together, we derive the integration formula (3.2.34).

3.2.4   Computation of the element advection matrix

When the advection velocity $\mathbf{u}$ is constant and equal to $\mathbf{U} = (U_x, U_y)$, the element advection matrix simplifies to

$$C_{ij}^{(l)} = 2A \, \mathbf{U} \cdot \iint \psi_i \, \nabla \psi_j \, \mathrm{d}\xi \, \mathrm{d}\eta \equiv \alpha \, \beta_j, \tag{3.2.48}$$

```
function [eam] = eam3 (ux,uy,x1,y1,x2,y2,x3,y3)

%==========================================
% Evaluation of the element advection matrix
% for a 3-node triangle
%==========================================

d32x = x3-x2; d32y = y3-y2;
d13x = x1-x3; d13y = y1-y3;
d21x = x2-x1; d21y = y2-y1;

eam(1,1) = (-ux*d32y + uy*d32x)/6.0;
eam(1,2) = (-ux*d13y + uy*d13x)/6.0;
eam(1,3) = (-ux*d21y + uy*d21x)/6.0;

eam(2,1) = eam(1,1);
eam(2,2) = eam(1,2);
eam(2,3) = eam(1,3);

eam(3,1) = eam(1,1);
eam(3,2) = eam(1,2);
eam(3,3) = eam(1,3);

%-----
% done
%-----

return;
```

Function eam3: Evaluation of the element advection matrix from the coordinates of the vertices of a 3-node triangle.

where α is a numerical coefficient given by

$$\alpha = \iint \psi_1 \, d\xi \, d\eta = \iint \psi_2 \, d\xi \, d\eta = \iint \psi_3 \, d\xi \, d\eta = \frac{1}{6}, \qquad (3.2.49)$$

β_j are dimensional coefficients given by

$$\beta_1 = \mathbf{V} \cdot \mathbf{d}_{32}, \qquad \beta_2 = \mathbf{V} \cdot \mathbf{d}_{13}, \qquad \beta_3 = \mathbf{V} \cdot \mathbf{d}_{21}, \qquad (3.2.50)$$

the vector $\mathbf{V}$ is defined as $\mathbf{V} = (U_y, -U_x)$, and the node distances $\mathbf{d}_{ij}$ are defined in (3.2.32). Note that, because the rows of the element advection matrix are identical, the determinant is zero and the matrix is rank-zero singular, that is, it has three null eigenvalues.

FSELIB function eam3, listed in the text, computes the element advection matrix, based on the preceding expressions. The input includes the velocity components, u_x and u_y, and the coordinates of the three nodes.

PROBLEMS

3.2.1 *Gradient of the element interpolation functions.*

Prove the geometrical interpretation of the gradients of the element interpolation functions shown in (3.2.30), as discussed in the text.

3.2.2 *The element diffusion matrix is singular.*

Show that the determinant of the element diffusion matrix displayed in (3.2.31) is zero, which means that the matrix is singular.

3.2.3 *Gradient of a function.*

Consider a function, $f(x, y)$, defined over the 3-node triangle. The gradient of this function, $\nabla f = (\partial f / \partial x, \partial f / \partial y)$, can be computed in terms of ξ and η derivatives using the counterpart of (3.2.25),

$$
\begin{bmatrix}
\frac{\partial x}{\partial \xi} & \frac{\partial y}{\partial \xi} \\[2mm]
\frac{\partial x}{\partial \eta} & \frac{\partial y}{\partial \eta}
\end{bmatrix}
\cdot \nabla f =
\begin{bmatrix}
\frac{\partial f}{\partial \xi} \\[2mm]
\frac{\partial f}{\partial \eta}
\end{bmatrix}.
\tag{3.2.51}
$$

Solving this system by Cramer's rule, we find

$$
\frac{\partial f}{\partial x} = \frac{1}{h_S} \left(\frac{\partial f}{\partial \xi} \frac{\partial y}{\partial \eta} - \frac{\partial f}{\partial \eta} \frac{\partial y}{\partial \xi} \right),
$$

$$
\frac{\partial f}{\partial y} = \frac{1}{h_S} \left(\frac{\partial f}{\partial \eta} \frac{\partial x}{\partial \xi} - \frac{\partial f}{\partial \xi} \frac{\partial x}{\partial \eta} \right).
\tag{3.2.52}
$$

Applying (3.2.52) first for $f = \xi$, and then for $f = \eta$, we find

$$
\frac{\partial \xi}{\partial x} = \frac{1}{h_S} \frac{\partial y}{\partial \eta}, \qquad \frac{\partial \xi}{\partial y} = -\frac{1}{h_S} \frac{\partial x}{\partial \eta},
$$

$$
\frac{\partial \eta}{\partial x} = -\frac{1}{h_S} \frac{\partial y}{\partial \xi}, \qquad \frac{\partial \eta}{\partial y} = \frac{1}{h_S} \frac{\partial x}{\partial \xi}.
\tag{3.2.53}
$$

(*a*) Confirm that the matrix

$$
\mathbf{J}^{-1} \equiv
\begin{bmatrix}
\frac{\partial \xi}{\partial x} & \frac{\partial \xi}{\partial y} \\[2mm]
\frac{\partial \eta}{\partial x} & \frac{\partial \eta}{\partial y}
\end{bmatrix}
\tag{3.2.54}
$$

is the inverse of the Jacobian.

(*b*) Consider a vector function, $\mathbf{F} = (F_x, F_y)$. The divergence of this function is defined as the scalar

$$
\nabla \cdot \mathbf{F} = \frac{\partial F_x}{\partial x} + \frac{\partial F_y}{\partial y}.
\tag{3.2.55}
$$

Show that the divergence also can be evaluated from the expression

$$
\nabla \cdot \mathbf{F} = \frac{1}{h_S} \left[\frac{\partial}{\partial \xi} \left(F_x \frac{\partial y}{\partial \eta} - F_y \frac{\partial x}{\partial \eta} \right) + \frac{\partial}{\partial \eta} \left(- F_x \frac{\partial y}{\partial \xi} + F_y \frac{\partial x}{\partial \xi} \right) \right]. \quad (3.2.56)
$$

(*c*) Show that the divergence of the vector function **F** defined in (*b*) can be evaluated from the expression

$$
\nabla \cdot \mathbf{F} = \frac{1}{h_S} \left[\frac{\partial}{\partial \xi} \left(h_S \, \mathbf{F} \cdot \nabla \xi \right) + \frac{\partial}{\partial \eta} \left(h_S \, \mathbf{F} \cdot \nabla \eta \right) \right], \quad (3.2.57)
$$

where $\nabla \xi = (\partial \xi / \partial x, \partial \xi / \partial y)$, and $\nabla \eta = (\partial \eta / \partial x, \partial \eta / \partial y)$.

3.3 Grid generation

To prepare the grounds for the finite element expansion, we develop a module for domain discretization into the 3-node triangles described in Section 3.2. First, we discuss a triangulation based on the successive subdivision of a parental structure, where a group of ancestral elements are hard-coded and smaller descendant elements arise by subdivision. Second, we discuss a method of grid generation based on the Delaunay triangulation implemented in a MATLAB function. Advanced methods of automatic domain triangulation and public domain codes are reviewed in Appendix E.

3.3.1 Successive subdivisions

FSELIB function `trgl3_disk`, listed in the text, performs the triangulation of the unit disk in the xy plane. The algorithm successively subdivides each of four ancestral triangles located at the four quadrants, shown in the first frame of Figure 3.3.1, into four descendant triangles. The vertices of the descendant triangles lie at the vertices and edge mid-points of their parental elements. In the process of triangulation, boundary nodes are projected in the radial direction onto the unit circle. The level of refinement is determined by the input flag *ndiv*, as follows:

- *ndiv* = 0 produces the four ancestral elements.

- *ndiv* = 1 produces sixteen first-generation elements.

- *ndiv* = 2 produces sixty-four second-generation elements.

- Each time a subdivision is carried out, the number of elements is increased by a factor of four.

Triangulations for discretization levels *ndiv* = 0, 1, 2, 3 are shown in Figure 3.3.1. These shapes can be subsequently deformed or mapped by a transformation to produce other simply-connected, disk-like domains.

```
function [ne,ng,p,c,efl,gfl] = trgl3_disk(ndiv)

%==========================================================
% Triangulation of the unit disk into 3-node elements
% by the successive subdivision of four
% ancestral elements
%
% ndiv: discretization level
%       0 yields the ancestral set
% ne:   number of elements
% ng:   number of global nodes
% p:    coordinates of global nodes
% c:    connectivity table
% efl:  element-node boundary flag
% gfl:  global-node boundary flag
%==========================================================

%-----------------------------------------
% ancestral structure with four elements
%-----------------------------------------

ne = 4;

x(1,1) = 0.0; y(1,1) = 0.0; efl(1,1)=0;  % first element
x(1,2) = 1.0; y(1,2) = 0.0; efl(1,2)=1;
x(1,3) = 0.0; y(1,3) = 1.0; efl(1,3)=1;

x(2,1) = 0.0; y(2,1) = 0.0; efl(2,1)=0;  % second element
x(2,2) = 0.0; y(2,2) = 1.0; efl(2,2)=1;
x(2,3) =-1.0; y(2,3) = 0.0; efl(2,3)=1;

x(3,1) = 0.0; y(3,1) = 0.0; efl(3,1)=0;  % third element
x(3,2) =-1.0; y(3,2) = 0.0; efl(3,2)=1;
x(3,3) = 0.0; y(3,3) =-1.0; efl(3,3)=1;

x(4,1) = 0.0; y(4,1) = 0.0; efl(4,1)=0;  % fourth element
x(4,2) = 0.0; y(4,2) =-1.0; efl(4,2)=1;
x(4,3) = 1.0; y(4,3) = 0.0; efl(4,3)=1;

if(ndiv > 0)       % refinement loop

for i=1:ndiv

  nm = 0; % count the new elements arising in each refinement;
          % four elements will be generated in each pass

  for j=1:ne    % loop over current elements

  % edge mid-nodes will become vertex nodes

    x(j,4) = 0.5*(x(j,1)+x(j,2)); y(j,4) = 0.5*(y(j,1)+y(j,2));
    x(j,5) = 0.5*(x(j,2)+x(j,3)); y(j,5) = 0.5*(y(j,2)+y(j,3));
    x(j,6) = 0.5*(x(j,3)+x(j,1)); y(j,6) = 0.5*(y(j,3)+y(j,1));
```

Function trgl3_disk: Continuing $\longrightarrow$

```
% set the flag of the mid-nodes; a mid-node is a boundary node
% only if the vertex lies on the boundary

    efl(j,4) = 0; efl(j,5) = 0; efl(j,6) = 0;

    if(efl(j,1)==1 & efl(j,2)==1) efl(j,4) = 1; end
    if(efl(j,2)==1 & efl(j,3)==1) efl(j,5) = 1; end
    if(efl(j,3)==1 & efl(j,1)==1) efl(j,6) = 1; end

% assign vertex nodes to sub-elements;
% these will be the "new" elements

    nm = nm+1; % first sub-element
    xn(nm,1) = x(j,1); yn(nm,1) = y(j,1); efln(nm,1) = efl(j,1);
    xn(nm,2) = x(j,4); yn(nm,2) = y(j,4); efln(nm,2) = efl(j,4);
    xn(nm,3) = x(j,6); yn(nm,3) = y(j,6); efln(nm,3) = efl(j,6);

    nm = nm+1; % second sub-element
    xn(nm,1) = x(j,4); yn(nm,1) = y(j,4); efln(nm,1) = efl(j,4);
    xn(nm,2) = x(j,2); yn(nm,2) = y(j,2); efln(nm,2) = efl(j,2);
    xn(nm,3) = x(j,5); yn(nm,3) = y(j,5); efln(nm,3) = efl(j,5);

    nm = nm+1; % third sub-element
    xn(nm,1) = x(j,6); yn(nm,1) = y(j,6); efln(nm,1) = efl(j,6);
    xn(nm,2) = x(j,5); yn(nm,2) = y(j,5); efln(nm,2) = efl(j,5);
    xn(nm,3) = x(j,3); yn(nm,3) = y(j,3); efln(nm,3) = efl(j,3);

    nm = nm+1; % fourth sub-element
    xn(nm,1) = x(j,4); yn(nm,1) = y(j,4); efln(nm,1) = efl(j,4);
    xn(nm,2) = x(j,5); yn(nm,2) = y(j,5); efln(nm,2) = efl(j,5);
    xn(nm,3) = x(j,6); yn(nm,3) = y(j,6); efln(nm,3) = efl(j,6);

end     % end of loop over current elements

ne = 4*ne;  % number of elements has increased
            % by a factor of four

for k=1:ne     % relabel the new points
    for l=1:3  % and put them in the master list

      x(k,l) = xn(k,l); y(k,l) = yn(k,l); efl(k,l) = efln(k,l);

      % project the boundary nodes onto the unit circle:

      if(efl(k,l)==1)
        rad = sqrt(x(k,l)^2+y(k,l)^2);
        x(k,l) = x(k,l)/rad; y(k,l) = y(k,l)/rad;
      end

    end
 end

end % end do of refinement loop
end % end if of refinement loop
```

Function trgl3_disk: $\longrightarrow$ Continuing $\longrightarrow$

```
%--------------------
% plotting (optional)
%--------------------

for i=1:ne
  xp(1) = x(i,1); xp(2) = x(i,2);
  xp(3) = x(i,3); xp(4) = x(i,1);
  yp(1) = y(i,1); yp(2) = y(i,2);
  yp(3) = y(i,3); yp(4) = y(i,1);
  plot(xp, yp,'-o'); hold on;
  xlabel('x'); ylabel('y');
end

%---------------------------------------------------
% define the global nodes and the connectivity table
%---------------------------------------------------

% 3 nodes of the first element are entered manually

p(1,1) = x(1,1); p(1,2) = y(1,1); gfl(1,1) = efl(1,1);
p(2,1) = x(1,2); p(2,2) = y(1,2); gfl(2,1) = efl(1,2);
p(3,1) = x(1,3); p(3,2) = y(1,3); gfl(3,1) = efl(1,3);

c(1,1) = 1;  % first  node of first element is global node 1
c(1,2) = 2;  % second node of first element is global node 2
c(1,3) = 3;  % third  node of first element is global node 3

ng = 3;

% loop over further elements

eps = 0.000001;

% Iflag=0 will signal a new global node

for i=2:ne         % loop over elements
  for j=1:3          % loop over element nodes

  Iflag=0;           % initialize

  for k=1:ng

   if(abs(x(i,j)-p(k,1)) < eps)
    if(abs(y(i,j)-p(k,2)) < eps)
      Iflag = 1;   % the node has been recorded previously
      c(i,j) = k;  % the jth local node of element i
                   % is the kth global node
    end
   end

  end
```

Function trgl3_disk: $\longrightarrow$ Continuing $\longrightarrow$

```
if(Iflag==0)   % record the node
  ng = ng+1;

  p(ng,1) = x(i,j);
  p(ng,2) = y(i,j);

  gfl(ng,1) = efl(i,j);
  c(i,j) = ng;    % the jth local node of the element
                  % is the new global node
end

end % end of loop over element nodes
end  % end of loop over elements

%-----
% done
%-----

return;
```

Function trgl3_disk: ($\longrightarrow$ Continued.) Triangulation of the unit disk into 3-node elements by the successive subdivision of four ancestral elements.

Distinguishing features of the triangulation algorithm include the following:

- In the first part of the algorithm, only the element nodes are generated. The unique global nodes and the connectivity matrix are defined in the second stage, after the triangulation has been completed.

- The array efl is the boundary flag of the element nodes, as described in Section 3.1; $efl(i,j)$ is set to zero at the interior nodes, and to unity at the boundary nodes.

- The flag efl of a new node is set according to the flags of the two vertex nodes defining the subdivided side. If both nodes are boundary nodes, then the new node is also a boundary node.

- The global array $p(i,j)$, where $j = 1, 2$, contains the x and y coordinates of the global nodes, for $i = 1, 2, \ldots, N_G$.

- The array $c(i,j)$ hosts the connectivity matrix described in Section 3.1.

- The array gfl is the boundary flag of the global nodes, as described in Section 3.1.

Triangulation of other domains can be performed in the spirit of the function trgl3_disk, beginning with a different set of ancestral triangles, as will be discussed later in this chapter. As an example, the triangulation of a square

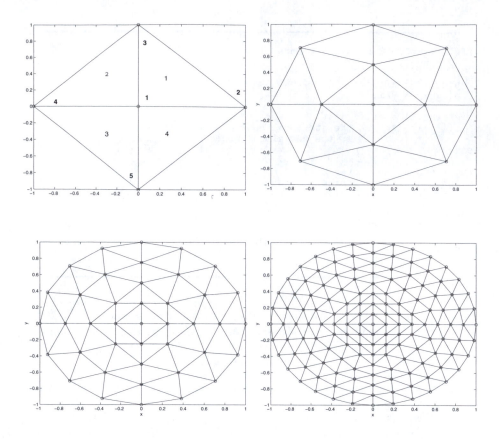

Figure 3.3.1 Discretization of a disk into 3-node triangles produced by the
FSELIB function `trgl3_disk`, for *ndiv* = 0 (ancestral set), 1, 2, and 3.
The labels of the global nodes are printed in bold in the first frame.

patch is discussed in Problem 3.3.1. In all cases, at least one element node in
the ancestral set must be an interior node, otherwise the flag $efl(i, 1)$ will be
erroneously set to unity for all generated element nodes.

3.3.2   Delaunay triangulation

In the inverse approach, we first specify the location of the global nodes defining
the triangle vertices, and then produce the elements by mapping global nodes to
element nodes. Element definition from a given global set can be done by various
methods, including the method of Delaunay triangulation. One limitation of
this approach is that the standard algorithm triangulates the convex hull of the
specified set of points, which is defined as the area enclosed by a rubber band
that is stretched and then allowed to relax toward the global nodes.

One way of carrying out the Delaunay triangulation is by means of the Dirichlet-Voronoi-Thiessen tessellation (DVT). In this approach, the discretization domain is divided into polygons, subject to the condition that each polygon contains only one global node, and the distance of an arbitrary point inside a polygon from the native global node is smaller than the distance from any other node. It can be shown that the sides of the polygons thus produced are perpendicular bisectors of the straight segments connecting pairs of nodes. Once the DVT tessellation has been completed, the Delaunay triangulation emerges by connecting a node to all other nodes that share a polygon side. Direct methods of Delaunay triangulation are reviewed by Rebay [47]. A wealth of information, algorithms, and useful links, can be found at the Internet site: `http://www.voronoi.com` .

The MATLAB function `delaunay` performs the Delaunay triangulation according to the function call:

```
c = delaunay(x,y)
```

where the input vectors x and y hold the coordinates of the specified global nodes, and the familiar $N_E \times 3$ global connectivity matrix, c, holds the global labels of the N_E 3-node triangles arising from the Delaunay triangulation.

FSELIB function `trgl3_delaunay`, listed in the text, performs the triangulation based on a specified set of global nodes tabulated in the file *points.dat*. The first column in this file holds the x coordinates of the nodes, the second column holds the y coordinates of the nodes, and the third column holds the global boundary flag, gfl, in the format of the sample file:

```
.6375 .2625 1
.4950 .3000 1
...
.4725 .4875 0
.5475 .5325 0
```

A triangulation produced by this function is shown in Figure 3.3.2(*a*), and the corresponding DVD tessellation produced by the MATLAB function `voronoi` is shown in Figure 3.3.2(*b*).

As a further application, we perform the triangulation of a square. FSELIB function `trgl3_delaunay_sqr`, listed in the text, generates a uniform $N \times M$ grid of global nodes, randomizes the coordinates of the interior nodes using the MATLAB random-number generator `rand`, performs the Delaunay triangulation, displays the triangulation using the MATLAB graphics function `trimesh`, and prepares a graph of the Voronoi tessellation. A triangulation produced by this function is shown in Figure 3.3.3(*a*), and the corresponding DVT tessellation is shown in Figure 3.3.3(*b*).

```
function [ne,ng,p,c,efl,gfl] = trgl3_delaunay

%====================================================
% Discretization of an arbitrary domain into 3-node
% elements by the Delaunay triangulation
%====================================================

%----------------------------------------
% Read the global nodes and boundary flags
% from file: points.dat
% The three columns in this file are the
% x and y coordinates and the boundary
% flag of the global nodes:
%
%   x(1) y(1) gfl(1,1)
%   x(2) y(2) gfl(2,1)
%   ... ... ...
%   x(n) y(n) gfl(n,1)
%----------------------------------------

file1 = fopen('points.dat');
 points = fscanf(file1,'%f');
fclose(file1);

%--------------------------
% number of global nodes, ng
%--------------------------

sp = size(points); ng = sp(1)/3;

%-------------------------------------------------
% coordinates and boundary flag of global nodes
%-------------------------------------------------

for i=1:ng
 p(i,1) = points(3*i-2);
 p(i,2) = points(3*i-1);
 gfl(i,1) = points(3*i);
 xdel(i) = p(i,1); ydel(i) = p(i,2);  % input to delaunay
end

%----------------------
% delaunay triangulation
%----------------------

c = delaunay (xdel,ydel);

%---------------------------------
% extract the number of elements, ne
%---------------------------------

sc = size(c); ne = sc(1,1);
```

Function trgl3_delaunay: Continuing $\longrightarrow$

```
%------------------------------------
% set the element-node boundary flags
%------------------------------------

for i=1:ne
 efl(i,1) = gfl(c(i,1),1);
 efl(i,2) = gfl(c(i,2),1);
 efl(i,3) = gfl(c(i,3),1);
end

%-----
% plot
%-----

for i=1:ne

  % plot the elements:

  xp(1) = p(c(i,1),1); yp(1) = p(c(i,1),2);
  xp(2) = p(c(i,2),1); yp(2) = p(c(i,2),2);
  xp(3) = p(c(i,3),1); yp(3) = p(c(i,3),2);
  xp(4) = p(c(i,1),1); yp(4) = p(c(i,1),2);

  plot(xp, yp); hold on

  % mark the boundary nodes:

  if(efl(i,1)==1)
     plot(xp(1), yp(1), 'o');
  end;
  if(efl(i,2)==1)
     plot(xp(2), yp(2), '+');
  end;
  if(efl(i,3)==1)
    plot(xp(3), yp(3), 'x');
  end;

end

xlabel('x'); ylabel('y');

%---------------------
% Voronoi tessellation
%---------------------

figure; voronoi(xdel,ydel)

%-----
% done
%-----

return
```

Function trgl3_delaunay: ($\longrightarrow$ Continued.) Triangulation of a domain with pre-assigned global nodes using the Delaunay triangulation.

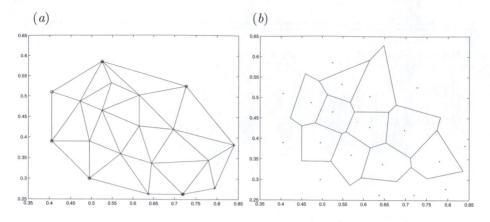

Figure 3.3.2 (*a*) Discretization of a domain into 3-node triangles by the Delaunay triangulation, produced by the FSELIB function `trgl3_delaunay`. The boundary nodes are indicated by filled circles. (*b*) Corresponding Voronoi tessellation.

```
function [ne,ng,p,c,efl,gfl] = trgl3_delaunay_sqr

%---------------------
% window and grid size
%---------------------

X1 = -1.0; X2 = 1.0;
Y1 = -1.0; Y2 = 1.0;

N = 8; M = 8;

%--------
% prepare
%--------

Dx = (X2-X1)/N;
Dy = (Y2-Y1)/N;

%-----------------------------
% arrange points on a mesh grid
% set the boundary flag
% and count the nodes (ng)
%-----------------------------

ng = 0;
```

Function trgl3_delaunay_sqr: Continuing $\longrightarrow$

```
for j=1:M+1

  for i=1:N+1
   ng = ng+1;
   p(ng,1) = X1+(i-1.0)*Dx;
   p(ng,2) = Y1+(j-1.0)*Dy;
   gfl(ng,1) = 0;
   if(i==1 | i==N+1 | j==1 | j==M+1) gfl(ng,1)=1; end;
  end

end

%-----------------------------
% randomize the interior nodes
%-----------------------------

Ic = N+2;

for j=2:M
 for i=2:N
   Ic=Ic+1;
   p(Ic,1) = p(Ic,1)+(rand-1.0)*0.5*Dx;
   p(Ic,2) = p(Ic,2)+(rand-1.0)*0.4*Dy;
 end
 Ic=Ic+2;
end

%----------------------
% Delaunay triangulation
%----------------------

for i=1:ng
 xdel(i) = p(i,1); ydel(i) = p(i,2);
end

c = delaunay (xdel,ydel);

%-----------------------------
% extract the number of elements
%-----------------------------

sc = size(c); ne = sc(1,1);

%-------------------------------------
% set the element-node boundary flags
%-------------------------------------

for i=1:ne
 efl(i,1) = gfl(c(i,1),1);
 efl(i,2) = gfl(c(i,2),1);
 efl(i,3) = gfl(c(i,3),1);
end
```

Function trgl3_delaunay_sqr: Continuing $\longrightarrow$

```
%---------------------------
% display the triangulation
%---------------------------

trimesh (c,p(:,1),p(:,2),zeros(ng,1));

%--------------------
% Voronoi tessellation
%--------------------

figure;
voronoi (xdel,ydel);

%-----
% done
%-----

return
```

Function trgl3_delaunay_sqr: ($\longrightarrow$ Continued.) Triangulation of a square with global nodes generated by a randomly perturbed square grid, using the Delaunay triangulation.

(a)   (b)

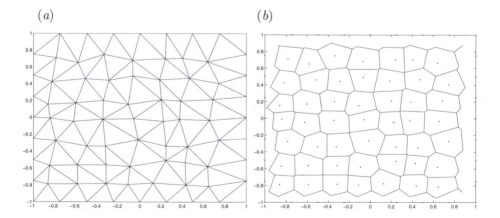

Figure 3.3.3 (a) Discretization of a square with randomized interior nodes into 3-node triangles using the Delaunay triangulation, produced by the FSELIB function `trgl3_delaunay_sqr`. (b) Corresponding Voronoi tessellation.

3.3.3 Generalized connectivity matrices

Both discretization methods discussed earlier in this section generate the connectivity matrix, $c(i, j)$, which relates the element nodes to the global nodes. The index i runs over the elements labels, $i = 1, 2, \ldots, N_E$, the index j runs over the three element node labels, $j = 1, 2, 3$, and the components of the connectivity matrix, $c(i, j)$, take values in the range: $1, 2, \ldots, N_G$.

Further connectivity tables can be defined to facilitate various computations, as required, including the following:

- The connectivity matrix $cen(i, j)$ is defined such that:

  - $cen(i, 1)$ is the number of elements sharing the ith global node, for $i = 1, 2, \ldots, N_G$.

  - $cen(i, j)$, for $j = 2, 3, \ldots, ce(i, 1) + 1$, are the corresponding element labels.

  This matrix can be used to compute nodal values as averages of neighboring element values.

- The connectivity matrix $ces(i, j)$ is defined such that $ces(i, j)$ is the label of the element sharing the jth edge (side) of the ith element, where $j = 1, 2, 3$, and $i = 1, 2, \ldots, N_E$. Side 1 is subtended between nodes 1 and 2, side 2 is subtended between nodes 2 and 3, and side 3 is subtended between nodes 3 and 1. If $ces(i, j) = 0$, the jth side of the ith element does not have a neighbor, which means that the node lies on the boundary.

  The matrix ces can be used to compute element side values as averages of neighboring element values.

The connectivity matrix cen can be computed with the help of the connectivity matrix, c, by running over all global nodes, and then running over all element nodes, according to the FSELIB function cen3, listed in the text. The connectivity matrix ces can be computed in a similar fashion according to the FSELIB function ces3, listed in the text. As an application, the FSELIB script see_elm3, listed in the text, scans the global nodes and draws neighboring elements.

3.3.4 Element and node numbering

Given an element layout, there are many ways to label the elements, the element nodes, and the unique global nodes. For example, the labels of two unique global nodes can be interchanged, provided that appropriate modifications are made to the connectivity matrix. Although the protocol of element and node numbering has no effect on the finite element solution itself, it may have a profound influence on the structure of the global diffusion, mass, and advection matrices, and thus on the efficiency of the finite element code.

```
function cen = cen3(ne,ng,c)

%=============================================
% Evaluation of the element-to-node
% connectivity matrix "cen"
%
% cen(i,1) is the number of elements sharing
% the ith global node, where i=1,2, ..., ng
%
% cen(i,j), for j=2,3, ..., ce(i,1)+1
% are the corresponding element labels
%
% ne: number of elements
% ng: number of global nodes
%=============================================

  for i=1:ng    % scan the global nodes

   cen(i,1) = 0;
   Icount = 1;

   for j=1:ne    % scan the elements
     for k=1:3

       if(c(j,k)==i)              % the ith global node and the
         cen(i,1)=cen(i,1)+1;     % test element node are identical
         Icount = Icount +1;
         cen(i,Nodeplace) = j;
       end

     end

   end
  end

%-----
% done
%-----

return
```

Function cen3: FSELIB function for computing the element-to-node connectivity matrix *cen* for a 3-node triangle.

Ideally, elements and nodes should be labeled so that the global matrices are as close to being diagonal as possible, that is, they have the smallest possible bandwidth. The objective is to economize storage and expedite the solution of the final system of algebraic or ordinary differential equations, as discussed in Appendix C. Methods of optimal node labeling are reviewed by Schwarz ([54], pp. 167–185).

```
function ces = ces3(ne,ng,c)

%====================================
% Evaluation of the element-to-sides
% connectivity matrix ces
%====================================

%--------------------
% wrap the first node
%--------------------

  for i=1:ne
    c(i,4) = c(i,1)
  end

%----------------------
% run over the elements
%----------------------

  for i=1:ne
   for j=1:3

    ces(i,j) = 0;

    for k=1:ne

       if(k ~= i) % skip the self-element
%----
          for l=1:3
            if(c(k,l)==c(i,j) &   c(k,l+1)==c(i,j+1) )
              ces(i,j) = k;
            end
            if(c(k,l)==c(i,j+1) &   c(k,l+1)==c(i,j) )
              ces(i,j) = k;
            end
          end
%----
       end

    end

   end
  end

%-----
% done
%-----

return
```

Function ces3: FSELIB function for computing the element-to-side connectivity matrix, *ces*.

```
ndiv=1;   % specify the discretization level

[ne,ng,p,c,efl,gfl] = trgl3_disk (ndiv);   % triangulate the disk

cen = cen3(ne,ng,c);   % element to node connectivity

for node=1:ng    % run over nodes; or else specify "node"

 for i=2:cen(node,1)+1

  l = cen(node,i);    % element number

  j = c(l,1); xp(1) = p(j,1); yp(1) = p(j,2);
  j = c(l,2); xp(2) = p(j,1); yp(2) = p(j,2);
  j = c(l,3); xp(3) = p(j,1); yp(3) = p(j,2);
  j = c(l,1); xp(4) = p(j,1); yp(4) = p(j,2);

  cp(1) = 1.0; cp(2) = 0.0; cp(3) = 0.0; cp(4) = 1.0;   % color indices

  patch(xp,yp,cp); hold on;

 end

end
```

Script see_elm3**:** This script scans the global nodes and displays the host neighboring elements.

PROBLEMS

3.3.1 *Triangulation of a square.*

FSELIB function trgl3_sqr triangulates a square using an algorithm that is similar to that implemented in the function trgl3_disk, as discussed in the text. In the case of the square, we begin by hard-coding eight ancestral triangles, two in each quadrant, as shown in Figure 3.3.4(*a*), and then generate different levels of triangulation by element subdivision. In the end, the grid may be deformed to yield a desired shape that is homologous to the square. A deformed grid for discretization level $ndiv = 2$ is displayed in Figure 3.3.4(*b*).

Execute the function trgl3_sqr, and display the grid layout for discretization levels $ndiv = 0$, 1, and 2.

3.3.2 *Delaunay triangulation.*

Execute the FSELIB function trgl3_delaunay with a set of global nodes of your choice, and display the grid layout and associated Voronoi diagram.

(*a*)   (*b*)

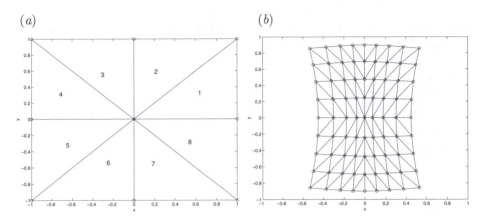

Figure 3.3.4 (*a*) Ancestral 8-element structure for triangulating a square, and (*b*) discretization of a deformed square for *ndiv* = 2, produced by the FSELIB function `trgl3_sqr`.

3.3.3 *Element removal.*

Write a function that subtracts a specified element from the triangulation produced by the FSELIB function `trgl3_delaunay_sqr`, possibly also removing a global node. Run the function until all elements have been removed one by one, and display intermediate stages of the triangulation.

3.4 Code for Laplace's equation with the Dirichlet boundary condition in a disk-like domain

As a first case study, we consider steady conduction in the absence of a distributed source. The unknown temperature field, $f(x, y)$, satisfies Laplace's equation,

$$\nabla^2 f = 0, \tag{3.4.1}$$

subject to the Dirichlet boundary condition around the contour of a simply connected, disk-like domain. Carrying out the Galerkin projection, we find that the solution vector, $\mathbf{f}$, containing the N_G values of the function, f, at the global nodes, satisfies the homogeneous linear algebraic system, $\mathbf{D} \cdot \mathbf{f} = \mathbf{0}$, where $\mathbf{D}$ is the global diffusion matrix. Implementing the Dirichlet boundary condition, as discussed in Section 3.1.4, we obtain an altered inhomogeneous system with a non-zero right-hand side,

$$\widehat{\mathbf{D}} \cdot \mathbf{f} = \mathbf{b}. \tag{3.4.2}$$

The complete finite element FSELIB code `lapl3_d` is listed in the text.

```
%================================================
% CODE: lapl3_d
%
% Finite element solution of Laplace's equation
% in a disk-like domain with the
% Dirichlet boundary condition
%================================================

%-----------
% input data
%-----------

ndiv = 3;    % level of triangulation

%------------
% triangulate
%------------

[ne,ng,p,c,efl,gfl] = trgl3_disk (ndiv);

%-------
% deform
%-------

defx = 0.6;                    % sample deformation

for i=1:ng
 p(i,1)=p(i,1)*(1.0-defx*p(i,2)^2 );
end

%--------------------------------------------
% specify the Dirichlet boundary condition
%--------------------------------------------

for i=1:ng
  if(gfl(i,1)==1)
    gfl(i,2) = p(i,1)*sin(0.5*pi*p(i,2));
  end
end

%------------------------------------
% assemble the global diffusion matrix
%------------------------------------

gdm = zeros(ng,ng); % initialize

for l=1:ne    % loop over the elements

% compute the element diffusion matrix:

 j=c(l,1); x1=p(j,1); y1=p(j,2);
 j=c(l,2); x2=p(j,1); y2=p(j,2);
 j=c(l,3); x3=p(j,1); y3=p(j,2);
```

Code lapl3_d: Continuing $\longrightarrow$

```
 edm_elm = edm3 (x1,y1,x2,y2,x3,y3);

   for i=1:3
     i1=c(1,i);
     for j=1:3
       j1=c(1,j);
       gdm(i1,j1) = gdm(i1,j1) + edm_elm(i,j);
     end
   end

end

%------------------------------
% compute the right-hand side
%------------------------------

for m=1:ng
  b(m)=0.0;    % initialize
end

for m=1:ng

  if(gfl(m,1)==1)
    for i=1:ng
      b(i) = b(i) - gdm(i,m) * gfl(m,2);
      gdm(i,m)=0; gdm(m,i)=0;
    end
    gdm(m,m)=1.0; b(m)=gfl(m,2);
  end

end

%-----------------------
% solve the linear system
%-----------------------

f = b/gdm';

%---------
% graphics
%---------

plot_3 (ne,ng,p,c,f); trimesh (c,p(:,1),p(:,2),f);
```

Code lapl3_d: ($\longrightarrow$ Continued.) Finite element solution of Laplace's equation with the Dirichlet boundary condition.

The basic modules of the code are the following:

- *Specify the input data.*

- *Perform the grid generation.*

- *Implement the Dirichlet boundary condition at the global nodes.*

If $gfl(i, 1) = 1$, the ith global node is a boundary node, and the second entry $gfl(i, 2)$ hosts the prescribed boundary data.

- *Assemble the global diffusion matrix from the element diffusion matrices according to Algorithm 2.1.2.*

The element diffusion matrix is produced by the FSELIB function edm3. For the lowest level of discretization, $ndiv = 0$, illustrated in the first frame of Figure 3.3.1, the 5×5 global diffusion matrix reads

$$
\mathbf{D} = \begin{bmatrix} 4 & -1 & -1 & -1 & -1 \\ -1 & 1 & 0 & 0 & 0 \\ -1 & 0 & 1 & 0 & 0 \\ -1 & 0 & 0 & 1 & 0 \\ -1 & 0 & 0 & 0 & 1 \end{bmatrix}. \tag{3.4.3}
$$

For the next level of discretization, $ndiv = 1$, illustrated in the second frame of Figure 3.3.1, the 13×13 global diffusion matrix is shown in Table 3.4.1.

- *Implement the Dirichlet boundary condition by modifying the equations corresponding to the boundary nodes, as discussed in Section 3.1.7.*

For the lowest level of discretization, $ndiv = 0$, the modified global diffusion matrix takes the diagonal form

$$
\widehat{\mathbf{D}} = \begin{bmatrix} 4 & 0 & 0 & 0 & 0 \\ 0 & 1 & 0 & 0 & 0 \\ 0 & 0 & 1 & 0 & 0 \\ 0 & 0 & 0 & 1 & 0 \\ 0 & 0 & 0 & 0 & 1 \end{bmatrix}. \tag{3.4.4}
$$

Note that all rows, except for the first row, implement the Dirichlet boundary condition. For the next level of discretization, $ndiv = 1$, the modified global diffusion matrix truncated at the first decimal place is shown in Table 3.4.2.

- *Solve the linear system using the* MATLAB *linear solver.*

- *Graphics display.*

Prepare a color graph of the finite element solution using (a) the MATLAB graphics function patch with color coding, as implemented in the FSELIB function plot_3, listed in the text, and (b) the MATLAB graphics function trimesh. Note that trimesh accepts as input the connectivity matrix, c, the x and y coordinates of the global nodes, and the solution vector.

The graphics display of a computation with the sinusoidal boundary condition implemented in the code is shown in Figure 3.4.1.

4.0	−1.0	−1.0	0	0	0	−1.0	0	0	−1.0	0	0	0
−1.0	3.9557	−0.2050	−1.4214	−0.5621	0	0	0	0	−0.2050	0	0	−0.5621
−1.0	−0.2050	3.7610	0	−0.8030	−0.7448	−0.2050	−0.8030	0	0	0	0	0
0	−1.4214	01.4143	0.0036	0	0		0	0	0	0	0	0.0036
0	−0.5621	−0.8030	0.0036	1.5708	−0.2092	0	0	0	0	0	0	0
0	0	−0.7448	0	−0.2092	1.1633	0	−0.2092	0	0	0	0	0
−1.0	0	−0.2050	0	0	0	3.9557	−0.5621	−1.4214	−0.2050	−0.5621	0	0
0	0	−0.8030	0	0	−0.2092	−0.5621	1.5708	0.0036	0	0	0	0
0	0	0	0	0	0	−1.4214	0.0036	1.4143	0	0.0036	0	0
−1.0	−0.2050	0	0	0	0	−0.2050	0	0	3.7610	−0.8030	−0.7448	−0.8030
0	0	0	0	0	0	−0.5621	0	0.0036	−0.8030	1.5708	−0.2092	0
0	0	0	0	0	0	0	0	0	−0.7448	−0.2092	1.1633	−0.2092
0	−0.5621	0	0.0036	0	0	0	0	0	−0.8030	0	−0.2092	1.5708

$\longrightarrow$

Table 3.4.1 The global diffusion matrix, **D**, for discretization level, $ndiv = 1$. The element layout is shown in the second frame of Figure 3.3.1.

$$
\begin{bmatrix}
4.0 & -1.0 & -1.0 & 0 & 0 & 0 & -1.0 & 0 & 0 & -1.0 & 0 & 0 & 0 \\
-1.0 & 4.0 & -0.2 & 0 & 0 & 0 & 0 & 0 & 0 & -0.2 & 0 & 0 & 0 \\
-1.0 & -0.2 & 3.8 & 0 & 0 & 0 & -0.2 & 0 & 0 & 0 & 0 & 0 & 0 \\
0 & 0 & 0 & 1.0 & 0 & 0 & 0 & 0 & 0 & 0 & 0 & 0 & 0 \\
0 & 0 & 0 & 0 & 1.0 & 0 & 0 & 0 & 0 & 0 & 0 & 0 & 0 \\
0 & 0 & 0 & 0 & 0 & 1.0 & 0 & 0 & 0 & 0 & 0 & 0 & 0 \\
-1.0 & 0 & -0.2 & 0 & 0 & 0 & 4.0 & 0 & 0 & -0.2 & 0 & 0 & 0 \\
0 & 0 & 0 & 0 & 0 & 0 & 0 & 1.0 & 0 & 0 & 0 & 0 & 0 \\
0 & 0 & 0 & 0 & 0 & 0 & 0 & 0 & 1.0 & 0 & 0 & 0 & 0 \\
-1.0 & -0.2 & 0 & 0 & 0 & 0 & -0.2 & 0 & 0 & 3.8 & 0 & 0 & 0 \\
0 & 0 & 0 & 0 & 0 & 0 & 0 & 0 & 0 & 0 & 1.0 & 0 & 0 \\
0 & 0 & 0 & 0 & 0 & 0 & 0 & 0 & 0 & 0 & 0 & 1.0 & 0 \\
0 & 0 & 0 & 0 & 0 & 0 & 0 & 0 & 0 & 0 & 0 & 0 & 1.0
\end{bmatrix}
$$

Table 3.4.2 The global diffusion matrix shown in Table 3.4.1, after the Dirichlet boundary condition has been implemented.

```
function plot_3 (ne,ng,p,c,f);

%===================================================
% Color-mapped visualization of a function f
% in a domain discretized into 3-node triangles.
%===================================================

% compute the maximim and minimum of the function f

fmax =-100.0;  % initialize
fmin = 100.0;  % initialize

for i=1:ng
 if(f(i) > fmax) fmax = f(i); end
 if(f(i) < fmin) fmin = f(i); end
end

range = 1.2*(fmax-fmin); shift = fmin;

%-------------------------------------------
% shift the color index in the range (0, 1)
% and plot 4-point patches
%-------------------------------------------
```

Function plot_3: Continuing ⟶

```
for l=1:ne
 j=c(1,1); xp(1)=p(j,1); yp(1)=p(j,2); cp(1)=(f(j)-shift)/range;
 j=c(1,2); xp(2)=p(j,1); yp(2)=p(j,2); cp(2)=(f(j)-shift)/range;
 j=c(1,3); xp(3)=p(j,1); yp(3)=p(j,2); cp(3)=(f(j)-shift)/range;
 j=c(1,1); xp(4)=p(j,1); yp(4)=p(j,2); cp(4)=(f(j)-shift)/range;
 patch (xp, yp,cp); hold on;
 xlabel('x'); ylabel('y'); zlabel('f');
end
%-----
% done
%-----
```

Function plot_3: Color-mapped visualization of a function in a domain discretized into 3-node triangles using the MATLAB function `patch`.

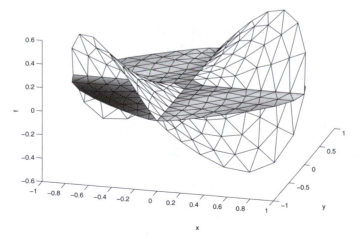

Figure 3.4.1 Finite element solution of Laplace's equation subject to the Dirichlet boundary condition, produced by code `lapl3_d`.

After the finite element solution has been concluded, the boundary flux can be calculated based on the Galerkin projection of the global interpolation functions corresponding to the boundary nodes, expressed by the following redacted form of the general projection (3.1.17),

$$\oint_C \phi_i \, q \, \mathrm{d}l = k \sum_{i=j}^{N_G} D_{ij} \, f_j, \qquad (3.4.5)$$

with the understanding that the ith node is a boundary node. Note that (3.4.5) is the counterpart of the one-dimensional algebraic form (1.1.44), in the absence of a source.

Applying (3.4.5) for the global interpolation functions of all boundary nodes, we obtain a system of integral equations for q whose solution can be found by standard numerical methods. The numerical implementation is similar to that discussed in Sections 3.2 and 3.3.

PROBLEMS

3.4.1 *Constant and linear boundary conditions.*

Run the FSELIB code `lapl3_d` for a circular domain, with boundary values that are (a) uniform (constant), (b) linear in x, and (c) linear in y, and discuss the results of your computation in each case.

3.4.2 *Deformed domain.*

Repeat Problem 3.4.1 with a deformed disk-like domain of your choice.

3.4.3 *Delaunay triangulation.*

Replace the discretization function `trgl3_disk` with the Delaunay-based FSELIB function `trgl3_delaunay` discussed in Section 3.3. Run the code for a domain geometry and boundary conditions of your choice, and discuss the results of your computation.

3.5 Code for steady convection–diffusion with the Dirichlet boundary condition

As a second case study, we consider the steady linear convection–diffusion equation

$$u_x \frac{\partial f}{\partial x} + u_y \frac{\partial f}{\partial y} = \kappa \, \nabla^2 f, \qquad (3.5.1)$$

where u_x and u_y are specified constant velocities along the x and y axes, and κ is the medium diffusivity. The Dirichlet boundary condition is prescribed around

the contiguous contour of a simply connected domain arising from the deformation of a circular disk. The importance of convection relative to diffusion is
expressed by the dimensionless Péclet number,

$$Pe \equiv \frac{UR}{\kappa}, \qquad (3.5.2)$$

where U is the maximum of $|u_x|$ and $|u_y|$, and R is the radius of the undeformed
disk. When $Pe = 0$, we recover the problem of steady diffusion governed by
Laplace's equation, as discussed in Section 3.4. As the Péclet number is raised,
convective transport becomes increasingly important everywhere except near
the boundaries where thin boundary layers arise.

Carrying out the Galerkin projection, we find that the solution vector, $\mathbf{f}$,
containing the N_G nodal values of the requisite function at the global nodes,
satisfies the homogeneous algebraic system,

$$(\kappa \, \mathbf{D} + \mathbf{N}) \cdot \mathbf{f} = \mathbf{0}, \qquad (3.5.3)$$

where $\mathbf{D}$ is the global diffusion matrix, and $\mathbf{N}$ is the global advection matrix.
Implementing the Dirichlet boundary condition, as discussed in Section 3.1.4,
we derive an altered inhomogeneous system with a non-zero right-hand side,

$$(\kappa \, \widehat{\mathbf{D}} + \widehat{\mathbf{N}}) \cdot \mathbf{f} = \mathbf{b}, \qquad (3.5.4)$$

where the hat designates modified diffusion and advection matrices. The finite
element implementation is a straightforward modification of that described in
Section 3.4 for Laplace's equation. The complete finite element code scd3_d,
incorporating the assembly of the global diffusion and advection matrices, is
listed in the text.

Figure 3.5.1 displays the finite element solution for convection along the
x axis, $u_x > 0$ and $u_y = 0$, at four values of the Péclet number, $Pe = 0$, 2,
5, and 10. The boundary conditions specify that the temperature, f, varies
linearly in the x coordinate around the boundary, as implemented in the code.
The numerical results confirm that, as the Péclet number is raised, convection
becomes increasingly important and steep gradients arise near the right half of
the boundary in agreement with physical intuition.

PROBLEMS

3.5.1 *Code* scd3_d.

(*a*) Run the code scd3_d for a circular domain with parameter values and
boundary conditions of your choice, and discuss the results of your computation.

(*b*) Repeat (*a*) for a deformed disk-like domain of your choice.

```
%=========================================================
% CODE scd3_d
%
% Finite element code for steady convection--diffusion
% subject to the Dirichlet boundary condition
%
% gdm: global diffusion matrix
% gam: global advection matrix
% lsm: linear system matrix
%=========================================================

%-----------
% input data
%-----------

k=1.0; rho=1.0; cp=1.0; ux=5.0; uy=0.0;

ndiv = 3;

%----------
% constants
%----------

kappa = k/(rho*cp);

%----------------------
% triangulate and deform
%----------------------

[ne,ng,p,c,efl,gfl] = trgl3_disk (ndiv);

for i=1:ng
 p(i,1) = p(i,1)*(1.0-0.5*p(i,2)^2 );
end

%------------------------------------------
% specify the Dirichlet boundary condition
%------------------------------------------

for i=1:ng

  if(gfl(i,1)==1)
    gfl(i,2) = p(i,1);  % example
  end

end

%----------------------------
% assemble the global diffusion
% and advection matrices
%----------------------------
```

Code scd3_d: Continuing $\longrightarrow$

```
gdm = zeros(ng,ng);    %  initialize

gam = zeros(ng,ng);    %  initialize

for l=1:ne          % loop over the elements
% compute the element diffusion
% and advection matrices

   j=c(l,1);
       x1=p(j,1); y1=p(j,2);
   j=c(l,2);
       x2=p(j,1); y2=p(j,2);
   j=c(l,3);
       x3=p(j,1); y3=p(j,2);

   [edm_elm] = edm3 (x1,y1,x2,y2,x3,y3);

   [eam_elm] = eam3 (ux,uy,x1,y1,x2,y2,x3,y3);

    for i=1:3
      i1 = c(l,i);
      for j=1:3
        j1 = c(l,j);
        gdm(i1,j1) = gdm(i1,j1) + edm_elm(i,j);
        gam(i1,j1) = gam(i1,j1) + eam_elm(i,j);
      end
    end

end

%--------------------------------
% compute the coefficient matrix
%--------------------------------

lsm = kappa*gdm-gam;

%---------------------------------------------
% compute the right-hand side and implement
% the Dirichlet boundary condition
%---------------------------------------------

for i=1:ng
  b(i) = 0.0;
end
```

Code scd3_d: Continuing $\longrightarrow$

3.5.2 *Delaunay triangulation.*

Replace the discretization function `trgl3_disk` implemented in code `scd3_d` with the Delaunay-based FSELIB function `trgl3_delaunay` discussed in Section 3.3.2. Run the modified code for a domain geometry, convection velocity

```
for j=1:ng

 if(gfl(j,1)==1)
    for i=1:ng
       b(i) = b(i) - lsm(i,j) * gfl(j,2);
       lsm(i,j) = 0.0;
       lsm(j,i) = 0.0;
    end
    lsm(j,j) = 1.0;
    b(j) = gfl(j,2);
 end

end
%------------------------
% solve the linear system
%------------------------

f = b/lsm';

%------------------
% plot the solution
%------------------

plot_3 (ne,ng,p,c,f);

trimesh (c,p(:,1),p(:,2),f);

%-----
% done
%-----
```

Code scd3_d: ($\longrightarrow$ Continued.)   Finite element solution of the steady convection–diffusion equation with the Dirichlet boundary condition around the contour of a disk-like domain.

and other parameter values of your choice, and discuss the results of your computation.

3.5.3 *Convection-dominated transport.*

Assess by numerical experimentation the performance of the numerical method implemented in the code scd3_d as the Péclet is raised to high values, for a domain geometry, parameter values, and boundary conditions of your choice.

3.5.4 *Convective transport.*

Run the code code scd3_d for purely convective transport corresponding to the limit $Pe \to \infty$. Discuss and explain on physical grounds the MATLAB response.

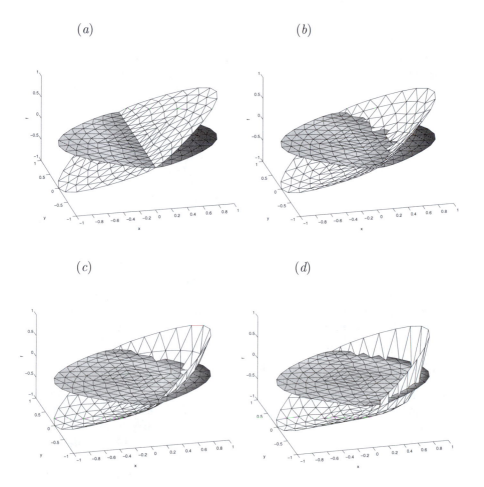

Figure 3.5.1 Solution of the steady convection–diffusion equation with a
Dirichlet boundary condition that is linear in x, produced by the FSELIB
code scd3_d. A uniform convection velocity U is imposed from left to
right. (a) $Pe = UR/\kappa = 0$, (b) 2, (c) 5, and (d) 10, where R is the radius
of the undeformed disk.

3.6 Code for Helmholtz's equation with the Neumann boundary condition

Helmholtz's equation can be viewed as a special case of the steady diffusion equation, arising when the source term is proportional to the *a priori* unknown solution, $s = \alpha\, f/k$, yielding

$$\nabla^2 f + \alpha\, f = 0, \tag{3.6.1}$$

where α is a specified constant with dimensions of inverse squared length. We will assume that the Neumann boundary condition specifying the flux, q, is prescribed around the boundary of a simply connected solution domain arising from the deformation of a circular disk.

Carrying out the Galerkin projection, we find that the master Galerkin finite element system (3.1.21) simplifies to the algebraic system

$$\Big(\mathbf{D} - \alpha\,\mathbf{M}\Big)\cdot \mathbf{f} = \mathbf{b}, \tag{3.6.2}$$

where the ith component of the vector $\mathbf{b}$ on the right-hand side is given by

$$b_i \equiv \frac{1}{k} \oint_C \phi_i\, q\, \mathrm{d}l. \tag{3.6.3}$$

The flux integral on the right-hand side (3.6.3) is nonzero only if the ith node associated with the global interpolation function, ϕ_i, corresponding to the ith Galerkin projection equation, is a boundary node. In the finite element implementation, this flux integral can be assembled in terms of *element edge integrals*, in a process that is similar to that used to assemble the coefficient matrix, as follows:

- Loop over the elements, and then loop over the three element vertex nodes.

- If the jth element node is a boundary node, for $j = 1, 2, 3$, examine whether the next element node, numbered $j + 1$, is also a boundary node. If both element nodes are boundary nodes, then the corresponding element edge lies on the boundary of the solution domain. In that case, perform the computation discussed in the next step. To facilitate the logistics, we wrap the element nodes by introducing a ghost fourth node, which is identical to the first node.

- Evaluate the contour integral on the right-hand side of (3.6.3) along the element edge for the element interpolation functions corresponding to the first and second end-nodes. An elementary computation shows that, consistent with the piecewise linear interpolation, these integrals are given, respectively, by

$$I_j = \left(\frac{1}{3}\, q_j + \frac{1}{6}\, q_{j+1}\right) \Delta s, \qquad I_{j+1} = \left(\frac{1}{6}\, q_j + \frac{1}{3}\, q_{j+1}\right) \Delta s, \tag{3.6.4}$$

where Δs is the element edge length. Note that the coefficients on the right-hand sides derive from the one-dimensional mass-matrix with linear elements, as shown in (1.1.65).

- Add these integrals to the right-hand sides of the appropriate Galerkin equations identified with the help of the connectivity matrix.

The complete FSELIB code, hlm3_n, is listed in the text. The individual modules are:

- *Specify the input data.*

- *Perform the grid generation.*

- *Assign the Neumann boundary conditions to the global nodes.*

- *Assemble the global diffusion and mass matrices from the element diffusion matrices according to Algorithm 2.1.2.*

   The element diffusion matrices are evaluated by the FSELIB function edm3, listed previously in this chapter, and the element mass matrices are evaluated by the FSELIB function emm3, listed in the text.

- *Assemble the right-hand side of the linear system.*

- *Solve the linear system using the intrinsic* MATLAB *solver.*

- *Prepare a graph of the results using the* FSELIB *function* **plot_3**, *listed previously in this chapter, and the* MATLAB *function* **trimesh**.

The graphics display of a computation with the constant flux boundary condition, $q = 1$, as implemented in the code, is shown in Figure 3.6.1 for $\alpha R^2 = 1$ and 5. As this coefficient becomes smaller, the solution tends to become more uniform. A singular behavior is encountered in the limit $\alpha = 0$ due to the compatibility condition for the Poisson equation, as discussed in Problem 3.6.2.

PROBLEMS

3.6.1 *Code* hlm3_n.

Run the FSELIB code hlm3_n for a deformed shape, parameter values, and Neumann boundary conditions of your choice, and discuss the results of your computation.

```
%================================================
% CODE hlm3_n
%
% Code for Helmholtz's equation in a disk-like
% domain subject to the Neumann
% boundary condition
%================================================

%-----
% input: conductivity,
%         Helmholtz coefficient,
%         discretization level
%-----

k = 1.0; alpha = 4.0; ndiv = 3;

%------------
% triangulate
%------------

[ne,ng,p,c,efl,gfl] = trgl3_disk (ndiv);

%-------
% deform
%-------

def = 0.20;

for i=1:ng
 p(i,1) = p(i,1)*(1.0-def*p(i,2)^2 );
end

%-------------------------------------
% specify the Neumann boundary condition
%-------------------------------------

for i=1:ng

 if(gfl(i,1)==1)
  gfl(i,2) = 1.0;       % example
 end

end

%---------------------------------------------
% assemble the global diffusion and mass matrix
%---------------------------------------------

gdm = zeros(ng,ng);  % initialize

gmm = zeros(ng,ng); % initialize
```

Code hlm3_n: Continuing $\longrightarrow$

```
for l=1:ne      % loop over the elements

% compute the element matrices

j = c(l,1); x1 = p(j,1); y1 = p(j,2);
j = c(l,2); x2 = p(j,1); y2 = p(j,2);
j = c(l,3); x3 = p(j,1); y3 = p(j,2);

[edm_elm] = edm3(x1,y1,x2,y2,x3,y3);
[emm_elm] = emm3(x1,y1,x2,y2,x3,y3);

    for i=1:3
      i1 = c(l,i);
      for j=1:3
        j1 = c(l,j);
        gdm(i1,j1) = gdm(i1,j1) + edm_elm(i,j);
        gmm(i1,j1) = gmm(i1,j1) + emm_elm(i,j);
      end
    end

end

%-------------------------
% set up the right-hand side
%-------------------------

for i=1:ng       % initialize
 b(i) = 0.0;
end

%-----------------------------------------
% Neumann integral on the right-hand side
%-----------------------------------------

for i=1:ne       % loop over the elements

 efl(i,4) = efl(i,1);  % wrap around the element
 c(i,4) = c(i,1);

  for j=1:3             % run over the element edges
   j1=c(i,j); j2=c(i,j+1);
   if( gfl(j1,1)==1 & gfl(j2,1)==1 )
     xe1 = p(j1,1); ye1 = p(j1,2); xe2 = p(j2,1); ye2 = p(j2,2);
     edge = sqrt((xe2-xe1)^2+(ye2-ye1)^2);
     int1 = edge * ( gfl(j1,2)/3 + gfl(j2,2)/6 );
     int2 = edge * ( gfl(j1,2)/6 + gfl(j2,2)/3 );
     b(j1) = b(j1)+int1/k;
     b(j2) = b(j2)+int2/k;
   end
  end

end
```

Code hlm3_n: $\longrightarrow$ Continuing $\longrightarrow$

```
%------------------
% coefficient matrix
%------------------

lsm = gdm-alpha*gmm;

%----------------------
% solve the linear system
%----------------------

f = b/lsm';

%---------
% plotting
%---------

plot_3 (ne,ng,p,c,f);
trimesh (c,p(:,1),p(:,2),f);

%-----
% done
%-----
```

Code hlm3_n: ($\longrightarrow$ Continued.) Code for Helmholtz's equation in a disk-like domain subject to the Neumann boundary condition.

(a) (b)

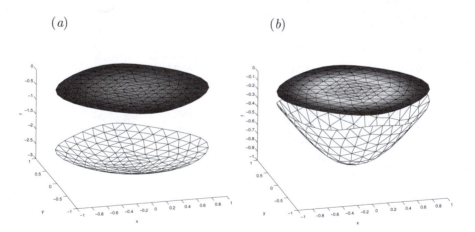

Figure 3.6.1 Solution of the Helmholtz equation subject to the Neumann boundary condition specifying a uniform boundary flux, $q = 1$, produced by the FSELIB code hlm3_n, for (a) $\alpha R^2 = 1$, and (b) 5.

```
function emm = emm3(x1,y1,x2,y2,x3,y3)

%=====================================
% Evaluation of the element mass matrix
% from the coordinates of the vertices
% of a 3-node triangle
%=====================================

 d23x = x2-x3; d23y = y2-y3;
 d31x = x3-x1; d31y = y3-y1;
 d12x = x1-x2; d12y = y1-y2;

 A = 0.5*(d31x*d12y - d31y*d12x);

 fc = A/12.0; fs = 2.0*fc;

 emm(1,1) = fs; emm(1,2) = fc; emm(1,3) = fc;
 emm(2,1) = fc; emm(2,2) = fs; emm(2,3) = fc;
 enm(3,1) = fc; emm(3,2) = fc; emm(3,3) = fs;

%-----
% done
%-----

return;
```

Function emm3: Evaluation of the element mass matrix from the coordinates of the vertices of a 3-node triangle.

3.6.2 *Compatibility condition for Poisson's equation.*

The compatibility condition for Poisson's equation,

$$\nabla^2 f + s = 0, \tag{3.6.5}$$

requires that the areal integral of the source function, s, over the solution domain, must be equal to the line integral of the flux across the boundary. Physically, the total rate of production must balance the total outward flow to prevent accumulation. If this condition is not met, a solution cannot be found. If this condition is met, an infinite number of solutions are possible differing by an arbitrary constant. In the case of Laplace's equation, $s = 0$, the compatibility condition requires that the line integral of the boundary flux is zero.

Run the code hlm3_d with $\alpha = 0$, whereupon Helmholtz's equation reduces to Laplace's equation, using other parameter values and boundary conditions of your choice, and confirm that the computation fails. In particular, MATLAB will produce the warning: "Matrix is close to singular or badly scaled." Compute the determinant of the coefficient matrix using the MATLAB function det, and explain the failure of the numerical method from a mathematical viewpoint.

3.7 Code for Laplace's equation with Dirichlet and Neumann boundary conditions

In the most demanding application, we solve Laplace's equation

$$\nabla^2 f = 0, \qquad (3.7.1)$$

in a disk-like domain centered at the origin, subject to the Dirichlet boundary condition on the left part of the boundary, $x < 0$, and the Neumann boundary condition on the right part of the boundary, $x > 0$. Under these circumstances, the master Galerkin finite element system (3.1.21) simplifies to the algebraic system

$$\mathbf{D} \cdot \mathbf{f} = \mathbf{b}, \qquad (3.7.2)$$

where the ith component of the vector $\mathbf{b}$ on the right-hand side is given by

$$b_i \equiv \frac{1}{k} \oint_C \phi_i \, q \, dl, \qquad (3.7.3)$$

and k is the medium conductivity. The flux integral on the right-hand side of (3.7.3) is nonzero only if the ith node associated with the global interpolation function, ϕ_i, corresponding to the ith Galerkin projection, is a Neumann boundary node.

The complete FSELIB code `lapl3_dn` is listed in the text. Highlights of the algorithm include the following:

- After the solution domain, D, has been triangulated, element edges that lie on the boundary are identified and counted sequentially using the element edge counter, Ic. An element edge is a boundary edge only if both element nodes defining the edge are boundary nodes. The total number of boundary edges is denoted as nbe in the code.

- The variable $face1(i)$ is the label of the global node of the first point of the ith boundary edge, and the variable $face2(i)$ is the label of the global node of the second point of the ith boundary edge, where $i = 1, 2, \ldots, nbe$.

  The flag $face3(i) = 1$ or 2 indicates, respectively, that the Dirichlet or Neumann boundary condition is applied at the ith boundary edge, where $i = 1, 2, \ldots, nbe$; the default value is $face3(i) = 1$. If $face3(i) = 2$, the vector $face4(i)$ holds the prescribed Neumann value at the first point of the ith edge, and the vector $face5(i)$ holds the prescribed Neumann value at the second point of the ith edge.

- The flag $gfl(i, 2) = 1$ indicates that the ith global node is a Dirichlet boundary node; the default value is zero. This flag is set by running over the boundary edges and raising the flags of the first and second edge

```
%===============================================
% CODE lapl3_dn
%
% Finite element solution of Laplace's equation
% in a disk-like domain, with Dirichlet
% and Neumann boundary conditions
%===============================================

%-----------
% input data
%-----------

k = 1.0;       % conductivity
ndiv = 3;      % level of triangulation

%----------------------
% triangulate and deform
%----------------------

[ne,ng,p,c,efl,gfl] = trgl3_disk (ndiv);

for i=1:ng
 p(i,1)=p(i,1)*(1.0-0.30*p(i,2)^2 );
end

%-------------------------------------
% find the element edges on the boundary
%-------------------------------------

Ic = 0;   % initialize the element edge counter

for i=1:ne

 efl(i,4) = efl(i,1); % wrap around the element
 c(i,4) = c(i,1);

 for j=1:3    % run over the element edges

  if( efl(i,j)==1 & efl(i,j+1)==1 )  % new edge side
   Ic= Ic+1;
   face1(Ic)=c(i,j); face2(Ic)=c(i,j+1);
  end

 end

end

nbe = Ic;   % number of boundary edges

%---------------------------------------
% specify the Neumann boundary condition on
% the right side of the solution domain
%---------------------------------------
```

Code lapl3_dn: Continuing $\longrightarrow$

```
for i=1:nbe  % run along the boundary sides

  m = face1(i);
  l = face2(i);

  face3(i)=1;  % default Dirichlet flag

  if( p(m,1) > 0.00001 | p(l,1)> 0.00001 )     % Neumann side
     face3(i)= 2;        % Neumann flag
     face4(i)= 1.0;      % Neumann condition at first  edge node
     face5(i)= 1.0;      % Neumann condition at second edge node
  end

end

%-------------------------------------------
% specify the Dirichlet boundary condition
% on the left side
%-------------------------------------------

for i=1,ng          % initialize node flag
  gfl(i,2) = 0;     % gfl(i,2)=1 will indicate a Dirichlet node
end

for i=1:nbe    % run over the boundary edges

  m = face1(i); l = face2(i);

  if(p(m,1)<-0.00001 | p(l,1)<-0.00001 )    % Dirichlet edge
     gfl(m,2)=1;   gfl(l,2)=1;          % Dirichlet flags
     gfl(m,3)=0.0;     % Dirichlet condition at first  edge node
     gfl(l,3)=0.0;     % Dirichlet condition at second edge node
  end

end

%-------------------------------------
% assemble the global diffusion matrix
%-------------------------------------

gdm = zeros(ng,ng); % initialize
```

Code lapl3_dn: $\longrightarrow$ Continuing $\longrightarrow$

nodes. If $gfl(i,2) = 1$, in which case the ith global node is a Dirichlet boundary node, the entry $gfl(i,3)$ is assigned the prescribed boundary value.

- The contour integral on the right-hand side of (3.7.3) is computed by running over the *nbe* boundary element edges, while making additive contributions to the equations corresponding to the first and second end-node, using formulas (3.6.4).

```
for l=1:ne      % loop over the elements
                %to compute the global diffusion matrix

j=c(l,1); x1=p(j,1); y1=p(j,2);
j=c(l,2); x2=p(j,1); y2=p(j,2);
j=c(l,3); x3=p(j,1); y3=p(j,2);

[edm_elm] = edm3(x1,y1,x2,y2,x3,y3);

   for i=1:3
     i1 = c(l,i);
     for j=1:3
       j1 = c(l,j);
       gdm(i1,j1) = gdm(i1,j1) + edm_elm(i,j);
     end
   end
end

%--------------------------------
% initialize the right-hand side
%--------------------------------

for i=1:ng
 b(i) = 0.0;
end

%-----------------------------------------
% Neumann integral on the right-hand side
%-----------------------------------------

for i=1:nbe

 if(face3(i)==2)

  m = face1(i); l = face2(i);

  xe1 = p(m,1); ye1 = p(m,2);
  xe2 = p(l,1); ye2 = p(l,2);

  edge = sqrt((xe2-xe1)^2+(ye2-ye1)^2);

  int1 = edge*( face4(i)/3 + face5(i)/6 );
  int2 = edge*( face4(i)/6 + face5(i)/3 );

  b(m) = b(m)+int1/k;   b(l) = b(l)+int2/k;

 end

end

%---------------------------------------------
% implement the Dirichlet boundary condition
%---------------------------------------------
```

Code lapl3_dn: $\longrightarrow$ Continuing $\longrightarrow$

```
for j=1:ng

 if(gfl(j,2)==1)

    for i=1:ng
      b(i) = b(i) - gdm(i,j) * gfl(j,3);
      gdm(i,j) = 0; gdm(j,i) = 0;
    end
    gdm(j,j) = 1.0; b(j) = gfl(j,3);

 end

end

%------------------------
% solve the linear system
%------------------------

f = b/gdm';

%---------
% plotting
%--------

plot_3 (ne,ng,p,c,f);
trimesh (c,p(:,1),p(:,2),f);

%-----
% done
%-----
```

Code lapl3_dn: ($\longrightarrow$ Continued.) Finite element solution of Laplace's equation in a disk-like domain produced by the FSELIB code lapl3_dn. The Dirichlet boundary condition is prescribed on the right part of the boundary, $x < 0$, and the Neumann boundary condition is prescribed on the left part of the boundary, $x > 0$.

- The Dirichlet boundary condition is implemented after the linear system has been compiled using a familiar method.

Implicit in the algorithm is the splitting of each of the two boundary nodes located at the junction of the Dirichlet and Neumann contours, $x = 0$, into a pair of companion nodes.

Results of a computation with the uniform flux boundary condition along the Neumann portion of the boundary, $q = 1$, and the homogeneous Dirichlet condition on the Dirichlet portion of the boundary, $f = 0$, as implemented in the code, are displayed in Figure 3.7.1.

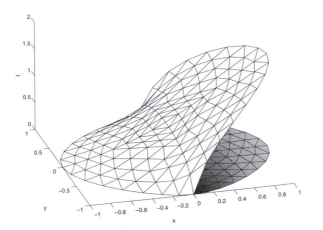

Figure 3.7.1 Finite element solution of Laplace's equation, subject to the Dirichlet boundary condition on the left half of the boundary, $x < 0$, $f = 0$, and the uniform flux Neumann boundary condition on the right half of the boundary, $x > 0$, $q = 1$, produced by the code `lapl3_dn`.

PROBLEMS

3.7.1 *Code* `lapl3_dn` .

Run the FSELIB code `lapl3_dn` for a deformed disk-like domain of your choice, and discuss the results of your computation.

3.7.2 *Modified code* `lapl3_dn`.

Modify the FSELIB code `lapl3_dn` so that the Dirichlet boundary condition is specified all around the boundary of the undeformed disk, except at the first quadrant where the Neumann boundary condition is specified. Run the modified code for a deformed shape of your choice, and discuss the results of your computation.

3.7.3 *Solution of Laplace's equation in a deformed square.*

FSELIB code `lapl3_dn_sqr` (not listed in the text) solves Laplace's equation in a deformed square, subject to the uniform Dirichlet boundary condition at the top, left, and bottom sides, $f = 0$, and the uniform Neumann boundary condition on the right side, $q = 1$. The implementation of the finite element method is similar to that described in the text for the corresponding code `lapl3_dn`. The graphics output for a deformed rectangular shape resembling a slab is displayed in Figure 3.7.2.

Execute code `lapl3_dn_sqr` for a deformed shape of your choice, and discuss the results of your computation.

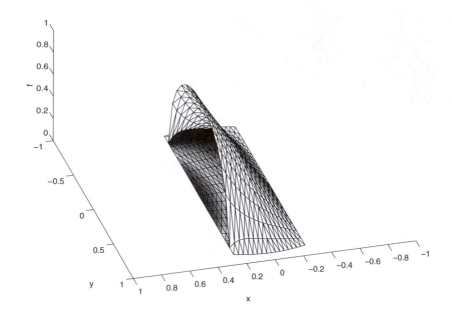

Figure 3.7.2 Finite element solution of Laplace's equation in a deformed
square resembling a slab, produced by the FSELIB code `lapl3_dn_sqr`.
The homogeneous Dirichlet boundary condition, $f = 0$, is applied along
three sides, and the uniform Neumann boundary condition, $q = 1$, is
applied along the fourth side.

3.8   Bilinear quadrilateral elements

In certain applications, it is convenient to use quadrilateral elements with four
straight or curved sides. The finite element methodology for these elements is
similar to that for the three-sided triangular element discussed previously in
this chapter.

The simplest quadrilateral element has four straight edges defined by four
vertices. To describe this element, we map it from the physical xy plane to a
standard square in the parametric $\xi\eta$ plane confined between $-1 \leq \xi \leq 1$ and
$-1 \leq \eta \leq 1$, as shown on the right of Figure 3.8.1(a). The first node is mapped
to the point $\xi = -1, \eta = -1$, the second to the point $\xi = 1, \eta = -1$, the third to
the point $\xi = -1, \eta = 1$, and the fourth to the point $\xi = 1, \eta = 1$. The mapping
from the xy to the $\xi\eta$ plane is mediated by the function

$$\mathbf{x} = \sum_{i=1}^{4} \mathbf{x}_i^E \, \psi_i(\xi, \eta), \qquad (3.8.1)$$

where $\mathbf{x}_i^E$ are the element nodes. The element-node cardinal interpolation func-

(a)

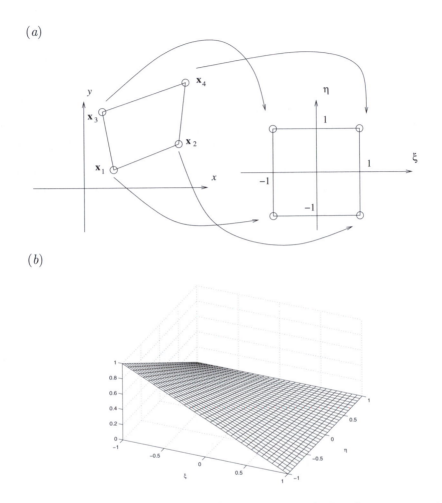

(b)

Figure 3.8.1 (a) A 4-node quadrilateral element in the xy plane is mapped to a square element in the parametric $\xi\eta$ plane. (b) Graph of the bilinear interpolation function corresponding to the southwestern node labeled 1.

tions, $\psi_i(\xi, \eta)$, are required to be bilinear: for a given ξ, $\psi_i(\xi, \eta)$ is linear in η, and for a given η, $\psi_i(\xi, \eta)$ is linear in ξ,

$$\psi_i(\xi, \eta) = (a\,\xi + b)\,(c\,\eta + d), \qquad (3.8.2)$$

where a–d are expansion coefficients. Expanding out the product on the right-hand side, we obtain an incomplete quadratic expansion, displayed in the deliberate pictorial form

$$
\begin{aligned}
\psi_i(\xi, \eta) \;=\; & a_{00} \\
& +a_{10}\,\xi + a_{01}\,\eta \\
& +a_{11}\,\xi\eta,
\end{aligned}
\tag{3.8.3}
$$

where $a_{00} = bd$, $a_{10} = ad$, $a_{01} = bc$, and $a_{11} = ac$. Thus, the interpolation functions are incomplete quadratics in (ξ, η), handicapped by the absence of the pure quadratic terms ξ^2 and η^2. Consequently, the element is spatially anisotropic.

To compute the four coefficients a–d, or the equivalent set of coefficients a_{00}, a_{10}, a_{01}, and a_{11}, we require the cardinal interpolation property demanding that $\psi_i(\xi, \eta) = 1$ at the ith node, and $\psi_i(\xi, \eta) = 0$ at the other three nodes. Working in the familiar way, we find

$$
\psi_1 = \frac{1}{4}\,(1 - \xi)\,(1 - \eta), \qquad \psi_2 = \frac{1}{4}\,(1 + \xi)\,(1 - \eta),
$$

$$
\tag{3.8.4}
$$

$$
\psi_3 = \frac{1}{4}\,(1 - \xi)\,(1 + \eta), \qquad \psi_4 = \frac{1}{4}\,(1 + \xi)\,(1 + \eta).
$$

Figure 3.8.1(b) shows a graph of the interpolation function, ψ_1, corresponding to the southwestern vertex node, generated by the FSELIB function `psi_q4` (not listed in the text). The interpolation functions for the other three nodes have similar shapes.

In the isoparametric interpolation, the requisite solution over the element is expressed in a form that is analogous to that shown in (3.8.1),

$$
f(x, y, t) = \sum_{i=1}^{4} f_i^E(t)\, \psi_i(\xi, \eta),
\tag{3.8.5}
$$

where $f_i^E(t)$ are the element nodal values. With this choice, the lth-element diffusion matrix is given by

$$
A_{ij}^{(l)} = \iint_{E_l} \nabla\psi_i \cdot \nabla\psi_j \, \mathrm{d}x\, \mathrm{d}y = \int_{-1}^{1}\int_{-1}^{1} \nabla\psi_i \cdot \nabla\psi_j \, h_S \, \mathrm{d}\xi\, \mathrm{d}\eta,
\tag{3.8.6}
$$

for $i, j = 1$–4, where h_S is the *surface metric coefficient*, given by

$$
h_S = \left| \frac{\partial \mathbf{x}}{\partial \xi} \times \frac{\partial \mathbf{x}}{\partial \eta} \right|.
\tag{3.8.7}
$$

Note that h_S is not constant, as in the case of the 3-node triangle, but depends on position inside the element. The element mass matrix is given by

$$
B_{ij}^{(l)} = \iint_{E_l} \psi_i\psi_j \, \mathrm{d}x\, \mathrm{d}y = \int_{-1}^{1}\int_{-1}^{1} \psi_i\psi_j \, h_S \, \mathrm{d}\xi\, \mathrm{d}\eta,
\tag{3.8.8}
$$

and the element advection matrix is given by

$$
C_{ij}^{(l)} = \iint_{E_l} \psi_i \, \mathbf{u} \cdot \nabla \psi_j \, \mathrm{d}x \, \mathrm{d}y = \int_{-1}^{1} \int_{-1}^{1} \psi_i \, \mathbf{u} \cdot \nabla \psi_j \, h_S \, \mathrm{d}\xi \, \mathrm{d}\eta, \qquad (3.8.9)
$$

for $i, j = 1$–4. To compute the interpolation function gradient, $\nabla \psi_i$, we work as discussed in Section 2.3.1.

Since the integration limits in the $\xi\eta$ plane are fixed, the element diffusion, mass, and advection matrices can be computed by the dual application of a one-dimensional integration quadrature. The Lobatto quadrature shown in (2.3.13) provides us with the approximation

$$
\int_{-1}^{1} \int_{-1}^{1} f(\xi, \eta) \, \mathrm{d}\xi \, \mathrm{d}\eta \simeq \sum_{p_1=1}^{k_1+1} \sum_{p_2=1}^{k_2+1} f(t_{p_1}, t_{p_2}) \, w_{p_1} \, w_{p_2}, \qquad (3.8.10)
$$

for some regular function $f(\xi, \eta)$, where k_1 and k_2 are specified quadrature orders.

PROBLEMS

3.8.1 *Surface metric coefficient.*

Derive an expression for the surface metric coefficient of the 4-node rectangular element, h_S, in terms of the node coordinates and derivatives of the element interpolation functions.

3.8.2 *Computation of an integral over a quadrilateral element.*

Write a MATLAB function that evaluates the integral of a given function, $f(\xi, \eta)$, over a quadrilateral element,

$$
\iint f(\xi, \eta) \, \mathrm{d}x \, \mathrm{d}y, \qquad (3.8.11)
$$

using the double Lobatto quadrature. The input to this function should include the position of the four vertices, $\mathbf{x}_i$, for $i=1$–4. Run the function for $f(\xi, \eta) = 1$, and confirm that the answer is equal to the area of the quadrilateral element in the xy plane.

Quadratic and spectral elements in two dimensions

4

In Chapter 3, we discussed the implementation of the finite element method in two dimensions with linear triangular elements defined by three nodes. To improve the accuracy of the interpolation and also account for the boundary curvature, we may discretize the solution domain into triangular or quadrilateral elements with curved edges defined by a higher number of nodes. In addition, we may approximate the solution over the individual elements isoparametrically or super-parametrically, using quadratic or high-order polynomial expansions expressed in nodal or modal form. Spectral element methods arise by judiciously deploying the element interpolation nodes at positions corresponding to optimal interpolation sets that are specific to the adopted element type.

In this chapter, we discuss the implementation of the finite element method with quadratic triangular elements, develop high-order and spectral element expansions with nodal and modal basis functions, and review high-order and spectral methods on the rectangle. The discussion of the spectral element method in the second and third parts assumes some familiarity with the theory of function interpolation and orthogonal polynomials reviewed in Appendices A and B. Readers who are not presently interested in high-order and spectral element methods may skip the corresponding sections and proceed without penalty to Chapters 5 and 6.

4.1 6-node triangular elements

Planar 6-node triangular elements with straight or curved edges are defined by three *vertex* nodes and three *edge* nodes, as illustrated in Figure 4.1.1(a). To describe an element in parametric form, we map it from the physical xy plane to the familiar right isosceles triangle in the $\xi\eta$ plane, as shown in Figure 4.1.1(a), so that:

- The first node is mapped to the origin, $\xi = 0$, $\eta = 0$.

- The second node is mapped to the point $\xi = 1, \eta = 0$ on the ξ axis.

- The third node is mapped to the point $\xi = 0, \eta = 1$ on the η axis.

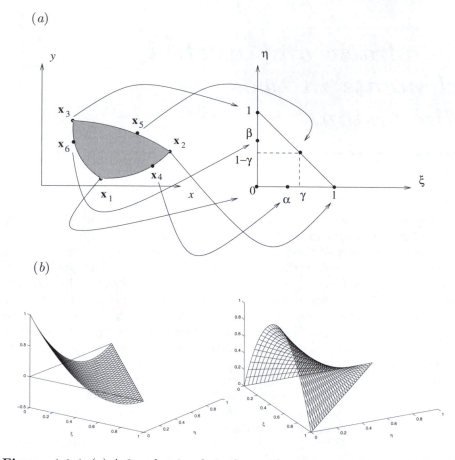

Figure 4.1.1 (*a*) A 6-node triangle in the *xy* plane is mapped to a right isosceles triangle in the parametric $\xi\eta$ plane. (*b*) Graphs of the element node interpolation functions ψ_1 and ψ_4 associated, respectively, with a vertex node and an edge node.

- The fourth node is mapped to the point $\xi = \alpha, \eta = 0$ on the ξ axis.

- The fifth node is mapped to the point $\xi = \gamma, \eta = 1 - \gamma$ on the hypotenuse.

- The sixth node is mapped to the point $\xi = 0, \eta = \beta$ on the η axis.

The dimensionless geometrical mapping coefficients, α, β, and γ are defined as

$$\alpha = \frac{1}{1 + \frac{|\mathbf{x}_4 - \mathbf{x}_2|}{|\mathbf{x}_4 - \mathbf{x}_1|}}, \qquad \beta = \frac{1}{1 + \frac{|\mathbf{x}_6 - \mathbf{x}_3|}{|\mathbf{x}_6 - \mathbf{x}_1|}}, \qquad \gamma = \frac{1}{1 + \frac{|\mathbf{x}_5 - \mathbf{x}_2|}{|\mathbf{x}_5 - \mathbf{x}_3|}}. \qquad (4.1.1)$$

FSELIB function `elm6_abc`, listed in the text, evaluates these parameters from the coordinates of the six vertices using expressions (4.1.1). In finite element

```
function [al, be, ga] = elm6_abc ...
 ...
    (x1,y1, x2,y2, x3,y3, x4,y4, x5,y5, x6,y6)

%======================================
% Computation of the (xi, eta) mapping
% coefficients alpha, beta, gamma
% for a 6-node triangle
%======================================

D42 = sqrt( (x4-x2)^2 + (y4-y2)^2 );
D41 = sqrt( (x4-x1)^2 + (y4-y1)^2 );
D63 = sqrt( (x6-x3)^2 + (y6-y3)^2 );
D61 = sqrt( (x6-x1)^2 + (y6-y1)^2 );
D52 = sqrt( (x5-x2)^2 + (y5-y2)^2 );
D53 = sqrt( (x5-x3)^2 + (y5-y3)^2 );

al = 1.0/(1.0+D42/D41);
be = 1.0/(1.0+D63/D61);
ga = 1.0/(1.0+D52/D53);

%-----
% done
%-----

return;
```

Function elm6_abc: Evaluation of the mapping coefficient α, β, and γ, for a 6-node triangle.

implementations, it is a common practice to set $\alpha = \beta = \gamma = 1/2$, which amounts to mapping the edge nodes 4, 5, and 6, to the edge mid-points. However, we will see that the scaling embedded in (4.1.1) ensures the regularity of the mapping when the edge nodes are located near the vertex nodes in the xy plane.

The mapping from the physical to the parametric space is mediated by the function

$$\mathbf{x} = \sum_{i=1}^{6} \mathbf{x}_i^E \, \psi_i(\xi, \eta), \tag{4.1.2}$$

whose Cartesian components are

$$x = \sum_{i=1}^{6} x_i^E \, \psi_i(\xi, \eta), \qquad y = \sum_{i=1}^{6} y_i^E \, \psi_i(\xi, \eta). \tag{4.1.3}$$

The quadratic element interpolation functions, $\psi_i(\xi, \eta)$, are required to satisfy the familiar cardinal interpolation properties requiring that $\psi_i = 1$ at the ith element node, and $\psi_i = 0$ at the other five nodes. In terms of Kronecker's delta, δ_{ij},

$$\psi_i(\xi_j, \eta_j) = \delta_{ij}, \qquad (4.1.4)$$

for $i, j = 1, 2, \ldots, 6$, where

$$(\xi_1, \eta_1) = (0, 0), \qquad (\xi_2, \eta_2) = (1, 0), \qquad (\xi_3, \eta_3) = (0, 1),$$

$$(4.1.5)$$

$$(\xi_4, \eta_4) = (\alpha, 0), \qquad (\xi_5, \eta_5) = (\gamma, 1 - \gamma), \qquad (\xi_6, \eta_6) = (0, \beta),$$

are the coordinates of the six nodes in the $\xi\eta$ plane.

To derive the node interpolation functions, we write

$$\psi_i(\xi, \eta) = a_i + b_i\, \xi + c_i\, \eta + d_i\, \xi^2 + e_i\, \xi\eta + f_i\, \eta^2, \qquad (4.1.6)$$

and compute the six coefficients a_i–f_i to satisfy the aforementioned interpolation conditions. The result is

$$\psi_2 = \frac{1}{1 - \alpha}\, \xi \left(\xi - \alpha + \frac{\alpha - \gamma}{1 - \gamma}\, \eta \right),$$

$$\psi_3 = \frac{1}{1 - \beta}\, \eta \left(\eta - \beta + \frac{\beta + \gamma - 1}{\gamma}\, \xi \right),$$

$$\psi_4 = \frac{1}{\alpha\,(1 - \alpha)}\, \xi\,\zeta,$$

$$\psi_5 = \frac{1}{\gamma\,(1 - \gamma)}\, \xi\,\eta, \qquad (4.1.7)$$

$$\psi_6 = \frac{1}{\beta\,(1 - \beta)}\, \eta\,\zeta,$$

$$\psi_1 = 1 - \psi_2 - \psi_3 - \psi_4 - \psi_5 - \psi_6,$$

where

$$\zeta \equiv 1 - \xi - \eta \qquad (4.1.8)$$

is the third barycentric coordinate. As in the case of the 3-node triangle, $\zeta = 0$ along the hypotenuse where $\eta = 1 - \xi$ and $\xi = 1 - \eta$. Figure 4.1.1(b) shows graphs of the interpolation functions ψ_1 and ψ_4 associated with a vertex node and an edge node, generated by the FSELIB function psi_t6 (not listed in the text).

If the edge nodes are fortuitously, intentionally, or capriciously mapped to the edge mid-points in the $\xi\eta$ plane, $\alpha = 1/2$, $\beta = 1/2$, and $\gamma = 1/2$, then the node interpolation functions take the simpler forms

$$\psi_1 = \zeta\,(2\,\zeta - 1), \qquad \psi_2 = \xi\,(2\,\xi - 1), \qquad \psi_3 = \eta\,(2\,\eta - 1),$$

$$(4.1.9)$$

$$\psi_4 = 4\,\xi\,\zeta, \qquad \psi_5 = 4\,\xi\,\eta, \qquad \psi_6 = 4\,\eta\,\zeta.$$

(a)   (b)

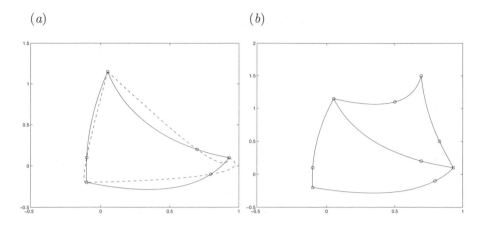

Figure 4.1.2 (*a*) Element contour generated by the general mapping (4.1.7)
(solid line), and by the simplified mapping (4.1.9) (broken line). (*b*) Two
elements join seamlessly to form a patch.

In Section 4.4, these formulas will be recovered as special cases of more general
expressions for high-order expansions.

Substituting the interpolation functions (4.1.7) in (4.1.2), we obtain a representation in terms of a *complete* quadratic function in ξ and η consisting of six
terms, as shown in (4.1.6). For a specified location in the triangle in the physical plane, (x, y), the corresponding location in the triangle in the parametric
space, (ξ, η), can be found by solving the system of two quadratic equations
(4.1.3), and retaining the physically meaningful solution.

The actual geometry of the curved edges of the element in the physical xy
plane is determined implicitly by the quadratic mapping expressed by (4.1.2).
An element has a straight edge only if the three nodes defining the edge are
collinear in the physical plane. More generally, since each edge is a quadratic
function of ξ or η with coefficients depending on the location of the three corresponding nodes, shared edges of adjacent elements coincide.

Figure 4.1.2(*a*) depicts the contour of a triangle computed with the general
mapping (4.1.7) (solid lines), and with the simplified mapping (4.1.9) (broken
lines), generated by the FSELIB function `edges6` (not listed in the text). This
example demonstrates that the general approach ensures the regularity of the
mapping by producing smooth element edges. Under extreme conditions, the
simplified mapping may lead to edge crossing, whereupon a singular point is
mapped to two pairs of barycentric coordinates (ξ, η), resulting in loss of analyticity. Figure 4.1.2(*b*) depicts two adjacent elements seamlessly joining to
form a two-element patch.

4.1.1   Integral over the triangle

The Jacobian matrix of the mapping from the physical xy plane to the parametric $\xi\eta$ plane was defined in (3.2.11) as

$$
\mathbf{J} \equiv
\begin{bmatrix}
\frac{\partial x}{\partial \xi} & \frac{\partial x}{\partial \eta} \\[2mm]
\frac{\partial y}{\partial \xi} & \frac{\partial y}{\partial \eta}
\end{bmatrix}.
\tag{4.1.10}
$$

Substituting expressions (4.1.3), we find that the components of the Jacobian matrix are linear functions of ξ and η. Thus, unlike the 3-node triangle discussed in Chapter 3, the 6-node triangle has a position-dependent Jacobian. A singularity of the Jacobian matrix at a point signals the failure of the quadratic mapping and the loss of analyticity.

The determinant of the Jacobian matrix is the *surface metric coefficient*, defined in (3.2.12) as

$$
h_S \equiv \mathrm{Det}[\mathbf{J}] = \left| \frac{\partial \mathbf{x}}{\partial \xi} \times \frac{\partial \mathbf{x}}{\partial \eta} \right|.
\tag{4.1.11}
$$

Substituting expressions (4.1.3), we find that h_S is a nonlinear function of ξ and η. Thus, unlike the 3-node triangle, the 6-node triangle has a position-dependent metric coefficient that cannot be extracted from an integral.

The integral of a function, $f(x, y)$, over the area of the triangle in the xy plane can be expressed as an integral over the area of the triangle in the parametric $\xi\eta$ plane, as

$$
\iint f(x, y)\, dx\, dy = \iint f(\xi, \eta)\, h_S\, d\xi\, d\eta.
\tag{4.1.12}
$$

The integral over the standard triangle on the right-hand side is routinely computed by numerical methods, as will be discussed later in this section.

4.1.2   Isoparametric interpolation and element matrices

In the isoparametric interpolation, a function of interest, $f(\xi, \eta)$, defined over the parametric triangle, is expressed by the counterpart of the geometrical expansion (4.1.2), as

$$
f(\xi, \eta) = \sum_{j=1}^{6} f(\xi_j, \eta_j)\, \psi_j(\xi, \eta),
\tag{4.1.13}
$$

where $f(\xi_j, \eta_j)$ are the element nodal values. The lth-element diffusion matrix is given by

$$
A_{ij}^{(l)} = \iint_{E_l} \nabla \psi_i \cdot \nabla \psi_j\, dx\, dy = \iint \nabla \psi_i \cdot \nabla \psi_j\, h_S\, d\xi\, d\eta,
\tag{4.1.14}
$$

the corresponding element mass matrix is given by

$$B_{ij}^{(l)} = \iint_{E_l} \psi_i \psi_j \, dx \, dy = \iint \psi_i \psi_j \, h_S \, d\xi \, d\eta, \tag{4.1.15}$$

and the corresponding element advection matrix is given by

$$C_{ij}^{(l)} = \iint_{E_l} \psi_i \, \mathbf{u} \cdot \nabla \psi_j \, dx \, dy = \iint \psi_i \, \mathbf{u} \cdot \nabla \psi_j \, h_S \, d\xi \, d\eta, \tag{4.1.16}$$

for $i, j = 1, 2, \ldots, 6$, where the integrals on the right-hand sides are performed over the area of the triangle in the parametric plane.

To compute the element diffusion and advection matrices, we require the gradients of the element interpolation functions, $\nabla \psi_i$. These can be found readily using relations (3.2.23), which amount to

$$\sum_{j=1}^{6} \frac{\partial \psi_j}{\partial \xi} \, \mathbf{x}_j^E \cdot \nabla \psi_i = \frac{\partial \psi_i}{\partial \xi},$$

$$\sum_{j=1}^{6} \frac{\partial \psi_j}{\partial \eta} \, \mathbf{x}_j^E \cdot \nabla \psi_i = \frac{\partial \psi_i}{\partial \eta}. \tag{4.1.17}$$

The associated matrix form is

$$\mathbf{J}^T \cdot \nabla \psi_i = \begin{bmatrix} \frac{\partial \psi_i}{\partial \xi} \\ \frac{\partial \psi_i}{\partial \eta} \end{bmatrix}, \tag{4.1.18}$$

where $\mathbf{J}^T$ is the transpose of the Jacobian matrix,

$$\mathbf{J}^T = \left[\begin{array}{c|c} \sum_{j=1}^{6} \frac{\partial \psi_j}{\partial \xi} x_j^E & \sum_{j=1}^{6} \frac{\partial \psi_j}{\partial \xi} y_j^E \\ \hline \sum_{j=1}^{6} \frac{\partial \psi_j}{\partial \eta} x_j^E & \sum_{j=1}^{6} \frac{\partial \psi_j}{\partial \eta} y_j^E \end{array} \right]. \tag{4.1.19}$$

The determinant of $\mathbf{J}^T$ is equal to the determinant of $\mathbf{J}$, which is equal to the surface metric coefficient, h_S. Applying Cramer's rule, we find that the solution of (4.1.18) is given by

$$\nabla \psi_i = \frac{1}{h_S} \left[\begin{array}{c|c} \sum_{j=1}^{6} \frac{\partial \psi_j}{\partial \eta} y_j^E & -\sum_{j=1}^{6} \frac{\partial \psi_j}{\partial \xi} y_j^E \\ \hline -\sum_{j=1}^{6} \frac{\partial \psi_j}{\partial \eta} x_j^E & \sum_{j=1}^{6} \frac{\partial \psi_j}{\partial \xi} x_j^E \end{array} \right] \cdot \begin{bmatrix} \frac{\partial \psi_i}{\partial \xi} \\ \frac{\partial \psi_i}{\partial \eta} \end{bmatrix}, \tag{4.1.20}$$

for $i = 1, 2, \ldots, 6$.

FSELIB function `elm6_interp`, listed in the text, computes the node interpolation functions and their gradients, and evaluates the surface metric coefficient,

```
function [psi, gpsi, hs] = elm6_interp ...
...
           (x1,y1, x2,y2, x3,y3, x4,y4, x5,y5, x6,y6 ...
           ,al,be,ga, xi,eta)

%==========================================================
% Evaluation of the surface metric coefficient, hs, and
% computation of the interpolation functions and their
% gradients over a 6-node triangle
%==========================================================

%--------
% prepare
%--------

alc = 1.0-al; bec = 1.0-be; gac = 1.0-ga;
alalc = al*alc; bebec = be*bec; gagac = ga*gac;

%-------------------------------------
% compute the interpolation functions
%-------------------------------------

psi(2) = xi*(xi-al+eta*(al-ga)/gac)/alc;
psi(3) = eta*(eta-be+xi*(be+ga-1.0)/ga)/bec;
psi(4) = xi*(1.0-xi-eta)/alalc;
psi(5) = xi*eta/gagac;
psi(6) = eta*(1.0-xi-eta)/bebec;
psi(1) = 1.0-psi(2)-psi(3)-psi(4)-psi(5)-psi(6);

%----------------------------------------------------------
% compute the xi derivatives of the interpolation functions
%----------------------------------------------------------

dps2 =  (2.0*xi-al+eta*(al-ga)/gac)/alc;
dps3 =  eta*(be+ga-1.0)/(ga*bec);
dps4 =  (1.0-2.0*xi-eta)/alalc;
dps5 =  eta/gagac; dps6 = -eta/bebec;
dps1 = -dps2-dps3-dps4-dps5-dps6;

%-----------------------------------------------------------
% compute the eta derivatives of the interpolation functions
%-----------------------------------------------------------

pps2 =  xi*(al-ga)/(alc*gac);
pps3 =  (2.0*eta-be+xi*(be+ga-1.0)/ga)/bec;
pps4 = -xi/alalc;
pps5 =  xi/gagac;
```

Function elm6_interp: Continuing $\longrightarrow$

h_S, at a position corresponding to a specified pair of barycentric coordinates, (ξ, η). The input to this function includes the coordinates of the six vertices and the mapping coefficients α, β, and γ. The latter are evaluated by the FSELIB function elm6_abc, listed previously in this section.

```
pps6 =  (1.0-xi-2.0*eta)/bebec;
pps1 = -pps2-pps3-pps4-pps5-pps6;

%----------------------------------------
% compute the xi and eta derivatives of x
%----------------------------------------

DxDxi = x1*dps1 + x2*dps2 + x3*dps3 ...
        +x4*dps4 + x5*dps5 + x6*dps6;
DyDxi = y1*dps1 + y2*dps2 + y3*dps3 ...
        +y4*dps4 + y5*dps5 + y6*dps6;
DxDet = x1*pps1 + x2*pps2 + x3*pps3 ...
        +x4*pps4 + x5*pps5 + x6*pps6;
DyDet = y1*pps1 + y2*pps2 + y3*pps3 ...
        +y4*pps4 + y5*pps5 + y6*pps6;

%---------------------------
% compute the surface metric hs
%---------------------------

vnz = DxDxi * DyDet - DxDet * DyDxi;

hs = sqrt(vnz^2);

%--------------------------------------------
% compute the gradient of the six interpolation
% functions by solving two linear equations:
%
% dx/dxi . grad = d psi/dxi
% dx/det . grad = d psi /det
%
% The solution is found by Cramer's rule
%--------------------------------------------

A11 = DxDxi; A12 = DyDxi;
A21 = DxDet; A22 = DyDet;
Det =   A11*A22-A21*A12;

%--- first

B1 = dps1; B2 = pps1;
Det1 =   B1*A22 - B2*A12;
Det2 = - B1*A21 + B2*A11;
gpsi(1,1) = Det1/Det; gpsi(1,2) = Det2/Det;

%--- second

B1 = dps2; B2 = pps2;
Det1 =   B1*A22 - B2*A12;
Det2 = - B1*A21 + B2*A11;
gpsi(2,1) = Det1/Det;
gpsi(2,2) = Det2/Det;
```

Function elm6_interp: $\longrightarrow$ Continuing $\longrightarrow$

```
%--- third

B1 = dps3; B2 = pps3;
Det1 =   B1*A22 - B2*A12;
Det2 = - B1*A21 + B2*A11;
gpsi(3,1) = Det1/Det; gpsi(3,2) = Det2/Det;

%--- fourth

B1 = dps4; B2 = pps4;
Det1 =   B1*A22 - B2*A12;
Det2 = - B1*A21 + B2*A11;

gpsi(4,1) = Det1/Det; gpsi(4,2) = Det2/Det;

%--- fifth

B1 = dps5; B2 = pps5;
Det1 =   B1*A22 - B2*A12;
Det2 = - B1*A21 + B2*A11;

gpsi(5,1) = Det1/Det; gpsi(5,2) = Det2/Det;

%--- sixth

B1 = dps6; B2 = pps6;
Det1 =   B1*A22 - B2*A12;
Det2 = - B1*A21 + B2*A11;

gpsi(6,1) = Det1/Det; gpsi(6,2) = Det2/Det;

%-----
% done
%-----

return;
```

Function elm6_interp: ($\longrightarrow$ Continued.) Evaluation the surface metric coefficient, h_S, and computation of the interpolation functions and their gradients over a 6-node triangle.

4.1.3 Element matrices and integration quadratures

Substituting the preceding expressions in (4.1.14), (4.1.15), and (4.1.16), we obtain lengthy integrands for the element diffusion, mass, and advection matrices. In practice, the integrals over the parametric triangle in the $\xi\eta$ plane are computed most efficiently by numerical integration, using a triangle integration quadrature (e.g., [43], Chapter 7).

Applying the integration quadrature amounts to approximating the integral of a function, $g(\xi, \eta)$, over the area of the triangle with a weighted sum,

```
function [xi, eta, w] = gauss_trgl(m)

%==========================================================
% Abscissas (xi, eta) and weights (w) for Gaussian
% integration over a flat triangle in the xi-eta plane
%
%
% The integration is performed with respect
% to the triangle barycentric coordinates
%
% m: number of base points (NQ)
%    choose from 1,3,4,6,7,9,12,13; default is 7
%==========================================================

%-----
% trap
%-----

if( (m ~= 1) & (m ~= 3) & (m ~= 4) & (m ~= 6) & (m ~= 7) ...
  & (m ~= 9) & (m ~= 12) & (m ~= 13) )
  disp('');disp(' Gauss_trgl: Chosen number of points');
           disp('   is not available; Will take m=7');
  m=7;
end
```

Function gauss_trgl: Continuing $\longrightarrow$

$$\iint g(\xi, \eta) \, \mathrm{d}\xi \, \mathrm{d}\eta \simeq \frac{1}{2} \sum_{i=1}^{N_Q} g(\xi_i, \eta_i) \, w_i, \qquad (4.1.21)$$

where N_Q is a specified number of quadrature base points. The beginning and tail-end of the FSELIB function **gauss_trgl** that assigns numerical values to the integration base points, (ξ_i, η_i), and corresponding weights, w_i, is listed in the text. Note that the sum of the weights is equal to unity for any value of N_Q. Consequently, if $g(\xi, \eta) = 1$, the integral is equal to the area of the triangle in the $\xi\eta$ plane, which is equal to $1/2$.

FSELIB function **edm6**, listed in the text, evaluates the element diffusion matrix using the Gauss triangle quadrature. The code first calls the FSELIB function **edm6_abc**, listed in the text, to evaluate the mapping coefficients, and then the function **edm6_interp**, also listed in the text, to evaluate the gradients of the interpolation functions at the integration base points.

FSELIB function **edam6**, listed in the text, evaluates the diffusion and advection matrices by a similar method. Note that the velocity at the quadrature base points is computed by interpolation in terms of the element-node interpolation functions and the values of the velocity at the six element nodes passed in the function call.

```
%-------
if(m==1)
%-------

xi(1) = 1.0/3.0; eta(1) = 1.0/3.0; w(1) = 1.0;

%-----------
elseif(m==3)
%-----------

xi(1) = 1.0/6.0; eta(1) = 1.0/6.0; w(1) = 1.0/3.0;
xi(2) = 2.0/3.0; eta(2) = 1.0/6.0; w(2) = w(1);
xi(3) = 1.0/6.0; eta(3) = 2.0/3.0; w(3) = w(1);

%-----------
elseif(m==4)
%-----------

xi(1) = 1.0/3.0; eta(1) = 1.0/3.0; w(1) = -27.0/48.0;
xi(2) = 1.0/5.0; eta(2) = 1.0/5.0; w(2) =  25.0/48.0;
xi(3) = 3.0/5.0; eta(3) = 1.0/5.0; w(3) =  25.0/48.0;
xi(4) = 1.0/5.0; eta(4) = 3.0/5.0; w(4) =  25.0/48.0;

%-----------
elseif(m==6)
%-----------

al = 0.816847572980459; be = 0.445948490915965;
ga = 0.108103018168070; de = 0.091576213509771;
o1 = 0.109951743655322; o2 = 0.223381589678011;

xi(1) = de; xi(2) = al; xi(3) = de; xi(4) = be;
xi(5) = ga; xi(6) = be;

eta(1) = de; eta(2) = de;eta(3) = al; eta(4) = be;
eta(5) = be; eta(6) = ga;

w(1) = o1; w(2) = o1; w(3) = o1; w(4) = o2;
w(5) = o2; w(6) = o2;

. . . . . .

%--
end
%--

return;
```

Function gauss_trgl: ($\longrightarrow$ Continued.) Base points and weights for integrating over a triangle in the $\xi\eta$ plane; the six dots at the end of the listing indicate additional code.

```
function [edm, arel] = edm6 ...
 ...
    (x1,y1, x2,y2, x3,y3, x4,y4, x5,y5, x6,y6, NQ)

%=================================================
% Evaluation of the element diffusion matrix for a
% 6-node triangle, using an integration quadrature
%=================================================

%---------------------------------
% compute the mapping coefficients
%---------------------------------

[al, be, ga] = elm6_abc...
 ...
    (x1,y1, x2,y2, x3,y3, x4,y4, x5,y5, x6,y6)

%---------------------------
% read the triangle quadrature
%---------------------------

[xi, eta, w] = gauss_trgl(NQ);

%---------------------------------------
% initialize the element diffusion matrix
%---------------------------------------

for k=1:6
 for l=1:6
  edm(k,l) = 0.0
 end
end

%----------------------
% perform the quadrature
%----------------------

arel = 0.0;  % element area

for i=1:NQ

 [psi, gpsi, hs] = elm6_interp ...
 ...
     (x1,y1, x2,y2, x3,y3, x4,y4, x5,y5, x6,y6 ...
     ,al,be,ga, xi(i),eta(i));

 cf = 0.5*hs*w(i);
```

Function edm6: Continuing ⟶

```
for k=1:6
 for l=1:6
  edm(k,l) = edm(k,l) + (gpsi(k,1)*gpsi(l,1)   ...
                      +  gpsi(k,2)*gpsi(l,2) )*cf;
 end
end

arel = arel + cf;

end

%-----
% done
%-----

return;
```

Function edm6: ($\longrightarrow$ Continued.) Evaluation of the element diffusion matrix for a 6-node triangle, using a Gauss integration quadrature.

```
function [edm, eam, arel] = edam6 ...
 ...
   (x1,y1, x2,y2, x3,y3, x4,y4, x5,y5, x6,y6  ...
   ,u1,v1, u2,v2, u3,v3, u4,v4, u5,v5, u6,v6, ...
   ,NQ)

%====================================================
% Evaluation of the element diffusion and advection
% matrices and element area for a 6-node triangle
%
% The element integrals are computed using
% the NQ-point Gauss-triangle quadrature
%
% u and v are the x and y velocity components
%
% The quadrature base points and weights are
% read from the function gauss_trgl
%====================================================

%-----------------------------------
% compute the mapping coefficients
%-----------------------------------
```

Function edam6: Continuing $\longrightarrow$

```
[al, be, ga] = elm6_abc ...
      ...
         (x1,y1, x2,y2, x3,y3, x4,y4, x5,y5, x6,y6);

%------------------------------
% read the triangle quadrature
%------------------------------

[xi, eta, w] = gauss_trgl(NQ);

%--------------------------------
% initialize the element diffusion
% and advection matrices
%--------------------------------

for k=1:6
 for l=1:6
   edm(k,l) = 0.0;
   eam(k,l) = 0.0;
 end
end

%----------------------
% perform the quadrature
%----------------------

arel = 0.0; % element area

for i=1:NQ

[psi, gpsi, hs] = elm6_interp ...
 ...
     (x1,y1, x2,y2, x3,y3, x4,y4, x5,y5, x6,y6 ...
     ,al,be,ga, xi(i),eta(i));

% interpolate the velocity at the base points:

 u = u1*psi(1) + u2*psi(2) + u3*psi(3) ...
   + u4*psi(4) + u5*psi(5) + u6*psi(6)
 v = v1*psi(1) + v2*psi(2) + v3*psi(3) ...
   + v4*psi(4) + v5*psi(5) + v6*psi(6)

 cf = 0.5*hs*w(i);

 for k=1:6
  for l=1:6

   edm(k,l) = edm(k,l) + (gpsi(k,1)*gpsi(l,1)    ...
                       +  gpsi(k,2)*gpsi(l,2) )*hs*w(i);
   prj = u*gpsi(l,1)+v*gpsi(l,1);
   eam(k,l) = eam(k,l) + psi(k)*prj*cf;

  end
 end
```

Function edam6: Continuing $\longrightarrow$

```
  arel = arel + cf;
end       % end of the quadrature loop

%-----
% done
%-----

return;
```

Function edam6: ($\longrightarrow$ Continued.) Evaluation of the element diffusion and advection matrices over a 6-node triangle.

PROBLEM

4.1.1 *Numerical integration over a triangle.*

Write a code that integrates a specified function, $g(\xi, \eta)$, over the standard triangle, using the integration quadrature discussed in the text. Run the code for $N_Q = 6$, and discuss whether the quadrature is able to produce the exact answer when the integrand is: (a) a linear function, $g(\xi) = \xi$ or $g(\eta) = \eta$, (b) a quadratic monomial product, $g(\xi, \eta) = \xi^p \eta^q$, where $p + q = 2$, (c) a cubic monomial product, $g(\xi, \eta) = \xi^p \eta^q$, where $p + q = 3$, and (d) a quartic monomial product, $g(\xi, \eta) = \xi^p \eta^q$, where $p + q = 4$.

4.2   Grid generation and finite element codes

FSELIB function `trig6_disk`, listed in the text, discretizes the unit disk into 6-node triangles, by successively subdividing an ancestral set of four triangles located in the four quadrants into four descendant elements. In this implementation, the vertices of the emerging triangles are placed at the mid-points of the parental triangles, which are computed by linear interpolation. In the process of triangulation, new boundary nodes are projected in the radial direction onto the unit circle.

The refinement level is determined by the input flag $ndiv$, which is defined so that $ndiv = 0$ produces the four ancestral elements, $ndiv = 1$ produces sixteen first-generation elements, and $ndiv = 2$ produces sixty-four descendant elements. Each time a subdivision is carried out, the number of elements increases by a factor of four. The methodology is similar to that discussed in Section 3.3.1 for 3-node triangles.

We recall, in particular, that the array efl is the boundary flag of the element nodes, as described in Section 3.1: $efl(i, j)$ is set to zero at the interior nodes, and to unity at the boundary nodes. The flag efl of a new node is set according to the flags of the two nodes defining the subdivided edge. If both nodes are boundary nodes, the new node is also a boundary node, otherwise the new node is an interior node.

Triangulations produced by the function trgl6_disk are shown in Figure 4.2.1(a) for triangulation levels $ndiv = 0$ (ancestral), 1, 2, and 3. These shapes can be subsequently deformed or mapped by a transformation to produce other disk-like, simply connected domains.

```
function [ne,ng,p,c,efl,gfl] = trgl6_disk(ndiv)

%==========================================================
% Triangulation of the unit disk into 6-node elements by
% the successive subdivision of four ancestral elements
%==========================================================

%----------------------------------------
% ancestral structure with four elements
%----------------------------------------

ne = 4;

x(1,1) = 0.0; y(1,1) = 0.0; efl(1,1)=0;  % first element
x(1,2) = 1.0; y(1,2) = 0.0; efl(1,2)=1;
x(1,3) = 0.0; y(1,3) = 1.0; efl(1,3)=1;
x(1,4) = 0.5*(x(1,1)+x(1,2));
y(1,4) = 0.5*(y(1,1)+y(1,2)); efl(1,4)=0;
x(1,5) = 0.5*(x(1,2)+x(1,3));
y(1,5) = 0.5*(y(1,2)+y(1,3)); efl(1,5)=1;
x(1,6) = 0.5*(x(1,3)+x(1,1));
y(1,6) = 0.5*(y(1,3)+y(1,1)); efl(1,6)=0;

x(2,1) = 0.0; y(2,1) = 0.0; efl(2,1)=0;  % second element
x(2,2) = 0.0; y(2,2) = 1.0; efl(2,2)=1;
x(2,3) =-1.0; y(2,3) = 0.0; efl(2,3)=1;
x(2,4) = 0.5*(x(2,1)+x(2,2));
y(2,4) = 0.5*(y(2,1)+y(2,2)); efl(2,4)=0;
x(2,5) = 0.5*(x(2,2)+x(2,3));
y(2,5) = 0.5*(y(2,2)+y(2,3)); efl(2,5)=1;
x(2,6) = 0.5*(x(2,3)+x(2,1));
y(2,6) = 0.5*(y(2,3)+y(2,1)); efl(2,6)=0;

x(3,1) = 0.0; y(3,1) = 0.0; efl(3,1)=0;  % third element
x(3,2) =-1.0; y(3,2) = 0.0; efl(3,2)=1;
x(3,3) = 0.0; y(3,3) =-1.0; efl(3,3)=1;
x(3,4) = 0.5*(x(3,1)+x(3,2));
y(3,4) = 0.5*(y(3,1)+y(3,2)); efl(3,4)=0;
x(3,5) = 0.5*(x(3,2)+x(3,3));
y(3,5) = 0.5*(y(3,2)+y(3,3)); efl(3,5)=1;
x(3,6) = 0.5*(x(3,3)+x(3,1));
y(3,6) = 0.5*(y(3,3)+y(3,1)); efl(3,6)=0;
```

Function trgl6_disk: Continuing $\longrightarrow$

```
x(4,1) = 0.0; y(4,1) = 0.0; efl(4,1)=0;   % fourth element
x(4,2) = 0.0; y(4,2) =-1.0; efl(4,2)=1;
x(4,3) = 1.0; y(4,3) = 0.0; efl(4,3)=1;
x(4,4) = 0.5*(x(4,1)+x(4,2));
y(4,4) = 0.5*(y(4,1)+y(4,2)); efl(4,4)=0;
x(4,5) = 0.5*(x(4,2)+x(4,3));
y(4,5) = 0.5*(y(4,2)+y(4,3)); efl(4,5)=1;
x(4,6) = 0.5*(x(4,3)+x(4,1));
y(4,6) = 0.5*(y(4,3)+y(4,1)); efl(4,6)=0;

if(ndiv > 0)

for i=1:ndiv

 nm = 0; % count the new elements arising by each refinement loop
         % four elements will be generated in each pass

 for j=1:ne    % loop over current elements

  % assign vertex nodes to sub-elements;
  % these will become the "new" elements

   nm = nm+1;                      %  first sub-element

   xn(nm,1)=x(j,1); yn(nm,1)=y(j,1); efln(nm,1)=efl(j,1);
   xn(nm,2)=x(j,4); yn(nm,2)=y(j,4); efln(nm,2)=efl(j,4);
   xn(nm,3)=x(j,6); yn(nm,3)=y(j,6); efln(nm,3)=efl(j,6);

   xn(nm,4) = 0.5*(xn(nm,1)+xn(nm,2));
   yn(nm,4) = 0.5*(yn(nm,1)+yn(nm,2));
   xn(nm,5) = 0.5*(xn(nm,2)+xn(nm,3));
   yn(nm,5) = 0.5*(yn(nm,2)+yn(nm,3));
   xn(nm,6) = 0.5*(xn(nm,3)+xn(nm,1));
   yn(nm,6) = 0.5*(yn(nm,3)+yn(nm,1));

   efln(nm,4) = 0; efln(nm,5) = 0; efln(nm,6) = 0;

   if(efln(nm,1)==1 & efln(nm,2)==1) efln(nm,4) = 1; end
   if(efln(nm,2)==1 & efln(nm,3)==1) efln(nm,5) = 1; end
   if(efln(nm,3)==1 & efln(nm,1)==1) efln(nm,6) = 1; end

   nm = nm+1;                      %  second sub-element

   xn(nm,1)=x(j,4); yn(nm,1)=y(j,4); efln(nm,1)=efl(j,4);
   xn(nm,2)=x(j,2); yn(nm,2)=y(j,2); efln(nm,2)=efl(j,2);
   xn(nm,3)=x(j,5); yn(nm,3)=y(j,5); efln(nm,3)=efl(j,5);

   xn(nm,4) = 0.5*(xn(nm,1)+xn(nm,2));
   yn(nm,4) = 0.5*(yn(nm,1)+yn(nm,2));
   xn(nm,5) = 0.5*(xn(nm,2)+xn(nm,3));
   yn(nm,5) = 0.5*(yn(nm,2)+yn(nm,3));
   xn(nm,6) = 0.5*(xn(nm,3)+xn(nm,1));
   yn(nm,6) = 0.5*(yn(nm,3)+yn(nm,1));
```

Function trgl6_disk: $\longrightarrow$ Continuing $\longrightarrow$

```
efln(nm,4) = 0; efln(nm,5) = 0; efln(nm,6) = 0;

if(efln(nm,1)==1 & efln(nm,2)==1) efln(nm,4) = 1; end
if(efln(nm,2)==1 & efln(nm,3)==1) efln(nm,5) = 1; end
if(efln(nm,3)==1 & efln(nm,1)==1) efln(nm,6) = 1; end

nm = nm+1; %  third sub-element

xn(nm,1)=x(j,6); yn(nm,1)=y(j,6); efln(nm,1)=efl(j,6);
xn(nm,2)=x(j,5); yn(nm,2)=y(j,5); efln(nm,2)=efl(j,5);
xn(nm,3)=x(j,3); yn(nm,3)=y(j,3); efln(nm,3)=efl(j,3);

xn(nm,4) = 0.5*(xn(nm,1)+xn(nm,2));
yn(nm,4) = 0.5*(yn(nm,1)+yn(nm,2));
xn(nm,5) = 0.5*(xn(nm,2)+xn(nm,3));
yn(nm,5) = 0.5*(yn(nm,2)+yn(nm,3));
xn(nm,6) = 0.5*(xn(nm,3)+xn(nm,1));
yn(nm,6) = 0.5*(yn(nm,3)+yn(nm,1));

efln(nm,4) = 0; efln(nm,5) = 0; efln(nm,6) = 0;
if(efln(nm,1)==1 & efln(nm,2)==1) efln(nm,4) = 1; end
if(efln(nm,2)==1 & efln(nm,3)==1) efln(nm,5) = 1; end
if(efln(nm,3)==1 & efln(nm,1)==1) efln(nm,6) = 1; end

nm = nm+1; %  fourth sub-element

xn(nm,1)=x(j,4); yn(nm,1)=y(j,4); efln(nm,1)=efl(j,4);
xn(nm,2)=x(j,5); yn(nm,2)=y(j,5); efln(nm,2)=efl(j,5);
xn(nm,3)=x(j,6); yn(nm,3)=y(j,6); efln(nm,3)=efl(j,6);

xn(nm,4) = 0.5*(xn(nm,1)+xn(nm,2));
yn(nm,4) = 0.5*(yn(nm,1)+yn(nm,2));
xn(nm,5) = 0.5*(xn(nm,2)+xn(nm,3));
yn(nm,5) = 0.5*(yn(nm,2)+yn(nm,3));
xn(nm,6) = 0.5*(xn(nm,3)+xn(nm,1));
yn(nm,6) = 0.5*(yn(nm,3)+yn(nm,1));

efln(nm,4) = 0; efln(nm,5) = 0; efln(nm,6) = 0;
if(efln(nm,1)==1 & efln(nm,2)==1) efln(nm,4) = 1; end
if(efln(nm,2)==1 & efln(nm,3)==1) efln(nm,5) = 1; end
if(efln(nm,3)==1 & efln(nm,1)==1) efln(nm,6) = 1; end

end % end of loop over current elements

ne = 4*ne;  % number of elements has increased
            % by a factor of four
for k=1:ne     % relabel the new points
               % and put them in the master list
               % project boundary nodes onto the circle
```

Function trgl6_disk: $\longrightarrow$ Continuing $\longrightarrow$

```
    for l=1:6
     x(k,l)=xn(k,l); y(k,l)=yn(k,l);  efl(k,l)=efln(k,l);
     if(efl(k,l) == 1)
        rad = sqrt(x(k,l)^2+y(k,l)^2);
        x(k,l)=x(k,l)/rad;y(k,l)=y(k,l)/rad;
     end
    end

 end

 for k=1:ne      % reposition the mid-nodes

    if(efl(k,4) == 0)
       x(k,4)=0.5*(x(k,1)+x(k,2));
       y(k,4)=0.5*(y(k,1)+y(k,2)); end

    if(efl(k,5) == 0)
       x(k,5)=0.5*(x(k,2)+x(k,3));
       y(k,5)=0.5*(y(k,2)+y(k,3)); end

    if(efl(k,6) == 0)
       x(k,6)=0.5*(x(k,3)+x(k,1));
       y(k,6)=0.5*(y(k,3)+y(k,1)); end

 end

end % end do of refinement loop
end % end if of refinement loop

% 6 nodes of the first element are entered manually

p(1,1)=x(1,1); p(1,2)=y(1,1); gfl(1,1)=efl(1,1);
p(2,1)=x(1,2); p(2,2)=y(1,2); gfl(2,1)=efl(1,2);
p(3,1)=x(1,3); p(3,2)=y(1,3); gfl(3,1)=efl(1,3);

p(4,1)=x(1,4); p(4,2)=y(1,4); gfl(4,1)=efl(1,4);
p(5,1)=x(1,5); p(5,2)=y(1,5); gfl(5,1)=efl(1,5);
p(6,1)=x(1,6); p(6,2)=y(1,6); gfl(6,1)=efl(1,6);

c(1,1) = 1;  % first  node of first element is
             % global node 1; similarly:
c(1,2) = 2;
c(1,3) = 3; c(1,4) = 4;
c(1,5) = 5; c(1,6) = 6;

% loop over further elements;
% Iflag=0 will signal a new global node

eps = 0.000001;
```

Function trgl6_disk: $\longrightarrow$ Continuing $\longrightarrow$

```
for i=2:ne        % loop over elements

 for j=1:6        % loop over element nodes

 Iflag=0;

  for k=1:ng

   if( abs(x(i,j)-p(k,1)) < eps)
    if(abs(y(i,j)-p(k,2)) < eps)
      Iflag = 1;    % the node has been recorded previously
      c(i,j) = k;   % the jth local node of element i
                    % is the kth global node
    end
   end

  end

  if(Iflag==0)  % record the node

    ng = ng+1;
    p(ng,1)=x(i,j);
    p(ng,2)=y(i,j);

    gfl(ng,1)=efl(i,j);

    c(i,j) = ng;   % the jth local node of this element
                   % is the new global node
  end

 end

end  % end of loop over elements
%-----
% done
%-----

return;
```

Function trgl6_disk: ($\longrightarrow$ Continued.) Triangulation of the unit disk into 6-node elements generated by successively subdividing four ancestral elements located in the four quadrants. Each time a subdivision occurs, the number of elements increases by a factor of four.

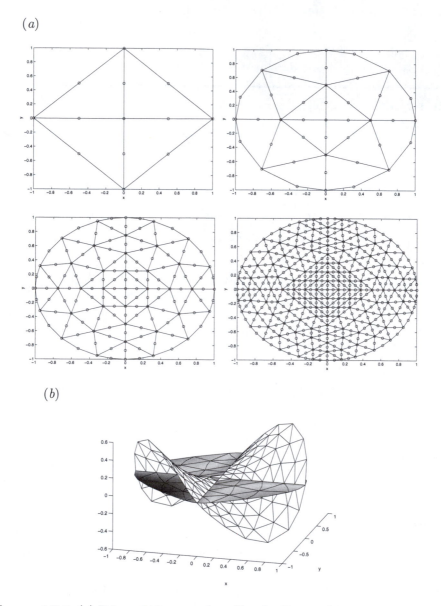

Figure 4.2.1 (*a*) Triangulations produced by the FSELIB function `trgl6_disk`, for discretization levels $ndiv = 0, 1, 2, 3$. Each triangle is defined by six nodes, including three vertex nodes and three edge nodes. (*b*) Finite element solution of Laplace's equation subject to the Dirichlet boundary condition, produced by the FSELIB code `lapl6_d`, for discretization level $ndiv = 2$.

4.2.1   Triangulation of an L-shaped domain

FSELIB includes the function trgl6_L that triangulates an L-shaped domain
in the spirit of the function trgl6_disk described in the previous section. In
the case of the L-shaped domain, we begin by defining six ancestral triangles
in the first, third, and fourth quadrants, as shown in Figure 4.2.2(*a*), and then
produce different levels of triangulation by element subdivision. The beginning
of the code showing the ancestral triangulation is listed in the text. After the
triangulation has been completed, the grid can be deformed to yield another
desired shape that is homologous to the L-shaped domain.

The triangulation for discretization level $ndiv = 2$ is shown in Figure 4.2.2(*b*)
along with the finite element solution of Laplace's equation, as will be discussed
later in this section.

4.2.2   Triangulation of a square with a square or circular hole

FSELIB includes the function trgl6_ss that triangulates a unit square pierced
by a square hole with half-side length equal to a, where $a < 0.5$. The companion
function trgl6_sc triangulates a unit square pierced by a circular hole of radius
a, where $a < 0.5$.

Both triangulations are performed in the spirit of the functions trgl6_disk
and trgl6_L discussed earlier in this section. In both cases, we begin by manually triangulating the domain into twelve ancestral triangles, as shown in Figures 4.2.3(*a*), and 4.2.4(*a*), and then produce different levels of triangulation
by element subdivision. The triangulation of a square with a square hole for
discretization level $ndiv = 1$ is shown in Figure 4.2.3(*b*). The triangulation of a
square with a circular hole for discretization level $ndiv = 2$ is shown in Figure
4.2.4(*b*). Superposed on these graphs are finite element solutions of Laplace's
equation, as will be discussed later in this section. The beginning of both
triangulation codes showing the ancestral triangular structure is listed in the
text.

A new feature of functions trgl6_ss and trgl6_sc is that, because the
solution domain is doubly connected, the element-node flags $efl(i,1)$ and corresponding global-node flags $gfl(i,1)$ may take the values 0, 1, or 2. The value
0 indicates that a node is an interior node, the value 1 indicates that a node lies
on the outer boundary, and the value 2 indicates that a node lies on the interior
boundary. The flag of a new node generated during element subdivision is set
accordingly, depending on whether the new node lies on a segment defined by
two end-nodes with identical flags.

4.2.3   Triangulation of a rectangle with a circular hole

FSELIB includes the function trgl6_rc (not listed in the text) that triangulates
a rectangular domain pierced by a circular hole. In this case, we begin by
manually triangulating the domain into twenty ancestral triangles, as shown in
Figures 4.2.5(*a*), and then produce different levels of triangulation by element

```
function [ne,ng,p,c,efl,gfl] = trgl6_L(ndiv)

%==============================================================
% Triangulation of an L-shaped domain into 6-node elements
% by the successive subdivision of six ancestral elements
%==============================================================

%--------------------------------------
% ancestral structure with six elements
%--------------------------------------

ne = 6;

x(1,1) = 0.0; y(1,1) = 0.0; efl(1,1)=0;   % first element
x(1,2) = 1.0; y(1,2) = 0.0; efl(1,2)=1;
x(1,3) = 0.0; y(1,3) = 1.0; efl(1,3)=1;

x(1,4) = 0.5*(x(1,1)+x(1,2));
y(1,4) = 0.5*(y(1,1)+y(1,2)); efl(1,4)=0;
x(1,5) = 0.5*(x(1,2)+x(1,3));
y(1,5) = 0.5*(y(1,2)+y(1,3)); efl(1,5)=0;
x(1,6) = 0.5*(x(1,3)+x(1,1));
y(1,6) = 0.5*(y(1,3)+y(1,1)); efl(1,6)=1;

x(2,1) = 1.0; y(2,1) = 0.0; efl(2,1)=1;   % second element
x(2,2) = 1.0; y(2,2) = 1.0; efl(2,2)=1;
x(2,3) = 0.0; y(2,3) = 1.0; efl(2,3)=1;

x(2,4) = 0.5*(x(2,1)+x(2,2));
y(2,4) = 0.5*(y(2,1)+y(2,2)); efl(2,4)=1;
x(2,5) = 0.5*(x(2,2)+x(2,3));
y(2,5) = 0.5*(y(2,2)+y(2,3)); efl(2,5)=1;
x(2,6) = 0.5*(x(2,3)+x(2,1));
y(2,6) = 0.5*(y(2,3)+y(2,1)); efl(2,6)=0;

x(3,1) = 0.0; y(3,1) = 0.0; efl(3,1)=1;   % third element
x(3,2) =-1.0; y(3,2) = 0.0; efl(3,2)=1;
x(3,3) = 0.0; y(3,3) =-1.0; efl(3,3)=1;

x(3,4) = 0.5*(x(3,1)+x(3,2));
y(3,4) = 0.5*(y(3,1)+y(3,2)); efl(3,4)=1;
x(3,5) = 0.5*(x(3,2)+x(3,3));
y(3,5) = 0.5*(y(3,2)+y(3,3)); efl(3,5)=0;
x(3,6) = 0.5*(x(3,3)+x(3,1));
y(3,6) = 0.5*(y(3,3)+y(3,1)); efl(3,6)=0;
```

Function trgl6_L: Continuing $\longrightarrow$

subdivision. The element-node flag $efl(i, 1)$ and the corresponding global-node flag $gfl(i, 1)$ may take the values 0, 1, or 2, respectively, for interior nodes, exterior boundary nodes, and interior boundary nodes, as discussed earlier in this section for code trgl6_sc.

```
x(4,1) =-1.0; y(4,1) = 0.0; efl(4,1)=1;  % fourth element
x(4,2) =-1.0; y(4,2) =-1.0; efl(4,2)=1;
x(4,3) = 0.0; y(4,3) =-1.0; efl(4,3)=1;

x(4,4) = 0.5*(x(4,1)+x(4,2));
y(4,4) = 0.5*(y(4,1)+y(4,2)); efl(4,4)=1;
x(4,5) = 0.5*(x(4,2)+x(4,3));
y(4,5) = 0.5*(y(4,2)+y(4,3)); efl(4,5)=1;
x(4,6) = 0.5*(x(4,3)+x(4,1));
y(4,6) = 0.5*(y(4,3)+y(4,1)); efl(4,6)=0;

x(5,1) = 0.0; y(5,1) = 0.0; efl(5,1)=0;  % fifth element
x(5,2) = 0.0; y(5,2) =-1.0; efl(5,2)=1;
x(5,3) = 1.0; y(5,3) = 0.0; efl(5,3)=1;

x(5,4) = 0.5*(x(5,1)+x(5,2));
y(5,4) = 0.5*(y(5,1)+y(5,2)); efl(5,4)=1;
x(5,5) = 0.5*(x(5,2)+x(5,3));
y(5,5) = 0.5*(y(5,2)+y(5,3)); efl(5,5)=1;
x(5,6) = 0.5*(x(5,3)+x(5,1));
y(5,6) = 0.5*(y(5,3)+y(5,1)); efl(5,6)=0;

x(6,1) = 1.0; y(6,1) = 0.0; efl(6,1)=1;  % sixth element
x(6,2) = 0.0; y(6,2) =-1.0; efl(6,2)=1;
x(6,3) = 1.0; y(6,3) =-1.0; efl(6,3)=1;

x(6,4) = 0.5*(x(6,1)+x(6,2));
y(6,4) = 0.5*(y(6,1)+y(6,2)); efl(6,4)=1;
x(6,5) = 0.5*(x(6,2)+x(6,3));
y(6,5) = 0.5*(y(6,2)+y(6,3)); efl(6,5)=1;
x(6,6) = 0.5*(x(6,3)+x(6,1));
y(6,6) = 0.5*(y(6,3)+y(6,1)); efl(6,6)=0;

%----------------
% refinement loop
%----------------

. . . . . .
```

Function trgl6_L: ($\longrightarrow$ Continued.) Triangulation of an L-shaped domain into 6-node elements generated by the successive subdivision of six ancestral elements. The six dots at the end of the listing indicate additional code included in FEMLIB.

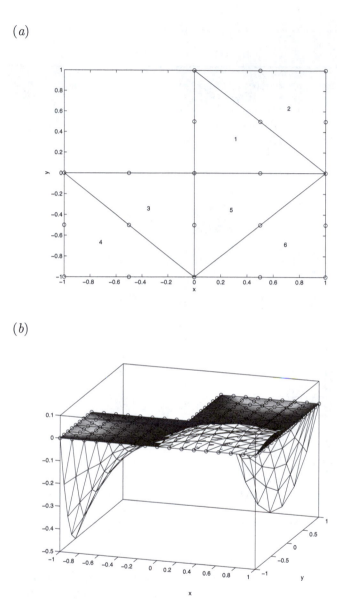

Figure 4.2.2 (*a*) Ancestral structure for triangulating an L-shaped domain, and (*b*) finite element solution of Laplace's equation subject to the Dirichlet boundary condition produced by the FSELIB code `lapl6_d_L`, for discretization level $ndiv = 2$.

```
function [ne,ng,p,c,efl,gfl] = trgl6_ss(a, ndiv)

%=========================================================
% Triangulation of a unit square with a square hole of
% radius ''a'' into 6-node elements by the successive
% subdivision of twelve ancestral elements
%=========================================================

%-----------------------------------------
% ancestral structure with twelve elements
%-----------------------------------------

ne = 12;

x(1,1) =    a;y(1,1) =   -a; efl(1,1)=2;  % first element
x(1,2) = 1.0;y(1,2) =-1.0; efl(1,2)=1;
x(1,3) = 1.0;y(1,3) = 0.0; efl(1,3)=1;

x(1,4) = 0.5*(x(1,1)+x(1,2));
y(1,4) = 0.5*(y(1,1)+y(1,2)); efl(1,4)=0;
x(1,5) = 0.5*(x(1,2)+x(1,3));
y(1,5) = 0.5*(y(1,2)+y(1,3)); efl(1,5)=1;
x(1,6) = 0.5*(x(1,3)+x(1,1));
y(1,6) = 0.5*(y(1,3)+y(1,1)); efl(1,6)=0;

x(2,1) =    a;y(2,1) =   -a; efl(2,1)=2;  % second element
x(2,2) = 1.0;y(2,2) = 0.0; efl(2,2)=1;
x(2,3) =    a;y(2,3) =    a; efl(2,3)=2;

x(2,4) = 0.5*(x(2,1)+x(2,2));
y(2,4) = 0.5*(y(2,1)+y(2,2)); efl(2,4)=0;
x(2,5) = 0.5*(x(2,2)+x(2,3));
y(2,5) = 0.5*(y(2,2)+y(2,3)); efl(2,5)=0;
x(2,6) = 0.5*(x(2,3)+x(2,1));
y(2,6) = 0.5*(y(2,3)+y(2,1)); efl(2,6)=2;

x(3,1) =    a;y(3,1) =    a; efl(3,1)=2;  % third element
x(3,2) = 1.0;y(3,2) = 0.0; efl(3,2)=1;
x(3,3) = 1.0;y(3,3) = 1.0; efl(3,3)=1;

x(3,4) = 0.5*(x(3,1)+x(3,2));
y(3,4) = 0.5*(y(3,1)+y(3,2)); efl(3,4)=0;
x(3,5) = 0.5*(x(3,2)+x(3,3));
y(3,5) = 0.5*(y(3,2)+y(3,3)); efl(3,5)=1;
x(3,6) = 0.5*(x(3,3)+x(3,1));
y(3,6) = 0.5*(y(3,3)+y(3,1)); efl(3,6)=0;

x(4,1) =    a;y(4,1) =    a; efl(4,1)=2;  % fourth element
x(4,2) = 1.0;y(4,2) = 1.0; efl(4,2)=1;
x(4,3) = 0.0;y(4,3) = 1.0; efl(4,3)=1;
```

Function trgl6_ss: Continuing $\longrightarrow$

```
x(4,4) = 0.5*(x(4,1)+x(4,2));
y(4,4) = 0.5*(y(4,1)+y(4,2)); efl(4,4)=0;
x(4,5) = 0.5*(x(4,2)+x(4,3));
y(4,5) = 0.5*(y(4,2)+y(4,3)); efl(4,5)=1;
x(4,6) = 0.5*(x(4,3)+x(4,1));
y(4,6) = 0.5*(y(4,3)+y(4,1)); efl(4,6)=0;

x(5,1) =  0.0;y(5,1) = 1.0; efl(5,1)=1;  % fifth element
x(5,2) =  -a;y(5,2) =   a; efl(5,2)=2;
x(5,3) =   a;y(5,3) =   a; efl(5,3)=2;

x(5,4) = 0.5*(x(5,1)+x(5,2));
y(5,4) = 0.5*(y(5,1)+y(5,2)); efl(5,4)=0;
x(5,5) = 0.5*(x(5,2)+x(5,3));
y(5,5) = 0.5*(y(5,2)+y(5,3)); efl(5,5)=2;
x(5,6) = 0.5*(x(5,3)+x(5,1));
y(5,6) = 0.5*(y(5,3)+y(5,1)); efl(5,6)=0;

%-------------------------------------
% rest of the elements by reflection
%-------------------------------------

for i=1:6
 for j=1:6

  x(6+i,j)=-x(i,j);
  y(6+i,j)=-y(i,j);
  efl(6+i,j)=efl(i,j);

 end
end

%----------------
% refinement loop
%----------------

. . . . . .
```

Function trgl6_ss: ($\longrightarrow$ Continued.) Triangulation of the unit square pierced by a square hole into 6-node elements generated by the successive subdivision of an ancestral structure with twelve elements. The six dots at the end of the listing indicate unprinted lines of code.

```
function [ne,ng,p,c,efl,gfl] = trgl6_sc(a, ndiv)

%==============================================================
% Triangulation of a unit square with a circular hole of
% radius ''a'' into 6-node elements by the successive
% subdivision of twelve ancestral elements
%==============================================================

%--------------------------------------------
% ancestral structure with twelve elements
%--------------------------------------------

ne = 12;

x(1,1) =    a; y(1,1) =  -a; efl(1,1)=2;  % first element
x(1,2) = 1.0; y(1,2) =-1.0; efl(1,2)=1;
x(1,3) = 1.0; y(1,3) = 0.0; efl(1,3)=1;
x(1,4) = 0.5*(x(1,1)+x(1,2));
y(1,4) = 0.5*(y(1,1)+y(1,2)); efl(1,4)=0;
x(1,5) = 0.5*(x(1,2)+x(1,3));
y(1,5) = 0.5*(y(1,2)+y(1,3)); efl(1,5)=1;
x(1,6) = 0.5*(x(1,3)+x(1,1));
y(1,6) = 0.5*(y(1,3)+y(1,1)); efl(1,6)=0;

x(2,1) =    a; y(2,1) =  -a; efl(2,1)=2;  % second element
x(2,2) = 1.0; y(2,2) = 0.0; efl(2,2)=1;
x(2,3) =    a; y(2,3) =    a; efl(2,3)=2;
x(2,4) = 0.5*(x(2,1)+x(2,2));
y(2,4) = 0.5*(y(2,1)+y(2,2)); efl(2,4)=0;
x(2,5) = 0.5*(x(2,2)+x(2,3));
y(2,5) = 0.5*(y(2,2)+y(2,3)); efl(2,5)=0;
x(2,6) = 0.5*(x(2,3)+x(2,1));
y(2,6) = 0.5*(y(2,3)+y(2,1)); efl(2,6)=2;

x(3,1) =    a; y(3,1) =    a; efl(3,1)=2;  % third element
x(3,2) = 1.0; y(3,2) = 0.0; efl(3,2)=1;
x(3,3) = 1.0; y(3,3) = 1.0; efl(3,3)=1;
x(3,4) = 0.5*(x(3,1)+x(3,2));
y(3,4) = 0.5*(y(3,1)+y(3,2)); efl(3,4)=0;
x(3,5) = 0.5*(x(3,2)+x(3,3));
y(3,5) = 0.5*(y(3,2)+y(3,3)); efl(3,5)=1;
x(3,6) = 0.5*(x(3,3)+x(3,1));
y(3,6) = 0.5*(y(3,3)+y(3,1)); efl(3,6)=0;

x(4,1) =    a;y(4,1) =    a; efl(4,1)=2;  % fourth element
x(4,2) = 1.0;y(4,2) = 1.0; efl(4,2)=1;
x(4,3) = 0.0;y(4,3) = 1.0; efl(4,3)=1;
x(4,4) = 0.5*(x(4,1)+x(4,2));
y(4,4) = 0.5*(y(4,1)+y(4,2)); efl(4,4)=0;
x(4,5) = 0.5*(x(4,2)+x(4,3));
y(4,5) = 0.5*(y(4,2)+y(4,3)); efl(4,5)=1;
x(4,6) = 0.5*(x(4,3)+x(4,1));
y(4,6) = 0.5*(y(4,3)+y(4,1)); efl(4,6)=0;
```

Function trgl6_sc: Continuing $\longrightarrow$

```
x(5,1) = 0.0; y(5,1) = 1.0; efl(5,1)=1;  % fifth element
x(5,2) =   -a; y(5,2) =    a; efl(5,2)=2;
x(5,3) =    a; y(5,3) =    a; efl(5,3)=2;
x(5,4) = 0.5*(x(5,1)+x(5,2));
y(5,4) = 0.5*(y(5,1)+y(5,2)); efl(5,4)=0;
x(5,5) = 0.5*(x(5,2)+x(5,3));
y(5,5) = 0.5*(y(5,2)+y(5,3)); efl(5,5)=2;
x(5,6) = 0.5*(x(5,3)+x(5,1));
y(5,6) = 0.5*(y(5,3)+y(5,1)); efl(5,6)=0;

x(6,1) =   -a; y(6,1) =    a; efl(6,1)=2;  % sixth element
x(6,2) = 0.0; y(6,2) = 1.0; efl(6,2)=1;
x(6,3) =-1.0; y(6,3) = 1.0; efl(6,3)=1;
x(6,4) = 0.5*(x(6,1)+x(6,2));
y(6,4) = 0.5*(y(6,1)+y(6,2)); efl(6,4)=0;
x(6,5) = 0.5*(x(6,2)+x(6,3));
y(6,5) = 0.5*(y(6,2)+y(6,3)); efl(6,5)=1;
x(6,6) = 0.5*(x(6,3)+x(6,1));
y(6,6) = 0.5*(y(6,3)+y(6,1)); efl(6,6)=0;

%-----------------------------------
% rest of the elements by reflection
%-----------------------------------

for i=1:6
 for j=1:6
  x(6+i,j)=-x(i,j); y(6+i,j)=-y(i,j); efl(6+i,j)=efl(i,j);
 end
end

%------------------------------------------------------------
% project the interior-boundary nodes onto a circle of radius "a"
%------------------------------------------------------------

for i=1:12
  for j=1:6
    if (efl(i,j)==2)
     rad = a/sqrt(x(i,j)^2+y(i,j)^2);
     x(i,j) = x(i,j)*rad;y(i,j) = y(i,j)*rad;
    end
 end
end

%----------------
% refinement loop
%----------------
. . . . . .
```

Function trgl6_sc: ($\longrightarrow$ Continued.) Triangulation of the unit square pierced
by a circular hole into 6-node elements generated by the successive subdivision of an ancestral structure with twelve elements. The six dots at
the end of the listing indicate additional lines of unprinted code.

(a)

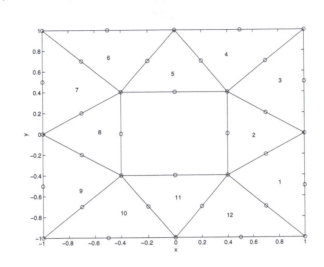

(b)

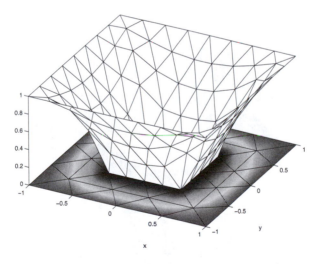

Figure 4.2.3 (a) Ancestral structure for triangulating a square with a square hole, and (b) finite element solution of Laplace's equation subject to the Dirichlet boundary condition produced by the FSELIB code `lapl6_d_ss` for discretization level $ndiv = 1$.

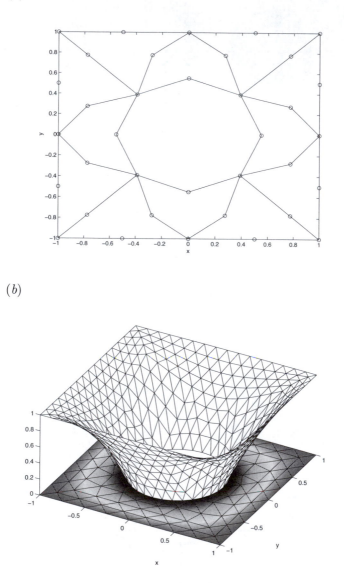

Figure 4.2.4 (*a*) Ancestral structure for triangulating a square with a circular hole, and (*b*) finite element solution of Laplace's equation subject to Dirichlet boundary condition produced by the FSELIB code `lapl6_d_sc` for discretization level *ndiv* = 2.

(a)

(b)

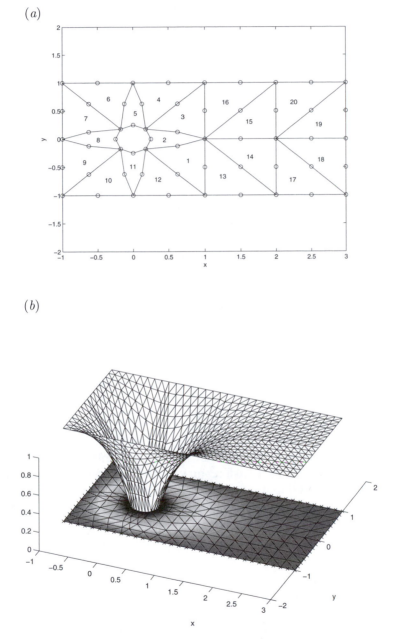

Figure 4.2.5 Continuing $\longrightarrow$

(*c*)

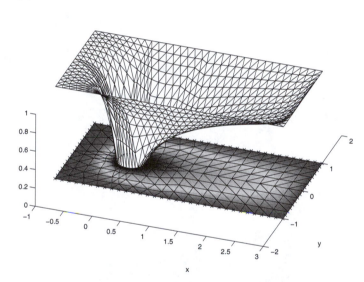

Figure 4.2.5 (⟶ Continued.) (*a*) Ancestral structure for triangulating a rectangular domain with a circular hole. (*b, c*) Finite element solution of the steady convection–diffusion equation subject to the Dirichlet boundary condition produced by the FSELIB code **scd6_d_rc**, for discretization level *ndiv* = 2, and (*b*) *Pe* = 0.0 (pure diffusion), and (*c*) 2.5 (diffusion combined with convection).

4.2.4 Code for Laplace's equation with the Dirichlet boundary condition

The layout of the finite element code for solving Laplace's equation,

$$\nabla^2 f = 0, \qquad (4.2.1)$$

subject to the Dirichlet boundary condition applied all around the boundary of the solution domain, is a straightforward modification of that discussed in Section 3.4 for the 3-node triangle. The input-output (I/O) and computational modules are repeated here for convenient reference:

- *Specify the input data.*

- *Perform the grid generation.*

- *Specify the Dirichlet boundary condition at the boundary nodes.*

- *Assemble the global diffusion matrix from the element diffusion matrices aided by the connectivity matrix.*

- *Assemble the right-hand side of the linear system.*

- *Implement the Dirichlet boundary condition, as discussed in Section 3.1.4.*

- *Solve the linear system.*

- *Prepare a graph of the solution using the* MATLAB *graphics functions* patch *and* trimesh.

Disk-like domain

FSELIB code `lap16_d`, listed in the text, solves Laplace's equation in a disk-like domain. The triangulation is performed by the function `trgl6_disk` discussed earlier in this section. The graphics module is implemented in the FSELIB function `plot_6` listed in the text, based on the MATLAB graphics function `patch`. An additional graphics display is based on the MATLAB function `trimesh`, which receives as input an extended connectivity matrix pertaining to 3-node sub-triangles that arise by dividing each 6-node triangle into four sub-elements. The graphics output of a finite element solution for discretization level $ndiv = 2$ with the boundary condition implemented in the code is displayed in Figure 4.2.1(b).

L-shaped domain

FSELIB code `lap16_d_L` (not listed in the text) solves Laplace's equation in the L-shaped domain shown in Figure 4.2.2, subject to the Dirichlet boundary condition around the polygonal boundary, $f = x\,y\,(x-1)\,(y+1)$, The implementation is nearly identical to that described earlier in this section for code `lap16_d`. The graphics output is displayed in Figure 4.2.2(b).

```
%===============================================
% CODE lapl6_d
%
%
% Finite element code for Laplace's equation
% in a disk-like domain with the Dirichlet
% boundary condition, using 6-node triangles
%===============================================

%-----------
% input data
%-----------

ndiv = 1;  % discretization level

NQ = 6;     % gauss-triangle quadrature

%------------
% triangulate
%------------

[ne,ng,p,c,efl,gfl] = trgl6_disk (ndiv);

%-------
% deform
%-------

for i=1:ng
   p(i,1)=p(i,1)*(1.0-0.60*p(i,2)^2 );
end

%----------------------------------------
% specify the Dirichlet boundary condition
%----------------------------------------

for i=1:ng

  if(gfl(i,1)==1)
    gfl(i,2) = p(i,1)*sin(0.5*pi*p(i,2));
  end

end

%-------------------------------------------------
% assemble the global diffusion matrix
% and compute the domain surface area (optional)
%-------------------------------------------------

gdm = zeros(ng,ng);  % initialize
area = 0.0;          % initialize
```

Code lapl6_d: Continuing $\longrightarrow$

```
for l=1:ne                % loop over the elements
                          % compute the element diffusion matrix
 j=c(l,1); x1=p(j,1); y1=p(j,2);
 j=c(l,2); x2=p(j,1); y2=p(j,2);
 j=c(l,3); x3=p(j,1); y3=p(j,2);
 j=c(l,4); x4=p(j,1); y4=p(j,2);
 j=c(l,5); x5=p(j,1); y5=p(j,2);
 j=c(l,6); x6=p(j,1); y6=p(j,2);

  [edm_elm, arel] = edm6 ...
    (x1,y1, x2,y2, x3,y3, x4,y4, x5,y5, x6,y6 ,NQ);

  area = area + arel;

   for i=1:6
     i1 = c(l,i);
     for j=1:6
       j1 = c(l,j);
       gdm(i1,j1) = gdm(i1,j1) + edm_elm(i,j);
     end
   end

end

%------------------------------------------------
% compute the right-hand side
% and implement the Dirichlet boundary condition
%------------------------------------------------

for i=1:ng
 b(i) = 0.0;
end

for j=1:ng

 if(gfl(j,1)==1)

   for i=1:ng
     b(i) = b(i) - gdm(i,j) * gfl(j,2);
     gdm(i,j) = 0; gdm(j,i) = 0;
   end

   gdm(j,j) = 1.0; b(j) = gfl(j,2);

 end

end

%-----------------------
% solve the linear system
%-----------------------

  f = b/gdm';
```

Code lapl6_d: $\longrightarrow$ Continuing $\longrightarrow$

```
%-----
% plot
%-----

plot_6 (ne,ng,p,c,f);

%-----
% Connectivity matrix for 3-node sub-triangles
%-----

Ic=0;

for i=1:ne
  Ic=Ic+1;
  c3(Ic,1) = c(i,1); c3(Ic,2) = c(i,4); c3(Ic,3) = c(i,6);
  Ic=Ic+1;
  c3(Ic,1) = c(i,4); c3(Ic,2) = c(i,2); c3(Ic,3) = c(i,5);
  Ic=Ic+1;
  c3(Ic,1) = c(i,5); c3(Ic,2) = c(i,3); c3(Ic,3) = c(i,6);
  Ic=Ic+1;
  c3(Ic,1) = c(i,4); c3(Ic,2) = c(i,5); c3(Ic,3) = c(i,6);
end

%-----
% plot
%-----

trimesh (c3,p(:,1),p(:,2),f);

%-----
% done
%-----
```

Code lapl6_d: ($\longrightarrow$ Continued.) Finite element code for solving Laplace's equation in a disk-like domain subject to the Dirichlet boundary condition, using 6-node triangles.

Square with a square or circular hole

FSELIB codes lapl6_d_ss and lapl6_d_sc (not listed in the text) implement the finite element method for Laplace's equation in a square with a square or circular hole, subject to the Dirichlet boundary condition specifying that $f = 1$ around the exterior boundary, and $f = 0$ around the interior boundary. Physically, the solution domain may be identified with the cross-section of a chimney wall.

The implementation of the finite element method is similar to that described earlier for code lapl6_d. The triangulation is performed by the functions trgl6_ss and trgl6_sc discussed earlier in this section. The graphics output of the codes is shown in Figures 4.2.3(*a*) and 4.2.4(*b*). The results show that the shape of the inner boundary is significant only in its proximity.

```
function plot_6 (ne,ng,p,c,f);

%===============================================
% Color mapped visualization of a function f in a
% domain discretized into 6-node triangles
% using the patch graphics function
%===============================================

%----------------------------------------------------
% compute the maximim and minimum of the function f
%----------------------------------------------------

fmax = -100.0;   fmin= 100; % trial values

for i=1:ng
 if(f(i) > fmax) fmax = f(i); end
 if(f(i) < fmin) fmin = f(i); end
end

range = 1.2*(fmax-fmin); shift = fmin;

%---------------------------------------------
% shift the color index in the range (0, 1)
% and paint seven-point patches
%---------------------------------------------

for l=1:ne

 j=c(1,1); xp(1)=p(j,1); yp(1)=p(j,2); cp(1)=(f(j)-shift)/range;
 j=c(1,4); xp(2)=p(j,1); yp(2)=p(j,2); cp(2)=(f(j)-shift)/range;
 j=c(1,2); xp(3)=p(j,1); yp(3)=p(j,2); cp(3)=(f(j)-shift)/range;
 j=c(1,5); xp(4)=p(j,1); yp(4)=p(j,2); cp(4)=(f(j)-shift)/range;
 j=c(1,3); xp(5)=p(j,1); yp(5)=p(j,2); cp(5)=(f(j)-shift)/range;
 j=c(1,6); xp(6)=p(j,1); yp(6)=p(j,2); cp(6)=(f(j)-shift)/range;
 j=c(1,1); xp(7)=p(j,1); yp(7)=p(j,2); cp(7)=(f(j)-shift)/range;

 patch(xp, yp,cp); hold on; xlabel('x'); ylabel('y');

end

%----------------------
% define the axes range
%----------------------

axis([-1 1 -1 1]);

%-----
% done
%-----

return;
```

Function plot_6: Color mapped visualization of a function in a domain discretized into 6-node triangles.

4.2.5 Code for steady convection–diffusion

FSELIB code scd6_d_rc, listed in the text, solves the steady convection–diffusion equation

$$u_x \frac{\partial f}{\partial x} + u_y \frac{\partial f}{\partial y} = \kappa \nabla^2 f, \qquad (4.2.2)$$

where $u_x(x, y)$ and $u_y(x, y)$ are the x and y components of the advection velocity, and κ is the diffusivity. The solution domain is a rectangle with a circular hole of radius a, physically representing the contour of a circular cylinder embedded in a channel streaming flow, as shown in Figure 4.2.5(a). The Dirichlet boundary condition specifies the distribution of the function f around the outer rectangular boundary and along the inner circular boundary. The triangulation is performed using the FSELIB function *trgl6_rc* discussed earlier in this section (not listed in the text).

The advection velocity prescribed in the code corresponds to streaming potential flow with velocity U past a cylinder, given by

$$u_x = U \left[1 + \frac{a^2}{r^2} (1 - 2 \frac{x^2}{r^2}) \right],$$

$$\qquad (4.2.3)$$

$$v_y = -2 U a^2 \frac{xy}{r^4},$$

where r is the distance from the cylinder center (e.g., [42]). The importance of convection relative to diffusion is determined by the Péclet number, defined as $Pe = Ua/\kappa$.

A new module of the finite element code is the interpolation of the x and y velocity components at the element integration base points from corresponding values at the six element nodes. This computation is performed inside the function edam6 that evaluates the global diffusion and advection matrices, listed in Section 4.1. The nodal velocities themselves are evaluated by the FSELIB function scd6_vel, listed in the text.

The graphics visualization of the finite element solution is displayed in Figure 4.2.5(b, c), respectively, for $Pe = 0.0$ and 2.5. In these computations, the Dirichlet boundary condition requires that $f = 1$ around the outer rectangular boundary, and $f = 0$ around the inner circular boundary. Physically, the results illustrate the temperature field established around an isothermal cold cylinder in a channel confined between two isothermal parallel plane walls, in the presence of a convective stream. As $Pe \to 0$, heat transport occurs due to conduction alone. As Pe is raised, convective transport becomes increasingly important, and the temperature field extends in the downstream direction, in agreement with physical intuition.

```
%=======================================================
% CODE scd6_d_rc
%
% Code for the steady convection--diffusion equation
% with the Dirichlet boundary condition in a
% rectangular domain with a circular hole
%=======================================================

%-----------
% input data
%-----------

k=1.0; rho=1.0; cp=1.0;
U=0.5;
NQ = 6;
ndiv = 2;
a = 0.25;  % cylinder radius

%----------
% constants
%----------

kappa = k/(rho*cp);

%------------
% triangulate
%------------

[ne,ng,p,c,efl,gfl] = trgl6_rc (a, ndiv);

%-------------------------------------------
% specify Dirichlet the boundary condition
%-------------------------------------------

for i=1:ng

 if(gfl(i,1)==1)
   gfl(i,2) = 1.0; % disp (gfl(i,1));
 end
 if(gfl(i,1)==2)
   gfl(i,2) = 0.0;%  disp (gfl(i,1));
 end

end

%------------------------------
% assemble the global diffusion
% and advection matrices
%------------------------------

gdm = zeros(ng,ng); % initialize
gam = zeros(ng,ng); % initialize
```

Code scd6_d_rc: Continuing $\longrightarrow$

```
for l=1:ne

% compute the element diffusion
% and advection matrices

   j=c(l,1); x1=p(j,1); y1=p(j,2); % node positions
   j=c(l,2); x2=p(j,1); y2=p(j,2);
   j=c(l,3); x3=p(j,1); y3=p(j,2);
   j=c(l,4); x4=p(j,1); y4=p(j,2);
   j=c(l,5); x5=p(j,1); y5=p(j,2);
   j=c(l,6); x6=p(j,1); y6=p(j,2);

% define the nodal velocities (ux=u, uy=v)

   [u1, v1] = scd6_vel(U,a,x1,y1); % node velocities
   [u2, v2] = scd6_vel(U,a,x2,y2);
   [u3, v3] = scd6_vel(U,a,x3,y3);
   [u4, v4] = scd6_vel(U,a,x4,y4);
   [u5, v5] = scd6_vel(U,a,x5,y5);
   [u6, v6] = scd6_vel(U,a,x6,y6);

% element matrices

[edm_elm, eam_elm, arel] = edam6 ...
...
   (x1,y1, x2,y2, x3,y3, x4,y4, x5,y5, x6,y6 ...
   ,u1,v1, u2,v2, u3,v3, u4,v4, u5,v5, u6,v6 ...
   ,NQ);

   for i=1:6
     i1 = c(l,i);
     for j=1:6
       j1 = c(l,j);
       gdm(i1,j1) = gdm(i1,j1) + edm_elm(i,j);
       gam(i1,j1) = gam(i1,j1) + eam_elm(i,j);
     end
   end
end

%-------------------
% coefficient matrix
%-------------------

lsm = kappa*gdm+gam;

%------------------------------------
% compute the right-hand side and
% implement the boundary conditions
%------------------------------------

for i=1:ng
 b(i) = 0.0;
end
```

Code scd6_d_rc: ⟶ Continuing ⟶

```
for j=1:ng

  if(gfl(j,1)==1 | gfl(j,1)==2)   % outer and inner
                                  % boundary nodes
     for i=1:ng
       b(i) = b(i) - lsm(i,j)*gfl(j,2);
       lsm(i,j) = 0;
       lsm(j,i) = 0;
     end

     lsm(j,j) = 1.0;
     b(j) = gfl(j,2);

  end

end

%------------------------
% solve the linear system
%------------------------

f = b/lsm';

%---------
% plotting
%---------

plot_6 (ne,ng,p,c,f);

%-----------------------------
% extended connectivity matrix
% for 3-node sub-triangles
%-----------------------------

Ic=0;

for i=1:ne

  Ic=Ic+1;
  c3(Ic,1)=c(i,1); c3(Ic,2)=c(i,4); c3(Ic,3)=c(i,6);
  Ic=Ic+1;
  c3(Ic,1)=c(i,4); c3(Ic,2)=c(i,2); c3(Ic,3)=c(i,5);
  Ic=Ic+1;
  c3(Ic,1)=c(i,5); c3(Ic,2)=c(i,3); c3(Ic,3)=c(i,6);
  Ic=Ic+1;
  c3(Ic,1)=c(i,4); c3(Ic,2)=c(i,5); c3(Ic,3)=c(i,6);

end
```

Code scd6_d_rc: Continuing $\longrightarrow$

```
%-----------------
% trimesh plotting
%-----------------

trimesh (c3,p(:,1),p(:,2),f);

%-----
% done
%-----
```

Code scd6_d_rc: ($\longrightarrow$ Continued.) Code for steady convection–diffusion with the Dirichlet boundary condition in a rectangular domain with a circular hole, as illustrated in Figure 4.2.6.

```
function [u, v] = scd6_vel (U,a,x,y)

%===================================
% velocity evaluation corresponding
% to streaming potential flow past
% a cylinder of radius "a"
%===================================

   as = a^2;
   rs = x^2+y^2;
   rq = rs^2;

   u = U*(1.0+as/rs - 2.0*as*x^2/rq);
   v = U*( -2.0*as*x*y/rq);

%-----
% done
%-----

return;
```

Function scd6_vel: Nodal velocity evaluation required by code scd6_d_rc.

PROBLEMS

4.2.1 *Laplace's equation in an L-shaped domain.*

(a) Execute the FSELIB triangulation function `trgl6_L`, and display the element layout for $ndiv = 0$, 1, and 2.

(b) Execute the FSELIB code `lapl6_d_L` with boundary conditions of your choice, and discuss the results of your computation.

4.2.2 *Laplace's equation in a square with a square hole.*

(a) Execute the FSELIB triangulation function `trgl6_ss`, and display the grid layout for $ndiv = 0$, 1, and 2.

(b) Execute the FSELIB code `lapl6_d_ss` with boundary conditions of your choice, and discuss the results of your computation.

4.2.3 *Laplace's equation in a square with a circular hole.*

(a) Execute the FSELIB triangulation function `trgl6_sc`, and display the element layout for $ndiv = 0$, 1, and 2.

(b) Execute the FSELIB code `lapl6_d_sc` with boundary conditions of your choice, and discuss the results of your computation.

4.2.4 *Convection–diffusion equation.*

(a) Execute the FSELIB triangulation function `trgl6_rc`, and display the element layout for $ndiv = 0$, 1, and 2.

(b) Execute the FSELIB code `scd6_d_rc` for $Pe = 0$, 1.0, 5.0, and 10.0, and discuss the effect of convection on the temperature distribution.

4.3   High-order triangle expansions

Equation (3.2.5) expresses a complete linear expansion, and equation (4.1.6) expresses a complete quadratic expansion over the area of the standard triangle in the parametric $\xi\eta$ plane. A complete mth-order polynomial expansion of any suitable function, $f(\xi, \eta)$, takes the triangular form:

$$
\begin{aligned}
f(\xi, \eta) = \quad & a_{00} \\
& + a_{10}\,\xi + a_{01}\,\eta \\
& + a_{20}\,\xi^2 + a_{11}\,\xi\eta + a_{02}\,\eta^2 \\
& + a_{30}\,\xi^3 + a_{21}\,\xi^2\eta + a_{12}\,\xi\eta^2 + a_{03}\,\eta^3 \\
& \cdots \quad \cdots \quad \cdots \quad \cdots \quad \cdots \quad \cdots \quad \cdots \quad \cdots \quad \cdots \\
& + a_{m,0}\,\xi^m + a_{m-1,1}\,\xi^{m-1}\eta + \ldots + a_{1,m-1}\,\xi\eta^{m-1} + a_{0,m}\,\eta^m.
\end{aligned}
\tag{4.3.1}
$$

Note that the sum of the indices $i + j$ of the coefficients a_{ij} is constant across each row of (4.3.1). The total number of coefficients is

$$N = 1 + 2 + 3 + \cdots m + (m + 1) = \frac{(m + 1)(m + 2)}{2} = \left(\begin{array}{c} m + 2 \\ 2 \end{array} \right), \quad (4.3.2)$$

where

$$\left(\begin{array}{c} l \\ k \end{array} \right) = \frac{l!}{k! \, (l - k)!} \quad (4.3.3)$$

is the combinatorial, expressing the number of possible ways by which k objects can be lifted from a set of l identical objects, leaving $l - k$ objects behind, where $k \leq l$. The exclamation mark denotes the factorial, $l! = 1 \cdot 2 \dots \cdot l$, with the understanding that $0! = 1$. When $k = l - 1$, we obtain

$$\left(\begin{array}{c} l \\ l - 1 \end{array} \right) = \frac{l!}{(l - 1)! \, 1!} = l, \quad (4.3.4)$$

which is true, since we have l choices for leaving one object behind. As expected, the total number of coefficients counted in (4.3.2) is an integer, irrespective of whether the expansion order m is odd or even.

An important advantage of the complete mth-order expansion, as compared with an incomplete expansion where some of the terms in (4.3.1) would be missing, is that the function, f, is an mth-order polynomial with respect to distance along any straight line in the $\xi \eta$ plane. More important, the function is a complete mth-order polynomial of ξ or η along the edges of the triangle. Consequently, C^0 continuity across an edge is guaranteed by the presence of $m + 1$ shared element nodes, including vertex and edge nodes.

Pascal triangle

The triangular structure displayed in (4.3.1) can be compared to Pascal's triangle,

$$\begin{array}{ccccccccccc}
 & & & & & 1 & & & & & \\
 & & & & 1 & & 1 & & & & \\
 & & & 1 & & 2 & & 1 & & & \\
 & & 1 & & 3 & & 3 & & 1 & & \\
 & 1 & & 4 & & 6 & & 4 & & 1 & \\
1 & & 5 & & 10 & & 10 & & 5 & & 1 \\
\cdots & & \cdots & & \cdots & & \cdots & & \cdots & & \cdots
\end{array} \qquad (4.3.5)$$

Each entry is the sum of the two entries immediately above it, while the outermost entries are equal to unity. The entries of the pth row provide us with the coefficients of the binomial expansion $(\xi + \eta)^{p-1}$, where p is an integer. For example, when $p = 3$, the third row provides us with the coefficients of the quadratic expansion,

$$(\xi + \eta)^2 = \mathbf{1} \, \xi^2 + \mathbf{2} \, \xi \eta + \mathbf{1} \, \eta^2.$$

More generally,

$$(\xi + \eta)^m = \sum_{k=0}^{m} \binom{m}{k} \xi^{m-k} \eta^k, \tag{4.3.6}$$

where the tall parentheses designate the combinatorial.

Let $f(x)$ and $g(x)$ be two functions of the independent variable x. Applying the rules of product function differentiation, we derive the Leibniz high-derivative product differentiation rule

$$\frac{\mathrm{d}^m(fg)}{\mathrm{d}x^m} = \sum_{k=0}^{m} \binom{m}{k} \frac{\mathrm{d}^{m-k}f}{\mathrm{d}x^{m-k}} \frac{\mathrm{d}^k g}{\mathrm{d}x^k}. \tag{4.3.7}$$

Polynomial forms and interpolation functions

It is instructive to recast the complete mth-order polynomial into the form

$$
\begin{aligned}
f(\xi, \eta) = \quad & d_1 \, \eta^m \\
+ \quad & (d_2 + d_3 \, \xi) \, \eta^{m-1} \\
+ \quad & (d_4 + d_5 \, \xi + d_6 \, \xi^2) \, \eta^{m-2} \\
+ \quad & (d_7 + d_8 \, \xi + d_9 \, \xi^2 + d_{10} \, \xi^3) \, \eta^{m-3} \\
& \cdots \\
+ \quad & (d_{N-m} + d_{N-m+1} \, \xi + \ldots + d_{N-1} \, \xi^{m-1} + d_N \, \xi^m),
\end{aligned}
\tag{4.3.8}
$$

where d_1–d_N is a new set of coefficients. This expression demonstrates that the coefficient multiplying the monomial η^p is an $(m-p)$-degree polynomial in ξ. For example, when $p = m$, the coefficient is a zeroth degree polynomial identified with a constant, d_1. Rearranging, we obtain the complementary form

$$
\begin{aligned}
f(\xi, \eta) = \quad & e_1 \, \xi^m \\
+ \quad & (e_2 + e_3 \, \eta) \, \xi^{m-1} \\
+ \quad & (e_4 + e_5 \, \eta + e_6 \, \eta^2) \, \xi^{m-2} \\
+ \quad & (e_7 + e_8 \, \eta + e_9 \, \eta^2 + e_{10} \, \eta^3) \, \xi^{m-3} \\
& \cdots \\
+ \quad & (e_{N-m} + e_{N-m+1} \, \eta + \ldots + e_{N-1} \, \eta^{m-1} + e_N \, \eta^m),
\end{aligned}
\tag{4.3.9}
$$

where e_1–e_N is a new set of coefficients. This expression demonstrates that the coefficient multiplying the monomial ξ^p is an $(m-p)$-degree polynomial in η.

To formalize the mth-order expansion, we denote the monomial products on the right-hand side of (4.3.1) as

$$\mathcal{M}_{ij}(\xi, \eta) \equiv \xi^i \, \eta^j, \tag{4.3.10}$$

and write

$$f(\xi, \eta) = \sum_{j=0}^{m} \left[\sum_{i=0}^{m-j} a_{ij} \, \mathcal{M}_{ij}(\xi, \eta) \right], \tag{4.3.11}$$

corresponding to (4.3.8), or

$$f(\xi, \eta) = \sum_{i=0}^{m} \left[\sum_{j=0}^{m-i} a_{ij} \, \mathcal{M}_{ij}(\xi, \eta) \right], \qquad (4.3.12)$$

corresponding to (4.3.9).

To implement the finite element method, we select the polynomial order, m, and introduce N interpolation nodes over the area and along the edges of the triangle, (ξ_i, η_i), $i = 1, 2, \ldots, N$, where the number of nodes, N, is related to the polynomial order, m, by (4.3.2). By definition, the cardinal interpolation function for the ith node, $\psi_i(\xi, \eta)$, is a complete mth-degree polynomial in ξ and η, required to satisfy the N interpolation conditions

$$\psi_i(\xi_j, \eta_j) = \delta_{ij}, \qquad (4.3.13)$$

for $j = 1, 2, \ldots, N$, where δ_{ij} is Kronecker's delta. The nodal expansion over the parametric triangle takes the familiar form

$$f(\xi, \eta) = \sum_{j=1}^{N} f(\xi_j, \eta_j) \, \psi_i(\xi_j, \eta_j), \qquad (4.3.14)$$

involving the specified or *a priori* unknown nodal values, $f(\xi_j, \eta_j)$.

4.3.1   Computation of the node interpolation functions

In certain special cases, the node interpolation functions, $\psi_i(\xi, \eta)$, can be deduced by inspection aided by geometrical intuition, based on a simple rule. Consider the interpolation function of the ith node, and assume that m straight lines can be found passing through all nodes, except for the ith node. If these lines are described by the linear functions

$$\mathcal{F}_j(\xi, \eta) = A_j \, \xi + B_j \, \eta + C_j = 0, \qquad (4.3.15)$$

for $j = 1, 2, \ldots, m$, where A_j, B_j, and C_j are constant coefficients, then the ith-node interpolation function is given by

$$\psi_i(\xi, \eta) = \prod_{j=1}^{m} \frac{\mathcal{F}_j(\xi, \eta)}{\mathcal{F}_j(\xi_i, \eta_i)}. \qquad (4.3.16)$$

A specific example will be given in Section 4.4.2. It is a simple matter to show that (4.3.16) satisfies the cardinal interpolation condition (4.3.13).

Polynomial expansions

More generally, the cardinal interpolation function corresponding to a node can be expressed in the polynomial form shown in (4.3.12) or (4.3.11), and the

N coefficients, a_{ij}, can be computed by solving a system of linear equations originating from the cardinal interpolation condition expressed by (4.3.13).

In the most general approach, the interpolation functions are expressed as linear combinations of a set of N independent polynomials that form a complete base of the mth-order expansion in the $\xi\eta$ plane, $\phi_j(\xi, \eta)$, for $j = 1, 2, \ldots, N$,

$$\psi_i(\xi, \eta) = c_N \, \phi_1(\xi, \eta) + c_{N-1} \, \phi_2(\xi, \eta)$$
$$+ \ldots + c_2 \, \phi_{N-1}(\xi, \eta) + c_1 \, \phi_N(\xi, \eta), \qquad (4.3.17)$$

where c_j comprise a set of $N + 1$ expansion coefficients for the ith node. For example, identifying the basis function with the monomial products shown in (4.3.10), we obtain

$$\phi_1 = \mathcal{M}_{00} = 1, \qquad \phi_2 = \mathcal{M}_{10} = \xi, \qquad \phi_3 = \mathcal{M}_{01} = \eta,$$
$$\ldots, \qquad (4.3.18)$$
$$\phi_{N-1} = \mathcal{M}_{1,m-1} = \xi \, \eta^{m-1}, \qquad \phi_N = \mathcal{M}_{0,m} = \eta^m.$$

Another possible polynomial base is provided by the nodal interpolation functions corresponding to a nodal set that is different than the set under consideration. Better choices are provided by the Appell and Proriol families of semi-orthogonal and orthogonal polynomials discussed later in this section.

Enforcing the cardinal interpolation condition (4.3.13) for any choice of basis functions, we obtain the linear system

$$\mathbf{V}_\phi^T \cdot \mathbf{c} = \mathbf{e}_i, \qquad (4.3.19)$$

where the superscript T denotes the matrix transpose,

$$\mathbf{V}_\phi \equiv \begin{bmatrix} \phi_1(\xi_1, \eta_1) & \phi_1(\xi_2, \eta_2) & \cdots & \phi_1(\xi_N, \eta_N) \\ \phi_2(\xi_1, \eta_1) & \phi_2(\xi_2, \eta_2) & \cdots & \phi_2(\xi_N, \eta_N) \\ \cdots & \cdots & \cdots & \cdots \\ \phi_{N-1}(\xi_1, \eta_1) & \phi_{N-1}(\xi_2, \eta_2) & \cdots & \phi_{N-1}(\xi_N, \eta_N) \\ \phi_N(\xi_1, \eta_1) & \phi_N(\xi_2, \eta_2) & \cdots & \phi_N(\xi_N, \eta_N) \end{bmatrix} \qquad (4.3.20)$$

is the $N \times N$ *generalized Vandermonde matrix* with components $V_{\phi_{ij}} = \phi_i(\xi_j, \eta_j)$,

$$\mathbf{c} \equiv \begin{bmatrix} c_N \\ c_{N-1} \\ \vdots \\ c_2 \\ c_1 \end{bmatrix}, \qquad (4.3.21)$$

is the basis functions coefficient vector, and

$$\mathbf{e}_i \equiv \begin{bmatrix} 0 \\ \vdots \\ 1 \\ \vdots \\ 0 \end{bmatrix}, \qquad (4.3.22)$$

is the unit vector of the N-dimensional space associated with the ith node; the unity on the right-hand side of (4.3.22) appears in the ith entry. Accordingly, $\mathbf{c} = \mathbf{V}_\phi^{T^{-1}} \cdot \mathbf{e}_i$, and the ith nodal interpolation function is given by

$$\psi_i(\xi, \eta) = \boldsymbol{\phi}(\xi, \eta) \cdot \mathbf{V}_\phi^{T^{-1}} \cdot \mathbf{e}_i, \qquad (4.3.23)$$

where $\boldsymbol{\phi}(\xi, \eta)$ is the vector of basis functions. Compiling the expressions for all nodes, we obtain the nodal function interpolation vector

$$\boldsymbol{\psi}(\xi, \eta) = \mathbf{V}_\phi^{-1} \cdot \boldsymbol{\phi}(\xi, \eta), \qquad (4.3.24)$$

which can be rearranged into the linear system

$$\mathbf{V}_\phi \cdot \boldsymbol{\psi}(\xi, \eta) = \boldsymbol{\phi}(\xi, \eta). \qquad (4.3.25)$$

Alternatively, this linear system can be derived immediately following the discussion of Section 2.1.4: using the Lagrange interpolation formula (see Appendix A), we write the exact representation

$$\phi_i(\xi, \eta) = \sum_{j=1}^{N} \phi_i(\xi_j, \eta_j) \, \psi_j(\xi, \eta), \qquad (4.3.26)$$

for $i = 1, 2, \ldots, N$, which is equivalent to (4.3.25).

Computing the solution of the linear system by Cramer's rule, we find that the ith-node element interpolation function is given by

$$\psi_i(\xi, \eta) = \frac{\mathrm{Det}[\mathbf{V}_\phi(\xi_1, \eta_1, \xi_2, \eta_2, \ldots, \xi_{i-1}, \eta_{i-1}, \xi, \eta, \xi_{i+1}, \eta_{i+1}, \ldots \xi_N, \eta_N)]}{\mathrm{Det}[\mathbf{V}_\phi(\xi_1, \eta_1, \xi_2, \eta_2, \ldots, \xi_N, \eta_N)]}, \qquad (4.3.27)$$

for $i = 1, 2, \ldots, N$. To confirm that this is an acceptable representation, we make three key observations:

1. Since the basis functions $\phi_i(\xi, \eta)$ involved in the generalized Vandermonde matrix form a complete basis for the mth-order expansion in (ξ, η), the numerator in (4.3.27) is an mth-degree polynomial in (ξ, η). This can be readily demonstrated by considering the Laplace expansion of the determinant with respect to the column involving ξ and η.

2. The numerator in (4.3.27) is zero when $\xi = \xi_j$ and $\eta = \eta_j$, where $j \neq i$, as two columns of the generalized Vandermonde matrix become identical.

3. The right-hand side of (4.3.27) is clearly equal to unity when $\xi = \xi_i$ and $\eta = \eta_i$.

For low polynomial orders, m, the linear system can be solved analytically by elementary methods. As an example, we consider the linear expansion corresponding to $m = 1$ and $N = 3$, and choose

$$\phi_1 = 1, \qquad \phi_2 = \xi, \qquad \phi_3 = \eta. \tag{4.3.28}$$

The nodal set is comprised of the three vertex nodes with barycentric coordinates $\xi_1 = 0, \eta_1 = 0$ for the first node, $\xi_2 = 1, \eta_2 = 0$ for the second node, and $\xi_3 = 0, \eta_3 = 1$ for the third node, yielding the Vandermonde matrix

$$\mathbf{V}_\phi \equiv \begin{bmatrix} 1 & 1 & 1 \\ 0 & 1 & 0 \\ 0 & 0 & 1 \end{bmatrix}, \tag{4.3.29}$$

whose inverse is

$$\mathbf{V}_\phi^{-1} \equiv \begin{bmatrix} 1 & -1 & -1 \\ 0 & 1 & 0 \\ 0 & 0 & 1 \end{bmatrix}. \tag{4.3.30}$$

The nodal interpolation function vector is thus given by

$$\psi = \begin{bmatrix} 1 & -1 & -1 \\ 0 & 1 & 0 \\ 0 & 0 & 1 \end{bmatrix} \cdot \begin{bmatrix} 1 \\ \xi \\ \eta \end{bmatrix}, \tag{4.3.31}$$

in agreement with previously obtained expressions shown in (3.2.7).

For higher polynomial orders, the solution of the linear system for the nodal interpolation functions is found by numerical methods. To ensure that the coefficient matrix is well-conditioned, it is beneficial to employ basis functions that are partially or entirely orthogonal, such as those provided by the Appell and Proriol polynomials discussed in the next two sections. The linear system may then be solved without difficulty using standard methods, such as the method of Gauss elimination discussed in Appendix C.

The set of nodes that maximizes the magnitude of the determinant of the generalized Vandermonde matrix within the confines of the triangle is the highly desirable Fekete set. Because any two polynomials bases are linearly dependent, the Fekete set is independent of the working base using the carry out the optimization.

4.3.2   Appell polynomial base

A function of interest defined over the standard triangle in the $\xi\eta$ plane can be approximated with a complete mth-degree polynomial in ξ and η. The approximating polynomial can be expressed in the form

$$f(\xi, \eta) = \sum_{k=0}^{m} \left(\sum_{l=0}^{m-k} \hat{a}_{kl} \, \mathcal{A}_{kl}(\xi, \eta) \right), \tag{4.3.32}$$

where $\hat{a}_{kl}$ are appropriate expansion coefficients, $\mathcal{A}_{kl}$ are the Appell polynomials defined as

$$\mathcal{A}_{kl}(\xi, \eta) \equiv \frac{\partial^{k+l}}{\partial \xi^k \partial \eta^l} \left(\xi^k \eta^l \zeta^{k+l} \right), \tag{4.3.33}$$

and $\zeta = 1 - \xi - \eta$ is the third barycentric coordinate [2]. In the literature, the definition (4.3.33) is sometimes called the *Rodrigues formula*.

Carrying out the differentiation with respect to ξ and η with the help of Leibniz's rule, we find that $\mathcal{A}_{kl}$ is, in fact, a $(k+l)$-degree polynomial in ξ and η, and may thus be expressed in the form

$$\mathcal{A}_{kl}(\xi, \eta) = \sum_{i=0}^{k+l} \left(\sum_{j=0}^{k+l-i} a_{kl,ij} \, \xi^i \, \eta^j, \right), \tag{4.3.34}$$

where $a_{kl,ij}$ are appropriate coefficients evaluated later in this section.

To derive the explicit form of the Appell polynomials, we apply Leibniz's rule (4.3.7) to compute the η derivatives, finding

$$\mathcal{A}_{kl}(\xi, \eta) = \frac{\partial^k}{\partial \xi^k} \left[\xi^k \sum_{j=0}^{l} \binom{l}{j} \frac{\mathrm{d}^{l-j} \eta^l}{\mathrm{d}\eta^{l-j}} \frac{\partial^j \zeta^{k+l}}{\partial \eta^j} \right]. \tag{4.3.35}$$

Carrying out the differentiations with respect to η, we obtain

$$\mathcal{A}_{kl}(\xi, \eta) = \frac{\partial^k}{\partial \xi^k} \left[\xi^k \sum_{j=0}^{l} \binom{l}{j} \frac{l!}{j!} \eta^j \times (-1)^j \frac{(k+l)!}{(k+l-j)!} \zeta^{k+l-j} \right], \tag{4.3.36}$$

which can be recast into the form

$$\mathcal{A}_{kl}(\xi, \eta) = \sum_{j=0}^{l} \eta^j \, (-1)^j \, \frac{(l!)^2 \, (k+l)!}{(j!)^2 \, (l-j)! \, (k+l-j)!} \frac{\partial^k}{\partial \xi^k} \left(\xi^k \, \zeta^{k+l-j} \right). \tag{4.3.37}$$

Next, we perform the differentiation with respect to ξ using once again Leibniz's rule, and simplify to obtain the explicit formula

$$\mathcal{A}_{kl}(\xi, \eta) = k! \, l! \sum_{i=0}^{k} \sum_{j=0}^{l} (-1)^{i+j} \frac{(k)_i \, (l)_j \, (k+l)_{i+j}}{i!^2 \, j!^2} \, \xi^i \, \eta^j \, \zeta^{k+l-i-j}, \tag{4.3.38}$$

subject to the definitions

$$(p)_0 \equiv 1,$$

$$(p)_q \equiv (-p)(-p+1)(-p+2) \cdots (-p+q-1), \quad \text{for} \quad q \geq 1,$$

(4.3.39)

for a pair of integers, p and q ([60], p. 65). For example, $(-1)_q = q!$.

Further manipulation shows that the coefficients $a_{kl,ij}$ defined in (4.3.34) are given by

$$a_{kl,ij} = (-1)^{i+j} \frac{(k+j)!\,(l+i)!}{i!^2\,j!^2} \frac{(k+l)!}{(k+l-i-j)!}.$$

(4.3.40)

Note that, as required by symmetry, $a_{kl,ij} = a_{lk,ji}$.

Applying these formulas, or else using the Rodrigues formula, we derive the first few Appell polynomials,

$$\begin{aligned}
\mathcal{A}_{00} &= 1, \\
\mathcal{A}_{10} &= \zeta - \xi = 1 - 2\xi - \eta, \\
\mathcal{A}_{01} &= \zeta - \eta = 1 - 2\eta - \xi, \\
\mathcal{A}_{20} &= 2\zeta\,(\zeta - 4\xi) + 2\xi^2, \\
\mathcal{A}_{11} &= \zeta\,(\zeta - 2\xi - 2\eta) + 2\xi\eta, \\
\mathcal{A}_{02} &= 2\zeta\,(\zeta - 4\eta) + 2\eta^2.
\end{aligned}$$

(4.3.41)

Note that $\mathcal{A}_{kl}$ is a complete $(k+l)$-degree polynomial in ξ and η.

It can be shown that the Appell polynomials satisfy the homogeneous partial differential equation

$$\xi\,(\xi - 1)\,\mathcal{A}_{\xi\xi} + 2\,\xi\,\eta\,\mathcal{A}_{\xi\eta} + \eta\,(\eta - 1)\,\mathcal{A}_{\eta\eta}$$

(4.3.42)

$$+ (3\xi - 1)\,\mathcal{A}_\xi + (3\eta - 1)\,\mathcal{A}_\eta = (k+l)(k+l+2)\,\mathcal{A},$$

where a subscript denotes a partial derivative with respect to the corresponding variable; for simplicity, we have denoted $\mathcal{A} = \mathcal{A}_{kl}$ ([60], p. 75). Equation (4.3.42) has $k+l+1$ independent eigensolutions, which are the Appell polynomials of the same total degree, $k+l$. For example, when $k+l = 1$, the two eigensolutions are $\mathcal{A}_{10}$ and $\mathcal{A}_{01}$ shown in (4.3.41).

Incomplete orthogonality

The Appell polynomials are biorthogonal against the lower class of monomials, $\mathcal{M}_{pq} = \xi^p \eta^q$, in the sense that

$$\iint \mathcal{A}_{kl}(\xi, \eta)\,\xi^p \eta^q \, d\xi \, d\eta = 0,$$

(4.3.43)

for $p+q < k+l \equiv m$, and the integration is performed over the area of the right isosceles parametric triangle. Consequently, $\mathcal{A}_{kl}(\xi, \eta)$ is orthogonal against any polynomial of lower total degree in ξ and η, that is, any polynomial involving monomials $\xi^p \eta^q$ with $p + q < k + l$. Biorthogonality implies the incomplete self-orthogonality property

$$\iint \mathcal{A}_{kl}(\xi, \eta) \, \mathcal{A}_{pq}(\xi, \eta) \, \mathrm{d}\xi \, \mathrm{d}\eta = 0, \qquad (4.3.44)$$

for $p + q \neq k + l$. Complete orthogonality would require that the projection expressed by the integral on the left-hand side of (4.3.44) is zero for $k \neq p$ and $l \neq q$, which is not true. Thus, the Appell polynomial $\mathcal{A}_{10}$ is orthogonal against any other Appell polynomial, except for the same-row polynomial $\mathcal{A}_{01}$. A corollary of (4.3.44) is that

$$\iint \mathcal{A}_{kl}(\xi, \eta) \, \mathrm{d}\xi \, \mathrm{d}\eta = 0, \qquad (4.3.45)$$

for $k > 0$ or $l > 0$, which states that the mean value of all polynomials, except for the very first polynomial, is zero over the area of the parametric triangle.

One way of demonstrating the biorthogonality property (4.3.43) is by directly integrating in ξ and η,

$$\iint \mathcal{A}_{kl}(\xi, \eta) \, \xi^p \, \eta^q \, \mathrm{d}\xi \, \mathrm{d}\eta = \int_0^1 \left(\int_0^{1-\eta} \frac{\partial^{k+l}}{\partial \xi^k \partial \eta^l} \left(\xi^k \eta^l \zeta^{k+l} \right) \xi^p \, \mathrm{d}\xi \right) \eta^q \, \mathrm{d}\eta. \qquad (4.3.46)$$

Focusing on the inner integral on the right-hand side, we integrate by parts to find

$$\int_0^{1-\eta} \frac{\partial^{k+l}}{\partial \xi^k \partial \eta^l} \left(\xi^k \eta^l \zeta^{k+l} \right) \xi^p \, \mathrm{d}\xi = \left[\frac{\partial^{k+l-1}}{\partial \xi^{k-1} \partial \eta^l} \left(\xi^k \eta^l \zeta^{k+l} \right) \xi^p \right]_{\xi=0}^{\xi=1-\eta}$$

$$- \int_0^{1-\eta} \frac{\partial^{k+l-1}}{\partial \xi^{k-1} \partial \eta^l} \left(\xi^k \eta^l \zeta^{k+l} \right) \frac{\partial \xi^p}{\partial \xi} \, \mathrm{d}\xi. \qquad (4.3.47)$$

Recalling the definition $\zeta = 1 - \xi - \eta$, we find that the first term on the right-hand side is zero. Repeating the integration by parts, we obtain

$$\int_0^{1-\eta} \frac{\partial^{k+l}}{\partial \xi^k \partial \eta^l} \left(\xi^k \eta^l \zeta^{k+l} \right) \xi^p \, \mathrm{d}\xi = (-1)^k \int_0^{1-\eta} \frac{\partial^l}{\partial \eta^l} \left(\xi^k \eta^l \zeta^{k+l} \right) \frac{\partial^k \xi^p}{\partial \xi^k} \, \mathrm{d}\xi, \qquad (4.3.48)$$

which is zero if $p < k$. This conclusion may also be reached independently using the explicit form shown in (4.3.37). Working in a similar manner, we write

$$\iint \mathcal{A}_{kl}(\xi, \eta) \, \xi^p \, \eta^q \, \mathrm{d}\xi \, \mathrm{d}\eta = \int_0^1 \left(\int_0^{1-\xi} \frac{\partial^{k+l}}{\partial \xi^k \partial \eta^l} \left(\xi^k \eta^l \zeta^{k+l} \right) \eta^q \, \mathrm{d}\eta \right) \xi^p \, \mathrm{d}\xi, \qquad (4.3.49)$$

and find that this integral is zero when $q < l$, which completes the proof.

A detailed calculation shows that, when $p \geq k$ and $q \geq l$,

$$\iint \mathcal{A}_{kl}(\xi, \eta) \, \xi^p \, \eta^q \, \mathrm{d}\xi \, \mathrm{d}\eta$$

$$= (-1)^{k+l} \frac{(p!)^2 \, (q!)^2 \, (k+l)!}{(p-k)! \, (q-l)! \, (k+l+p+q+2)!}. \qquad (4.3.50)$$

Weighted orthogonality

The Appell polynomials defined in (4.3.33) can be generalized into a broader family of polynomials parametrized by the three exponents $\alpha > -1$, $\beta > -1$, and $\gamma > -1$. The generalized polynomials are

$$\mathcal{A}_{kl}^{(\alpha, \beta, \gamma)}(\xi, \eta) = \frac{1}{\xi^\alpha \, \eta^\eta \, \zeta^\gamma} \frac{\partial^{k+l}}{\partial \xi^k \partial \eta^l} \left(\xi^{k+\alpha} \eta^{l+\beta} \zeta^{k+l+\gamma} \right), \qquad (4.3.51)$$

where $\zeta = 1 - \xi - \eta$. The standard Appell polynomials shown in (4.3.33) correspond to $\alpha = 0$, $\beta = 0$, and $\gamma = 0$.

Repeating the preceding analysis, we find that the generalized Appell polynomials satisfy the weighted orthogonality properties

$$\iint \mathcal{A}_{kl}(\xi, \eta) \, \xi^p \, \eta^q \, w(\xi, \eta) \, \mathrm{d}\xi \, \mathrm{d}\eta = 0, \qquad (4.3.52)$$

for $p + q < k + l \equiv m$, and

$$\iint \mathcal{A}_{kl}(\xi, \eta) \, \mathcal{A}_{pq}(\xi, \eta) \, w(\xi, \eta) \, \mathrm{d}\xi \, \mathrm{d}\eta = 0, \qquad (4.3.53)$$

for $p + q \neq k + l$, with weighting function

$$w(\xi, \eta) = \xi^\alpha \, \eta^\beta \, \zeta^\gamma. \qquad (4.3.54)$$

When $p \geq k$ and $q \geq l$, we find

$$\iint \mathcal{A}_{kl}(\xi, \eta) \, \xi^p \, \eta^q \, w(\xi, \eta) \, \mathrm{d}\xi \, \mathrm{d}\eta \qquad (4.3.55)$$

$$= (-1)^{k+l} \frac{p! \, q!}{(p-k)! \, (q-l)!} \frac{\Gamma(p+\alpha+1) \, \Gamma(q+\beta+1) \, \Gamma(k+l+\gamma+1)}{\Gamma(p+q+k+l+\alpha+\beta+\gamma+3)},$$

where Γ is the Gamma function; if m is an integer, then $\Gamma(m+1) = m!$.

4.3.3 Proriol polynomial base

The most desirable base consists of Proriol's polynomials, which are entirely orthogonal over the area of the triangle [46]. To introduce these polynomials, we map the standard triangle from the $\xi\eta$ plane to the standard square in the

$\xi'\eta'$ parametric plane, as shown in Figure 3.2.2. Proriol's polynomials are then given by

$$\mathcal{P}_{kl} = L_k(\xi') \left(\frac{1-\eta'}{2}\right)^k J_l^{(2k+1,0)}(\eta')$$
$$= L_k(\xi') (1-\eta)^k J_l^{(2k+1,0)}(\eta'), \qquad (4.3.56)$$

where the variables ξ' and η' are defined in (3.2.40), repeated here for convenience,

$$\xi' = \frac{2\xi}{1-\eta} - 1, \qquad \eta' = 2\eta - 1, \qquad (4.3.57)$$

L_k are the Legendre polynomials, and $J_l^{(2k+1,0)}$ are the Jacobi polynomials discussed in Section B.8 of Appendix B. Note that the factor $(1-\eta)^k$ in (4.3.56) cancels the denominator of $L_k(\xi')$ arising from the fraction on the right-hand side of the transformation rule for ξ' shown in (4.3.57). Accordingly, the Proriol polynomial, $\mathcal{P}_{kl}$, involves products, $\xi^p \eta^{(k-p+q)}$, with combined order $k+q$, where $p = 1, 2, \ldots, k$ and $q = 1, 2, \ldots, l$.

We can derive explicit expressions for the first few Proriol polynomials by recalling the following expressions for the first few Jacobi polynomials,

$$J_0^{(\alpha,0)}(t) = 1,$$
$$J_1^{(\alpha,0)}(t) = \frac{1}{2}\left[(\alpha+2)t + \alpha\right], \qquad (4.3.58)$$
$$J_2^{(\alpha,0)}(t) = \frac{1}{8}(\alpha+3)(\alpha+4)t^2 + \frac{1}{4}\alpha(\alpha+3)t + \frac{1}{8}(\alpha^2 - \alpha - 4).$$

Using these formulas together with corresponding formulas for the Legendre polynomials, we find the constant polynomial

$$\mathcal{P}_{00} = L_0(\xi') J_0^{(1,0)}(\eta') = 1, \qquad (4.3.59)$$

the linear polynomials

$$\mathcal{P}_{10} = L_1(\xi')(1-\eta) J_0^{(3,0)}(\eta') = \xi'(1-\eta) = 2\xi + \eta - 1 = \xi - \zeta,$$

$$\mathcal{P}_{01} = L_0(\xi') J_1^{(1,0)}(\eta') = \frac{1}{2}(3\eta'+1) = 3\eta - 1, \qquad (4.3.60)$$

and the quadratic polynomials

$$\mathcal{P}_{20} = L_2(\xi')(1-\eta)^2 J_0^{(5,0)}(\eta')$$
$$= \frac{1}{2}(3\xi'^2 - 1)(1-\eta)^2 = 6\xi^2 + 6\xi\eta + \eta^2 - 6\xi - 2\eta + 1,$$

(a)   (b)

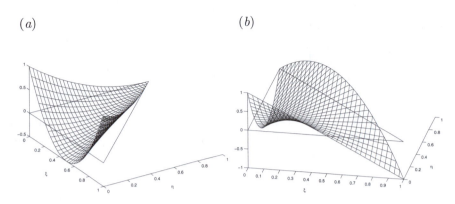

Figure 4.3.1 Graphs of the first two quadratic Proriol polynomials (a) $\mathcal{P}_{20}$, and (b) $\mathcal{P}_{11}$. The third quadratic polynomial, $\mathcal{P}_{02}$, is only a function of η.

$$\mathcal{P}_{11} = L_1(\xi')\,(1-\eta)\,J_1^{(3,0)}(\eta') = \xi'\,(1-\eta)\,\frac{1}{2}\,(5\,\eta'+3)$$
$$= (2\,\xi+\eta-1)\,(5\,\eta-1), \tag{4.3.61}$$

$$\mathcal{P}_{02} = L_0(\xi')\,J_2^{(1,0)}(\eta') = \frac{5}{2}\,\eta'^2 + \eta' - \frac{1}{2} = 10\,\eta^2 - 8\,\eta + 1.$$

Note that $\mathcal{P}_{00}$, $\mathcal{P}_{01}$, $\mathcal{P}_{02}$, ... are pure polynomials in η, whereas $\mathcal{P}_{10}$, $\mathcal{P}_{20}$, $\mathcal{P}_{30}$, ... are polynomials in both ξ and η.

FSELIB includes the function `proriol` (not listed in the text) that evaluates the Proriol polynomials based on the FSELIB function `jacobi` (not listed in the text). Graphs of two quadratic polynomials produced by the FSELIB function `proriol_graph` (not listed in the text) are shown in Figure 4.3.1.

Orthogonality

Using the integration formula (3.2.43), we find that the integral of the product of the ij and kl Proriol polynomials over the area of the standard triangle is given by

$$B_{ij,kl} \equiv \iint \mathcal{P}_{ij}\,\mathcal{P}_{kl}\,\mathrm{d}\xi\,\mathrm{d}\eta = \frac{1}{8}\int_{-1}^{1}\int_{-1}^{1}\mathcal{P}_{ij}\,\mathcal{P}_{kl}\,(1-\eta')\,\mathrm{d}\xi'\,\mathrm{d}\eta'$$

$$= \frac{1}{4}\int_{-1}^{1}\int_{-1}^{1}L_i(\xi')\,L_k(\xi')\,\left(\frac{1-\eta'}{2}\right)^{i+k+1} \tag{4.3.62}$$

$$\times\,J_j^{(2i+1,0)}(\eta')\,J_l^{(2k+1,0)}(\eta')\,\mathrm{d}\xi'\,\mathrm{d}\eta',$$

which can be recast into the form

$$B_{ij,kl} = \frac{1}{4} \int_{-1}^{1} \left(\int_{-1}^{1} L_i(\xi') \, L_k(\xi') \, \mathrm{d}\xi' \right) \tag{4.3.63}$$

$$\times \left(\frac{1 - \eta'}{2} \right)^{i+k+1} J_j^{(2i+1,0)}(\eta') \, J_l^{(2k+1,0)}(\eta') \, \mathrm{d}\eta'.$$

Using the orthogonality properties of the Legendre and Jacobi polynomials discussed in Appendix B, we find that $B_{ij,kl} = 0$ if $i \neq k$ or $j \neq l$, which concludes the proof of orthogonality. The self-projection integral is given by

$$\mathcal{G}_{ij} \equiv B_{ij,ij} = \frac{1}{(2i+1)(2i+2j+2)}. \tag{4.3.64}$$

For example, $\mathcal{G}_{00} = 1/2$, which is equal to the area of the standard triangle in the $\xi\eta$ plane. FSELIB includes the scripts `proriol_ortho` and `proriol_ortho1` that confirm the orthogonality of the Proriol polynomials by direct integration using a Gaussian quadrature or integration rule.

Orthogonal expansion

Proriol's polynomials provide us with a complete orthogonal basis. Any function, $f(\xi, \eta)$, defined over the standard triangle in the $\xi\eta$ plane can be approximated with a complete mth-degree polynomial in ξ and η, expressed in the form

$$f(\xi, \eta) \simeq \sum_{k=0}^{m} \left(\sum_{l=0}^{m-k} a_{kl} \, \mathcal{P}_{kl}(\xi, \eta) \right). \tag{4.3.65}$$

Multiplying (4.3.65) by $\mathcal{P}_{ij}$, integrating over the surface of the triangle, and using the orthogonality property, we derive the expansion coefficients

$$a_{kl} = \frac{1}{\mathcal{G}_{kl}} \iint f(\xi, \eta) \, \mathcal{P}_{kl}(\xi, \eta) \, \mathrm{d}\xi \, \mathrm{d}\eta, \tag{4.3.66}$$

where $\mathcal{G}_{kl}$ is defined in (4.3.64).

To convert a monomial product series to an equivalent Proriol series, we may use the expressions

$$1 = \mathcal{P}_{00},$$

$$\xi = \frac{1}{6} \left(3 \, \mathcal{P}_{10} - \mathcal{P}_{01} + 2 \right),$$

$$\eta = \frac{1}{3} \left(\mathcal{P}_{01} + 1 \right),$$

$$\xi^2 = \frac{1}{30} \left(5 \mathcal{P}_{20} - 3 \mathcal{P}_{11} + 3 \mathcal{P}_{02} - 4 \mathcal{P}_{01} + 12 \mathcal{P}_{10} + 5 \right),$$

$$\xi \eta = \frac{1}{60} \left(-3 \mathcal{P}_{02} + 6 \mathcal{P}_{11} + 6 \mathcal{P}_{10} + 2 \mathcal{P}_{01} + 5 \right),$$

$$\eta^2 = \frac{1}{30} \left(3 \mathcal{P}_{02} + 8 \mathcal{P}_{01} + 5 \right). \tag{4.3.67}$$

...

4.3.4   The Lebesgue constant

A complete nodal set is comprised on N interpolation nodes, where N is related to a specified polynomial order, m, through (4.3.2). To assess the relative merits of two candidate nodal sets, we require a measure of success mediated by an objective function. A heuristic generalization of the theory of one-dimensional interpolation discussed in Appendix A leads us to the Lebesgue function in two dimensions,

$$\mathcal{L}_N(\xi, \eta) \equiv \sum_{i=1}^{N} \left| \psi_i(\xi, \eta) \right|, \qquad (4.3.68)$$

and associated Lebesgue constant

$$\Lambda_N \equiv \text{Max}\left[\mathcal{L}(\xi, \eta) \right], \qquad (4.3.69)$$

where $\text{Max}[\bullet]$ signifies the maximum of the enclosed function over the area of the triangle. The theory of interpolation suggests that, the lower the value of Λ_N, the better the nodal point distribution. Other objective functions can be defined by exercising common sense and physical intuition, as will be discussed in Section 4.4.

4.3.5   Node condensation

In a typical nodal set, three nodes are placed at the triangle vertices, some nodes are placed along the triangle edges, and other nodes are placed in the triangle interior. When node condensation is applied, the unknowns associated with the interior nodes are eliminated from the finite element equations arising from the Galerkin projection of the global interpolation functions, and the final system is condensed into a system of smaller size involving only the shared vertex and edge nodes. The process of elimination is similar to that for one-dimensional problems discussed in Sections 1.3 and 2.4 [67]. Condensation techniques are reviewed by Schwarz ([54], pp. 185–201).

PROBLEMS

4.3.1 *Appell polynomials.*

Verify property (4.3.45) for the polynomials $\mathcal{A}_{10}$ and $\mathcal{A}_{01}$ listed in the text. *Hint:* Use the integration formula (3.2.34).

4.3.2 *Orthogonality of the Proriol polynomials.*

(*a*) Confirm the orthogonality of the first three Proriol polynomials shown in (4.3.60).

(*b*) Confirm formula (4.3.64) for the first three Proriol polynomials shown in (4.3.60).

4.3.3 *Orthogonal expansion.*

(*a*) Expand each of the three linear interpolation functions shown in (3.2.7) in a series of Proriol polynomials.

(*b*) Repeat (*a*) for the quadratic functions shown in (4.1.9).

4.4   High-order node distributions

In Section 4.3, we discussed the computation of the element node interpolation functions, but evaded the important issue of how the nodes can be distributed over the area of the triangle in the physical xy or parametric $\xi\eta$ plane. The objectives are that the global matrices should be well-conditioned, and the accuracy of the interpolation should be highest for a given polynomial order, m.

With these considerations in mind, we now work in two stages. First, we develop a general policy of node distribution with the goal of ensuring that the number of available interpolation nodes is exactly equal to that required by the complete mth-order polynomial expansion in ξ and η. Second, we seek to optimize the node distribution guided by the theory of polynomial interpolation and orthogonal polynomials summarized in Appendices A and B.

4.4.1   Node distribution based on a 1D master grid

One way to ensure that the number of interpolation nodes, N, is equal to the number of coefficients in the complete mth-order expansion, is to deploy the nodes as shown in Figure 4.4.1. To generate this construction, we introduce a one-dimensional *master grid* defined by a set of $m + 1$ points, v_i, where $i = 1, 2, \ldots, m + 1$, subject to the end-node constraints

$$v_1 = 0, \qquad v_{m+1} = 1. \tag{4.4.1}$$

If the master grid is symmetric with respect to the mid-point, $v = 0.5$,

$$v_{m+2-j} = 1 - v_j \tag{4.4.2}$$

for $j = 1, 2, \ldots, m+1$. Nodes on the ξ and η axes are then deployed at positions

$$\xi_i = v_i, \qquad \eta_j = 1 - v_{m+2-j}, \tag{4.4.3}$$

for $i, j = 1, 2, \ldots, m + 1$, and nodes along the hypotenuse of the triangle are identified by moving either vertically upward from the nodes along the ξ axis, or horizontally to the right from the nodes along the η axis.

The frames on the left and right columns of Figure 4.4.1 illustrate three possible ways of locating the interior nodes. The left column corresponds to an arbitrary one-dimensional master grid, and the right column corresponds to a symmetric master grid. The interior nodes at the top frames of Figure 4.4.1

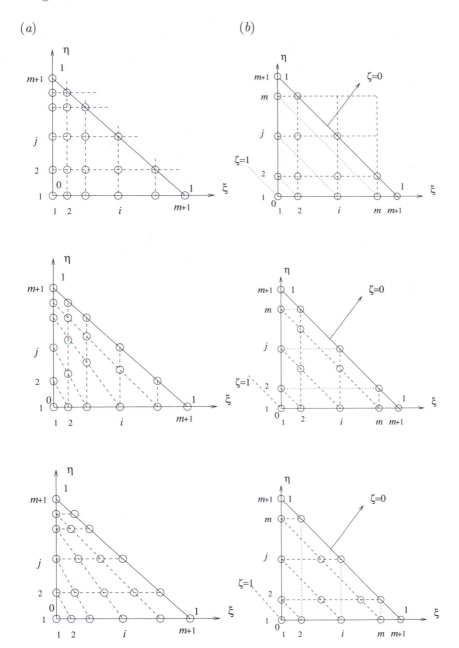

Figure 4.4.1 Distribution of interpolation nodes corresponding to a complete mth-order polynomial expansion over the standard triangle, generated by an (a) arbitrary, and (b) a symmetric one-dimensional master grid.

are located at intersections of vertical and horizontal grid lines, yielding pairs (ξ_i, η_j), where $i = 1, 2, \ldots, m + 1$; for each value of the index i, the index j takes values in the range $j = 1, 2, \ldots, m + 2 - i$. The nodes in the mid frames of Figure 4.4.1 are located at the intersections of vertical and diagonal lines, and the nodes in the bottom frames are located at the intersections of horizontal and diagonal lines,

When $m = 1$, we obtain a 3-node triangle with three vertex nodes, as discussed in Section 2.2. When $m = 2$, we obtain a 6-node triangle with three vertex nodes and three edge nodes, as discussed in Section 3.2. When $m = 3$, we obtain a 10-node triangle with three vertex nodes, six edge nodes, and one interior node.

The nodes can be labeled sequentially moving horizontally along the ξ axis and then vertically or diagonally, or *vice versa*, within the confines of the triangle. In all cases, to enforce C^0 continuity of the interpolated function over the entire solution domain consisting of the union of the elements, we require that the nodal values of the function at the vertex and edge nodes are shared by neighboring elements at corresponding positions.

4.4.2   Uniform grid

When the one-dimensional master grid is uniform, corresponding to evenly spaced vertical and horizontal lines,

$$v_i = \frac{i - 1}{m}, \tag{4.4.4}$$

for $i = 1, 2, \ldots, m + 1$, the three node distributions shown on the left and right columns of Figure 4.4.1 coincide, and the interior nodes lie along diagonal lines, drawn as dotted lines in Figure 4.2.2(*a*).

The diagonal lines correspond to constant values of the third barycentric coordinate $\zeta = 1 - \xi - \eta$, ranging from $\zeta = 0$ along the hypotenuse, to $\zeta = 1$ at the first vertex node labeled 1. Each diagonal line is identified by the index

$$k = m + 3 - i - j, \tag{4.4.5}$$

which increases from the value of 1 along the hypotenuse, to the value of $m + 1$ at the origin, $\xi = 0$, $\eta = 0$. Bos [5] observed that the nodes are also deployed around the perimeters of

$$\left[\frac{m - 1}{3}\right] + 1 \tag{4.4.6}$$

nested triangles, where the square brackets denote the integral part of the enclosed fraction. An example for $m = 10$ is shown in Figure 4.4.2(*b*).

Node interpolation functions

The cardinal node interpolation functions for the uniform grid can be constructed using formula (4.3.16), on the observation that for each node, all other

(a) (b)

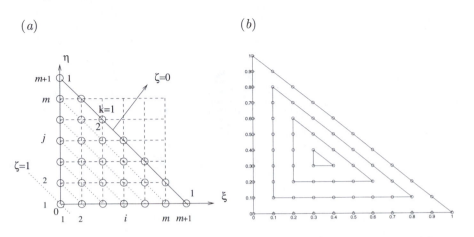

Figure 4.4.2 (*a*) Illustration of a uniform grid over the triangle. (*b*) The interpolation nodes fall on a nested sequence of concentric triangles.

nodes lie on preceding horizontal and vertical lines, and on subsequent diagonal lines. Specifically, the interpolation function corresponding to the (i, j) node can be expressed as the product of three polynomials, as

$$\psi_{ij}(\xi, \eta) = \Xi_i^{(i-1)}(\xi) \times H_j^{(j-1)}(\eta) \times Z_k^{(k-1)}(\zeta), \tag{4.4.7}$$

where:

- $\Xi_i^{(i-1)}(\xi)$ is an $(i-1)$-degree polynomial, defined such that

$$\Xi_1^0(\xi) = 1,$$

$$\Xi_i^{(i-1)}(\xi) = \frac{(\xi - v_1)(\xi - v_2) \ldots (\xi - v_{i-2})(\xi - v_{i-1})}{(v_i - v_1)(v_i - v_2) \ldots (v_i - v_{i-2})(v_i - v_{i-1})}, \tag{4.4.8}$$

  for $i = 2, 3, \ldots, m+1$.

- $H_j^{(j-1)}(\eta)$ is a $(j-1)$-degree polynomial, defined such that

$$H_1^0(\eta) = 1,$$

$$H_j^{(j-1)}(\eta) = \frac{(\eta - v_1)(\eta - v_2) \ldots (\eta - v_{j-2})(\eta - v_{j-1})}{(v_j - v_1)(v_j - v_2) \ldots (v_j - v_{j-2})(v_j - v_{j-1})}, \tag{4.4.9}$$

  for $j = 2, 3, \ldots, m+1$.

- $Z_k^{(j-1)}(\zeta)$ is a $(k-1)$-degree polynomial, defined such that

$$Z_1^0(\zeta) = 1,$$

$$(4.4.10)$$

$$Z_k^{(k-1)}(\zeta) = \frac{(\zeta - v_1)(\zeta - v_2) \ldots (\zeta - v_{k-2})(\zeta - v_{k-1})}{(v_k - v_1)(v_k - v_2) \ldots (v_k - v_{k-2})(v_k - v_{k-1})},$$

for $k = 2, 3, \ldots, m + 1$.

It is a straightforward exercise to verify that $\psi_{ij}(\xi, \eta)$ is a polynomial of degree

$$(i - 1) + (j - 1) + (k - 1) = i + j + k - 3 = m \qquad (4.4.11)$$

with respect to ξ and η, as required by (4.4.5), and that all cardinal interpolation conditions are met.

As an example, we consider a 6-node triangle corresponding to $m = 2$, $v_1 = 0$, $v_2 = 1/2$, and $v_3 = 1$. The interpolation function of the first vertex node corresponding to $i = 1$, $j = 1$, $k = 3$, is

$$\begin{aligned} \psi_{11}(\xi, \eta) &= \Xi_1^{(0)}(\xi) \times H_1^{(0)}(\eta) \times Z_3^{(2)}(\zeta) \\ &= 1 \times 1 \times \frac{(\zeta - v_1)(\zeta - v_2)}{(v_3 - v_1)(v_3 - v_2)} = \zeta\,(2\,\zeta - 1), \end{aligned} \qquad (4.4.12)$$

which is consistent with the expression for ψ_1 given in (4.1.9). Working similarly with the mid-side node, $i = 2$, $j = 1$, $k = 2$, we find

$$\begin{aligned} \psi_{21}(\xi, \eta) &= \Xi_2^{(1)}(\xi) \times H_1^{(0)}(\eta) \times Z_2^{(1)}(\zeta) \\ &= \frac{\xi - v_1}{v_2 - v_1} \times 1 \times \frac{\zeta - v_1}{v_2 - v_1} = 4\,\xi\,\zeta, \end{aligned} \qquad (4.4.13)$$

which is consistent with the expression for ψ_4 given in (4.1.9).

Figure 4.4.3 displays graphs of the node interpolation functions for $m = 3$, generated by the FSELIB function psi_tm (not listed in the text). Frame (a) shows the interpolation function corresponding to the interior node, $i = 2$ and $j = 2$. For obvious reasons, this function is described as a *bubble mode*. Frame (b) displays the interpolation function corresponding to the edge node, $i = 3$ and $j = 2$. Note that this function is zero along the two perpendicular edges described by $\xi = 0$ and $\eta = 0$.

Unfortunately, as the expansion order m is raised, the accuracy of the interpolation does not necessarily improve uniformly due to the two-dimensional version of the Runge effect discussed in Section 2.2, manifested by growing oscillations near the triangle edges. Specifically, numerical computations show that the Lebesgue constant increases rapidly with the polynomial order, m, or size of the nodal set, N [5]. Thus, the uniform node distribution is recommended only for low-order polynomial expansions, typically $m \leq 3$.

To circumvent this difficulty, Bos [5] modified the uniform node distribution by adjusting the sizes of the individual nested triangles illustrated in Figure (*a*)   (*b*)

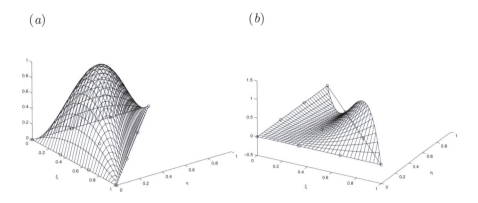

Figure 4.4.3 Cardinal interpolation functions on a uniform grid for $m = 3$: (*a*) interior (bubble) mode corresponding to $i = 2$ and $j = 2$, and (*b*) edge node corresponding to $i = 3$ and $j = 2$.

4.4.2(*b*), while maintaining the uniformity of the point distribution around the perimeters. Although the optimized distributions considerably improve the accuracy and convergence properties of the interpolation, they do not represent the optimal layout.

4.4.3   Lobatto grid on the triangle

In Section 2.6, we saw that the Lobatto node distribution is optimal for one-dimensional interpolation, subject to the constraint that one interpolation node is placed at each element end-node. Motivated by this result, we employ a one-dimensional master grid with $v_1 = 0$, $v_{m+1} = 1$, and interior nodes v_i, $i = 2, 3, \ldots, m$, positioned at the scaled zeros of the $(m - 1)$-degree Lobatto polynomial. Specifically, the Lobatto master grid is defined as

$$v_1 = 0,$$
$$v_2 = \frac{1}{2}\,(1 + t_1),$$
$$v_3 = \frac{1}{2}\,(1 + t_2),$$
$$\ldots$$
$$v_m = \frac{1}{2}\,(1 + t_{m-1}),$$
$$v_{m+1} = 1.$$

(4.4.14)

where t_i, for $i = 1, 2, \ldots, m - 1$, are the zeros of the $(m - 1)$-degree Lobatto polynomial, $Lo_{m-1}(t)$, distributed in the interval $(-1, 1)$. Figure 4.4.4(*a*) shows the node distribution generated by the intersection of horizontal and vertical

(a)   (b)

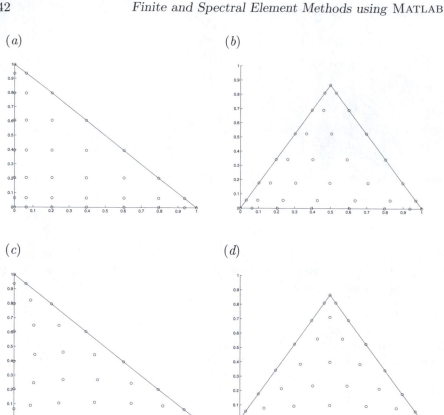

(c)   (d)

Figure 4.4.4 Node distribution for $m = 7$, generated by the intersection of horizontal and vertical Lobatto grid lines over (a) the right triangle in the $\xi\eta$ plane, and (b) the corresponding equilateral triangle in the $\hat{\xi}\hat{\eta}$ plane. (c, d) Improved point distribution with rotational symmetry over the (c) right, and (d) equilateral triangle.

grid lines, corresponding to the top frames of Figure 4.4.1, for $m = 7$. The nodes are identified by the coordinates

$$(\xi_i, \eta_j) = (v_i, v_j), \tag{4.4.15}$$

where $i = 1, 2, \ldots, m+1$, and $j = 1, 2, \ldots, m+2-i$.

Figure 4.4.4(b) shows the node distribution after the standard right isosceles triangle in the $\xi\eta$ plane has been mapped to an equilateral triangle in the $\hat{\xi}\hat{\eta}$ plane, using the transformation rules

$$\hat{\xi} = \xi + \frac{1}{2}\,\eta, \qquad \hat{\eta} = \frac{\sqrt{3}}{2}\,\eta. \tag{4.4.16}$$

The inverse transformation rules are

$$\xi = \hat{\xi} - \frac{1}{\sqrt{3}}\,\hat{\eta}, \qquad \eta = \frac{2}{\sqrt{3}}\,\hat{\eta}. \qquad (4.4.17)$$

The asymmetry of the distribution with respect to the vertices, evidenced in Figure 4.4.4(*a, b*), is due to the choice of horizontal and vertical intersections for identifying the interior nodes. Alternatives would be the horizontal and diagonal intersections, and the vertical and diagonal intersections, as shown in the middle and bottom frames of Figure 4.4.1. Because the three vertices in the physical *xy* plane have been labeled arbitrarily, 1, 2, and 3, subject to the counterclockwise convention, the lack of three-fold symmetry is unacceptable.

To eliminate this deficiency, we compute all three node distributions shown in the top, middle, and bottom frames of Figure 4.4.1, and then average the coordinates of the interior nodes over the three realizations. The result is an improved node distribution where the desirable three-fold symmetry is satisfied. The nodal coordinates are given by

$$\xi_{ij} = \frac{1}{3}\,(1 + 2\,v_i - v_j - v_k), \qquad \eta_{ij} = \frac{1}{3}\,(1 - v_i + 2\,v_j - v_k), \qquad (4.4.18)$$

for $i = 1, 2, \ldots, m+1$, and $j = 1, 2, \ldots, m+2-i$, where $k = m+3-i-j$. The node distributions for $m = 7$ over the right isosceles and equilateral triangles are shown in Figure 4.4.4(*c, d*).

Figure 4.4.5(*a*) shows the node distribution for $m = 6$ over the right isosceles triangle. Drawing a mesh of dotted lines connecting the edge nodes on the equilateral triangle reveals that the interior nodes are situated at the centroids of smaller inner triangles surrounding them, as shown in Figure 4.4.5(*b*). Figure 4.4.5(*c-f*) presents an assortment of node interpolation functions generated by the FSELIB function psi_lob (not listed in the text), computed using the Vandermonde matrix method discussed in Section 4.3.1, where the basis functions, ϕ_i, are identified with the Proriol polynomials.

Numerical investigation has shown that the Lobatto triangle distribution with three-fold symmetry described in this section enjoys interpolation convergence properties that compare favorably with those of the best-known distributions comprised of the Fekete nodes [4].

4.4.4 The Fekete set

An alternative method of distributing the interpolation nodes over the triangle is based on the notion of the Fekete points [5]. To introduce these points, we apply the polynomial expansion (4.3.1) at $N = \frac{1}{2}(m+1)(m+2)$ nodes over the area of the triangle, (ξ_i, η_i), as discussed earlier in this section, and consider the $N \times N$ generalized Vandermonde matrix defined in (4.3.20), regarded as a function of the nodal positions,

$$\mathbf{V}(\xi_1, \eta_1, \xi_2, \eta_2, \ldots, \xi_N, \eta_N). \qquad (4.4.19)$$

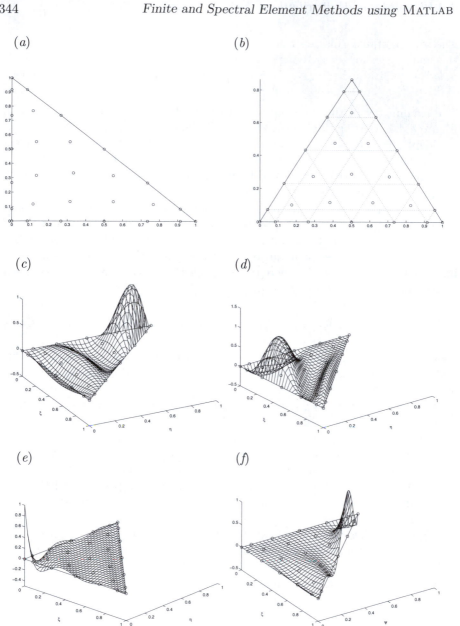

Figure 4.4.5 Lobatto triangle grid with three-fold symmetry for $m = 6$ over the (*a*) standard triangle in the parametric $\xi\eta$ plane, and (*b*) corresponding equilateral triangle. (*c-f*) An assortment of cardinal interpolation functions for sample nodes.

By definition, the Fekete set maximizes the magnitude of the determinant of this matrix within the confines of the triangle. Since the determinant of a matrix is multiplied by a constant factor when a multiple of a row is added to another to alter the polynomial base, the computed Fekete set is independent of the working base.

Invoking the expression of the node interpolation functions shown in (4.3.27), we find that, by construction of the Fekete points,

$$|\psi_i(\xi, \eta)| \leq 1, \tag{4.4.20}$$

with the equality holding only when $\xi = \xi_i$ and $\eta = \eta_i$. This property ensures a rapid interpolation convergence with respect to polynomial order, as discussed in Section A.5 of Appendix A.

The Fekete set over the triangle includes the three vertex nodes together with a group of $m - 1$ nodes along each edge distributed at the scaled zeros of the $(m - 1)$-degree Lobatto polynomial. Recall that the edge nodes are optimal for one-dimensional interpolation over a finite interval with fixed end-nodes, as discussed in Section A.5 of Appendix A. The remaining nodes are deployed inside the element at non-obvious positions. As an example, when $m = 3$, the Fekete set includes ten nodes, including the three vertex nodes, two nodes along each edge corresponding to the zeros of the Lo_2 Lobatto polynomial, and one interior node located at the element centroid. The Fekete point coordinates and associated Lebesgue constants for higher-order expansions are available in tabular form [7, 63].

Figure 4.4.6 shows the Fekete nodes over the equilateral triangle, marked as circles, and the corresponding symmetric Lobatto triangle nodes discussed in Section 4.4.3, marked as crosses, for expansion order $m = 9$ [4]. Both sets include fifty-five nodes, including three vertex nodes, twenty-four edge nodes, and twenty-eight interior nodes, one of which is located at the triangle centroid. It is interesting to observe that the Lobatto triangle interior nodes are displaced only slightly inward with respect to the corresponding Fekete nodes.

4.4.5 Further nodal distributions

Chen and Babuška [7] computed node distributions by minimizing the L_2 norm

$$\left(\iint \sum_{i=1}^{N} |\psi_i(\xi, \eta)|^2 \, d\xi \, d\eta \right)^{1/2}, \tag{4.4.21}$$

and found that the properties of the resulting sets comparable favorably with those of the Fekete nodes. Note that the Lebesgue function defined in (4.3.68) is the L_∞ norm. Unlike the Fekete sets, the sets corresponding to the L_2 norm do not include edge nodes corresponding to the zeros of Lobatto polynomials. Hesthaven [26] performed a similar calculation by minimizing instead a properly defined energy that depends on the mutual position of all pairs of nodes. Like the Fekete sets, the electrostatic sets include edge nodes corresponding

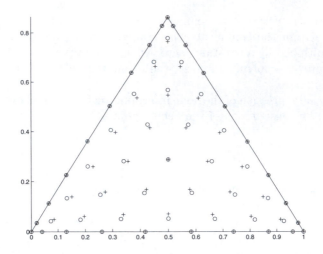

Figure 4.4.6 Comparison of the Fekete nodes (circles) with the symmetric
Lobatto triangle nodes (crosses), for polynomial order $m = 9$.

to the zeros of the Lobatto polynomials. In the absence of a rigorous theory
of two-dimensional polynomial interpolation, the relative merits of candidate
sets computed based on different objective functions can be assessed only by
numerical experimentation.

PROBLEMS

4.4.1 *Quadratic expansion.*

Verify that formula (4.4.7) reproduces the element node interpolation functions
given in (4.1.9).

4.4.2 *Cardinal functions with uniform node distribution.*

Run a properly modified version of the FSELIB code `psi_tm` to generate graphs
of the three vertex-node interpolation functions corresponding to a uniform
node distribution with $m = 4$.

4.5   Modal expansion on the triangle

Dubiner [17] proposed a modal expansion where a function of interest defined
over the standard triangle in the $\xi\eta$ plane is approximated with an mth-degree
polynomial, as

$$f(\xi, \eta) \simeq F_v + F_e + F_i, \tag{4.5.1}$$

where

$$F_v(\xi, \eta) = f_1 \, \zeta_1^v(\xi, \eta) + f_2 \, \zeta_2^v(\xi, \eta) + f_3 \, \zeta_3^v(\xi, \eta) \tag{4.5.2}$$

is the *vertex part*,

$$F_e(\xi, \eta) = \sum_{i=1}^{m-1} c_i^{12} \zeta_i^{12}(\xi, \eta) + \sum_{i=1}^{m-1} c_i^{13} \zeta_i^{13}(\xi, \eta) + \sum_{i=1}^{m-1} c_i^{23} \zeta_i^{23}(\xi, \eta)$$

$$\tag{4.5.3}$$

is the *edge part* arising when $m \geq 2$, and

$$F_i(\xi, \eta) = \sum_{i=1}^{m-2} \left(\sum_{j=1}^{m-i-1} c_{ij} \, \zeta_{ij}(\xi, \eta) \right) \tag{4.5.4}$$

is the *interior part* arising when $m \geq 3$. Each part is comprised of a set of modes multiplied by respective coefficients, sometimes called degrees of freedom.

The modal expansion includes $N_v = 3$ vertex modes, $N_e = 3(m-1)$ edge modes, and

$$N_i = (m-2) + (m-1) + \ldots + 1 = \frac{(m-1)(m-2)}{2} \tag{4.5.5}$$

interior modes. The total number of modes is

$$N = N_v + N_e + N_i$$
$$= 3 + 3(m-1) + \frac{(m-1)(m-2)}{2} = \frac{(m+1)(m+2)}{2}, \tag{4.5.6}$$

which is precisely equal to the number of terms in the complete mth-order polynomial expansion in ξ and η. In fact, the modes are designed so that the modal expansion expressed by (4.5.1) is a complete mth-degree polynomial in ξ and η. For example, when $m = 1$, we obtain the familiar three-term linear expansion involving only the vertex modes. When $m = 2$, we obtain the familiar six-term quadratic expansion involving three vertex modes and three edge modes. Interior modes arise for higher-order expansions.

To develop expressions for the three families of modes, we map the standard triangle in the $\xi\eta$ plane to the standard square in the $\xi'\eta'$ parametric plane, as shown in Figure 3.2.3, where $-1 < \xi' < 1$ and $-1 < \eta' < 1$. The barycentric coordinates, ξ and η, are related to the Cartesian coordinates, ξ' and η', and *vice versa*, by means of the transformation rules (3.2.37), (3.2.38), and (3.2.40), repeated here for convenience,

$$\xi = \frac{1+\xi'}{2} \frac{1-\eta'}{2}, \tag{4.5.7}$$

$$\eta = \frac{1+\eta'}{2}, \tag{4.5.8}$$

$$\zeta \equiv 1 - \xi - \eta = \frac{1-\xi'}{2} \frac{1-\eta'}{2}, \tag{4.5.9}$$

$$1 - \eta = 1 - \frac{1+\eta'}{2} = \frac{1-\eta'}{2}, \tag{4.5.10}$$

and

$$\xi' = \frac{2\xi}{1-\eta} - 1, \qquad \eta' = 2\eta - 1. \tag{4.5.11}$$

Vertex modes

The coefficients f_1, f_2, and f_3 in (4.5.2) are the values of the function $f(\xi, \eta)$ at the three vertex nodes labeled 1, 2, and 3. To ensure C^0 continuity of the finite element expansion, these values must be common to the neighboring elements sharing these nodes.

The corresponding cardinal interpolation functions, ζ_1^v, ζ_2^v, and ζ_3^v, are identical to those given in (3.2.7) for the linear triangle,

$$\zeta_1^v = \zeta = \frac{1-\xi'}{2}\frac{1-\eta'}{2},$$

$$\zeta_2^v = \xi = \frac{1+\xi'}{2}\frac{1-\eta'}{2}, \tag{4.5.12}$$

$$\zeta_3^v = \eta = \frac{1+\eta'}{2}.$$

Edge modes

The coefficients c_i^{12}, c_i^{13}, and c_i^{23} in (4.5.3) are associated with three corresponding families of edge interpolation functions ζ_i^{12}, ζ_i^{13}, and ζ_i^{23}, where $i = 1, 2, \ldots, m-1$. To ensure C^0 continuity of the finite element expansion, these coefficients must be common to the neighboring elements sharing an edge.

The interpolation functions ζ_i^{12} are zero along the two edges 13 and 23, corresponding to $\xi = 0$ and $\zeta = 0$. To satisfy this requirement, we write

$$\zeta_i^{12} = \xi \zeta \, \Psi_i^{12}(\xi', \eta'), \tag{4.5.13}$$

where $\Psi_i^{12}(\xi', \eta')$ is a new set of functions. Moreover, we set

$$\Psi_i^{12}(\xi', \eta') = Lo_{i-1}(\xi') \, \Phi_i^{12}(\eta'), \tag{4.5.14}$$

for $i = 1, 2, \ldots, m-1$, where $Lo_{i-1}(\xi')$ is a Lobatto polynomial, and $\Phi_i^{12}(\eta')$ is a new set of functions. The choice of Lobatto polynomials is motivated by the partial diagonalization of the element diffusion matrix, as discussed in Section 2.6. In summary, we have

$$\zeta_i^{12} = \xi \zeta \, Lo_{i-1}(\xi') \, \Phi_i^{12}(\eta'). \tag{4.5.15}$$

Next, we require that ζ_i^{12} is a polynomial in ξ and η. Noting the denominator of the fraction in the transformation rule for ξ' shown in the first equation of (4.5.11), we set

$$\Phi_i^{12}(\eta') = (1-\eta)^{i-1} = (\frac{1-\eta'}{2})^{i-1}, \tag{4.5.16}$$

and derive the *hierarchical* modal expansion

$$\zeta_i^{12} = \xi \, \zeta \, Lo_{i-1}(\xi') \left(\frac{1 - \eta'}{2}\right)^{i-1}, \tag{4.5.17}$$

or

$$\zeta_i^{12} = \frac{1 - \xi'}{2} \frac{1 + \xi'}{2} \left(\frac{1 - \eta'}{2}\right)^{i+1} Lo_{i-1}(\xi'), \tag{4.5.18}$$

for $i = 1, 2, \ldots, m-1$. The terminology "hierarchical" implies that the members of the function set can be arranged with respect to the polynomial order.

Working in a similar fashion with the other two families of edge modes, we find

$$\zeta_i^{13} = \eta \, \zeta \, Lo_{i-1}(\eta'), \tag{4.5.19}$$

or

$$\zeta_i^{13} = \frac{1 + \eta'}{2} \frac{1 - \xi'}{2} \frac{1 - \eta'}{2} Lo_{i-1}(\eta'), \tag{4.5.20}$$

for $i = 1, 2, \ldots, m - 1$, and

$$\zeta_i^{23} = \xi \, \eta \, Lo_{i-1}(\eta') \tag{4.5.21}$$

or

$$\zeta_i^{23} = \frac{1 + \xi'}{2} \frac{1 + \eta'}{2} \frac{1 - \eta'}{2} Lo_{i-1}(\eta'), \tag{4.5.22}$$

for $i = 1, 2, \ldots, m - 1$.

For example, when $i = 1$, which is possible only when $m \geq 2$, we note that $Lo_0 = 1$, and obtain the familiar quadratic edge-interpolation functions

$$\zeta_1^{12} = \xi \zeta, \qquad \zeta_1^{13} = \eta \zeta, \qquad \zeta_1^{23} = \xi \eta. \tag{4.5.23}$$

Interior modes

The coefficients c_{ij} in (4.5.4) correspond to the interior or bubble modes described by the interpolation functions ζ_{ij}. To ensure that these functions are zero all around the perimeter of the triangle, we express them in the form

$$\zeta_{ij} = \xi \, \eta \, \zeta \, X_{ij}(\xi', \eta')$$

$$= \frac{1 + \xi'}{2} \frac{1 - \xi'}{2} \left(\frac{1 - \eta'}{2}\right)^2 \frac{1 + \eta'}{2} X_{ij}(\xi', \eta'), \tag{4.5.24}$$

where $X_{ij}(\xi', \eta')$ is a new set of functions. Working as previously for the edge modes, we set

$$X_{ij}(\xi', \eta') = Lo_{i-1}(\xi') \left(\frac{1 - \eta'}{2}\right)^{i-1} J_{j-1}^{(2i+1,1)}(\eta'), \tag{4.5.25}$$

where $J_{j-1}^{(2i+1,1)}$ are the Jacobi polynomials discussed in Section B.8 of Appendix B. Substituting this expression in (4.5.24), we obtain the final form

$$\zeta_{ij} = \xi \, \eta \, \zeta \, Lo_{i-1}(\xi') \left(\frac{1 - \eta'}{2} \right)^{i-1} J_{j-1}^{(2i+1,1)}(\eta'), \tag{4.5.26}$$

or

$$\zeta_{ij} = \frac{1+\xi'}{2} \frac{1-\xi'}{2} \frac{1+\eta'}{2} \left(\frac{1-\eta'}{2} \right)^{i+1} Lo_{i-1}(\xi') J_{j-1}^{(2i+1,1)}(\eta'), \tag{4.5.27}$$

for $i = 1, 2, \ldots, m-2$, $j = 1, 2, \ldots, m-i-1$, and $m \geq 3$.

For example, when $m = 3$, we obtain a single bubble mode corresponding to $i = 1$ and $j = 1$, given by

$$\zeta_{11} = \xi \, \eta \, \zeta. \tag{4.5.28}$$

Note that this mode is identical to the interior-node interpolation function plotted in Figure 4.5.3(a) for $m = 3$, corresponding to $\xi = 1/3$ and $\eta = 1/3$, except that it is scaled by the factor $3^3 = 27$.

Partial orthogonality of the bubble modes

Applying the integration formula (3.2.43), we find that the integral of the product of the ij and kl interior modes over the triangle is given by the modal mass matrix component

$$B_{ij,kl} = \frac{1}{8} \int_{-1}^{1} \int_{-1}^{1} \zeta_{ij} \, \zeta_{kl} \, (1 - \eta') \, d\xi' \, d\eta'$$

$$= \frac{1}{8} \int_{-1}^{1} \int_{-1}^{1} \left(\frac{1-\xi'^2}{4} \frac{1+\eta'}{2} \right)^2 \left(\frac{1-\eta'}{2} \right)^{i+1} \left(\frac{1-\eta'}{2} \right)^{k+1}$$

$$\times Lo_{i-1}(\xi') \, Lo_{k-1}(\xi') \, J_{j-1}^{(2i+1,1)}(\eta') \, J_{l-1}^{(2k+1,1)}(\eta') \, (1 - \eta') \, d\xi' \, d\eta', \tag{4.5.29}$$

which can be rearranged into

$$B_{ij,kl} = \frac{1}{4} \int_{-1}^{1} \left(\frac{1-\xi'^2}{4} \right)^2 Lo_{i-1}(\xi') \, Lo_{k-1}(\xi') \, d\xi'$$

$$\times \int_{-1}^{1} \left(\frac{1-\eta'}{2} \right)^{i+k+3} \left(\frac{1+\eta'}{2} \right)^2 J_{j-1}^{(2i+1,1)}(\eta') \, J_{l-1}^{(2k+1,1)}(\eta') \, d\eta'. \tag{4.5.30}$$

The orthogonality of the Lobatto polynomials requires

$$\int_{-1}^{1} (1 - \xi'^2) \, P_q(\xi') \, Lo_r(\xi') \, d\xi' = 0, \tag{4.5.31}$$

where $\mathcal{P}_q(\xi')$ is a q-degree polynomial with $q < r$. Consequently, the first integral on the right-hand side of (4.5.30) is zero when either $2 + (i - 1) < k - 1$ or $2 + (k - 1) < i - 1$, that is, $i + 2 < k < i - 2$, which shows that the mass matrix has a finite bandwidth [55].

It is interesting to observe that the first integral on the right-hand side of (4.5.30) will be zero for $i \neq k$, provided that the Lobatto polynomials, Lo_{i-1}, are replaced by the Jacobi polynomials, $J_{i-1}^{(2,2)}$, discussed in Section B.8 of Appendix B. For similar reasons, the second integral on the right-hand side of (4.5.30) with $i = k$ will be zero, provided that $J_{j-1}^{(2i+1,1)}$ is replaced by $J_{j-1}^{(2i+3,2)}$. However, these modifications promote the coupling of the interior with the vertex and edge modes ([30], p. 91).

Implementation of the modal expansion

The Galerkin finite element method for the modal expansion is implemented as discussed previously in this chapter for the nodal expansion, the only new feature being that the element-node interpolation functions are replaced by the non-cardinal modal interpolation functions and the interior nodal values are replaced by the corresponding expansion coefficients (degrees of freedom). The numerical computation of the element mass and diffusion matrices is discussed in detail by Sherwin and Karniadakis [30, 55].

In concluding this discussion, we call attention to two important features of the modal expansion. First, we emphasize that the modal decomposition (4.5.1) is *not* entirely orthogonal, which means that the corresponding mass matrix is only *nearly* diagonal. The lack of complete orthogonality can be traced to the presence of vertex and edge modes, which must be included to facilitate enforcing the C^0 continuity of the finite element expansion. In the absence of this constraint, it is beneficial to use a modal expansion in terms of the entirely orthogonal Proriol polynomials, as discussed in Section 4.4.

Second, the aforementioned partial orthogonality of the modal expansion applies only to the super-parametric interpolation over the 3-node triangle in the physical xy plane. The reason is that this triangle has a constant surface metric coefficient, h_S, which can be extracted from the integrals defining the various element matrices in the $\xi\eta$ parametric plane. In contrast, because the surface metric coefficient of the 6-node triangle is generally a function of position, that is, it depends on ξ and η, bubble-mode orthogonality with a constant weighting function discussed in this section does not produce a sparse mass matrix.

PROBLEMS

4.5.1 *Cubic modal expansion.*

When $m = 3$, we obtain the ten-mode cubic expansion involving three vertex modes, six edge modes, and one bubble mode.

(*a*) Write out the explicit form of the six edge modes as functions of ξ and η, and prepare graphs over the area of the triangle.

(*b*) Repeat (*a*) for the bubble mode.

4.5.2 *Vandermonde matrix.*

Discuss the structure of the Vandermonde matrix, $V_{ij} = \zeta_i(\xi_j, \eta_j)$, for the Lobatto triangle set and for the Fekete set discussed in Section 4.4, where the modal functions are used as a polynomial base, ζ_i, and the nodes are ordered in the following sequence: vertex nodes, edge nodes, and interior nodes. Then explain why the determinant of the Vandermonde matrix is the product of three determinants defined with respect to the vertex nodes and modes, edge nodes and modes, and interior nodes and modes.

4.6   Surface elements

The mathematical modeling of a certain class of problems in science and engineering results in partial differential equations defined over three-dimensional, stationary or evolving surfaces. In physical applications, these surfaces are material boundaries, interfaces, and propagating or moving fronts. An example is the transport equation governing the evolution of the surface concentration of an insoluble surfactant residing at the interface between two immiscible fluids [45]. The surfactant molecules are advected and diffuse tangentially to the interface, while their number density changes because of interfacial in-plane stretching and expansion associated with normal motion.

A key notion in the formulation of such three-dimensional problems is the "surface gradient," which is a vector consisting of spatial derivatives in directions that are tangential to the surface. Specifically, the surface gradient operator is defined as

$$\nabla_s \equiv (\mathbf{I} - \mathbf{nn}) \cdot \nabla, \tag{4.6.1}$$

where

$$\nabla = \left(\frac{\partial}{\partial x}, \frac{\partial}{\partial y}, \frac{\partial}{\partial z} \right), \tag{4.6.2}$$

is the usual three-dimensional gradient, $\mathbf{I}$ is the unit matrix, and $\mathbf{n}$ is the unit vector normal to the surface. The matrix

$$\mathbf{P} \equiv \mathbf{I} - \mathbf{nn}, \tag{4.6.3}$$

with components $P_{ij} = \delta_{ij} - n_i\, n_j$, projects a vector onto the plane that is normal to $\mathbf{n}$, and therefore tangential to the surface, by eliminating the normal component. Thus, if $\mathbf{v}$ is three-dimensional vector defined over the surface, then the vector

$$\mathbf{P} \cdot \mathbf{v} = (\mathbf{I} - \mathbf{nn}) \cdot \mathbf{v} = \mathbf{v} - \mathbf{n}\,(\mathbf{n} \cdot \mathbf{v}) \tag{4.6.4}$$

is its tangential projection onto the surface. An identity allows us to write

$$\mathbf{P} \cdot \mathbf{v} = \mathbf{n} \times (\mathbf{v} \times \mathbf{n}) \tag{4.6.5}$$

(Problem 4.6.1). The right-hand side of (4.6.5) is the double rotation of the tangential component of $\mathbf{v}$.

The methodology described earlier in the text for planar 3-node and 6-node triangular elements may be extended in a straightforward fashion to accommodate surface elements in three-dimensional space, as illustrated in Figure 4.6.1. In the case of 6-node elements, the mapping from the physical to parameter space is described by (4.1.2), and the element interpolation functions are given in (4.1.7). The element diffusion and mass matrices are given by the right-hand sides of (4.1.14) and (4.1.15).

To compute the three-dimensional surface gradient of the element interpolation functions, ψ_i, we complement the counterparts of equations (4.1.17) for a 6-node triangle,

$$\sum_{j=1}^{6} \frac{\partial \psi_j}{\partial \xi} \mathbf{x}_j^E \cdot \nabla_s \psi_i = \frac{\partial \psi_i}{\partial \xi},$$

$$\tag{4.6.6}$$

$$\sum_{j=1}^{6} \frac{\partial \psi_j}{\partial \eta} \mathbf{x}_j^E \cdot \nabla_s \psi_i = \frac{\partial \psi_i}{\partial \eta},$$

with the condition

$$\mathbf{n} \cdot \nabla_s \psi_i = 0, \tag{4.6.7}$$

where $\mathbf{n} = (n_x, n_y, n_z)$ is the unit vector normal to the element. The resulting 3×3 linear system takes the form

$$\boldsymbol{\Gamma} \cdot \nabla_s \psi_i = \begin{bmatrix} \partial \psi_i / \partial \xi \\ \partial \psi_i / \partial \eta \\ 0 \end{bmatrix}, \tag{4.6.8}$$

where

$$\boldsymbol{\Gamma} = \begin{bmatrix} \sum_{j=1}^{6} \frac{\partial \psi_j}{\partial \xi} x_j^E & \sum_{j=1}^{6} \frac{\partial \psi_j}{\partial \xi} y_j^E & \sum_{j=1}^{6} \frac{\partial \psi_j}{\partial \xi} z_j^E \\ \sum_{j=1}^{6} \frac{\partial \psi_j}{\partial \eta} x_j^E & \sum_{j=1}^{6} \frac{\partial \psi_j}{\partial \eta} y_j^E & \sum_{j=1}^{6} \frac{\partial \psi_j}{\partial \eta} z_j^E \\ n_x & n_y & n_z \end{bmatrix}. \tag{4.6.9}$$

The solution of this system can be found readily by elementary analytical or numerical methods.

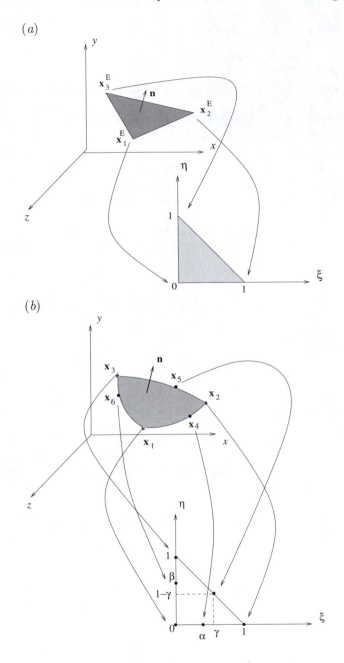

Figure 4.6.1 (*a*) A 3-node, and (*b*) a 6-node triangle in three-dimensional space is mapped to a right isosceles triangle in the $\xi\eta$ plane.

PROBLEM

4.6.1 *Tangential projection.*

To prove identity (4.6.5), we employ index notation and write

$$(\delta_{ij} - n_i\, n_j)\, v_j = \epsilon_{ijk}\, n_j\, (\epsilon_{klm}\, v_l\, n_m), \qquad (4.6.10)$$

where δ_{ij} is Kronecker's delta, ϵ_{ijk} is the alternating tensor, and summation is implied over the repeated indices j, k, l, and m. The alternating tensor is defined such that $\epsilon_{ijk} = 0$ if at least two of the indices take the same value, $\epsilon_{123} = \epsilon_{231} = \epsilon_{312} = 1$, and $\epsilon_{ijk} = -1$ otherwise, as discussed in Section D.3 of Appendix D. The product of two alternating tensors with one shared index satisfies the identity

$$\epsilon_{kij}\, \epsilon_{klm} = \delta_{il}\, \delta_{jm} - \delta_{im}\, \delta_{jl}, \qquad (4.6.11)$$

under the repeated index summation convention for k.

Based on this identity, and using the fundamental property of Kronecker's delta, $\delta_{ij}\, v_j = v_i$, complete the proof of (4.6.5).

4.7 High-order quadrilateral elements

In certain applications, it is desirable to use quadrilateral elements with four straight or curved edges instead of their triangular counterparts. The finite and spectral element methodology for these elements is similar to that for the triangular elements discussed in Chapter 3 and previously in this chapter.

To describe a quadrilateral element, we map it from the physical xy plane to the standard square in the parametric $\xi\eta$ plane confined between $-1 \le \xi \le 1$ and $-1 \le \eta \le 1$, as shown in Figure 4.7.1. The precise form of the mapping depends on the number of element nodes and chosen form of the polynomial expansion.

A complete mth-order polynomial expansion of a function, $f(\xi, \eta)$, over the standard square in the $\xi\eta$ plane, has the flat-base triangular form shown in (4.3.1), repeated here for convenience,

$$
\begin{aligned}
f(\xi, \eta) = \quad & a_{00} \\
& + a_{10}\, \xi + a_{01}\, \eta \\
& + a_{20}\, \xi^2 + a_{11}\, \xi\eta + a_{02}\, \eta^2 \\
& + a_{30}\, \xi^3 + a_{21}\, \xi^2\eta + a_{13}\, \xi\eta^2 + a_{03}\, \eta^3 \\
& \cdots \quad \cdots \quad \cdots \quad \cdots \quad \cdots\cdots \quad \cdots \quad \cdots \quad \cdots \\
& + a_{m0}\, \xi^m + a_{m-1,1}\, \xi^{m-1}\eta + \ldots + a_{1,m-1}\, \xi\eta^{m-1} + a_{0m}\, \eta^m,
\end{aligned}
\qquad (4.7.1)
$$

where a_{ij} are expansion coefficients. Quadrilateral elements with various numbers of interpolation nodes can accommodate different truncations of this triangular form, typically corresponding to incomplete polynomial expansions. One

(*a*)

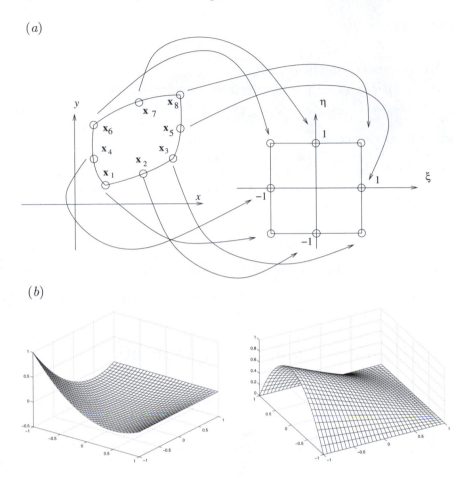

(*b*)

Figure 4.7.1 (*a*) An 8-node quadrilateral in the xy plane is mapped to the standard square in the parametric $\xi\eta$ plane. (*b*) Graphs of the interpolation functions corresponding to a vertex and a mid-node.

disadvantage of the incomplete expansion is that, in general, the solution will not be invariant with respect to the rotation of the global Cartesian axes.

The simplest quadrilateral element is the 4-node bilinear element discussed in Section 3.8. In the remainder of this section, we discuss higher-order and spectral elements described by a higher number of vertex, edge, and interior interpolation nodes.

4.7.1   8-node serendipity elements

The 8-node quadrilateral element is defined by four vertex nodes and four edge nodes, as shown in Figure 4.7.1(*a*). To describe the element in parametric form, we map it from the xy plane to the standard square in the $\xi\eta$ plane, so that the first node is mapped to the point $\xi = -1, \eta = -1$, the second to the point $\xi = 0, \eta = -1$, the third to the point $\xi = 1, \eta = -1$, the fourth to the point $\xi = -1, \eta = 0$, the fifth to the point $\xi = 1, \eta = 0$, the sixth to the point $\xi = -1, \eta = 1$, the seventh to the point $\xi = 0, \eta = 1$, and the eighth to the point $\xi = 1, \eta = 1$.

The mapping from the physical to the parametric space is mediated by the function

$$\mathbf{x} = \sum_{i=1}^{8} \mathbf{x}_i^E \, \psi_i(\xi, \eta), \tag{4.7.2}$$

where $\mathbf{x}_i^E$ are the element nodes. The element-node cardinal interpolation functions, $\psi_i(\xi, \eta)$, are biquadratic of the serendipity class, named after Horace Walpole's novel *"The Three Princes of Serendip,"* and are given by

$$
\begin{aligned}
\psi_i(\xi, \eta) \quad = \quad & (a + b\,\xi + c\,\xi^2) \\
& + (d + e\,\xi + f\,\xi^2)\,\eta \\
& + (g + h\,\xi)\,\eta^2,
\end{aligned} \tag{4.7.3}
$$

where a–h are expansion coefficients. Rearranging, we obtain an incomplete cubic expansion written in the deliberate pictorial form

$$
\begin{aligned}
\psi_i(\xi, \eta) = \quad & a_{00} \\
& + a_{10}\,\xi + a_{01}\,\eta, \\
& + a_{20}\,\xi^2 + a_{11}\,\xi\eta + a_{02}\,\eta^2 \\
& + a_{21}\,\xi^2\eta + a_{12}\,\xi\eta^2,
\end{aligned} \tag{4.7.4}
$$

where the new coefficients, a_{ij}, are reincarnations of the previous coefficients a–h. Note that, for a given ξ, $\psi_i(\xi, \eta)$ is quadratic in η, for a given η, $\psi_i(\xi, \eta)$ is quadratic in ξ, and only terms $\xi^m \eta^n$ with $m \le 2$ and $n \le 2$ appear. The expansion shown in (4.7.4) is a chopped flat-base Pascal triangle.

To compute the polynomial coefficients of the node interpolation functions, we require the cardinal interpolation property demanding that $\psi_i(\xi, \eta) = 1$ at the ith node, and $\psi_i(\xi, \eta) = 0$ at all other nodes. A detailed calculation yields

$$
\begin{aligned}
\psi_1 &= \frac{1}{4}\,(\xi - 1)\,(1 - \eta)\,(\xi + \eta + 1), \\
\psi_2 &= \frac{1}{2}\,(1 - \xi^2)\,(1 - \eta), \\
\psi_3 &= \frac{1}{4}\,(1 + \xi)\,(1 - \eta)\,(\xi - \eta - 1), \\
\psi_4 &= \frac{1}{2}\,(1 - \xi)\,(1 - \eta^2), \\
\psi_5 &= \frac{1}{2}\,(1 + \xi)\,(1 - \eta^2),
\end{aligned} \tag{4.7.5}
$$

$$\psi_6 = \frac{1}{4}\,(1-\xi)\,(1+\eta)\,(\eta-\xi-1),$$

$$\psi_7 = \frac{1}{2}\,(1-\xi^2)\,(1+\eta),$$

$$\psi_8 = \frac{1}{4}\,(1+\xi)\,(1+\eta)\,(\xi+\eta-1).$$

Graphs of the interpolation functions ψ_1 and ψ_2 corresponding to the south-western vertex node and adjacent edge node, produced by the FSELIB function `psi_q8` (not listed in the text) are shown in Figure 4.7.1(b).

In the isoparametric interpolation, the requisite solution over the element is expressed in a form that is analogous to that shown in (4.7.2), as

$$f(x,y) = \sum_{i=1}^{8} f_i^E \, \psi_i(\xi,\eta). \tag{4.7.6}$$

The element diffusion, mass, and advection matrices are computed as discussed in Section 3.8 for the 4-node quadrilateral element.

4.7.2   12-node serendipity elements

The 4-node element discussed in Section 3.8 and the 8-node element discussed previously in this section belong to the class of incomplete serendipity expansions. The next member in this family is the quartic serendipity element whose node interpolation functions have the general form

$$\begin{aligned}
\psi_i(\xi,\eta) = \quad & a_{00} \\
+ & a_{10}\,\xi + a_{01}\,\eta, \\
+ a_{20}\,\xi^2 + & a_{11}\,\xi\eta + a_{12}\,\eta^2 \\
+ a_{30}\,\xi^3 + a_{21}\,\xi^2\eta + a_{12}\,\xi\eta^2 + & a_{03}\,\eta^3 \\
+ a_{31}\,\xi^3\eta \quad\quad\quad + & a_{13}\,\xi\eta^3,
\end{aligned} \tag{4.7.7}$$

involving twelve coefficients. Four interpolation nodes are placed at the vertices of the quadrilateral, and four pairs of interpolation nodes are placed along the four edges, as shown in Figure 4.7.2(a).

To compute the polynomial coefficients, we require $\psi_i(\xi,\eta)=1$ at the ith node, and $\psi_i(\xi,\eta)=0$ at the other eleven nodes, finding

$$\psi_1 = -\frac{1}{32}\,(1-\xi)\,(1-\eta)\,(10-9\,\xi^2-9\,\eta^2),$$

$$\psi_2 = \frac{9}{32}\,(1-\xi^2)\,(1-\eta)\,(1-3\,\xi),$$

$$\psi_3 = \frac{9}{32}\,(1-\xi^2)\,(1-\eta)\,(1+3\,\xi),$$

$$\psi_4 = -\frac{1}{32}\,(1+\xi)\,(1-\eta)\,(10-9\,\xi^2-9\,\eta^2), \tag{4.7.8}$$

$$\psi_5 = \frac{9}{32} (1 - \xi)(1 - \eta^2)(1 - 3\,\eta),$$

$$\psi_6 = \frac{9}{32} (1 + \xi)(1 - \eta^2)(1 - 3\,\eta),$$

$$\psi_7 = \frac{9}{32} (1 - \xi)(1 - \eta^2)(1 + 3\,\eta),$$

$$\psi_8 = \frac{9}{32} (1 + \xi)(1 - \eta^2)(1 + 3\,\eta),$$

$$\psi_9 = \frac{1}{32} (1 - \xi)(1 + \eta)(10 - 9\,\xi^2 - 9\,\eta^2),$$

$$\psi_{10} = \frac{9}{32} (1 - \xi^2)(1 - 3\,\xi)(1 + \eta),$$

$$\psi_{11} = \frac{9}{32} (1 - \xi^2)(1 + 3\,\xi)(1 + \eta),$$

$$\psi_{12} = \frac{1}{32} (1 + \xi)(1 + \eta)(10 - 9\,\xi^2 - 9\,\eta^2).$$

Graphs of the interpolation functions ψ_1 and ψ_4 corresponding to the southwestern vertex node and adjacent edge node, produced by the FSELIB function psi_q12 (not listed in the text), are shown in Figure 4.7.2(*b*).

In the isoparametric interpolation, the requisite solution over the element is expressed in the form

$$f(x, y) = \sum_{i=1}^{12} f_i^E\, \psi_i(\xi, \eta). \tag{4.7.9}$$

The element diffusion, mass, and advection matrices are computed as discussed in Section 3.8 for 4-node quadrilateral elements.

4.7.3   Grid nodes via tensor-product expansions

In this class of quadrilateral elements, the interpolation nodes are distributed on a $(m_1 + 1) \times (m_2 + 1)$ Cartesian grid covering the standard square in the $\xi\eta$ plane, as shown in Figure 4.7.3. The nodes are identified by the coordinate doublet (ξ_i, η_j), where $i = 1, 2, \ldots, m_1 + 1$, and $j = 1, 2, \ldots, m_2 + 1$. To ensure that common nodes are introduced at the element vertices and along the edges, and thereby facilitate enforcing the C^0 continuity condition of the finite element expansion, we require

$$\begin{aligned} \xi_1 &= -1, & \xi_{m_1+1} &= 1, \\ \eta_1 &= -1, & \eta_{m_2+1} &= 1. \end{aligned} \tag{4.7.10}$$

The cardinal interpolation function of the (i, j) node is given by the tensor product

$$\psi_{ij}(\xi, \eta) = \mathcal{L}_i(\xi)\, \mathcal{M}_j(\eta), \tag{4.7.11}$$

(*a*)

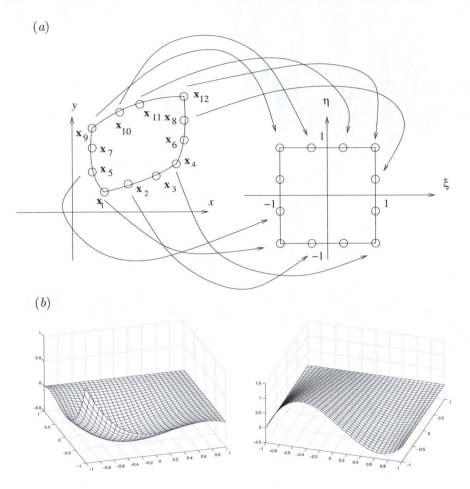

(*b*)

Figure 4.7.2 (*a*) A 12-node quadrilateral element in the xy plane is mapped to the standard square in the parametric $\xi\eta$ plane. (*b*) Graphs of the interpolation functions corresponding to a vertex node (left) and adjacent edge node (right).

where $\mathcal{L}_i(\xi)$ is an m_1-degree Lagrange interpolating polynomial defined with respect to the ξ grid lines,

$$\mathcal{L}_i(\xi) = \frac{(\xi - \xi_1)(\xi - \xi_2)\cdots(\xi - \xi_{i-1})(\xi - \xi_{i+1})\cdots(\xi - \xi_{m_1+1})}{(\xi_i - \xi_1)(\xi_i - \xi_2)\cdots(\xi_i - \xi_{i-1})(\xi_i - \xi_{i+1})\cdots(\xi_i - \xi_{m_1+1})},$$

$$(4.7.12)$$

and $\mathcal{M}_j(\eta)$ is an m_2-degree Lagrange interpolating polynomial defined with respect to the η grid lines,

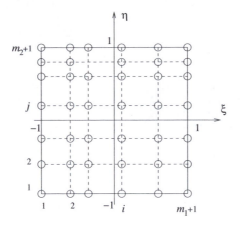

Figure 4.7.3 Distribution of interpolation nodes corresponding to a tensor-product expansion over the standard square.

$$M_j(\eta) = \frac{(\eta - \eta_1)(\eta - \eta_2) \cdots (\eta - \eta_{j-1})(\eta - \eta_{j+1}) \cdots (\eta - \eta_{m_2+1})}{(\eta_j - \eta_1)(\eta_j - \eta_2) \cdots (\eta_j - \eta_{j-1})(\eta_j - \eta_{j+1}) \cdots (\eta_j - \eta_{m_2+1})}.$$

$$(4.7.13)$$

Substituting these expressions in (4.7.11), we obtain a polynomial expansion in (ξ, η) involving $(m_1 + 1) \times (m_2 + 1)$ coefficients.

As an example, when $m_1 = 3$ and $m_2 = 1$, the element interpolation functions have the polynomial form

$$\psi_{ij}(\xi, \eta) = \quad a_{00}$$
$$+ a_{10}\,\xi + a_{01}\,\eta$$
$$+ a_{20}\,\xi^2 + a_{11}\,\xi\eta$$
$$+ a_{30}\,\xi^3 + a_{21}\,\xi^2\eta$$
$$+ a_{31}\,\xi^3\,\eta, \qquad (4.7.14)$$

involving eight coefficients. When $m_1 = m_2$, the expansion takes the symmetric diamond-like form:

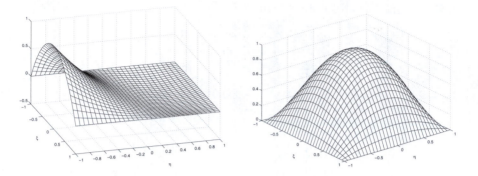

Figure 4.7.4 Graphs of the mid-edge (left) and bubble-mode (right) interpolation functions for a spectral node distribution with $m_1 = 2$ and $m_2 = 2$.

Spectral node distribution

For best interpolation accuracy and to ensure numerical stability, the vertical interior grid lines ξ_i, $i = 2, 3, \ldots, m_1$, should be positioned at the zeros of the $(m_1 - 1)$-degree Lobatto polynomial, and the horizontal interior grid lines η_j, $j = 2, 3, \ldots, m_2$, should be positioned at the zeros of the (m_2-1)-degree Lobatto polynomial. When this is done, the interpolation error behaves spectrally, that is, it decreases faster than algebraically with respect to the polynomial order m_1 or m_2. The spectral properties of the tensor product expansion in two dimensions derive from those of its one-dimensional constituents, as discussed in Chapter 2.

For example, when $m_1 = 2$ and $m_2 = 2$, we obtain the vertical grid lines

$$\xi_1 = -1, \qquad \xi_2 = 0, \qquad \xi_3 = 1, \qquad (4.7.15)$$

and the horizontal grid lines

$$\eta_1 = -1, \qquad \eta_2 = 0, \qquad \eta_3 = 1. \qquad (4.7.16)$$

Graphs of the node interpolation functions corresponding to the edge node $i = 1$, $j = 2$, and interior node $i = 2$, $j = 2$, produced by the FSELIB function `psi_q22` (not listed in the text), are shown in Figure 4.7.4.

Use of the dual Lobatto grid is further motivated by the realization that the resulting nodal set is, in fact, identical to the Fekete nodal set for the square [6]. By definition, the Fekete set maximizes the magnitude of the determinant of the generalized Vandermonde matrix within the confines of the square.

4.7.4   Modal expansion

Consider the standard square in the parametric $\xi\eta$ plane, and number the vertices as shown in Figure 3.8.1. In the modal expansion, a function of interest,

$f(\xi, \eta)$, defined over the standard square, is approximated with the sum of three parts comprised of expansion modes multiplied by corresponding coefficients, as

$$f(\xi, \eta) \simeq F_v + F_e + F_i, \qquad (4.7.17)$$

where

$$F_v(\xi, \eta) = f_1\, \zeta_1^v(\xi, \eta) + f_2\, \zeta_2^v(\xi, \eta) + f_3\, \zeta_3^v(\xi, \eta) + f_4\, \zeta_4^v(\xi, \eta)$$
$$(4.7.18)$$

is the *vertex part*,

$$F_e(\xi, \eta) = \sum_{i=1}^{m_1-1} c_i^{12}\zeta_i^{12}(\xi, \eta) + \sum_{i=1}^{m_2-1} c_i^{34}\zeta_i^{34}(\xi, \eta)$$
$$(4.7.19)$$
$$+ \sum_{i=1}^{m_1-1} c_i^{13}\zeta_i^{13}(\xi, \eta) + \sum_{i=1}^{m_2-1} c_i^{24}\zeta_i^{24}(\xi, \eta)$$

is the *edge part* arising when $m_1 > 1$ or $m_2 > 1$, and

$$F_i(\xi, \eta) = \sum_{i=1}^{m_1-1} \left(\sum_{j=1}^{m_2-1} c_{ij}\, \zeta_{ij}(\xi, \eta) \right), \qquad (4.7.20)$$

is the *interior part* arising when $m_1 > 1$ or $m_2 > 1$. The truncation limits m_1 and m_2 are specified polynomial orders.

Adding the $N_v = 4$ vertex modes, the $N_e = 2 \times (m_1 - 1) + 2 \times (m_2 - 1)$ edge modes, and the $N_i = (m_1 - 1)(m_2 - 1)$ interior nodes, we obtain the total number of modes,

$$N = 4 + 2\,(m_1 - 1) + 2\,(m_2 - 1) + (m_1 - 1)(m_2 - 1)$$
$$= (m_1 + 1)(m_2 + 1), \qquad (4.7.21)$$

which is precisely equal to that involved in the tensor-product expansion discussed in Section 4.7.3.

Vertex modes

The coefficients f_1, f_2, f_3, and f_4 in (4.7.18) are the values of the function $f(\xi, \eta)$ at the four vertex nodes. To ensure C^0 continuity, these values must be shared by neighboring elements at the common nodes. The corresponding cardinal interpolation functions, ζ_1, ζ_2, ζ_3, and ζ_4, are identical to the bilinear node interpolation functions for the 4-node quadrilateral element given in (3.8.4), repeated here for convenient reference,

$$\zeta_1 = \frac{1-\xi}{2}\frac{1-\eta}{2},$$

$$\zeta_2 = \frac{1+\xi}{2}\frac{1-\eta}{2},$$

$$\zeta_3 = \frac{1-\xi}{2}\frac{1+\eta}{2}, \qquad (4.7.22)$$

$$\zeta_4 = \frac{1+\xi}{2}\frac{1+\eta}{2}.$$

Edge modes

The coefficients c_i^{12}, c_i^{34}, c_i^{13}, and c_i^{24} in (4.7.19) are associated with the edge interpolation functions ζ_i^{12} and ζ_i^{34} for $i = 1, 2, \ldots, m_1 - 1$, and with the edge interpolation functions ζ_i^{13}, and ζ_i^{24}, for $i = 1, 2, \ldots, m_2 - 1$. To ensure C^0 continuity, these coefficients must be shared by neighboring elements with common edges.

The interpolation functions, ζ_i^{12}, are zero along the 13, 34, and 23 edges. To satisfy this requirement, we write

$$\zeta_i^{12} = \frac{1-\xi}{2}\frac{1+\xi}{2}\frac{1-\eta}{2}\, Lo_{i-1}(\xi), \qquad (4.7.23)$$

for $i = 1, 2, \ldots, m_1 - 1$. The choice of the Lobatto polynomial on the right-hand side ensures the orthogonality of the $\zeta_i^{12}(\xi)$ set,

$$\int_{-1}^{1} \zeta_i^{12}(\xi, \eta)\, \zeta_j^{12}(\xi, \eta)\, \mathrm{d}\xi = 0 \qquad (4.7.24)$$

if $i \neq j$.

Working in a similar fashion, we derive the remaining edge modes,

$$\zeta_i^{34} = \frac{1-\xi}{2}\frac{1+\xi}{2}\frac{1+\eta}{2}\, Lo_{i-1}(\xi), \qquad (4.7.25)$$

for $i = 1, 2, \ldots, m_1 - 1$,

$$\zeta_i^{13} = \frac{1-\xi}{2}\frac{1-\eta}{2}\frac{1+\eta}{2}\, Lo_{i-1}(\eta), \qquad (4.7.26)$$

for $i = 1, 2, \ldots, m_2 - 1$, and

$$\zeta_i^{24} = \frac{1+\xi}{2}\frac{1-\eta}{2}\frac{1+\eta}{2}\, Lo_{i-1}(\eta), \qquad (4.7.27)$$

for $i = 1, 2, \ldots, m_2 - 1$, which enjoy similar orthogonality properties.

As an example, we consider the modal expansion for $m_1 = m_2 = 2$, recall that the zeroth-degree Lobatto polynomial is constant, $Lo_0(\xi) = 1$, and obtain the cubic edge interpolation functions

$$\zeta_1^{12} = \frac{1-\xi}{2}\frac{1+\xi}{2}\frac{1-\eta}{2},$$

$$\zeta_1^{34} = \frac{1-\xi}{2}\frac{1+\xi}{2}\frac{1+\eta}{2},$$

$$\zeta_1^{13} = \frac{1-\eta}{2}\frac{1+\eta}{2}\frac{1-\xi}{2}, \qquad (4.7.28)$$

$$\zeta_1^{24} = \frac{1-\eta}{2}\frac{1+\eta}{2}\frac{1+\xi}{2}.$$

Note that these modes are proportional, respectively, to the mid-point edge interpolation functions of the 8-node element $\zeta_2, \zeta_7, \zeta_4$, and ζ_5, as listed in (4.7.5).

Interior modes

The coefficients c_{ij} in (4.7.20) correspond to the interior modes, also called bubble modes, described by the interpolation functions ζ_{ij}. To ensure that these functions are zero all around the perimeter of the standard square, we write

$$\zeta_{ij} = \frac{1-\xi}{2}\frac{1+\xi}{2}Lo_{i-1}(\xi)\,\frac{1-\eta}{2}\frac{1+\eta}{2}Lo_{j-1}(\eta), \qquad (4.7.29)$$

where $i = 1, 2, \ldots, m_1 - 1$, and $j = 1, 2, \ldots, m_2 - 1$. The choice of Lobatto polynomial on the right-hand side is motivated by the orthogonality of the set ζ_{ij},

$$\int_{-1}^{1}\int_{-1}^{1} \zeta_{ij}(\xi, \eta)\,\zeta_{kl}(\xi, \eta)\,\mathrm{d}\xi\,\mathrm{d}\eta \neq 0 \qquad (4.7.30)$$

only if $i = k$ and $j = l$. The proof follows immediately from the properties of the Lobatto polynomials discussed in Section B.6 of Appendix B.

As an example, we consider the quadratic modal expansion for $m_1 = m_2 = 2$, recall that the zeroth-degree Lobatto polynomial is constant, $Lo_0(\xi) = 1$ and $Lo_0(\eta) = 1$, and obtain the solitary quartic interior mode

$$\zeta_{11} = \frac{1-\xi}{2}\frac{1+\xi}{2}\frac{1-\eta}{2}\frac{1+\eta}{2}. \qquad (4.7.31)$$

PROBLEMS

4.7.1 *Element node interpolation functions.*

Prepare graphs of the element node interpolation functions corresponding to the Lobatto node distribution with $m_1 = 2$ and $m_2 = 2$.

4.7.2 *Interior modes.*

Prepare graphs and discuss the structure of the interior modes for $m_1 = 3$ and $m_2 = 3$.

Applications in solid and $\boldsymbol{5}$ fluid mechanics

The finite element method finds important practical applications in the field of computational mechanics, including solid and structural statics and dynamics, and hydrodynamics. An application concerning the bending and buckling of beams was already discussed in Sections 1.7 and 1.8. The general methodology and implementation of the finite element method in these applications is similar to that discussed in previous chapters for the convection–diffusion equation. Two new features are that the mathematical modeling typically results in a system of differential equations instead of a single equation, and the order of the individual differential equations may vary across the unknown functions.

In this chapter, we illustrate basic finite element formulations and solution procedures for selected applications in solid, structural, and fluid mechanics, and present integrated finite element codes complete with grid generation and data visualization.

5.1 Plane stress-strain analysis

Consider an elastic material deforming under the action of an external surface load or body force. The plane stress and plane strain problems are distinguished by the assumption that material point particles are deflected so that their displacement in the xy plane is independent of the transverse coordinate, z, as illustrated in Figure 5.1.1. (e.g., [37], Chapter 7). Physically, this may occur when an in-plane force that is parallel to the xy plane is exerted around the edges, at the upper surface, or at the lower surface of an elastic plate, or when an in-plane distributed body force, such as gravity, acts over the volume of the plate. The complementary situation where the plate exhibits transverse deformation normal to the xy plane under the action of a distributed or localized transverse load or body force directed along the z axis, will be discussed in Section 5.2 in the context of plate bending.

To develop the equilibrium equations governing the plate deformation after a surface load or body force has been applied, we temporarily broaden the scope of our discussion and review basic concepts from the theory of three-dimensional elasticity.

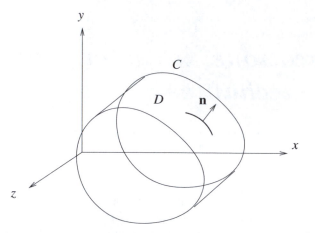

Figure 5.1.1 Sketch of a material confined by two parallel surfaces used to perform the two-dimensional (plane) stress or strain analysis.

5.1.1   Elements of elasticity theory

To describe the forces developing in a deformed three-dimensional material, we introduce the Cauchy stress tensor,

$$
\boldsymbol{\sigma} = \begin{bmatrix} \sigma_{xx} & \sigma_{xy} & \sigma_{xz} \\ \sigma_{yx} & \sigma_{yy} & \sigma_{yz} \\ \sigma_{zx} & \sigma_{zy} & \sigma_{zz} \end{bmatrix}. \tag{5.1.1}
$$

A tensor is a matrix whose components are physical entities endowed with special properties that allow us to compute values corresponding to one coordinate system from values corresponding to another coordinate system, using simple geometrical transformation rules (e.g., [42]).

   The components of the Cauchy stress tensor are defined such that the force surface density, defined as the force per unit area exerted on an infinitesimal surface element drawn inside or at the boundary of the material *in the deformed state*, is given by the traction vector

$$
\mathbf{f} = \mathbf{n} \cdot \boldsymbol{\sigma}, \tag{5.1.2}
$$

where $\mathbf{n} = (n_x, n_y, n_z)$ is the unit vector normal to the infinitesimal surface. By convention, the traction is exerted on the side of the material that lies behind the normal vector, $\mathbf{n}$. Newton's third law requires that the traction exerted on the other side pulls in the opposite direction with equal strength. Explicitly, the Cartesian components of the traction vector are given by

$$
\begin{aligned}
f_x &= n_x\,\sigma_{xx} + n_y\,\sigma_{yx} + n_z\,\sigma_{zx}, \\
f_y &= n_x\,\sigma_{xy} + n_y\,\sigma_{yy} + n_z\,\sigma_{zy}, \\
f_z &= n_x\,\sigma_{xz} + n_y\,\sigma_{yz} + n_z\,\sigma_{zz}.
\end{aligned} \tag{5.1.3}
$$

In the absence of an external torque field, moment equilibrium requires that the stress tensor is symmetric,

$$\sigma_{xy} = \sigma_{yx}, \qquad \sigma_{xz} = \sigma_{zx}, \qquad \sigma_{yz} = \sigma_{zy}. \tag{5.1.4}$$

A torque field would arise when a material made of small magnetized particles is subjected to an electrical field.

Performing a force balance over an infinitesimal parallelepiped whose sides are parallel to the x, y, and z axes, we derive the equilibrium equations

$$\frac{\partial \sigma_{xx}}{\partial x} + \frac{\partial \sigma_{yx}}{\partial y} + \frac{\partial \sigma_{zx}}{\partial z} + b_x = 0,$$

$$\frac{\partial \sigma_{xy}}{\partial x} + \frac{\partial \sigma_{yy}}{\partial y} + \frac{\partial \sigma_{zy}}{\partial z} + b_y = 0, \tag{5.1.5}$$

$$\frac{\partial \sigma_{xz}}{\partial x} + \frac{\partial \sigma_{yz}}{\partial y} + \frac{\partial \sigma_{zz}}{\partial z} + b_z = 0,$$

where $\mathbf{b} = (b_x, b_y, b_z)$ is the body force exerted on the material, defined as the force per unit volume of the material. In the case of the gravitational force, $\mathbf{b} = \rho \mathbf{g}$, where ρ is the material density and $\mathbf{g}$ is the acceleration of gravity. In compact vector notation, the three scalar equilibrium equations (5.1.5) combine into the vectorial equation

$$\nabla \cdot \boldsymbol{\sigma} + \mathbf{b} = \mathbf{0}, \tag{5.1.6}$$

where $\nabla = [\partial/\partial x, \partial/\partial y, \partial/\partial z]$ is the three-dimensional gradient. The first term on the left-hand side of (5.1.6) is the divergence of the stress tensor.

Deformation and constitutive equations

Consider a point particle of the elastic material in the undeformed, and then in the deformed configuration. The displacement of the point particle is denoted by the vector

$$\mathbf{v} = \begin{bmatrix} v_x \\ v_y \\ v_z \end{bmatrix}, \tag{5.1.7}$$

where v_x, v_y, and v_z are the scalar displacements along the x, y, and z axes. Physical arguments dictate that the components of the stress tensor developing in the material due to the deformation at the position of the point particle are functions of the components of the strain tensor,

$$\epsilon_{kl} = \frac{1}{2} \left(\frac{\partial v_k}{\partial x_l} + \frac{\partial v_l}{\partial x_k} \right), \tag{5.1.8}$$

for $k, l = 1, 2, 3$. To simplify the notation, we have denoted $x_1 = x$, $x_2 = y$, $x_3 = z$, and $v_1 = v_x, v_2 = v_y, v_3 = v_z$. This functional dependence is mediated by a constitutive equation that reflects the physical structure and mechanical properties of the material.

Explicitly, the diagonal components of the strain tensor are given by

$$\epsilon_{11} \equiv \epsilon_{xx} = \frac{\partial v_x}{\partial x}, \qquad \epsilon_{22} \equiv \epsilon_{yy} = \frac{\partial v_y}{\partial y}, \qquad \epsilon_{33} \equiv \epsilon_{zz} = \frac{\partial v_z}{\partial z}, \qquad (5.1.9)$$

and the off-diagonal components are given by

$$\epsilon_{12} = \epsilon_{21} \equiv \epsilon_{xy} = \epsilon_{yx} = \frac{1}{2}\left(\frac{\partial v_x}{\partial y} + \frac{\partial v_y}{\partial x}\right),$$

$$\epsilon_{13} = \epsilon_{31} \equiv \epsilon_{xz} = \epsilon_{zx} = \frac{1}{2}\left(\frac{\partial v_x}{\partial z} + \frac{\partial v_z}{\partial x}\right), \qquad (5.1.10)$$

$$\epsilon_{23} = \epsilon_{32} \equiv \epsilon_{yz} = \epsilon_{zx} = \frac{1}{2}\left(\frac{\partial v_y}{\partial z} + \frac{\partial v_z}{\partial y}\right).$$

These definitions suggest the compatibility conditions

$$\frac{\partial^2 \epsilon_{xx}}{\partial y^2} + \frac{\partial^2 \epsilon_{yy}}{\partial x^2} = 2\frac{\partial^2 \epsilon_{xy}}{\partial x \partial y},$$

$$\frac{\partial^2 \epsilon_{xx}}{\partial z^2} + \frac{\partial^2 \epsilon_{zz}}{\partial x^2} = 2\frac{\partial^2 \epsilon_{xz}}{\partial x \partial z}, \qquad (5.1.11)$$

$$\frac{\partial^2 \epsilon_{yy}}{\partial z^2} + \frac{\partial^2 \epsilon_{zz}}{\partial y^2} = 2\frac{\partial^2 \epsilon_{yz}}{\partial y \partial z},$$

which prevent us from arbitrarily specifying all components of the strain tensor.

Linear elasticity

If the material behaves like a linearly elastic medium, the constitutive equation relating stress to deformation takes the form

$$\sigma_{ij} = \lambda\, \delta_{ij}\, \alpha + 2\,\mu\, \epsilon_{ij}, \qquad (5.1.12)$$

where δ_{ij} is the Kronecker delta, λ and μ are the Lamé constants, and

$$\alpha \equiv \epsilon_{xx} + \epsilon_{yy} + \epsilon_{zz} = \frac{\partial v_k}{\partial x_k} = \nabla \cdot \mathbf{v} \qquad (5.1.13)$$

is the dilatation due to the deformation expressed by the divergence of the displacement vector field, which is equal to the trace of the strain tensor; summation of the repeated index k over $1, 2, 3$ or x, y, z is implied in (5.1.13). Explicitly, the linear constitutive equation reads

$$\sigma_{ij} = \lambda\, \delta_{ij} \frac{\partial v_k}{\partial x_k} + \mu\left(\frac{\partial v_i}{\partial x_j} + \frac{\partial v_j}{\partial x_i}\right). \qquad (5.1.14)$$

Note that, in agreement with our earlier discussion, the stress tensor is symmetric, $\sigma_{ij} = \sigma_{ji}$.

Alternatively, the components of the strain tensor can be related to the components of the stress tensor by a generalized Hooke's law for an isotropic medium,

$$
\begin{bmatrix} \epsilon_{xx} \\ \epsilon_{yy} \\ \epsilon_{zz} \\ \epsilon_{xy} \\ \epsilon_{xz} \\ \epsilon_{yz} \end{bmatrix} = \frac{1}{E} \begin{bmatrix} 1 & -\nu & -\nu & 0 & 0 & 0 \\ -\nu & 1 & -\nu & 0 & 0 & 0 \\ -\nu & -\nu & 1 & 0 & 0 & 0 \\ 0 & 0 & 0 & 1+\nu & 0 & 0 \\ 0 & 0 & 0 & 0 & 1+\nu & 0 \\ 0 & 0 & 0 & 0 & 0 & 1+\nu \end{bmatrix} \cdot \begin{bmatrix} \sigma_{xx} \\ \sigma_{yy} \\ \sigma_{zz} \\ \sigma_{xy} \\ \sigma_{xz} \\ \sigma_{yz} \end{bmatrix},
$$

$$(5.1.15)$$

where E is the Young modulus of elasticity, and ν is the Poisson ratio taking values in the range $[-1, 0.5]$; the extreme value $\nu = 0.5$ corresponds to an incompressible solid. While negative values of the Poisson ratio are theoretically possible and have been observed for wrinkled materials that unfold upon deformation, most common materials have a positive Poisson ratio in the range $(0, 0.5)$.

The constitutive equations (5.1.12) and (5.1.15) are, in fact, identical, provided that the Lamé constants are related to the modulus of elasticity, E, and Poisson's ratio, ν, by

$$
\mu = \frac{E}{2(1+\nu)}, \qquad \lambda = \frac{E\nu}{(1+\nu)(1-2\nu)}. \tag{5.1.16}
$$

As an application of (5.1.15), we write the third scalar component of Hooke's law,

$$
\epsilon_{zz} = \frac{1}{E}\left[\sigma_{zz} - \nu\left(\sigma_{xx} + \sigma_{yy}\right)\right], \tag{5.1.17}
$$

and note the coupling of the directional stresses and strain in three dimensions for non-zero Poisson ratio.

Substituting the constitutive equation (5.1.12) in the equilibrium equation (5.1.6), we derive Navier's equation,

$$
\mu \nabla^2 \mathbf{v} + (\mu + \lambda) \nabla \alpha + \mathbf{b} = \mathbf{0}. \tag{5.1.18}
$$

Using relations (5.1.16), we find

$$
\mu + \lambda = \frac{E}{2(1+\nu)(1-2\nu)} = \frac{\mu}{1-2\nu}, \tag{5.1.19}
$$

and derive the alternative form

$$
\mu \nabla^2 \mathbf{v} + \frac{\mu}{1-2\nu} \nabla \alpha + \mathbf{b} = \mathbf{0}, \tag{5.1.20}
$$

where ∇^2 is the three-dimensional Laplacian operators. Explicitly, the scalar components of (5.1.20) read

$$\mu \left(\frac{\partial^2 v_x}{\partial x^2} + \frac{\partial^2 v_x}{\partial y^2} + \frac{\partial^2 v_x}{\partial z^2} \right) + \frac{\mu}{1 - 2\nu} \frac{\partial \alpha}{\partial x} + b_x = 0,$$

$$\mu \left(\frac{\partial^2 v_y}{\partial x^2} + \frac{\partial^2 v_y}{\partial y^2} + \frac{\partial^2 v_y}{\partial z^2} \right) + \frac{\mu}{1 - 2\nu} \frac{\partial \alpha}{\partial y} + b_y = 0, \qquad (5.1.21)$$

$$\mu \left(\frac{\partial^2 v_z}{\partial x^2} + \frac{\partial^2 v_z}{\partial y^2} + \frac{\partial^2 v_z}{\partial z^2} \right) + \frac{\mu}{1 - 2\nu} \frac{\partial \alpha}{\partial z} + b_z = 0.$$

Plane stress and plane strain

In the plane stress and plane strain problems, all components of the stress tensor and all components of the strain tensor are independent of the z coordinate, as illustrated in Figure 5.1.1. These physical circumstances considerably simplify the analysis.

5.1.2 Plane stress

The plane stress problem is further distinguished by the stress conditions

$$\sigma_{iz} = \sigma_{zi} = 0, \qquad (5.1.22)$$

for $i = x, y, z$. Physically, the two parallel surfaces of the material normal to the z axis are free to deform in the z direction, but are unable to support traction. This situation occurs when a flat wrench, modeled as a plate, is pushed on its handle to loosen a bolt.

Under these assumptions, the master constitutive equation (5.1.15) simplifies to

$$\begin{bmatrix} \epsilon_{xx} \\ \epsilon_{yy} \\ \epsilon_{xy} \\ \epsilon_{zz} \end{bmatrix} = \frac{1}{E} \begin{bmatrix} 1 & -\nu & 0 \\ -\nu & 1 & 0 \\ 0 & 0 & 1 + \nu \\ -\nu & -\nu & 0 \end{bmatrix} \cdot \begin{bmatrix} \sigma_{xx} \\ \sigma_{yy} \\ \sigma_{xy} \end{bmatrix}, \qquad (5.1.23)$$

complemented by $\epsilon_{xz} = 0$ and $\epsilon_{yz} = 0$. Note that the right-hand side of (5.1.23) involves a 4×3 matrix.

Solving the first three equations in (5.1.23) for σ_{xx}, σ_{yy}, and σ_{xy} in terms of ϵ_{xx}, ϵ_{yy}, and ϵ_{xy}, we find

$$\begin{bmatrix} \sigma_{xx} \\ \sigma_{yy} \\ \sigma_{xy} \end{bmatrix} = \frac{E}{1 - \nu^2} \begin{bmatrix} 1 & \nu & 0 \\ \nu & 1 & 0 \\ 0 & 0 & 1 - \nu \end{bmatrix} \cdot \begin{bmatrix} \epsilon_{xx} \\ \epsilon_{yy} \\ \epsilon_{xy} \end{bmatrix}. \qquad (5.1.24)$$

The fourth equation in (5.1.23) is a simplification of (5.1.17),

$$\epsilon_{zz} = -\frac{\nu}{E} (\sigma_{xx} + \sigma_{yy}). \qquad (5.1.25)$$

This relation reveals that a material with a Poisson ratio equal to zero, such as a bottle cork, will not expand laterally when subjected to an in-plane load. On the other hand, when the Poisson ratio is negative, stretching of the material in the xy plane, $\sigma_{xx} + \sigma_{yy} > 0$, causes expansion in the lateral direction, $\epsilon_{zz} > 0$, which is counterintuitive.

Equilibrium equations

In the case of the plane stress problem, the x and y components of the equilibrium equations (5.1.5) simplify to

$$\frac{\partial \sigma_{xx}}{\partial x} + \frac{\partial \sigma_{yx}}{\partial y} + b_x = 0,$$

$$\frac{\partial \sigma_{xy}}{\partial x} + \frac{\partial \sigma_{yy}}{\partial y} + b_y = 0. \tag{5.1.26}$$

Taking the partial derivative of the first equation with respect to x and the partial derivative of the second equation with respect to y, adding the resulting expressions and rearranging, we find

$$\frac{\partial^2 \sigma_{xx}}{\partial x^2} + \frac{\partial^2 \sigma_{yy}}{\partial y^2} + 2 \frac{\partial^2 \sigma_{xy}}{\partial x \, \partial y} = -\left(\frac{\partial b_x}{\partial x} + \frac{\partial b_y}{\partial y}\right). \tag{5.1.27}$$

Now, the first of the compatibility conditions (5.1.11) in conjunction with (5.1.24) yields

$$\frac{\partial^2 \epsilon_{xx}}{\partial y^2} + \frac{\partial^2 \epsilon_{yy}}{\partial x^2} = \frac{1}{E}\left[\frac{\partial^2 (\sigma_{xx} - \nu \sigma_{yy})}{\partial y^2} + \frac{\partial^2 (\sigma_{yy} - \nu \sigma_{xx})}{\partial x^2}\right]$$

$$= 2 \frac{\partial^2 \epsilon_{xy}}{\partial x \partial y} = \frac{2\,(1+\nu)}{E} \frac{\partial^2 \sigma_{xy}}{\partial x \partial y}, \tag{5.1.28}$$

which can be rearranged into

$$2 \frac{\partial^2 \sigma_{xy}}{\partial x \partial y} = \frac{1}{1+\nu}\left[\frac{\partial^2 (\sigma_{xx} - \nu \sigma_{yy})}{\partial y^2} + \frac{\partial^2 (\sigma_{yy} - \nu \sigma_{xx})}{\partial x^2}\right]. \tag{5.1.29}$$

Combining (5.1.27) and (5.1.29) to eliminate $\partial^2 \sigma_{xy}/(\partial x \partial y)$, we derive a Poisson equation for the sum of the normal stresses, which is equal to the trace of the stress tensor in the xy plane,

$$\left(\frac{\partial^2}{\partial x^2} + \frac{\partial^2}{\partial y^2}\right)(\sigma_{xx} + \sigma_{yy}) = -(1+\nu)\left(\frac{\partial b_x}{\partial x} + \frac{\partial b_y}{\partial y}\right). \tag{5.1.30}$$

In compact notation,

$$\nabla^2 (\sigma_{xx} + \sigma_{yy}) = -(1+\nu)\, \nabla \cdot \mathbf{b}. \tag{5.1.31}$$

The solution is to be found subject to appropriate boundary conditions around the boundary.

Biharmonic equation for the Airy function

If the body force is conservative, its Cartesian components, b_x and b_y, derive from a scalar potential function, $\mathcal{V}$, as

$$b_x = \frac{\partial \mathcal{V}}{\partial x}, \qquad b_y = \frac{\partial \mathcal{V}}{\partial y}. \tag{5.1.32}$$

In vector notation,

$$\mathbf{b} = \nabla \mathcal{V}. \tag{5.1.33}$$

For example, if the body force is uniform in space, then

$$\mathcal{V} = b_x \, x + b_y \, y = \mathbf{b} \cdot \mathbf{x}. \tag{5.1.34}$$

Under these circumstances, the equilibrium equations (5.1.26) become

$$\frac{\partial(\sigma_{xx} + \mathcal{V})}{\partial x} + \frac{\partial \sigma_{yx}}{\partial y} = 0,$$

$$\frac{\partial \sigma_{xy}}{\partial x} + \frac{\partial(\sigma_{yy} + \mathcal{V})}{\partial y} = 0, \tag{5.1.35}$$

which are satisfied if the stress components derive from the Airy stress function, ϕ, as

$$\sigma_{xx} = \frac{\partial^2 \phi}{\partial y^2} - \mathcal{V}, \qquad \sigma_{yy} = \frac{\partial^2 \phi}{\partial x^2} - \mathcal{V},$$

$$\sigma_{xy} = \sigma_{yx} = -\frac{\partial^2 \phi}{\partial x \partial y}. \tag{5.1.36}$$

The trace of the stress tensor is given by

$$\sigma_{xx} + \sigma_{yy} = \nabla^2 \phi - 2 \, \mathcal{V}. \tag{5.1.37}$$

Substituting this expression along with (5.1.32) in (5.1.31), we find

$$\nabla^2 (\nabla^2 \phi - 2 \, \mathcal{V}) = -(1 + \nu) \, \nabla \cdot \mathbf{b} = -(1 + \nu) \, \nabla^2 \, \mathcal{V}. \tag{5.1.38}$$

Rearranging, we find that the Airy stress function satisfies the inhomogeneous biharmonic equation

$$\nabla^4 \phi \equiv \left(\frac{\partial^2}{\partial x^2} + \frac{\partial^2}{\partial y^2} \right)^2 \phi = \frac{\partial^4 \phi}{\partial x^4} + 2 \frac{\partial^4 \phi}{\partial x^2 \partial y^2} + \frac{\partial^4 \phi}{\partial y^4}$$

$$= (1 - \nu) \, \nabla^2 \, \mathcal{V}. \tag{5.1.39}$$

If the potential $\mathcal{V}$ is harmonic, which is true when the components of the body force, b_x and b_y, are constant in space, in which case $\mathcal{V}$ is linear in x and

y, the right-hand side of (5.1.40) is zero, yielding the homogeneous biharmonic equation

$$\nabla^4 \phi = 0. \tag{5.1.40}$$

This equation is a common point of departure for computing solutions of the plane stress problem by finite difference methods. The finite element formulation of the biharmonic equation will be discussed in Section 5.2 in the context of plate bending.

5.1.3 Plane strain

In the plane-strain problem, all material point particles lying on a line that is parallel to the z axis undergo the same displacement. Thus, the plane-strain problem is distinguished by the strain conditions

$$\epsilon_{iz} = \epsilon_{zi} = 0, \tag{5.1.41}$$

for $i = x, y, z$. Physically, the planar boundaries of the two-dimensional material normal to the z axis are anchored, as though they were sandwiched between two impenetrable walls that allow sliding motions. In practice, the plane-strain problem occurs when a water dam is immobilized by side-walls or when a slab of earth is held fixed between two hard rocks. Under these circumstances, the master constitutive equation (5.1.15) simplifies to

$$
\begin{bmatrix} \epsilon_{xx} \\ \epsilon_{yy} \\ 0 \\ \epsilon_{xy} \\ 0 \\ 0 \end{bmatrix} = \frac{1}{E} \begin{bmatrix} 1 & -\nu & -\nu & 0 & 0 & 0 \\ -\nu & 1 & -\nu & 0 & 0 & 0 \\ -\nu & -\nu & 1 & 0 & 0 & 0 \\ 0 & 0 & 0 & 1+\nu & 0 & 0 \\ 0 & 0 & 0 & 0 & 1+\nu & 0 \\ 0 & 0 & 0 & 0 & 0 & 1+\nu \end{bmatrix} \cdot \begin{bmatrix} \sigma_{xx} \\ \sigma_{yy} \\ \sigma_{zz} \\ \sigma_{xy} \\ \sigma_{xz} \\ \sigma_{yz} \end{bmatrix}. \tag{5.1.42}
$$

The last two equations require $\sigma_{xz} = 0$ and $\sigma_{yz} = 0$, whereupon the x and y components of the equilibrium equations (5.1.5) reduce to (5.1.26). Taking the partial derivative of the first equation in (5.1.42) with respect to x and the partial derivative of the second equation with respect to y, adding the resulting expressions and rearranging, we derive the second-order equation (5.1.27). The fourth equation in (5.1.42) requires

$$\sigma_{zz} = \nu \left(\sigma_{xx} + \sigma_{yy} \right). \tag{5.1.43}$$

Using this constraint to eliminate σ_{zz}, we obtain the simplified constitutive equation

$$
\begin{bmatrix} \epsilon_{xx} \\ \epsilon_{yy} \\ \epsilon_{xy} \end{bmatrix} = \frac{1+\nu}{E} \begin{bmatrix} 1-\nu & -\nu & 0 \\ -\nu & 1-\nu & 0 \\ 0 & 0 & 1 \end{bmatrix} \cdot \begin{bmatrix} \sigma_{xx} \\ \sigma_{yy} \\ \sigma_{xy} \end{bmatrix}. \tag{5.1.44}
$$

Solving for the stresses in terms of the strains, we finally arrive at the targeted constitutive equation

$$
\begin{bmatrix} \sigma_{xx} \\ \sigma_{yy} \\ \sigma_{xy} \end{bmatrix} = \frac{E}{(1+\nu)(1-2\nu)} \begin{bmatrix} 1-\nu & \nu & 0 \\ \nu & 1-\nu & 0 \\ 0 & 0 & 1-2\nu \end{bmatrix} \cdot \begin{bmatrix} \epsilon_{xx} \\ \epsilon_{yy} \\ \epsilon_{xy} \end{bmatrix}.
$$
(5.1.45)

Alternatively, equations (5.1.45) can be derived directly from (5.1.12), subject to relations (5.1.16) (Problem 5.1.2). Note that a singular behavior occurs when $\nu = 0.5$, corresponding to an incompressible material. When $\nu = 0$, and only then, the constitutive equations for plane strain reduce to those for plane stress.

PROBLEMS

5.1.1 *Linear constitutive equations.*

Show that the constitutive equations (5.1.12) and (5.1.15) are identical, provided that the Lamé constants are related to the modulus of elasticity, E, and Poisson's ratio, ν, according to (5.1.16).

5.1.2 *Plane strain analysis.*

(*a*) Derive equations (5.1.45) directly from (5.1.12), subject to relations (5.1.16).

(*b*) Write out the counterparts of equations (5.2.11) for plane strain analysis.

5.1.3 *Compression of a rectangular plate in plane stress analysis.*

(*a*) Consider a rectangular plate with two sides parallel to the x axis and two sides parallel to the y axis, and set the origin of the Cartesian axes at the center-point. Show that, when the plate is compressed normal to the two edges that are parallel to the y axes with a uniform force per unit length, F_x, while the other two edges are free, the displacement field is given by

$$
v_x = -\frac{F_x}{Eh}\, x, \qquad v_y = \frac{F_x}{Eh}\, \nu\, y, \tag{5.1.46}
$$

where h is the plate thickness, and derive the associated stress field. Note that, when the Poisson ratio is positive, compression induces expansion in the lateral direction.

(*b*) Repeat (*a*) for the case where the plate is compressed normal to the two edges that are parallel to the x axes with a uniform force per unit length, F_y, while the other two edges are free.

5.1.4 *Shear flow past a circular or elliptical membrane.*

(*a*) Consider shear flow along the x axis with velocity $U = G\, z$, past a circular membrane of radius a that is clamped around the rim on a plane wall, as shown

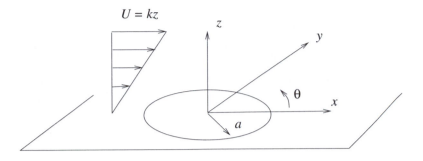

Figure 5.1.2 Illustration of shear flow past a circular membrane that is clamped around the rim on a plane wall.

in Figure 5.1.2, where G is the shear rate. The upper surface of the membrane is exposed to the hydrodynamic shear stress, $\tau = \eta G$, which amounts to an in-plane body force with components $b_x = \tau/h$ and $b_y = 0$, where η is the fluid viscosity, and h is the membrane thickness.

Verify that the displacement field is given by

$$v_x = \frac{\tau}{Eh} \frac{1-\nu^2}{3-\nu} (a^2 - x^2 - y^2), \qquad v_y = 0, \qquad (5.1.47)$$

and confirm that the associated stresses are given by

$$\sigma_{xx} = -\frac{2}{3-\nu} \frac{\eta k}{h} x, \qquad \sigma_{xy} = -\frac{1-\nu}{3-\nu} \frac{\eta k}{h} y,$$

$$\sigma_{yy} = \nu \, \sigma_{xx}. \qquad (5.1.48)$$

(*b*) Consider an elliptical membrane patch with principal axes in the x and y directions, whose contour is described by the equation $(x/a_x)^2 + (y/a_y)^2 = 1$, where a_x and a_y are the elliptical semi-axes. Verify that the displacement field for the problem described in (*a*) is given by

$$v_x = V \left(1 - \frac{x^2}{a_x^2} - \frac{y^2}{a_y^2} \right), \qquad v_y = 0, \qquad (5.1.49)$$

where

$$V = \frac{\tau}{Eh} \frac{(1-\nu^2) \, a_x^2 \, a_y^2}{(1-\nu) \, a_x^2 + 2 \, a_y^2} \qquad (5.1.50)$$

is a constant with dimensions of length, and confirm that the associated membrane stresses are given by

$$\sigma_{xx} = \frac{E}{1-\nu^2}\frac{\partial v_x}{\partial x} = -2\frac{EV}{(1-\nu^2)\,a_x^2}\,x = -\frac{\tau}{h}\frac{2\,a_y^2}{(1-\nu)\,a_x^2 + 2\,a_y^2}\,x,$$

$$\sigma_{xy} = \frac{E}{2(1+\nu)}\frac{\partial v_x}{\partial y} = -\frac{EV}{(1+\nu)\,a_y^2}\,y = -\frac{\tau}{h}\frac{(1-\nu)\,a_x^2}{(1-\nu)\,a_x^2 + 2\,a_y^2}\,y,$$

$$\sigma_{yy} = \nu\,\sigma_{xx}. \tag{5.1.51}$$

5.2 Finite element methods for plane stress/strain

We begin developing the finite element method for plane stress and plane strain analysis by discretizing the solution domain in the xy plane into finite elements, as discussed in Chapters 3 and 4 for the convection–diffusion equation. In the second step, we express the x and y components of the displacement field in terms of global interpolation functions, ϕ_j, in the familiar form

$$\mathbf{v}(x,y) = \sum_{j=1}^{N_G} \mathbf{v}_j\,\phi_j(x,y), \tag{5.2.1}$$

where N_G is the number of global interpolation nodes. The coefficient vector,

$$\mathbf{v}_j = \begin{bmatrix} v_{x_j} \\ v_{y_j} \end{bmatrix} \tag{5.2.2}$$

is the vectorial displacement at the position of the jth node.

Next, we carry out the Galerkin projection of the vectorial equilibrium equation (5.1.6) over the solution domain, D,

$$\iint_D \phi_i(x,y)\,(\nabla \cdot \boldsymbol{\sigma} + \mathbf{b})\,\mathrm{d}x\,\mathrm{d}y = \mathbf{0}, \tag{5.2.3}$$

and use the Gauss divergence theorem to remove the divergence of the stress tensor working as follows. First, we write

$$\iint_D \phi_i\,\nabla \cdot \boldsymbol{\sigma}\,\mathrm{d}x\,\mathrm{d}y = \iint_D \nabla \cdot (\phi_i\,\boldsymbol{\sigma})\,\mathrm{d}x\,\mathrm{d}y - \iint_D \nabla\phi_i \cdot \boldsymbol{\sigma}\,\mathrm{d}x\,\mathrm{d}y. \tag{5.2.4}$$

Second, we use the divergence theorem to convert the first integral on the right-hand side to a contour integral, obtaining

$$\iint_D \phi_i\,\nabla \cdot \boldsymbol{\sigma}\,\mathrm{d}x\,\mathrm{d}y = \oint_C \phi_i\,\mathbf{n} \cdot \boldsymbol{\sigma}\,\mathrm{d}l - \iint_D \nabla\phi_i \cdot \boldsymbol{\sigma}\,\mathrm{d}x\,\mathrm{d}y, \tag{5.2.5}$$

where **n** is the unit vector normal to the boundary of the solution domain, C, pointing *inward*, and l is the arc length along C. Substituting this expression in (5.2.3) and rearranging, we derive the final form

$$\iint_D \nabla \phi_i \cdot \boldsymbol{\sigma} \, dx \, dy = \oint_C \phi_i \, \mathbf{f} \, dl + \iint_D \phi_i \, \mathbf{b} \, dx \, dy, \qquad (5.2.6)$$

where $\mathbf{f} = \mathbf{n} \cdot \boldsymbol{\sigma}$ is the boundary traction.

Explicitly, the x and y components of the ith vectorial Galerkin equation (5.2.6) read

$$\iint_D \left(\frac{\partial \phi_i}{\partial x} \sigma_{xx} + \frac{\partial \phi_i}{\partial y} \sigma_{yx} \right) dx \, dy = F_{x_i} + B_{x_i},$$

$$\iint_D \left(\frac{\partial \phi_i}{\partial x} \sigma_{xy} + \frac{\partial \phi_i}{\partial y} \sigma_{yy} \right) dx \, dy = F_{y_i} + B_{y_i}, \qquad (5.2.7)$$

where

$$F_{x_i} \equiv \oint_C \phi_i \, f_x \, dl, \qquad F_{y_i} \equiv \oint_C \phi_i \, f_y \, dl, \qquad (5.2.8)$$

are the effective nodal *edge forces*, and

$$B_{x_i} \equiv \iint_D \phi_i \, b_x \, dx \, dy, \qquad B_{y_i} \equiv \iint_D \phi_i \, b_y \, dx \, dy, \qquad (5.2.9)$$

are the effective nodal *body forces*. Note that, since ϕ_i is zero around the boundary for an interior node, F_{x_i} and F_{y_i} are non-zero only if the ith node is a boundary node.

Further implementation of the finite element method involves the following steps:

1. Substitute the finite element expansion for the nodal displacement (5.2.1) in the definition (5.1.8) to obtain the components of the strain tensor in terms of the node displacements, as

$$\epsilon_{xx} = \sum_{j=1}^{N_G} v_{x_j} \frac{\partial \phi_j}{\partial x},$$

$$\epsilon_{yy} = \sum_{j=1}^{N_G} v_{y_j} \frac{\partial \phi_j}{\partial y}, \qquad (5.2.10)$$

$$\epsilon_{xy} = \frac{1}{2} \sum_{j=1}^{N_G} \left(v_{x_j} \frac{\partial \phi_j}{\partial y} + v_{y_j} \frac{\partial \phi_j}{\partial x} \right).$$

2. Substitute the strain components in (5.1.24) for the plane stress problem, or in (5.1.45) for the plane stress problem, to obtain the stresses in terms of the nodal displacements and their corresponding interpolation functions.

3. Substitute the stresses thus computed in (5.2.7) to derive a linear system of algebraic equations for the x and y components of the nodal displacements.

4. Implement the boundary conditions.

5. Solve the linear system for the nodal displacements.

6. Having recovered the displacements, compute the stresses using (5.1.24) for the plane stress problem, or (5.1.45) for the plane stress problem.

7. Visualize the displacement and stress fields.

5.2.1 Code for plane stress

To make the implementation more specific, we consider the most commonly encountered plane stress problem.

In Step 2, we find

$$
\sigma_{xx} = \frac{E}{1-\nu^2} \sum_{j=1}^{N_G} \left(v_{x_j} \frac{\partial \phi_j}{\partial x} + \nu\, v_{y_j} \frac{\partial \phi_j}{\partial y} \right),
$$

$$
\sigma_{yy} = \frac{E}{1-\nu^2} \sum_{j=1}^{N_G} \left(\nu\, v_{x_j} \frac{\partial \phi_j}{\partial x} + v_{y_j} \frac{\partial \phi_j}{\partial y} \right), \qquad (5.2.11)
$$

$$
\sigma_{xy} = \frac{E}{2\,(1+\nu)} \sum_{j=1}^{N_G} \left(v_{x_j} \frac{\partial \phi_j}{\partial y} + v_{y_j} \frac{\partial \phi_j}{\partial x} \right).
$$

In Step 3, we find

$$
E \sum_{j=1}^{N_G} \left(M_{ij}^{xx}\, v_{x_j} + M_{ij}^{xy}\, v_{y_j} \right) = F_{x_i} + B_{x_i},
$$

$$
E \sum_{j=1}^{N_G} \left(M_{ij}^{yx}\, v_{x_j} + M_{ij}^{yy}\, v_{y_j} \right) = F_{y_i} + B_{y_i}, \qquad (5.2.12)
$$

where

$$
M_{ij}^{xx} = \frac{1}{1-\nu^2}\, K_{ij}^{xx} + \frac{1}{2(1+\nu)}\, K_{ij}^{yy},
$$

$$
M_{ij}^{xy} = \frac{\nu}{1-\nu^2}\, K_{ij}^{xy} + \frac{1}{2(1+\nu)}\, K_{ij}^{yx},
$$

$$
M_{ij}^{yx} = \frac{\nu}{1-\nu^2}\, K_{ij}^{yx} + \frac{1}{2(1+\nu)}\, K_{ij}^{xy}, \qquad (5.2.13)
$$

$$
M_{ij}^{yy} = \frac{1}{1-\nu^2}\, K_{ij}^{yy} + \frac{1}{2(1+\nu)}\, K_{ij}^{xx},
$$

and we have introduced the stiffness matrices

$$K_{ij}^{xx} \equiv \iint_D \frac{\partial \phi_i}{\partial x} \frac{\partial \phi_j}{\partial x} \, dx \, dy, \qquad K_{ij}^{xy} \equiv \iint_D \frac{\partial \phi_i}{\partial x} \frac{\partial \phi_j}{\partial y} \, dx \, dy,$$

$$K_{ij}^{yx} \equiv \iint_D \frac{\partial \phi_i}{\partial y} \frac{\partial \phi_j}{\partial x} \, dx \, dy, \qquad K_{ij}^{yy} \equiv \iint_D \frac{\partial \phi_i}{\partial y} \frac{\partial \phi_j}{\partial y} \, dx \, dy. \tag{5.2.14}$$

Note the symmetry properties

$$K_{ij}^{xy} = K_{ji}^{yx}, \qquad\qquad M_{ij}^{xy} = M_{ji}^{yx}. \tag{5.2.15}$$

In Step 4, we derive the linear system

$$\mathbf{M} \cdot \begin{bmatrix} v_{x_1} \\ v_{x_2} \\ \vdots \\ v_{x_{N_G}} \\ --- \\ v_{y_1} \\ v_{y_2} \\ \vdots \\ v_{y_{N_G}} \end{bmatrix} = \begin{bmatrix} F_{x_1} \\ F_{x_1} \\ \vdots \\ F_{x_{N_G}} \\ --- \\ F_{y_1} \\ F_{y_2} \\ \vdots \\ F_{y_{N_G}} \end{bmatrix} + \begin{bmatrix} B_{x_1} \\ B_{x_1} \\ \vdots \\ B_{x_{N_G}} \\ --- \\ B_{y_1} \\ B_{y_2} \\ \vdots \\ B_{y_{N_G}} \end{bmatrix}, \tag{5.2.16}$$

where

$$\mathbf{M} = E \left[\begin{array}{c|c} \mathbf{M}^{xx} & \mathbf{M}^{xy} \\ \hline \mathbf{M}^{yx} & \mathbf{M}^{yy} \end{array} \right] \tag{5.2.17}$$

is the block global stiffness matrix. Note that, because of the symmetry properties (5.2.15), the matrix $\mathbf{M}$ is symmetric. The first vector on the right-hand side of (5.2.16) represents the effect of the edge force, and the second vector represents the effect of the body force.

FSELIB code psa6, listed in the text, solves the equations of plane stress analysis in a square possibly pierced by a circular hole, in the absence of a body force. In the case of the whole square, the domain is discretized into 6-node triangles using the FSELIB function trgl6_sqr, which is similar to the function trgl6_disk for a disk-like domain discussed in Section 4.2. In the case of a pierced square, the domain is discretized into 6-node triangles using the FSELIB function trgl6_sc discussed in Section 4.2.

The structure of the finite element code is similar to that discussed in Section 4.2 for Laplace's equation, involving the assembly of the stiffness matrices from corresponding element matrices evaluated using the FSELIB function esm6, listed in the text. One new feature concerns the computation of the nodal values of the surface traction components F_{x_i} and F_{y_i} defined in (5.2.8). In the code, the corresponding integrals are assembled from integrals around the element edges.

```
%===========
% CODE psa6
%
% Plane stress analysis in a square
% or a square with a circular hole
%
%=============

%-----------
% input data
%-----------

E = 1.0;    % modulus of elasticity
nu = 0.5;   % Poisson ratio
NQ = 6;     % gauss-triangle quadrature
ndiv = 2;   % discretization level

%------------
% triangulate
%------------

[ne,ng,p,c,efl,gfl] = trgl6_sqr(ndiv);    % square

% a= 0.5;    % hole radius
% [ne,ng,p,c,efl,gfl] = trgl6_sc(a, ndiv); % square with a hole

disp('Number of elements:'); ne

%--------
% prepare
%--------

ng2 = 2*ng;
nus = nu^2; % square of the Poisson ratio

%------------------------
% count the interior nodes
% (optional)
%------------------------

ic = 0;
```

Code psa6: Continuing $\longrightarrow$

Specifically, if the traction boundary condition is imposed at the element vertex nodes 1 and 2, then the element edge subtended between these nodes contributes to the integrals shown in (5.2.8). If the mid-side node 4 is located midway between vertex nodes 1 and 3, consideration of the element node interpolation functions shows that the contribution to the global node i, corresponding to the element vertex nodes 1 or 2, is given by a consistent lumping approximation that arises by applying the product integration rule, as

```
for j=1:ng
 if(gfl(j,1)==0) ic=ic+1; end
end

disp('Number of interior nodes:'); ic

%----------------------------------------
% specify the boundary condition on the
% top, bottom, and left side, and
% the traction on the right side
%----------------------------------------

for i=1:ng

 gfl(i,2) = 0;    % initialize

 if(gfl(i,1)==1)     % boundary node

% default:              traction boundary condition

  gfl(i,2) = 1;   gfl(i,3) = 0.0;   % fx = 0
                  gfl(i,4) = 0.0;   % fy = 0

% apply the load at the right edge:

  if(p(i,1) > 0.99)

    gfl(i,3) = 0.0;   % Fx
    gfl(i,4) = 0.01*(1.0-p(i,2)^2); % Fy

  end

% zero displacement at the left edge

  if(p(i,1) < -0.99)
    gfl(i,2) = 2;     % displacement boundary condition
    gfl(i,3) = 0.0;   % vx: x-displacement
    gfl(i,4) = 0.0;   % vy: y-displacement
  end

 end
end

%----------------------
% deform to a rectangle
%----------------------

defx=1.0;   % x deformation factors
defy=1.0;   % y deformation factors

for i=1:ng
  p(i,1)=p(i,1)*defx;
  p(i,2)=p(i,2)*defy;
end
```

Code psa6: Continuing $\longrightarrow$

```
%-----------------------
% plot the element nodes
%-----------------------

for i=1:ne
 i1=c(i,1); i2=c(i,2); i3=c(i,3); i4=c(i,4); i5=c(i,5); i6=c(i,6);
 xp(1)=p(i1,1); xp(2)=p(i4,1); xp(3)=p(i2,1); xp(4)=p(i5,1);
               xp(5)=p(i3,1); xp(6)=p(i6,1); xp(7)=p(i1,1);
 yp(1)=p(i1,2); yp(2)=p(i4,2); yp(3)=p(i2,2); yp(4)=p(i5,2);
               yp(5)=p(i3,2); yp(6)=p(i6,2); yp(7)=p(i1,2);
 plot(xp, yp,'+'); hold on;
end

%----------------------------------------
% assemble the global stiffness matrices
%----------------------------------------

gsm_xx = zeros(ng,ng);    % initialize
gsm_xy = zeros(ng,ng);
gsm_yy = zeros(ng,ng);

% compute the element stiffness matrices

j=c(1,1); x1=p(j,1); y1=p(j,2);
j=c(1,2); x2=p(j,1); y2=p(j,2);
j=c(1,3); x3=p(j,1); y3=p(j,2);
j=c(1,4); x4=p(j,1); y4=p(j,2);
j=c(1,5); x5=p(j,1); y5=p(j,2);
j=c(1,6); x6=p(j,1); y6=p(j,2);

[esm_xx, esm_xy, esm_yy, arel] = esm6 ...
...
    (x1,y1, x2,y2, x3,y3, x4,y4, x5,y5, x6,y6 ,NQ);

  for i=1:6

    i1 = c(1,i);

    for j=1:6
      j1 = c(1,j);
      gsm_xx(i1,j1) = gsm_xx(i1,j1) + esm_xx(i,j);
      gsm_xy(i1,j1) = gsm_xy(i1,j1) + esm_xy(i,j);
      gsm_yy(i1,j1) = gsm_yy(i1,j1) + esm_yy(i,j);
    end

  end

end

%------------------------
% form the transpose block
%------------------------

gsm_yx = gsm_xy';    % a prime denotes the transpose
```

Code psa6: $\longrightarrow$ Continuing $\longrightarrow$

```
%-------------------------
% define the block matrices
%-------------------------

Mxu =     gsm_xx/(1-nus) + 0.5*gsm_yy/(1+nu);
Mxv = nu*gsm_xy/(1-nus) + 0.5*gsm_yx/(1+nu);
Myu = Mxv';
Myv =     gsm_yy/(1-nus) + 0.5*gsm_xx/(1+nu);

%-------------------------
% compile the grand matrix
%-------------------------

for i=1:ng
 for j=1:ng
  Gm(i,    j)    = E*Mxu(i,j);
  Gm(i,    j+ng) = E*Mxv(i,j);
  Gm(ng+i,j)     = E*Myu(i,j);
  Gm(ng+i,j+ng)  = E*Myv(i,j);
 end
end

%---------------------------
% assemble the right-hand side
% of the linear system
%---------------------------

for i=1:ng2
 b(i) = 0.0;
end

for i=1:ne

  l1=c(i,1); l2=c(i,2); l3=c(i,4);

  if(gfl(l1,2)==1 & gfl(l2,2)==1 )

   side = sqrt( (p(l2,1)-p(l1,1))^2+(p(l2,2)-p(l1,2))^2 );

   b(l1) = b(l1) + gfl(l1,3)*side*1.0/6.0;
   b(l2) = b(l2) + gfl(l2,3)*side*1.0/6.0;
   b(l3) = b(l3) + gfl(l3,3)*side*2.0/3.0;

   k1 = l1+ng;
   k2 = l2+ng;
   k3 = l3+ng;

   b(k1) = b(k1) + gfl(k1,4)*side*1.0/6.0;
   b(k2) = b(k2) + gfl(k2,4)*side*1.0/6.0;
   b(k3) = b(k3) + gfl(k3,4)*side*2.0/3.0;
  end
```

Code psa6: $\longrightarrow$ Continuing $\longrightarrow$

```
   l1=c(i,2); l2=c(i,3); l3=c(i,5);

   if(gfl(l1,2)==1 & gfl(l2,2)==1 )
     side = sqrt( (p(l2,1)-p(l1,1))^2+(p(l2,2)-p(l1,2))^2 );
     b(l1) = b(l1) + gfl(l1,3)*side*1.0/6.0;
     b(l2) = b(l2) + gfl(l2,3)*side*1.0/6.0;
     b(l3) = b(l3) + gfl(l3,3)*side*2.0/3.0;
     k1 = l1+ng; k2 = l2+ng; k3 = l3+ng;
     b(k1) = b(k1) + gfl(l1,4)*side*1.0/6.0;
     b(k2) = b(k2) + gfl(l2,4)*side*1.0/6.0;
     b(k3) = b(k3) + gfl(l3,4)*side*2.0/3.0;
   end

   l1=c(i,3); l2=c(i,1); l3=c(i,6);

   if(gfl(l1,2)==1 & gfl(l2,2)==1 )
     side = sqrt( (p(l2,1)-p(l1,1))^2+(p(l2,2)-p(l1,2))^2 );
     b(l1) = b(l1) + gfl(l1,3)*side*1.0/6.0;
     b(l2) = b(l2) + gfl(l2,3)*side*1.0/6.0;
     b(l3) = b(l3) + gfl(l3,3)*side*2.0/3.0;
     k1 = l1+ng; k2 = l2+ng; k3 = l3+ng;
     b(k1) = b(k1) + gfl(k1,4)*side*1.0/6.0;
     b(k2) = b(k2) + gfl(k2,4)*side*1.0/6.0;
     b(k3) = b(k3) + gfl(k3,4)*side*2.0/3.0;
   end

end

%--------------------------------------------
% implement the Dirichlet boundary condition
%--------------------------------------------


for j=1:ng  % run over global nodes

%---
 if(gfl(j,2)==2)   % node with displacement boundary condition

    for i=1:ng2
      b(i) = b(i) - Gm(i,j) * gfl(j,3) - Gm(i,ng+j) * gfl(j,4);
      Gm(i,j) = 0; Gm(i,ng+j) = 0;
      Gm(j,i) = 0; Gm(ng+j,i) = 0;
    end

    Gm(j,j)       = 1.0;
    Gm(j+ng,j+ng) = 1.0;

    b(j)    = gfl(j,3);
    b(j+ng) = gfl(j,4);

 end
%---

end
```

Code psa6: $\longrightarrow$ Continuing $\longrightarrow$

```
%------------------------
% solve the linear system
%------------------------

f = b/Gm';

%------------------------
% assign the displacements
%------------------------

for i=1:ng
  vx(i) = f(i); vy(i) = f(ng+i);
end

%--------------------------
% plot the deformed elements
%--------------------------

for i=1:ng
  p(i,1) = p(i,1)+vx(i);
  p(i,2) = p(i,2)+vy(i);
end

for i=1:ne

  i1=c(i,1); i2=c(i,2); i3=c(i,3); i4=c(i,4); i5=c(i,5); i6=c(i,6);

  xp(1)=p(i1,1); xp(2)=p(i4,1); xp(3)=p(i2,1); xp(4)=p(i5,1);
                 xp(5)=p(i3,1); xp(6)=p(i6,1); xp(7)=p(i1,1);

  yp(1)=p(i1,2); yp(2)=p(i4,2); yp(3)=p(i2,2); yp(4)=p(i5,2);
                 yp(5)=p(i3,2); yp(6)=p(i6,2); yp(7)=p(i1,2);

  plot(xp, yp,'-o');

end

axis([-1.2 1.2 -1.2 1.2]);
xlabel('x'); ylabel('y');

%---
% reset
%---

for i=1:ng
  p(i,1) = p(i,1)-vx(i);
  p(i,2) = p(i,2)-vy(i);
end

%-------------------------------------------------
% average the element nodal stresses to compute
% the stresses at the global nodes
%-------------------------------------------------
```

Code psa6: $\longrightarrow$ Continuing $\longrightarrow$

```
%---
% initialize the global node stresses:
%---

for j=1:ng
  gsig_xx(j)=0.0; gsig_xy(j)=0.0; gsig_yy(j)=0.0;
  gitally(j)=0;
end

%---
% loop over the element nodes and compute
% the element nodal stresses
%---

for l=1:ne

  j=c(1,1); xe(1)=p(j,1); ye(1)=p(j,2); vxe(1)=u(j); vye(1)=v(j);
  j=c(1,2); xe(2)=p(j,1); ye(2)=p(j,2); vxe(2)=u(j); vye(2)=v(j);
  j=c(1,3); xe(3)=p(j,1); ye(3)=p(j,2); vxe(3)=u(j); vye(3)=v(j);
  j=c(1,4); xe(4)=p(j,1); ye(4)=p(j,2); vxe(4)=u(j); vye(4)=v(j);
  j=c(1,5); xe(5)=p(j,1); ye(5)=p(j,2); vxe(5)=u(j); vye(5)=v(j);
  j=c(1,6); xe(6)=p(j,1); ye(6)=p(j,2); vxe(6)=u(j); vye(6)=v(j);

  [sig_xx, sig_xy, sig_yy] = psa6_stress ...
   ...
      (xe,ye,vxe,vye,E,nu);

%---
% add element node to the global node and increase tally by one
%---

  for k=1:6
    j=c(1,k);
    gitally(j) = gitally(j) + 1;
    gsig_xx(j) = gsig_xx(j) + sig_xx(k);
    gsig_xy(j) = gsig_xy(j) + sig_xy(k);
    gsig_yy(j) = gsig_yy(j) + sig_yy(k);
  end

end

for j=1:ng
  gsig_xx(j)=gsig_xx(j)/gitally(j);
  gsig_xy(j)=gsig_xy(j)/gitally(j);
  gsig_yy(j)=gsig_yy(j)/gitally(j);
end

%-------------------------------
% extend the connectivity matrix
% for 3-node sub-triangles
%-------------------------------

Ic=0;
```

Code psa6: $\longrightarrow$ Continuing $\longrightarrow$

```
for i=1:ne

  Ic=Ic+1;
  c3(Ic,1) = c(i,1); c3(Ic,2) = c(i,4); c3(Ic,3) = c(i,6);
  Ic=Ic+1;
  c3(Ic,1) = c(i,4); c3(Ic,2) = c(i,2); c3(Ic,3) = c(i,5);
  Ic=Ic+1;
  c3(Ic,1) = c(i,5); c3(Ic,2) = c(i,3); c3(Ic,3) = c(i,6);
  Ic=Ic+1;
  c3(Ic,1) = c(i,4); c3(Ic,2) = c(i,5); c3(Ic,3) = c(i,6);

end

%----------------------------
% plot the stress field using
% the matlab function trimesh
%----------------------------

figure
%trimesh (c3,p(:,1),p(:,2),gsig_xx);
%trimesh (c3,p(:,1),p(:,2),gsig_yy);
trimesh (c3,p(:,1),p(:,2),gsig_xy);
hold on;

%----------------------------------
% plot the displacement vector field
%----------------------------------

quiver (p(:,1)',p(:,2)',vx,vy);
xlabel('x'); ylabel('y');
axis([-1.2 1.2 -1.2 1.2]);

%-----
% done
%-----
```

Code psa6: ($\longrightarrow$ Continued.) Code for plane stress analysis in a square possibly pierced by a circular hole.

$$\Delta F_{x_i} = \Delta l \int_0^1 \psi_1(\xi) f_x \, d\xi \simeq \Delta l \, \frac{1}{6} \, f_{x_i},$$

$$(5.2.18)$$

$$\Delta F_{y_i} = \Delta l \int_0^1 \psi_1(\xi) f_y \, d\xi \simeq \Delta l \, \frac{1}{6} \, f_{y_i},$$

where Δl is the length of the side. The contribution to the ith global node corresponding to the mid-side node 4 is

```
function [esm_xx, esm_xy, esm_yy, arel] = esm6 ...
 ...
      (x1,y1, x2,y2, x3,y3, x4,y4, x5,y5, x6,y6, NQ)

%=================================================
% Evaluation of the element stiffness matrices
% for plane stress analysis
%=================================================

%-----------------------------------
% compute the mapping coefficients
%-----------------------------------

[al, be, ga] = elm6_abc ...
 ...
   (x1,y1, x2,y2, x3,y3, x4,y4, x5,y5, x6,y6);
```

Function esm6: Continuing $\longrightarrow$

$$\Delta F_{x_i} = \Delta l \int_0^1 \psi_4(\xi) \, f_x \, \mathrm{d}\xi \simeq \Delta l \, \frac{2}{3} \, f_{x_i},$$

$$\Delta F_{y_i} = \Delta l \int_0^1 \psi_1(\xi) \, f_y \, \mathrm{d}\xi \simeq \Delta l \, \frac{2}{3} \, f_{y_i}.$$

(5.2.19)

Similar contributions are made by the element edges subtended by the element vertex nodes (1, 2) and (3, 1).

Once the finite solution has been concluded, post-processing is performed to extract various quantities of interest derived from the nodal displacements. In the *stress recovery* module, the element node stresses are evaluated directly, as implemented in the FSELIB function psa6_stress, listed in the text. In some finite element implementations with quadrilateral elements, the stresses are computed at Gaussian integration points and then extrapolated to the element interpolation nodes.

Because the solution for the nodal displacements is only C^0 continuous, the strains are discontinuous across the element edges, and so are the stresses. To construct continuous fields, we evaluate the stresses at the global nodes as straight averages of the corresponding element nodal values, as shown in code psa6. Alternatively, we may evaluate the global nodal stresses as weighted averages of the element nodal values, where the weights are adjusted according to the apertures of the corners subtended by the element edges. Finally, the stress field is visualized based on the values at the global nodes, as shown in code psa6.

```
%----------------------------
% read the triangle quadrature
%----------------------------

[xi, eta, w] = gauss_trgl(NQ);

%--------------------------------
% initialize the stiffness matrices
%--------------------------------

 for k=1:6
  for l=1:6
   esm_xx(k,l)=0.0; esm_xy(k,l)=0.0;
   esm_yy(k,l)=0.0;
  end
 end

%----------------------
% perform the quadrature
%----------------------

arel = 0.0;      % element area

for i=1:NQ

[psi, gpsi, hs] = elm6_interp ...
...
    (x1,y1, x2,y2, x3,y3, x4,y4, x5,y5, x6,y6 ...
    ,al,be,ga, xi(i),eta(i));

 cf = 0.5*hs*w(i);

 for k=1:6
  for l=1:6
   esm_xx(k,l) = esm_xx(k,l) + gpsi(k,1)*gpsi(l,1)*cf;
   esm_xy(k,l) = esm_xy(k,l) + gpsi(k,1)*gpsi(l,2)*cf;
   esm_yy(k,l) = esm_yy(k,l) + gpsi(k,2)*gpsi(l,2)*cf;
  end
 end

 arel = arel + cf;

end

% disp (arel)

%-----
% done
%-----

return;
```

Function esm6: ($\longrightarrow$ Continued.) Evaluation of the element stiffness matrices for plane stress analysis.

```
function [sig_xx, sig_xy, sig_yy] = psa6_stress ...
 ...
     (x,y,u,v,E,nu);

%=========================================
% Evaluation of the node element stresses
% in plane stress analysis
%=========================================

%----------------------------------------
% compute the mapping coefficients
%----------------------------------------

[al, be, ga] = elm6_abc ...
 ...
  (x(1),y(1), x(2),y(2), x(3),y(3), x(4),y(4), x(5),y(5), x(6),y(6) );
```

Function psa6_stress: Continuing $\longrightarrow$

Figures 5.2.2 and 5.2.3 show the graphics display of a computation with the following boundary conditions:

- Zero displacement boundary condition on the left side (fixed side).

- Zero traction boundary condition at the bottom and top sides (free sides).

- Zero traction boundary condition for the x component, and a quadratically varying boundary condition for the y component of the traction on the right side (loaded side), as implemented in the code.

- In the case of a square with a hole, the zero stress condition is applied around the inner contour.

The element shapes and nodal position after deformation are shown in frame (a) of each figure. Further graphics display includes a plot of the components of the stress tensor prepared using the MATLAB function `trimesh`, and an overlapping *quiver* (arrow) plot of the nodal displacements, **u**, as shown in frames (b- d) of each figure.

It is reassuring to observe that the computed boundary stress distribution is consistent with the specified boundary conditions. In particular, σ_{xx} and σ_{yy} are zero along the right edge of the plate, whereas σ_{xy} displays the specified parabolic dependence with respect to y, as implemented in the code.

```
%-----------------------------
% nodal (xi, eta) coordinates
%-----------------------------

xi(1)=0.0; eta(1)=0.0;
xi(2)=1.0; eta(2)=0.0;
xi(3)=0.0; eta(3)=1.0;
xi(4)=al;  eta(4)=0.0;
xi(5)=be ; eta(5)=1-be;
xi(6)=0.0; eta(6)=ga;

%--------------------------
% compute the node stresses
%--------------------------

nus = nu^2;

for l=1:6  % run over the nodes

  sig_xx(l)=0.0; sig_xy(l)=0.0; sig_yy(l)=0.0;

    [psi, gpsi, hs] = elm6_interp ...
    ...
     (x(1),y(1), x(2),y(2), x(3),y(3) ...
     ,x(4),y(4), x(5),y(5), x(6),y(6) ...
     ,al,be,ga, xi(l),eta(l));

  for k=1:6

    sig_xx(l) = sig_xx(l)...
        +(u(k)*gpsi(k,1)+nu*v(k)*gpsi(k,2) )/(1-nus);

    sig_xy(l) = sig_xy(l)...
        +0.5*(u(k)*gpsi(k,2)+v(k)*gpsi(k,1) )/(1+nu);

    sig_yy(l) = sig_yy(l)...
        +(nu*u(k)*gpsi(k,1)+v(k)*gpsi(k,2) )/(1-nus);
  end

  sig_xx(l)=E*sig_xx(l);
  sig_xy(l)=E*sig_xy(l);
  sig_yy(l)=E*sig_yy(l);

end

%-----
% done
%-----

return;
```

Function psa6_stress: ($\longrightarrow$ Continued.) Evaluation of the element node stresses for plane stress analysis.

(*a*)

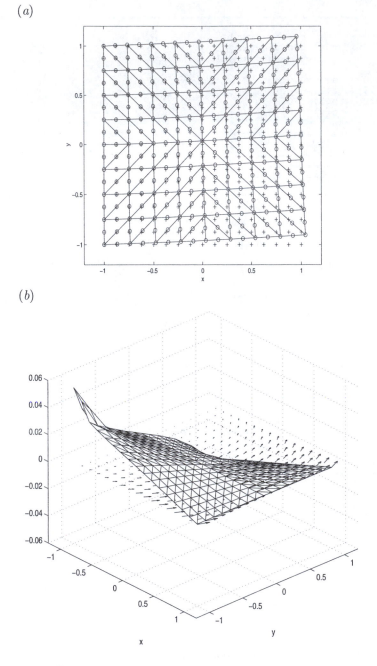

(*b*)

Figure 5.2.2 Continuing $\longrightarrow$

(c)

(d)

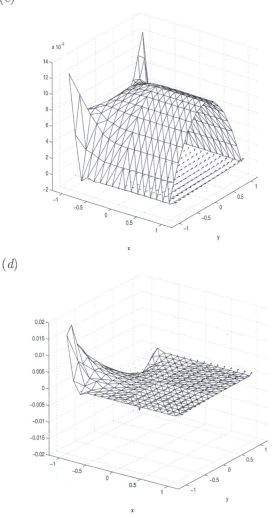

Figure 5.2.2 ($\longrightarrow$ Continued.) Plane-stress analysis for a square plate with the zero displacement boundary condition on the left edge (fixed edge), the zero traction boundary condition on the bottom and top edges (free edges), and a vertical traction boundary condition on the right side (loaded edge). (a) Element shapes and nodal positions before (crosses) and after (circles) deformation, and graphs of the (b) xx, (c) xy, and (d) yy, component of the stress tensor. The vectors in (b-d) illustrate the nodal displacements.

(a)

(b)

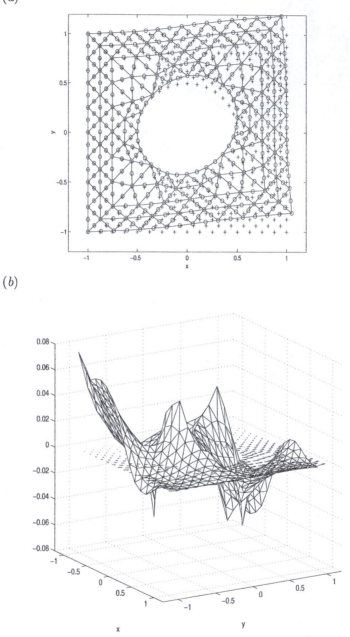

Figure 5.2.3 Continuing $\longrightarrow$

(c) (d)

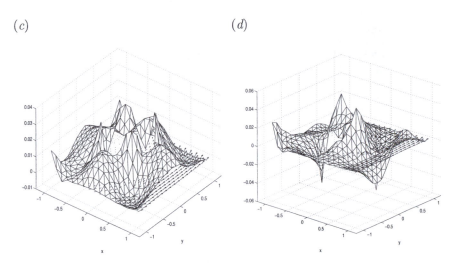

Figure 5.2.3 Same as Figure 5.2.2, except that the square plate is perforated with a circular hole where the zero-traction boundary condition is required.

PROBLEMS

5.2.1 *Code for plane stress analysis.*

(*a*) Run the code **psa6** for a rectangular plate with aspect ratio 1:4 and 4:1, and discuss the results of your computation.

(*b*) Repeat (*a*) for a rectangular plate with a circular hole.

(*c*) Modify the code **psa6** so that the boundary conditions specify the zero displacement boundary condition on the left and right sides (fixed sides), the zero traction boundary condition on the bottom side (free side), and a specified non-zero boundary condition for the x and y components of the traction on the upper side (loaded side). Run the modified code for a rectangular plate and an upper-side traction boundary condition of your choice, and discuss the results of your computation.

5.2.2 *Code for plane strain analysis.*

Modify the FSELIB code **psa6** into code **psta6** that performs the plane strain analysis. Run the modified code for a rectangular plate and boundary conditions of your choice, and discuss the results of your computation.

5.2.3 *Shear flow past a membrane.*

Consider shear flow past a membrane patch that is fixed on a plane wall, as discussed in Problem 5.1.4. FSELIB code **membrane** (not listed in the text) solves the pertinent plane stress problem.

(a) Run the code and confirm that the numerical results are consistent with the analytical solution for a circular membrane given in Problem 5.1.4(a).

(b) Run the code and confirm that the numerical results are consistent with the analytical solution for an elliptical membrane given in Problem 5.1.4(b).

(c) Modify the code to solve the corresponding problem for a square membrane, where the shear flow occurs along two parallel edges of the membrane, and discuss the results of your computation.

5.3   Plate bending

Consider a flat elastic plate with small thickness, h, that is parallel to the xy plane in the undeformed configuration, as shown in Figure 5.3.1. The plate now deforms under the influence of gravity or a mechanical load applied normally to the upper or lower surface, so that the position of the plate mid-surface is described by the transverse displacement function

$$z = f(x, y), \tag{5.3.1}$$

where $f = 0$ corresponds to the undeformed configuration. The complementary plane-stress problem where the plate deforms predominantly in its plane under the influence of an in-plane edge or body force that is parallel to the xy plane was discussed in Sections 5.1 and 5.2. The bending of a compressed plate under the influence of a combined transverse and in-plane load will be discussed later in this section.

Because of the predominantly transverse deformation, stresses develop over a cross-section of the plate, as shown in Figure 5.3.1. In the classical theory of plates, the cross-sectional stress is resolved into a transverse shear stress, σ_T, an in-plane normal stress, σ_N, and an in-plane shear stress, σ_S, as illustrated in Figure 5.3.1, defined as follows:

- The transverse shear stress points in the direction of the z axis.

- The in-plane normal stress is tangential to the plate and perpendicular to the cross-sectional contour, C.

- The in-plane shear stress is tangential to the plate and also tangential to the cross-sectional contour, C.

To develop the theory of plate bending, the transverse, normal, and shear stresses are integrated over the plate cross-section to yield corresponding tensions and bending moments, which are then used to formulate force and torque equilibrium equations and subsequently shape-governing equations.

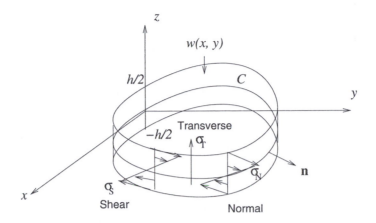

Figure 5.3.1 Illustration of an elastic plate deforming under the influence of a transverse load with surface density $w(x, y)$. The contour C corresponds to the plate mid-surface, and the unit vector, $\mathbf{n}$, is normal to C, pointing inward.

5.3.1 Tensions and bending moments

Integrating the transverse shear stress over a cross-section of the plate, we obtain the transverse shear tension,

$$q \equiv \int_{-h/2}^{h/2} \sigma_T \, dz, \tag{5.3.2}$$

where $h/2$ is the plate half-thickness. The transverse shear tension is the force per unit length exerted around the contour of the plate cross-section in the midplane, C, as shown in Figure 5.3.1. Integrating likewise the in-plane stresses over a cross-section of the plate, we derive expressions for the in-plane normal and shear tensions,

$$\tau_N \equiv \int_{-h/2}^{h/2} \sigma_N \, dz, \qquad \tau_S \equiv \int_{-h/2}^{h/2} \sigma_S \, dz. \tag{5.3.3}$$

Variations of the in-plane stresses over the cross-section of the plate are responsible for normal and tangential bending moments computed with respect to the undeformed mid-surface,

$$M_N \equiv \int_{-h/2}^{h/2} \sigma_N \, z \, dz, \qquad M_S \equiv \int_{-h/2}^{h/2} \sigma_S \, z \, dz. \tag{5.3.4}$$

We have mentioned that, in the classical theory of plates and shells, equilibrium equations are written with respect to tensions and bending moments

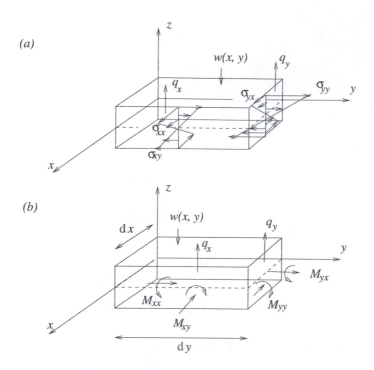

Figure 5.3.2 (*a*) Transverse shear tensions and in-plane stresses acting on a cross-section of the plate that is normal to the x and y axis. (*b*) Corresponding distribution of the bending moments.

instead of stresses. One important step in completing the theory and achieving closure is the derivation of constitutive equations relating tensions and bending moments to deformation. This challenging task must be done in a way that does not introduce mathematical incompatibilities and compromise fundamental physical laws.

5.3.2   Tension and bending moment tensors

To derive force and moment equilibrium equations, we consider two intersecting cross-sections of the plate that are normal to the x and y axes, as illustrated in Figure 5.3.2(*a*), and introduce:

- The transverse shear tensions q_x and q_y, acting on a cross-section that is normal to the x or y axis.

- The in-plane normal and shear stresses σ_{xx} and σ_{xy}, acting on a cross-section that is normal to the x axis, and their corresponding tensions τ_{xx} and τ_{xy}.

- The in-plane normal and shear stresses σ_{yy} and σ_{yx}, acting on a cross-section that is normal to the y axis, and their corresponding tensions τ_{yy} and τ_{yx}.

The transverse shear tensions can be collected into the tangential vector field

$$\mathbf{q} = (q_x, q_y), \tag{5.3.5}$$

which is defined such that the transverse shear tension exerted on a cross-section of the plate that is normal to the tangential unit vector, $\mathbf{n}$, as illustrated in Figure 5.3.1, is given by

$$q = \mathbf{n} \cdot \mathbf{q}. \tag{5.3.6}$$

For example, setting $\mathbf{n}$ parallel to the x axis, $\mathbf{n} = (1, 0)$, we find $q = q_x$.

The in-plane tensions can be collected into the tangential tension tensor field

$$\boldsymbol{\tau} = \begin{bmatrix} \tau_{xx} & \tau_{xy} \\ \tau_{yx} & \tau_{yy} \end{bmatrix}, \tag{5.3.7}$$

which is defined such that the in-plane tension exerted on a cross-section of the plate that is normal to the tangential unit vector, $\mathbf{n}$, is given by

$$\mathbf{f} = \mathbf{n} \cdot \boldsymbol{\tau}. \tag{5.3.8}$$

Figure 5.3.2(*b*) illustrates the tangential bending moments computed with respect to the undeformed mid-surface, defined as

$$M_{ij} \equiv \int_{-h/2}^{h/2} \sigma_{ij} \, z \, \mathrm{d}z, \tag{5.3.9}$$

for $i, j = x, y$. Recall that $z = 0$ marks the location of the undeformed mid-surface. The four scalar bending moments can be collected into the Cartesian bending moments tensor,

$$\mathbf{M} = \begin{bmatrix} M_{xx} & M_{xy} \\ M_{yx} & M_{yy} \end{bmatrix}. \tag{5.3.10}$$

For obvious reasons, the off-diagonal components of $\mathbf{M}$ are called the *twisting moments.*

The bending moments tensor, $\mathbf{M}$, has been defined such that the vector of bending moments acting on a cross-section of the plate that is normal to the tangential unit vector, $\mathbf{n}$, as illustrated in Figure 5.3.1, is given by

$$\mathbf{m} = \mathbf{e}_z \times (\mathbf{n} \cdot \mathbf{M}), \tag{5.3.11}$$

where $\mathbf{e}_z$ is the unit vector along the z axis, and $\times$ denotes the outer vector product. For example, setting $\mathbf{n}$ parallel to the x axis, $\mathbf{n} = (1, 0)$, we find

$$\mathbf{m} = \mathbf{e}_z \times \begin{bmatrix} M_{xx} \\ M_{xy} \end{bmatrix} = \begin{bmatrix} -M_{xy} \\ M_{xx} \end{bmatrix}, \tag{5.3.12}$$

which is consistent with the vector drawings on the front side of the rectangular section depicted in Figure 5.3.2(b). Similarly, setting $\mathbf{n}$ parallel to the y axis, $\mathbf{n} = (0, 1)$, we find

$$\mathbf{m} = \mathbf{e}_z \times \begin{bmatrix} M_{yx} \\ M_{yy} \end{bmatrix} = \begin{bmatrix} -M_{yy} \\ M_{yx} \end{bmatrix}, \tag{5.3.13}$$

which is consistent with the vector drawings on the right side of the rectangular section depicted in Figure 5.3.2(b).

The vector of bending moments can be further decomposed into (a) a normal twisting component, and (b) a tangential pure bending component denoted by $\mathbf{m}_b$,

$$\mathbf{m} = m_{twist}\, \mathbf{n} + \mathbf{m}_b, \tag{5.3.14}$$

where

$$m_{twist} \equiv \mathbf{m} \cdot \mathbf{n} = [\mathbf{e}_z \times (\mathbf{n} \cdot \mathbf{M})] \cdot \mathbf{n}, \tag{5.3.15}$$

and

$$\mathbf{m}_b = \mathbf{m} - m_{twist}\, \mathbf{n} = \mathbf{m} - (\mathbf{m} \cdot \mathbf{n})\, \mathbf{n} = (\mathbf{I} - \mathbf{nn}) \cdot \mathbf{m}. \tag{5.3.16}$$

The matrix $\mathbf{I} - \mathbf{nn}$ projects the in-plane vector $\mathbf{m}$ tangentially to the boundary contour C illustrated in Figure 5.3.1, where $\mathbf{I}$ is the identity matrix. In index notation, the ith component of the pure bending moment is given by

$$(\mathbf{m}_b)_i = (\delta_{ij} - n_i\, n_j)\, m_j, \tag{5.3.17}$$

where δ_{ij} is Kronecker's delta. For example, when $\mathbf{n} = (1, 0)$, we find

$$\mathbf{m}_b = \begin{bmatrix} 0 \\ M_{xx} \end{bmatrix}, \tag{5.3.18}$$

and when $\mathbf{n} = (0, 1)$, we find

$$\mathbf{m}_b = \begin{bmatrix} -M_{yy} \\ 0 \end{bmatrix}. \tag{5.3.19}$$

Both expressions are consistent with the vector drawings in Figure 5.3.2(b).

5.3.3 Equilibrium equations

To derive equilibrium equations, we perform a force balance in the z direction around the four edges of a rectangular section of the plate with infinitesimal dimensions dx and dy, as shown in Figure 5.3.2. Balancing the elastic forces exerted around the edges with the vertical load exerted on the rectangular surface area $dx\,dy$, we find

$$\Big[q_x(x+dx) - q_x(x)\Big]\,dy + \Big[q_y(y+dy) - q_y(y)\Big]\,dx - w\,dxdy = 0.$$

(5.3.20)

Dividing by $dx\,dy$, taking the limit as dx and dy tend to zero, and rearranging, we derive the differential equation

$$\frac{\partial q_x}{\partial x} + \frac{\partial q_y}{\partial y} = w.$$

(5.3.21)

In vector notation,

$$\nabla \cdot \mathbf{q} = w,$$

(5.3.22)

where $\mathbf{q} = (q_x, q_y)$ is the vector of transverse shear tensions.

Next, we perform a y moment balance about the center-point of the rectangular section, finding

$$\Big[M_{xx}(x+dx) - M_{xx}(x)\Big]\,dy$$
$$+ \Big[M_{yx}(y+dy) - M_{yx}(y)\Big]\,dx - (q_x\,dy)\,dx = 0.$$

(5.3.23)

Dividing by $dx\,dy$, taking the limit as dx and dy tend to zero, and rearranging, we derive the differential relation

$$q_x = \frac{\partial M_{xx}}{\partial x} + \frac{\partial M_{yx}}{\partial y}.$$

(5.3.24)

Performing a similar x moment balance, we derive the companion relation

$$q_y = \frac{\partial M_{xy}}{\partial x} + \frac{\partial M_{yy}}{\partial y}.$$

(5.3.25)

In vector notation, equations (5.3.24) and (5.3.25) assume the compact form

$$\mathbf{q} = \nabla \cdot \mathbf{M},$$

(5.3.26)

where $\mathbf{M}$ is the Cartesian tensor of bending moments introduced in (5.3.10).

Substituting (5.3.24) and (5.3.25) in (5.3.21), we obtain the desired equilibrium equation

$$\nabla \cdot (\nabla \cdot \mathbf{M}) = \frac{\partial^2 M_{xx}}{\partial x^2} + \frac{\partial^2 (M_{xy} + M_{yx})}{\partial x\,\partial y} + \frac{\partial^2 M_{yy}}{\partial y^2} = w.$$

(5.3.27)

Once this equation has been solved, the transverse shear tensions can be computed from (5.3.26).

(a) (b)

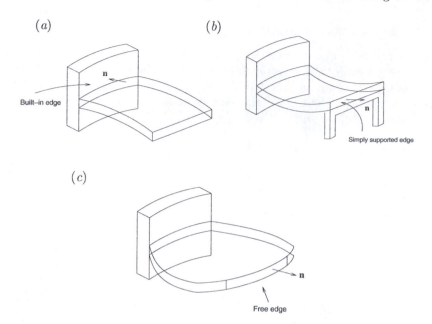

(c)

Figure 5.3.3 Illustration of a plate with (a) a clamped, built-in, or fixed edge,
   (b) a simply-supported edge, and (c) a free edge.

5.3.4   Boundary conditions

At the edge of the plate, C, we require different types of boundary conditions
according to the physical circumstances surrounding the problem under consideration. Common boundary conditions include the following:

- *Clamped, built-in, or fixed edge*: The deflection and its derivative normal
  to the edge are required to be zero,

$$f = 0, \qquad \mathbf{n} \cdot \nabla f \equiv \frac{\partial f}{\partial l_n} = 0, \qquad (5.3.28)$$

  where l_n is the arc length in the direction of the normal vector $\mathbf{n}$, as
  illustrated in Figure 5.3.3(a).

- *Simply supported edge*: The deflection and the pure bending moment
  around the edge are required to be zero,

$$f = 0, \qquad |\mathbf{m}_b| = 0, \qquad (5.3.29)$$

  as illustrated in Figure 5.3.3(b). Physically, the plate rests on a straight
  beam or curved bar.

- *Free edge*: The transverse shear tension and the pure bending moment
  around the edge are required to be zero,

$$q = 0, \qquad |\mathbf{m}_b| = 0, \qquad (5.3.30)$$

as illustrated in Figure 5.3.3(c). Using (5.3.5) and (5.3.26), we restate the first condition as

$$q \equiv \mathbf{n} \cdot \mathbf{q} = \mathbf{n} \cdot \nabla \cdot \mathbf{M}$$

$$= n_x \left(\frac{\partial M_{xx}}{\partial x} + \frac{\partial M_{yx}}{\partial y} \right) + n_y \left(\frac{\partial M_{xy}}{\partial x} + \frac{\partial M_{yy}}{\partial y} \right) = 0.$$

(5.3.31)

Physically, the edge may be identified with the edge of a balcony or with the far side of an aircraft wing.

It will be noted that, in all cases, two scalar boundary conditions are required along each edge, which is consistent with our impending discovery that plate deflection is governed by a fourth-order differential equation.

5.3.5 Constitutive and governing equations

To proceed further, we must introduce constitutive equations relating the bending moments to the deformation of the plate mid-surface, as described by the deflection function, $f(x, y)$. For this purpose, we introduce constitutive equations for the in-plane stresses. Multiplying these expressions by the distance from the plate mid-surface and integrating over the plate cross-section, we derive the requisite constitutive equations for the bending moments.

In the classical Kirchhoff theory of plates, material lines that are normal to the plate mid-surface before deformation are assumed to remain normal after deformation. Physically, this assumption is justified when the plate is structurally and geometrically symmetric with respect to the mid-surface, the thickness of the plate is either constant or varies slowly over the mid-surface, and the mid-surface remains virtually undeformed in the lateral plane after a transverse load has been applied.

Elementary differential geometry shows that the displacement of material points along the x or y axis, denoted by v_x and v_y, are given by

$$v_x = -z \frac{\partial f}{\partial x}, \qquad v_y = -z \frac{\partial f}{\partial y}.$$

(5.3.32)

Note that, at this level of approximation, material points at the mid-plane, located at $z = 0$, remain stationary at the undeformed position in the xy plane, as required. Using these equations, we find that the in-plane strain components are given by

$$\epsilon_{xx} \equiv \frac{\partial v_x}{\partial x} = -z \frac{\partial^2 f}{\partial x^2},$$

$$\epsilon_{xy} \equiv \frac{1}{2} \left(\frac{\partial v_x}{\partial y} + \frac{\partial v_y}{\partial x} \right) = -z \frac{\partial^2 f}{\partial x \partial y},$$

(5.3.33)

$$\epsilon_{yy} \equiv \frac{\partial v_y}{\partial y} = -z \frac{\partial^2 f}{\partial y^2}.$$

Substituting these expressions in the plane-stress constitutive equations given in (5.1.24), we find

$$
\begin{bmatrix} \sigma_{xx} \\ \sigma_{yy} \\ \sigma_{xy} \end{bmatrix} = -z \frac{E}{1-\nu^2} \begin{bmatrix} 1 & \nu & 0 \\ \nu & 1 & 0 \\ 0 & 0 & 1-\nu \end{bmatrix} \cdot \begin{bmatrix} \frac{\partial^2 f}{\partial x^2} \\ \frac{\partial^2 f}{\partial y^2} \\ \frac{\partial^2 f}{\partial x \partial y} \end{bmatrix}. \tag{5.3.34}
$$

Substituting further these expressions in (5.3.4) and carrying out the integration with respect to z, we find that the bending moments are given by

$$
M_{xx} = -E_B \left(\frac{\partial^2 f}{\partial x^2} + \nu \frac{\partial^2 f}{\partial y^2} \right),
$$

$$
M_{yy} = -E_B \left(\frac{\partial^2 f}{\partial y^2} + \nu \frac{\partial^2 f}{\partial x^2} \right), \tag{5.3.35}
$$

$$
M_{xy} = M_{yx} = -E_B \left(1 - \nu \right) \frac{\partial^2 f}{\partial x \, \partial y},
$$

where

$$
E_B \equiv \frac{Eh^3}{12 \left(1 - \nu^2 \right)} \tag{5.3.36}
$$

is the plate modulus of bending.

The trace of the bending moments tensor is given by

$$
M \equiv M_{xx} + M_{yy} = -E_B \left(1 + \nu \right) \nabla^2 f, \tag{5.3.37}
$$

where

$$
\nabla^2 f = \frac{\partial^2 f}{\partial x^2} + \frac{\partial^2 f}{\partial y^2} \tag{5.3.38}
$$

is the Laplacian of the transverse displacement. Rearranging (5.3.37), we obtain a Poisson equation,

$$
\nabla^2 f = -\frac{M}{E_B(1+\nu)}. \tag{5.3.39}
$$

To derive expressions for the transverse shear tensions in terms of the displacement, we substitute (5.3.35) in (5.3.24) and (5.3.25), finding

$$
q_x = -E_B \frac{\partial \nabla^2 f}{\partial x} = \frac{1}{1+\nu} \frac{\partial M}{\partial x},
$$

$$
\tag{5.3.40}
$$

$$
q_y = -E_B \frac{\partial \nabla^2 f}{\partial y} = \frac{1}{1+\nu} \frac{\partial M}{\partial y}.
$$

In vector notation, these equations combine into

$$\mathbf{q} = -E_B \, \nabla(\nabla^2 f) = \frac{1}{1+\nu} \, \nabla M. \qquad (5.3.41)$$

Finally, we substitute expressions (5.3.35) in (5.3.27), rearrange, and derive Lagrange's inhomogeneous biharmonic equation for the mid-surface deflection,

$$\nabla^4 f \equiv \nabla^2 \nabla^2 \, f = \frac{\partial^4 f}{\partial x^4} + 2 \, \frac{\partial^4 f}{\partial x^2 \, \partial y^2} + \frac{\partial^4 f}{\partial y^4} = -\frac{1}{E_B} \, w(x, y). \qquad (5.3.42)$$

This fourth-order partial differential equation is the counterpart of the fourth-order ordinary differential equation (1.7.12) governing the bending of a prismatic beam.

Because (5.3.42) is a fourth-order differential equation, the finite element solution must be C^1 continuous, which means that the transverse deflection, $f(x, y)$, and its gradient consisting of the first derivatives with respect to x and y, must be continuous at shared element nodes and along the element edges. To ensure C^1 continuity, we design special Hermitian bending elements, also called plate elements, as discussed in Section 5.4.

The requirement of C^1 continuity can be relaxed by converting the fourth-order equation into a system of two second-order equations. For example, substituting (5.3.39) in (5.3.42), we derive the Poisson equation

$$\nabla^2 M = (1 + \nu) \, w(x, y), \qquad (5.3.43)$$

which is to be solved together with the companion Poisson equation (5.3.39), subject to appropriate boundary conditions. Other ways of deriving Poisson-like equations will be discussed in Section 5.5.

Physical limitations

The limitations of the Kirchhoff theory of plates become evident by setting the displacement of material point particles normal to the plate equal to the deflection of the mid-surface,

$$v_z = f(x, y), \qquad (5.3.44)$$

whereupon the transverse strain components turn out to be zero,

$$\epsilon_{zz} \equiv \frac{\partial v_z}{\partial z} = 0,$$

$$\epsilon_{xz} \equiv \frac{1}{2} \left(\frac{\partial v_x}{\partial z} + \frac{\partial v_z}{\partial x} \right) = 0, \qquad (5.3.45)$$

$$\epsilon_{yz} \equiv \frac{1}{2} \left(\frac{\partial v_y}{\partial z} + \frac{\partial v_z}{\partial y} \right) = 0.$$

The vanishing of ϵ_{zz} implies that the plate is in a state of *plane strain* instead of the assumed state of *plane stress*, which cannot be true if the plate locally supports a transverse load. Moreover, the vanishing of the transverse shear strains

ϵ_{xz} and ϵ_{yz} implies the absence of corresponding transverse shear stresses σ_{xz} and σ_{yz}, at least for some types of materials, which contradicts the theoretical premise discussed earlier in this section. Notwithstanding these limitations, the Kirchhoff theory is an acceptable idealization when the in-plane stresses are larger than their transverse shear counterparts.

Improvements of the Kirchhoff model are built into the geometrically non-linear von Kármán model which reduces to the Kirchhoff model for small deflections, in the Reissner-Mindlin model which takes into consideration the transverse shear tensions, and in further models based on the theory of three-dimensional elasticity.

5.3.6   Buckling and wrinkling of a stressed plate

In the preceding analysis, we have assumed that the plate is not subjected to in-plane stresses other than those developing due to the deformation, that is, the plate is laterally unstressed. An applied in-plane edge or body force generates an independent in-plane stress field, presently denoted by Σ, as discussed in Section 5.1.

After the plate has buckled, the in-plane stresses point in tangential directions, thereby producing a net force component normal to the undeformed plate with a *downward* force density

$$
w(x,y) = -h \left[\frac{\partial}{\partial x} \left(\Sigma_{xx} \frac{\partial f}{\partial x} \right) + \frac{\partial}{\partial y} \left(\Sigma_{xy} \frac{\partial f}{\partial x} \right) \right.
$$
$$
\left. + \frac{\partial}{\partial x} \left(\Sigma_{xy} \frac{\partial f}{\partial y} \right) + \frac{\partial}{\partial x} \left(\Sigma_{yy} \frac{\partial f}{\partial y} \right) \right], \qquad (5.3.46)
$$

where h is the plate thickness (e.g., [64], p. 305). Expanding the derivatives on the right-hand side and using the equilibrium equations for Σ stated in (5.1.26), we find

$$
w(x,y) = -h \left(\Sigma_{xx} \frac{\partial^2 f}{\partial x^2} + 2 \Sigma_{xy} \frac{\partial^2 f}{\partial x \partial y} + \Sigma_{yy} \frac{\partial^2 f}{\partial y^2} - b_x \frac{\partial f}{\partial x} - b_y \frac{\partial f}{\partial y} \right),
$$
$$
(5.3.47)
$$

where $\mathbf{b}$ is the in-plane body force. In the absence of a body force, the last two terms on the right-hand side of (5.3.47) do not appear.

To derive (5.3.46), we project the tangential forces exerted around the edges of a rectangular section onto the z axis, as shown in Figure 5.3.4. The downward contribution of Σ_{xx} is

$$
-\left(\Sigma_{xx} \frac{\partial f}{\partial x} h \, \Delta y \right)_{x+\Delta x} + \left(\Sigma_{xx} \frac{\partial f}{\partial x} h \, \Delta y \right)_x
$$
$$
\simeq -\Delta x \Delta y \, h \, \frac{\left(\Sigma_{xx} \frac{\partial f}{\partial x} \right)_{x+\Delta x} - \left(\Sigma_{xx} \frac{\partial f}{\partial x} \right)_x}{\Delta x}. \qquad (5.3.48)
$$

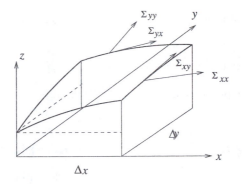

Figure 5.3.4 Derivation of equilibrium equations governing the vertical deflection of a compressed plate.

Taking the limit as Δx and Δy tend to zero, we derive the first term on the right-hand side of (5.3.46). The remaining terms are derived working in a similar fashion (Problem 5.3.2).

Substituting expression (5.3.47) in the equilibrium equation (5.3.42), we obtain the Saint Venant or linearized von Kármán equation

$$\nabla^4 f \equiv \nabla^2 \nabla^2 f = \frac{\partial^4 f}{\partial x^4} + 2 \frac{\partial^4 f}{\partial x^2 \, \partial y^2} + \frac{\partial^4 f}{\partial y^4} \tag{5.3.49}$$

$$= \frac{h}{E_B} \left(\Sigma_{xx} \frac{\partial^2 f}{\partial x^2} + 2 \, \Sigma_{xy} \frac{\partial^2 f}{\partial x \partial y} + \Sigma_{yy} \frac{\partial^2 f}{\partial y^2} - b_x \frac{\partial f}{\partial x} - b_y \frac{\partial f}{\partial y} \right).$$

In the absence of a body force, $b_x = 0$ and $b_y = 0$, this is the counterpart of the beam equation (1.8.4). The solution procedure involves determining the in-plane stresses Σ while neglecting the transverse deformation, and then solving (5.3.49) for the deflection, f. As an example, in the case of shear flow over a circular membrane discussed in Problem 5.1.3, equation (5.3.49) takes the specific form

$$\nabla^4 f = -\frac{2\tau}{E_B(3-\nu)} \left[x \frac{\partial^2 f}{\partial x^2} + (1-\nu) \, y \frac{\partial^2 f}{\partial x \partial y} + \nu \, x \frac{\partial^2 f}{\partial y^2} + \frac{3-\nu}{2} \frac{\partial f}{\partial x} \right].$$

$$\tag{5.3.50}$$

A trivial solution of (5.3.49) is $f = 0$. Nontrivial solutions are possible when the in-plane stresses reach a sequence of thresholds, physically representing critical in-plane loads for different modes of plate buckling. These critical loads, and the corresponding modes of deformation, are found by solving algebraic eigenvalue problems arising from the finite element formulation.

When the body force is conservative, we may use express the in-plane stresses, Σ, in terms of the Airy stress function ϕ defined in (5.1.36), to obtain an alternative statement of the linearized von Kármán equation,

$$\nabla^4 f = \frac{h}{E_B}\left[\frac{\partial^2 f}{\partial x^2}\frac{\partial^2 \phi}{\partial y^2} + \frac{\partial^2 f}{\partial y^2}\frac{\partial^2 \phi}{\partial x^2} - 2\frac{\partial^2 f}{\partial x\partial y}\frac{\partial^2 \phi}{\partial x\partial y} - \nabla\cdot(\nu\,\nabla f)\right], \quad (5.3.51)$$

which is a common point of departure in developing finite element solutions (e.g., [16]).

5.3.7 Buckling and wrinkling of a circular plate

In the case of a circular plate, it is convenient to work in plane polar coordinates, (r,θ), which are related to the Cartesian coordinates by

$$x = r\,\cos\theta, \qquad y = r\,\sin\theta, \qquad (5.3.52)$$

where r is the distance from the center of the plate, and θ is the associated polar angle measured in the counterclockwise direction. Using standard coordinate transformation rules, we find the following relations between the first partial derivatives of a function, f, with respect to Cartesian and plane polar coordinates,

$$\frac{\partial f}{\partial x} = \cos\theta\,\frac{\partial f}{\partial r} - \frac{\sin\theta}{r}\frac{\partial f}{\partial \theta},$$

$$(5.3.53)$$

$$\frac{\partial f}{\partial y} = \sin\theta\,\frac{\partial f}{\partial r} + \frac{\cos\theta}{r}\frac{\partial f}{\partial \theta}.$$

For the second partial derivatives, we find

$$\frac{\partial^2 f}{\partial x^2} = \cos^2\theta\,\frac{\partial^2 f}{\partial r^2} - \frac{\sin 2\theta}{r}\frac{\partial^2 f}{\partial r\partial\theta} + \frac{\sin^2\theta}{r}\frac{\partial f}{\partial r} + \frac{\sin 2\theta}{r^2}\frac{\partial f}{\partial\theta} + \frac{\sin^2\theta}{r^2}\frac{\partial^2 f}{\partial\theta^2},$$

$$\frac{\partial^2 f}{\partial y^2} = \sin^2\theta\,\frac{\partial^2 f}{\partial r^2} + \frac{\sin 2\theta}{r}\frac{\partial^2 f}{\partial r\partial\theta} + \frac{\cos^2\theta}{r}\frac{\partial f}{\partial r} - \frac{\sin 2\theta}{r^2}\frac{\partial f}{\partial\theta} + \frac{\cos^2\theta}{r^2}\frac{\partial^2 f}{\partial\theta^2},$$

$$\frac{\partial^2 f}{\partial x\partial y} = \frac{1}{2}\left(\sin 2\theta\,\frac{\partial^2 f}{\partial r^2} + 2\frac{\cos 2\theta}{r}\frac{\partial^2 f}{\partial r\partial\theta} - \frac{\sin 2\theta}{r}\frac{\partial f}{\partial r} - 2\frac{\cos 2\theta}{r^2}\frac{\partial f}{\partial\theta}\right.$$

$$\left. - \frac{\sin 2\theta}{r^2}\frac{\partial^2 f}{\partial\theta^2}\right). \qquad (5.3.54)$$

Using these relations, we find that the Laplacian of a function f is given by

$$\nabla^2 f = \frac{1}{r}\frac{\partial}{\partial r}\left(r\,\frac{\partial f}{\partial r}\right) + \frac{1}{r^2}\frac{\partial^2 f}{\partial\theta^2}. \qquad (5.3.55)$$

The biharmonic operator acting on f yields

$$\nabla^4 f = \frac{\partial^4 f}{\partial r^4} + \frac{2}{r}\frac{\partial^3 f}{\partial r^3} - \frac{1}{r^2}\frac{\partial^2 f}{\partial r^2} + \frac{1}{r^3}\frac{\partial f}{\partial r} + \frac{2}{r^2}\frac{\partial^4 f}{\partial r^2 \partial \theta^2}$$

$$-\frac{2}{r^3}\frac{\partial^3 f}{\partial r \partial \theta^2} + \frac{4}{r^4}\frac{\partial^2 f}{\partial \theta^2} + \frac{1}{r^4}\frac{\partial^4 f}{\partial \theta^4}. \qquad (5.3.56)$$

The general solution of the plate bending equation for the transverse displacement can be expressed as a Fourier series with respect to θ,

$$f(r,\theta) = \frac{1}{2}p_0(r) + \sum_{n=1}^{\infty}\left[\, p_n(r)\,\cos n\theta + q_n(r)\,\sin n\theta \,\right], \qquad (5.3.57)$$

where $p_n(r)$ and $q_n(r)$ are requisite functions (e.g., [35]). The equivalent complex form is

$$f(r,\theta) = \sum_{n=-\infty}^{\infty} \mathcal{F}_n(r)\,\exp(-in\theta), \qquad (5.3.58)$$

where i is the imaginary unit. We have adopted Euler's formula for the complex exponential,

$$\exp(-in\theta) = \cos(n\theta) - i\,\sin(n\theta), \qquad (5.3.59)$$

and we have introduced the complex function, $\mathcal{F}_n(r)$, defined such that

$$\mathcal{F}_n(r) \equiv \frac{1}{2}\left[\, p_n(r) + i\,q_n(r)\,\right] \qquad (5.3.60)$$

for $n \geq 0$, and

$$\mathcal{F}_n(r) \equiv \frac{1}{2}\left[\, p_n(r) - i\,q_n(r)\,\right] \qquad (5.3.61)$$

for $n < 0$. Note that $\mathcal{F}_n(r) = \mathcal{F}^*_{-n}(r)$, which guarantees that the right-hand side of (5.3.58) is real, where an asterisk denotes the complex conjugate.

Substituting the Fourier series in the plate bending governing equation, $\nabla^4 f = -w/E_B$, we obtain

$$\nabla^4 f = \sum_{n=-\infty}^{\infty}\left(\mathcal{F}_n'''' + \frac{2}{r}\mathcal{F}_n''' - \frac{1+2n^2}{r^2}\mathcal{F}_n'' + \frac{1+2n^2}{r^3}\mathcal{F}_n'\right.$$

$$\left. + n^2\frac{n^2-4}{r^4}\mathcal{F}_n\right)\exp(-in\theta) = -\frac{w}{E_B}. \qquad (5.3.62)$$

The general solution is composed of a particular solution and a homogeneous solution corresponding to $w = 0$. The Fourier coefficients of the homogeneous solution are found by setting the expression enclosed by the large parentheses

in (5.3.62) equal to zero, thereby obtaining a system of equidimensional linear ordinary differential equations. The solution can be found by elementary
methods as

$$\mathcal{F}_0 = A_0 \, r^2 + B_0 \, r^2 \, \ln r + C_0 \, \ln r + D_0,$$

$$\mathcal{F}_1 = A_1 \, r^3 + B_1 \, r \, \ln r + C_1 \, r + D_1 \, \frac{1}{r}, \qquad\qquad (5.3.63)$$

$$\mathcal{F}_n = A_n \, r^{n+2} + B_n \, r^n + C_n \, r^{-n+2} + D_n \, r^{-n}, \quad \text{for} \quad n \geq 2,$$

where the constants, A_i, B_i, C_i, D_i, are determined by the boundary conditions,
as discussed in Problem 5.3.3.

PROBLEMS

5.3.1 *Buckling of a rectangular plate.*

Consider a rectangular plate with two sides parallel to the x and y axes. The
plate is simply supported along the two edges that are parallel to the x axis
and along one edge that is parallel to the y axis, and has a free fourth edge, as
shown in the following illustration. Write down the boundary conditions to be
applied along each side.

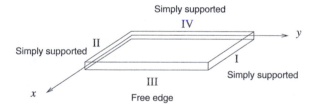

5.3.2 *Buckling of a plate under compression.*

Derive the second and third terms on the right-hand side of expression (5.3.46).

5.3.3 *Bending of a clamped circular plate subject to a uniform load.*

Derive an expression for the deflection of a clamped circular plate under the
action of a spatially uniform load, corresponding to a constant function, w.

5.4   Hermite triangles

Pursuant to our discussion of the plate bending problem formulated in the
preceding section, we require C^1 continuity of the finite element solution and
use polynomial expansions that are defined not only with respect to the nodal
values, but also with respect to their first and possibly higher-order derivatives
with respect to x and y.

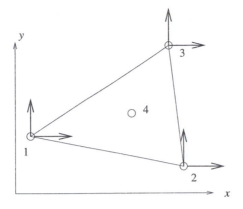

Figure 5.4.1 A Hermite triangle with three vertex nodes where the displacement and its gradient are defined, and one interior node where the displacement only is defined.

5.4.1 4-node, 10-dof triangle

A complete cubic expansion in x and y over an element takes the form

$$
\begin{aligned}
f(x,y) = \quad & a_{00} \\
& + a_{10}\, x + a_{01}\, y \\
& + a_{20}\, x^2 + a_{11}\, xy + a_{02}\, y^2 \\
& + a_{30}\, x^3 + a_{21}\, x^2 y + a_{12}\, xy^2 + a_{03}\, y^3,
\end{aligned}
\tag{5.4.1}
$$

involving ten coefficients, a_{ij}. To ensure an equal number of unknowns, we introduce a 4-node triangle defined by three vertex nodes and one interior node, as illustrated in Figure 5.4.1.

At the vertex nodes, we define the values of the function and its gradient composed of the x and y derivatives, denoted by

$$
f_i, \qquad f_i^x \equiv \left(\frac{\partial f}{\partial x}\right)_i, \qquad f_i^y \equiv \left(\frac{\partial f}{\partial y}\right)_i, \tag{5.4.2}
$$

for $i = 1, 2, 3$. At the interior node, labeled 4, we only define the value of the function, denoted by f_4. These unknown values are collected into the ten-dimensional vector

$$
\mathbf{h} \equiv
\begin{bmatrix}
f_1 \\
f_1^x \\
f_1^y \\
- \\
f_2 \\
f_2^x \\
f_2^y \\
- \\
f_3 \\
f_3^x \\
f_3^y \\
- \\
f_4
\end{bmatrix},
\tag{5.4.3}
$$

encapsulating ten *degrees of freedom*. The requirement of C^1 continuity demands that the function values and its derivatives at the vertex nodes are shared by neighboring elements at corresponding positions.

To compute the coefficients of the cubic expansion, a_{ij}, we require the interpolation conditions

$$
f(x_i, y_i) = f_i,
\tag{5.4.4}
$$

for $i = 1, 2, 3, 4$, and obtain four equations,

$$
\begin{aligned}
a_{00} + a_{10}\, x_i + a_{01}\, y_i + a_{20}\, x_i^2 + a_{11}\, x_i\, y_i + a_{02}\, y_i^2 \\
+ a_{30}\, x_i^3 + a_{21}\, x_i^2\, y_i + a_{12}\, x_i\, y_i^2 + a_{03}\, y_i^3 = f_i,
\end{aligned}
\tag{5.4.5}
$$

which can be expressed in the form of the vector product

$$
\begin{bmatrix} 1 & x_i & y_i & x_i^2 & x_i\, y_i & y_i^2 & x_i^3 & x_i^2\, y_i & x_i\, y_i^2 & y_i^3 \end{bmatrix} \cdot \mathbf{a} = f_i,
\tag{5.4.6}
$$

where

$$
\mathbf{a} \equiv
\begin{bmatrix}
a_{00} \\
a_{10} \\
a_{01} \\
a_{20} \\
a_{11} \\
a_{02} \\
a_{30} \\
a_{21} \\
a_{12} \\
a_{03}
\end{bmatrix}
\tag{5.4.7}
$$

is the expansion coefficient vector. Requiring further the interpolation conditions

$$
\left(\frac{\partial f}{\partial x} \right)_{(x_i, y_i)} = f_i^x,
\qquad
\left(\frac{\partial f}{\partial y} \right)_{(x_i, y_i)} = f_i^y,
\tag{5.4.8}
$$

at the vertex nodes, $i = 1, 2, 3$, we obtain six additional equations,

$$\begin{bmatrix} 0 & 1 & 0 & 2\,x_i & y_i & 0 & 3\,x_i^2 & 2\,x_i y_i & y_i^2 & 0 \end{bmatrix} \cdot \mathbf{a} = f_i^x,$$

$$\begin{bmatrix} 0 & 0 & 1 & 0 & x_i & 2\,y_i & 0 & x_i^2 & 2\,x_i y_i & 3\,y_i^2 \end{bmatrix} \cdot \mathbf{a} = f_i^y.$$

$$(5.4.9)$$

Compiling (5.4.6) and (5.4.9), we formulate a system of ten linear equations for the coefficient vector, $\mathbf{a}$,

$$\mathbf{D} \cdot \mathbf{a} = \mathbf{h}, \qquad (5.4.10)$$

where $\mathbf{D}$ is the grand coefficient matrix,

$$\mathbf{D} = \begin{bmatrix}
1 & x_1 & y_1 & x_1^2 & x_1\,y_1 & y_1^2 & x_1^3 & x_1^2\,y_1 & x_1\,y_1^2 & y_1^3 \\
0 & 1 & 0 & 2\,x_1 & y_1 & 0 & 3\,x_1^2 & 2\,x_1\,y_1 & y_1^2 & 0 \\
0 & 0 & 1 & 0 & x_1 & 2\,y_1 & 0 & x_1^2 & 2\,x_1\,y_1 & 3\,y_1^2 \\
1 & x_2 & y_2 & x_2^2 & x_2\,y_2 & y_2^2 & x_2^3 & x_2^2\,y_2 & x_2\,y_2^2 & y_2^3 \\
0 & 1 & 0 & 2\,x_2 & y_2 & 0 & 3\,x_2^2 & 2\,x_2\,y_2 & y_2^2 & 0 \\
0 & 0 & 1 & 0 & x_2 & 2\,y_2 & 0 & x_1^2 & 2\,x_1\,y_1 & 3\,y_1^2 \\
1 & x_3 & y_3 & x_3^2 & x_3\,y_3 & y_3^2 & x_3^3 & x_3^2\,y_3 & x_3\,y_3^2 & y_3^3 \\
0 & 1 & 0 & 2\,x_3 & y_3 & 0 & 3\,x_3^2 & 2\,x_3\,y_3 & y_3^2 & 0 \\
0 & 0 & 1 & 0 & x_3 & 2\,y_3 & 0 & x_3^2 & 2\,x_3\,y_3 & 3\,y_3^2 \\
1 & x_4 & y_4 & x_4^2 & x_4\,y_2 & y_4^2 & x_4^3 & x_4^2\,y_4 & x_4\,y_4^2 & y_4^3
\end{bmatrix}. \qquad (5.4.11)$$

Solving for $\mathbf{a}$ in terms of the inverse matrix, $\mathbf{D}^{-1}$, we find

$$\mathbf{a} = \mathbf{D}^{-1} \cdot \mathbf{h}, \qquad (5.4.12)$$

The grand coefficient matrix $\mathbf{D}$ and its inverse are completely defined by position of the four nodes. FSELIB function `psi_10_Dinv`, listed in the text, evaluates $\mathbf{D}$ and computes its inverse using the MATLAB function `inv`. Numerical methods for computing the inverse of an arbitrary matrix based on Gauss elimination and other algorithms are discussed in textbooks on scientific computing (e.g., [43], see also Appendix C).

The complete cubic expansion (5.4.1) can be expressed as the inner product of two vectors,

$$f(x, y) = \boldsymbol{\varphi} \cdot \mathbf{a}, \qquad (5.4.13)$$

where

$$\boldsymbol{\varphi} \equiv \begin{bmatrix} 1 & x & y & x^2 & xy & y^2 & x^3 & x^2 y & x y^2 & y^3 \end{bmatrix} \qquad (5.4.14)$$

is the horizontal vector of monomial products. Substituting the coefficient vector $\mathbf{a}$ from (5.4.12), we obtain

$$f(x, y) = \boldsymbol{\varphi} \cdot \mathbf{D}^{-1} \cdot \mathbf{h} \equiv \boldsymbol{\psi} \cdot \mathbf{h}. \qquad (5.4.15)$$

```
function Dinv = psi_10_Dinv (x1,y1, x2,y2, x3,y3, x4,y4)

%==============================================================
% Evaluation of the inverse of the grand coefficient matrix
% for a 4-node,10-dof Hermite triangle
%==============================================================

%------------------------------------
% define the coefficient matrix
%------------------------------------

D = [ ...
1   x1  y1  x1^2  x1*y1  y1^2  x1^3  x1^2*y1  x1*y1^2  y1^3   ; ...
0   1   0   2*x1  y1     0     3*x1^2  2*x1*y1  y1^2     0      ; ...
0   0   1   0     x1     2*y1  0     x1^2      2*x1*y1  3*y1^2 ; ...
1   x2  y2  x2^2  x2*y2  y2^2  x2^3  x2^2*y2  x2*y2^2  y2^3   ; ...
0   1   0   2*x2  y2     0     3*x2^2  2*x2*y2  y2^2     0      ; ...
0   0   1   0     x2     2*y2  0     x2^2      2*x2*y2  3*y2^2 ; ...
1   x3  y3  x3^2  x3*y3  y3^2  x3^3  x3^2*y3  x3*y3^2  y3^3   ; ...
0   1   0   2*x3  y3     0     3*x3^2  2*x3*y3  y3^2     0      ; ...
0   0   1   0     x3     2*y3  0     x3^2      2*x3*y3  3*y3^2 ; ...
1   x4  y4  x4^2  x4*y4  y4^2  x4^3  x4^2*y4  x4*y4^2  y4^3   ; ...
];

%--------------------
% compute the inverse
%--------------------

Dinv = inv(D);

%-----
% done
%-----

return
```

Function psi_10_Dinv: Computation of the inverse of the grand coefficient matrix, **D**, for a 4-node, 10-dof Hermite triangle.

The ten-dimensional vector function

$$\boldsymbol{\psi}(x,y) \equiv \boldsymbol{\varphi} \cdot \mathbf{D}^{-1}, \tag{5.4.16}$$

encapsulates the Hermite interpolation functions, also called expansion modes, associated with the degrees of freedom (dof) composing the vector **h**. Specifically, the ith interpolation function is given by

$$\psi_i(x,y) = \sum_{j=1}^{10} \varphi_j(x,y) \, D_{ji}^{-1}, \tag{5.4.17}$$

for $i = 1, 2, \ldots, 10$. Alternatively, the individual functions, ψ_i, arise by solving the linear system

$$\mathbf{D} \cdot \mathbf{a} = \mathbf{e}_i,$$ (5.4.18)

where the right-hand side is equal to the unit vector

$$\mathbf{e}_i = \begin{bmatrix} 0 & \cdots & 0 & 1 & 0 & \cdots & 0 \end{bmatrix}^T,$$ (5.4.19)

and the unity resides in the ith entry. Although the triangle can be mapped from the physical xy plane to the standard triangle in the $\xi\eta$ plane, and the independent variables x and y can be expressed in terms of ξ and η using relations (3.2.10), this is both cumbersome and unnecessary.

FSELIB code psi_10, listed in the text, generates graphs of the interpolation functions contained in the vector $\boldsymbol{\psi}$. Mesh lines are produced in the parametric $\xi\eta$ plane and then transferred to the physical xy plane using the transformation rules (3.2.10). Figure 5.4.2 shows graphs of four selected interpolation functions for an arrangement where the first vertex is located at $x_1 = 0$, $y_1 = 0$, the second vertex is located at $x_2 = 1$, $y_2 = 0$, the third at vertex is located at $x_3 = 0$, $y_3 = 1$, and the fourth vertex is located at the element centroid. Frames (a-c) show the interpolation functions ψ_1, ψ_2, and ψ_3, associated with the first node, corresponding to f_1, f_1^x, and f_1^y. Frame (d) shows the interpolation function associated with the interior node, ψ_{10}, corresponding to f_4.

The function ψ_1 shown in Figure 5.4.2(a) is zero along the triangle edge 2-3. This is because ψ_1 is a cubic polynomial with respect to arc length along this edge, with a double root at each vertex; accordingly, the cubic must be identically equal to zero. The distribution of ψ_1 along the 1-2 edge depends only on the position of the first and second nodes, and is independent of the position of the third and fourth nodes. To explain this, we introduce the arc length along the 1-2 edge measured from point 1, denoted by s, and require $\psi_1(0) = 1$, $(d\psi_1/ds)_0 = 0$, $\psi_1(l) = 0$, and $(d\psi_1/ds)_l = 0$, where l is the length of the 1-2 edge. These conditions completely determine the cubic function $\psi_1(s)$, as $\psi_1(s) = (2\hat{s} + 1)(1 - \hat{s})^2$, where $\hat{s} \equiv s/l$. Similar arguments can be made to show that the distribution of ψ_1 along the 1-3 edge depends only on the position of the first and third nodes, and is independent of the position of the second and fourth nodes.

The function ψ_2 shown in Figure 5.4.2(b) is zero along the 2-3 and 1-3 edges. This is because ψ_2 is a cubic polynomial with respect to arc length along each edge, with a double root at each vertex; accordingly, the cubics must be identically equal to zero. The distribution of ψ_2 along the 1-2 edge depends only on the position of the first and second nodes, and is independent of the position of the third and fourth nodes. To explain this, we introduce the arc length along the 1-2 edge measured from point 1, denoted by s, and require $\psi_2(0) = 0$, $(d\psi_1/ds)_0 = 1$, $\psi_2(l) = 0$, and $(d\psi_2/ds)_l = 0$, where l is the length of the 1-2 edge. These conditions completely determine the cubic function $\psi_1(s)$, as $\psi_2(s) = \hat{s}(1 - \hat{s})^2$, where $\hat{s} \equiv s/l$. A similar behavior is observed for the function ψ_3 shown in Figure 5.4.2(c).

```
%===========================================================
% psi_10
%
% prepare a graph of the element interpolation functions
% for a 4-node, 10-dof Hermite triangle
%===========================================================

%-----------------------------------
% define the vertices arbitrarily
%-----------------------------------

x1= 0.0; y1= 0.0;
x2= 1.0; y2= 0.0;
x3= 0.0; y3= 1.0;

%-----------------------------------------------------
% put the interior node at the centroid (arbitrary)
%-----------------------------------------------------

x4=(x1+x2+x3)/3.0;
y4=(y1+y2+y3)/3.0;

%-----------------------------------
% inquire on the mode to be graphed
%-----------------------------------

mode = input('Please enter the mode to plot: ')

%-----------------------------
% compute the inverse matrix
%-----------------------------

Dinv = psi_10_Dinv (x1,y1, x2,y2, x3,y3, x4,y4);

%-----------------------------------------------
% define the plotting grid in the xi-eta plane
%-----------------------------------------------

N=32; M=32;      % arbitrary

Dxi = 1.0/N; Deta = 1.0/M;

for i=1:N+1
   xi(i)=Dxi*(i-1.0);
end

for j=1:M+1
   eta(j)=Deta*(j-1.0);
end

%-----------------------------------------
% draw horizonal and vertical mesh lines
%-----------------------------------------
```

Code psi_10: Continuing $\longrightarrow$

```
%----------------------
% horizontal mesh lines
%----------------------

for i=1:N
 for j=1:M+2-i
  x = x1 + (x2-x1)*xi(i) + (x3-x1)*eta(j);
  y = y1 + (y2-y1)*xi(i) + (y3-y1)*eta(j);
  monvec = [1  x y  x^2 x*y y^2  x^3 x^2*y x*y^2 y^3];
  psi = monvec*Dinv;
  xp(j) = x; yp(j)=y; zp(j) = psi(mode);
 end

 plot3(xp,yp,zp); hold on;

end

%--------------------
% vertical mesh lines
%--------------------

for j=1:M
 for i=1:N+2-j
  x = x1 + (x2-x1)*xi(i) + (x3-x1)*eta(j);
  y = y1 + (y2-y1)*xi(i) + (y3-y1)*eta(j);
  monvec = [1  x y  x^2 x*y y^2  x^3 x^2*y x*y^2 y^3];
  psi = monvec*Dinv;
  xp(i) = x; yp(i)=y; zp(i) = psi(mode);
 end
 plot3(xp,yp,zp); hold on;
end

%----------------------------------
% plot the perimeter of the triangle
%----------------------------------

xx(1)=x1; yy(1)=y1; zz(1)=0.0;
xx(2)=x2; yy(2)=y2; zz(2)=0.0;
xx(3)=x3; yy(3)=y3; zz(3)=0.0;
xx(4)=x1; yy(4)=y1; zz(4)=0.0;

plot3(xx,yy,zz); hold on;

%----------------------
% plot the interior node
%----------------------

plot3(x4,y4,0.0,'o'); hold on; xlabel('x'); ylabel('y')

%-----
% done
%-----
```

Code psi_10: ($\longrightarrow$ Continued.) A script for generating graphs of the Hermite interpolation functions for a 4-node, 10-dof Hermite triangle.

(a)   (b)

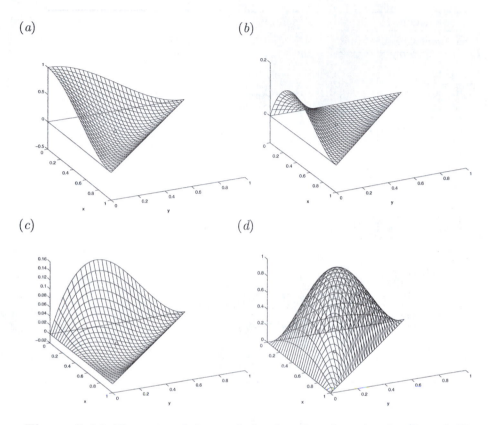

(c)   (d)

Figure 5.4.2 Element-node interpolation functions for a 4-node, 10-mode Hermite triangle: (*a*) ψ_1 corresponding to f_1, (*b*) ψ_2 corresponding to f_1^x, (*c*) ψ_3 corresponding to f_1^y, (*d*) ψ_{10}, corresponding to f_4.

Finally, we observe that the function ψ_{10} shown in Figure 5.4.2(*d*) is zero around all three edges, and may thus be classified as a bubble mode. This is because ψ_{10} is a cubic polynomial with respect to arc length along each triangle edge, with a double root at each vertex. Accordingly, all three cubics must be identically equal to zero.

Global interpolation functions

The union of the element interpolation functions discussed previously in this section defines the global interpolation functions, ϕ_i, corresponding to the deflection and its gradient at the position of the global nodes. The index i runs over an appropriate range of *global modes* associated with global degrees of freedom, that is typically far greater than the number of global nodes.

As an example, we consider the two-element arrangement shown in Figure 5.4.3(a), where the labels of the global nodes are circled and printed in bold. In this case, we have four global vertex nodes and two interior element nodes. The finite element expansion involves

$$N_G^m = 4 \times 3 + 2 \times 1 = 14 \tag{5.4.20}$$

global modes, three for each global vertex node and one for each interior node. Specifically, if $\psi_i^{(1)}$ are the Hermite interpolation functions of the first element and $\psi_i^{(2)}$ are the Hermite interpolation functions of the second element, then:

- ϕ_1 is identical to $\psi_1^{(1)}$, ϕ_2 is identical to $\psi_2^{(1)}$, and ϕ_3 is identical to $\psi_3^{(1)}$. All of these functions are associated with the first global node.

- ϕ_4 is the union of $\psi_4^{(1)}$ and $\psi_1^{(2)}$, ϕ_5 is the union of $\psi_5^{(1)}$ and $\psi_2^{(2)}$, and ϕ_6 is the union of $\psi_6^{(1)}$ and $\psi_3^{(2)}$. All of these functions are associated with the second global node.

- ...

Graphs of the functions ϕ_4, ϕ_5, and ϕ_6 associated with the second global node are shown in Figure 5.4.3(b).

For a general element arrangement, the total number of global interpolation functions is

$$N_G^m = 3\,N_V + N_I, \tag{5.4.21}$$

where N_V is the number of global vertex nodes, and N_I is the number of element interior nodes. Any suitable function can be expanded in the usual form

$$f(x,y) = \sum_{j=1}^{N_G^m} h_j^G\, \phi_j(x,y), \tag{5.4.22}$$

where the coefficients h_j^G represent the values of the solution and the Cartesian components of the gradient of the solution at appropriate global nodes. The Galerkin finite element equations are derived in the usual way based on this expansion.

Cursory inspection of the element interpolation functions composing the global interpolation functions reveals that expansion (5.4.22) represents a continuous function whose derivatives are continuous at the element node, but are not necessarily continuous across the entire length of the element edges. Elements yielding this undesirable property are called *incompatible* or *non-conforming*.

Some, but not all, non-conforming elements produce numerical solutions that do not converge to the true solution of the plate-bending governing equations as the element size becomes smaller. Unfortunately, the 4–node, 10–dof element discussed in this section falls in the category of unacceptable elements in the context of the biharmonic equation. The reason can be traced to the lack of polynomial invariance associated with the interior node ([10], p. 373).

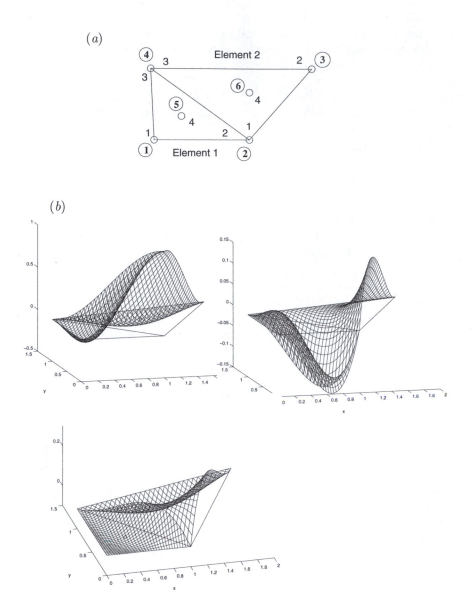

Figure 5.4.3 (*a*) A two-element arrangement used as a prototype for illustrating the relation between the element nodes, the global nodes (circled and bold faced), and the corresponding interpolation functions. (*b*) Graphs of the three global interpolation functions associated with the second global node.

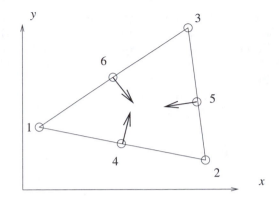

Figure 5.4.4 Illustration of the Morley element described by three vertex nodes where the value of the solution is defined, and three mid-side nodes where the normal inward derivative is defined.

5.4.2 The Morley element

Morley [36] introduced a simple non-conforming element defined by six nodes, including three vertex nodes and three mid-side nodes, as illustrated in Figure 5.4.4.

At the vertex nodes, the value of the function, f_i, is defined, for $i = 1, 2, 3$. At the mid-side nodes, the normal derivative is defined,

$$f_i^n \equiv \mathbf{n}_i \cdot \left(\nabla f \right)_{x_i, y_i}, \tag{5.4.23}$$

for $i = 4, 5, 6$, where $\mathbf{n}$ is the inward unit vector normal to the edge hosting the ith node. Using elementary geometry, we find

$$\mathbf{n}_4 = \frac{1}{|\mathbf{x}_2 - \mathbf{x}_1|} \begin{bmatrix} -(y_2 - y_1) \\ (x_2 - x_1) \end{bmatrix},$$

$$\mathbf{n}_5 = \frac{1}{|\mathbf{x}_3 - \mathbf{x}_2|} \begin{bmatrix} -(y_3 - y_2) \\ (x_3 - x_2) \end{bmatrix}, \tag{5.4.24}$$

$$\mathbf{n}_6 = \frac{1}{|\mathbf{x}_1 - \mathbf{x}_3|} \begin{bmatrix} -(y_1 - y_3) \\ (x_1 - x_3) \end{bmatrix}.$$

The unknowns can be collected into the six-dimensional vector

$$\mathbf{h} \equiv \begin{bmatrix} f_1 \\ f_2 \\ f_3 \\ - \\ f_4^n \\ f_5^n \\ f_6^n \end{bmatrix}, \tag{5.4.25}$$

encapsulating six degrees of freedom. The requirement of a C^0 continuous solution demands that the vertex values are shared by neighboring elements at corresponding positions. Similarly, the requirement of a C^1 continuous solution demands that the normal derivatives at the mid-side nodes are shared by neighboring elements at corresponding positions, provided that allowance is made for the opposite orientation of the normal vector.

Having available six degrees of freedom, we may introduce the complete quadratic expansion,

$$
\begin{aligned}
f(x, y) = \quad & a_{00} \\
& + a_{10}\, x + a_{01}\, y \\
& + a_{20}\, x^2 + a_{11}\, x\, y + a_{02}\, y^2,
\end{aligned}
\tag{5.4.26}
$$

involving six coefficients, a_{ij}. To compute these coefficients, we require the interpolation conditions $f(x_i, y_i) = f_i$, for $i = 1, 2, 3$, and obtain three equations,

$$
a_{00} + a_{10}\, x_i + a_{01}\, y_i + a_{20}\, x_i^2 + a_{11}\, x_i\, y_i + a_{02}\, y_i^2 = f_i,
\tag{5.4.27}
$$

which can be recast into the form of the vector product

$$
\begin{bmatrix} 1 & x_i & y_i & x_i^2 & x_i\, y_i & y_i^2 \end{bmatrix} \cdot \mathbf{a} = f_i,
\tag{5.4.28}
$$

where

$$
\mathbf{a} \equiv
\begin{bmatrix}
a_{00} \\
a_{10} \\
a_{01} \\
a_{20} \\
a_{11} \\
a_{02}
\end{bmatrix}
\tag{5.4.29}
$$

is the expansion coefficient vector. Requiring further the derivative interpolation conditions (5.4.23) at the mid-side nodes, we obtain three additional equations,

$$
\begin{bmatrix} 0 & n_{x_i} & n_{y_i} & 2\, x_i\, n_{x_i} & y_i\, n_{x_i} + x_i\, n_{y_i} & 2\, y_i\, n_{y_i} \end{bmatrix} \cdot \mathbf{a} = f_i^n,
\tag{5.4.30}
$$

for $i = 4, 5, 6$. Collecting (5.4.28) and (5.4.30), we formulate a system of six linear equations, $\mathbf{D} \cdot \mathbf{a} = \mathbf{h}$, where $\mathbf{D}$ is the grand coefficient matrix,

$$
\mathbf{D} =
\begin{bmatrix}
1 & x_1 & y_1 & x_1^2 & x_1\, y_1 & y_1^2 \\
1 & x_2 & y_2 & x_2^2 & x_2\, y_2 & y_2^2 \\
1 & x_3 & y_3 & x_3^2 & x_3\, y_3 & y_3^2 \\
0 & n_{x_4} & n_{y_4} & 2\, x_4\, n_{x_4} & y_4\, n_{x_4} + x_4\, n_{y_4} & 2\, y_4\, n_{y_4} \\
0 & n_{x_5} & n_{y_5} & 2\, x_5\, n_{x_5} & y_5\, n_{x_5} + x_5\, n_{y_5} & 2\, y_5\, n_{y_5} \\
0 & n_{x_6} & n_{y_6} & 2\, x_6\, n_{x_6} & y_6\, n_{x_6} + x_6\, n_{y_6} & 2\, y_6\, n_{y_6}
\end{bmatrix}.
\tag{5.4.31}
$$

Solving for the coefficient vector $\mathbf{a}$ in terms of the inverse matrix, $\mathbf{D}^{-1}$, we find $\mathbf{a} = \mathbf{D}^{-1} \cdot \mathbf{h}$. FSELIB function `morley_Dinv`, listed in the text, compiles the matrix $\mathbf{D}$ and computes its inverse using the MATLAB function `inv`.

```
function Dinv = morley_Dinv (x1,y1, x2,y2, x3,y3)

%=============================================
% Computation of the inverse of the matrix D
% for the Morley element.
%=============================================

%------------------------------------
% mid-side nodes and normal vectors
%------------------------------------

x4=0.5*(x1+x2); y4=0.5*(y1+y2);
x5=0.5*(x2+x3); y5=0.5*(y2+y3);
x6=0.5*(x3+x1); y6=0.5*(y3+y1);

d12 = sqrt((x2-x1)^2+(y2-y1)^2);

  nx4=-(y2-y1)/d12; ny4=(x2-x1)/d12;

d23 = sqrt((x3-x2)^2+(y3-y2)^2);

  nx5=-(y3-y2)/d23; ny5=(x3-x2)/d23;

d31 = sqrt((x1-x3)^2+(y1-y3)^2);

  nx6=-(y1-y3)/d31; ny6=(x1-x3)/d31;

%----------------------------
% define the coefficient matrix
%----------------------------

D = [ ...
1  x1  y1  x1^2  x1*y1  y1^2 ; ...
1  x2  y2  x2^2  x2*y2  y2^2 ; ...
1  x3  y3  x3^2  x3*y3  y3^2 ; ...
0 nx4 ny4 2*x4*nx4 y4*nx4+x4*ny4 2*y4*ny4 ; ...
0 nx5 ny5 2*x5*nx5 y5*nx5+x5*ny5 2*y5*ny5 ; ...
0 nx6 ny6 2*x6*nx6 y6*nx6+x6*ny6 2*y6*ny6 ...
];

%--------------------
% compute the inverse
%--------------------

Dinv = inv(D);

%-----
% done
%-----

return
```

Function morley_Dinv: Computation of the inverse of the grand coefficient
matrix for the Morley element.

(a) 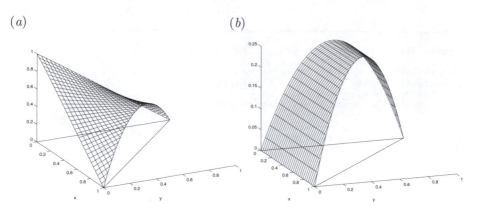 (b)

Figure 5.4.5 Element-node interpolation functions for the Morley element:
(a) ψ_1 corresponding to f_1, and (b) ψ_4 corresponding to f_4^n.

The complete quadratic expansion (5.4.26) can be expressed as the inner product of two vectors, $f(x, y) = \boldsymbol{\varphi} \cdot \mathbf{a}$, where

$$\boldsymbol{\varphi} \equiv \begin{bmatrix} 1 & x & y & x^2 & xy & y^2 \end{bmatrix} \tag{5.4.32}$$

is the horizontal vector of the monomial products. Substituting the expression for the coefficient vector, $\mathbf{a}$, we obtain

$$f(x, y) = \boldsymbol{\varphi} \cdot \mathbf{D}^{-1} \cdot \mathbf{h} \equiv \boldsymbol{\psi} \cdot \mathbf{h}. \tag{5.4.33}$$

The six-dimensional vector function $\boldsymbol{\psi}(x, y) \equiv \boldsymbol{\varphi} \cdot \mathbf{D}^{-1}$, hosts the Hermite interpolation functions associated with the components of the vector, $\mathbf{h}$, which are expansion *modes* associated with the available degrees of freedom. Specifically, the ith modal function is given by

$$\psi_i(x, y) = \sum_{j=1}^{6} \varphi_j(x, y) \, D_{ji}^{-1}, \tag{5.4.34}$$

for $i = 1, 2, \ldots, 6$.

FSELIB code `morley` (not listed in text) generates graphs of the interpolations functions contained in the vector $\boldsymbol{\psi}$. The script is similar to `psi_10`, listed previously in this section for the 10-mode element. Figure 5.4.5 shows graphs of the interpolation functions for an arrangement where the first vertex is located at $x_1 = 0$, $y_1 = 0$, the second node is located at $x_2 = 1$, $y_2 = 0$, and the third node is located at $x_3 = 0$, $y_3 = 1$. Frame (a) shows the interpolation function ψ_1 corresponding to f_1, and frame (b) shows the interpolation function ψ_4, corresponding to f_4^n.

Like the 4-node, 10-dof element discussed in Section 5.4.1, the Morley element does not guarantee C^1 continuity of the finite element expansion. In fact,

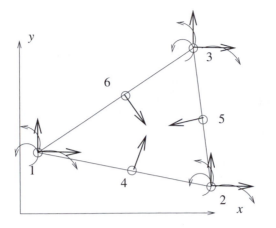

Figure 5.4.6 A Hermite triangle with 6 nodes and 21 degrees of freedom. The straight arrows indicate degrees of freedom associated with first derivatives, and the curved arrows indicate degrees of freedom associated with second derivatives.

the Morley element does not even guarantee C^0 continuity across the entire length of the element edges. In spite of this deficiency, the Morley element remarkably allows for convergent solutions of the plate bending problem ([10], pp. 374–376).

5.4.3   6-node, 21-dof triangle

To achieve C^1 continuity across the element edges, we employ conforming triangles with additional degrees of freedom. A complete quintic expansion in x and y takes the form

$$
\begin{aligned}
f(x,y) = \quad & a_{00} \\
+ & a_{10}\, x + a_{01}\, y \\
+ & a_{20}\, x^2 + a_{11}\, x\, y + a_{02}\, y^2 \\
+ & a_{30}\, x^3 + a_{21}\, x^2\, y + a_{12}\, x\, y^2 + a_{03}\, y^3 \\
+ & a_{40}\, x^4 + a_{31}\, x^3\, y + a_{22}\, x^2\, y^2 + a_{13}\, x\, y^3 + a_{04}\, x\, y^3 \\
+ & a_{50}\, x^5 + a_{41}\, x^4\, y + a_{32}\, x^3\, y^2 + a_{23}\, x^2\, y^3 + a_{14}\, x\, y^4 + a_{05} y^5,
\end{aligned}
\tag{5.4.35}
$$

involving twenty one coefficients, a_{ij}. To ensure an equal number of unknowns, we introduce a 6-node triangle with three vertex nodes and three edge nodes, as illustrated in Figure 5.4.6.

At the vertex nodes, we define the value of the function together with its first and second derivatives, denoted by

$$f_i, \qquad f_i^x \equiv \left(\frac{\partial f}{\partial x}\right)_i, \qquad f_i^y \equiv \left(\frac{\partial f}{\partial y}\right)_i,$$

$$\tag{5.4.36}$$

$$f_i^{xx} \equiv \left(\frac{\partial^2 f}{\partial x^2}\right)_i, \qquad f_i^{xy} \equiv \left(\frac{\partial^2 f}{\partial x \partial y}\right)_i, \qquad f_i^{yy} \equiv \left(\frac{\partial^2 f}{\partial y^2}\right)_i,$$

for $i = 1, 2, 3$. At the mid-side nodes, we only define the value of the normal derivative,

$$f_i^n \equiv \left(\mathbf{n} \cdot \nabla f\right)_{(x_i, y_i)}, \tag{5.4.37}$$

for $i = 4, 5, 6$, where $\mathbf{n}$ is the unit vector normal to the corresponding edge. The associated *modes* are computed as discussed previously for the Morley element.

The requirement of a C^1 continuous solution demands that the nodal values of the function and its derivatives are shared by neighboring elements at corresponding positions. The advantage of using this more complicated element is that the counterpart of the global expansion (5.4.22) produces a continuous function whose derivatives are also continuous across the entire length of the element edges. We describe this property by saying that the element is *compatible* or *conforming*. Conformity guarantees that the numerical solution of the governing plate bending equations converges to the exact solution as the finite element grid becomes finer ([10], Chapter 6).

5.4.4   The Hsieh-Clough-Tocher (HCT) element

To avoid introducing second derivatives as degrees of freedom, Clough and Felippa [11] suggested subdividing a triangular element into 3 sub-elements, and using individual cubic expansions over each sub-element, as illustrated in Figure 5.4.7, where the sub-elements labels are printed in bold. The resulting element is known as the Hsieh-Clough-Tocher (HCT) triangle. The parental element is sometimes called a macro-element.

A complete cubic expansion in x and y over the qth sub-element takes the form

$$
\begin{aligned}
f^{(q)}(x, y) = \quad & a_{00}^{(q)} \\
& + a_{10}^{(q)} \, x + a_{01}^{(q)} \, y \\
& + a_{20}^{(q)} \, x^2 + a_{11}^{(q)} \, x\,y + a_{02}^{(q)} \, y^2 \\
& + a_{30}^{(q)} \, x^3 + a_{21}^{(q)} \, x^2\,y + a_{12}^{(q)} \, x\,y^2 + a_{03}^{(q)} \, y^3,
\end{aligned}
\tag{5.4.38}
$$

for $q = 1, 2, 3$, involving thirty coefficients, $a_{ij}^{(q)}$. To ensure a sufficient number of degrees of freedom, we introduce the values of the function and its first derivatives at the vertex nodes,

$$f_i, \qquad f_i^x \equiv \left(\frac{\partial f}{\partial x}\right)_i, \qquad f_i^y \equiv \left(\frac{\partial f}{\partial y}\right)_i, \tag{5.4.39}$$

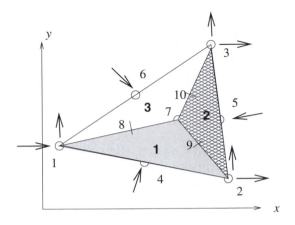

Figure 5.4.7 Illustration of the HCT element defined by three vertex nodes, three mid-side nodes, and several interior nodes defined with regard to three sub-triangles. The labels of the sub-triangles are printed in bold. The arrows indicate degrees of freedom associated with the x, y, and normal derivatives.

for $i = 1, 2, 3$, and the normal derivatives at the mid-side nodes,

$$f_i^n \equiv \left(\mathbf{n} \cdot \nabla f \right)_i, \tag{5.4.40}$$

for $i = 4, 5, 6$. These unknowns can be collected into the vector

$$\mathbf{h} \equiv \begin{bmatrix} f_1 \\ f_1^x \\ f_1^y \\ - \\ f_2 \\ f_2^x \\ f_2^y \\ - \\ f_3 \\ f_3^x \\ f_3^y \\ - \\ f_4^n \\ f_5^n \\ f_6^n \end{bmatrix}, \tag{5.4.41}$$

encapsulating twelve degrees of freedom.

To compute the thirty coefficients of the cubic expansions, $a_{ij}^{(q)}$, we require the vertex node interpolation conditions

$$f^{(q)}(x_i, y_i) = f_i,$$

$$\left(\frac{\partial f^{(q)}}{\partial y}\right)_{(x_i, y_i)} = f_i^y, \qquad \left(\frac{\partial f^{(q)}}{\partial x}\right)_{(x_i, y_i)} = f_i^x,$$

(5.4.42)

where $i = 1, 2$ for $q = 1$, $i = 2, 3$ for $q = 2$, and $i = 3, 1$ for $q = 3$, which provide us with $3 \times (2 \times 3) = 18$ equations. In addition, we require the mid-side node interpolation conditions

$$\mathbf{n}_4 \cdot \left(\nabla f^{(1)}\right)_{(x_4, y_4)} = f_4^n,$$

$$\mathbf{n}_5 \cdot \left(\nabla f^{(2)}\right)_{(x_5, y_5)} = f_5^n,$$

(5.4.43)

$$\mathbf{n}_6 \cdot \left(\nabla f^{(3)}\right)_{(x_6, y_6)} = f_6^n,$$

which provide us with three additional equations. The remaining nine equations express: (*a*) continuity of the three polynomial expansions and their gradients at the internal node labeled 7, and (*b*) continuity of the normal derivative at the interior mid-side nodes labeled 8, 9, 10, indicated by the tick marks in Figure 5.4.7. Compiling these equations, we formulate a system of thirty linear equations,

$$\mathbf{D} \cdot \begin{bmatrix} \mathbf{a}^{(1)} \\ \mathbf{a}^{(2)} \\ \mathbf{a}^{(3)} \end{bmatrix} = \mathbf{b},$$

(5.4.44)

where $\mathbf{D}$ is a grand coefficient matrix, and $\mathbf{b}$ is a suitable right-hand side defined in terms of the degrees of freedom. The last nine entries of the vector $\mathbf{b}$ are zero, reflecting the interior node continuity conditions.

FSELIB function HCT_sys, listed in the text, generates and solves the linear system for the polynomial coefficients. The twelve-dimensional vector $\mathbf{h}$ defined in (5.4.41) is contained in the input vector dof. The element interpolation functions can be generated by setting the value of a selected entry of this vector equal to unity, while holding all other values to zero, and then running over the available degrees of freedom.

FSELIB code HCT (not listed in text) generates graphs of the modes associated with the degrees of freedom. Figure 5.4.8 shows graphs of four modes for an arrangement where the first vertex is located at $x_1 = 0$, $y_1 = 0$, the second node is located at $x_2 = 1$, $y_2 = 0$, and the third node is located at $x_3 = 0$, $y_3 = 1$.

```
function [a] = HCT_sys (x1,y1, x2,y2, x3,y3, dof)

%===================================================
% Generate the modal (dof) polynomial coefficients
% for the HCT element
%
% dof: degrees of freedom
%===================================================

%-----------------------------------
% mid-side nodes and normal vectors
%-----------------------------------

x4=0.5*(x1+x2); y4=0.5*(y1+y2);
d4=sqrt((x2-x1)^2+(y2-y1)^2);
nx4=-(y2-y1)/d4; ny4=(x2-x1)/d4;

x5=0.5*(x2+x3); y5=0.5*(y2+y3);
d5=sqrt((x3-x2)^2+(y3-y2)^2);
nx5=-(y3-y2)/d5; ny5=(x3-x2)/d5;

x6 = 0.5*(x3+x1); y6=0.5*(y3+y1);
d6=sqrt((x1-x3)^2+(y1-y3)^2);
nx6=-(y1-y3)/d6; ny6=(x1-x3)/d6;

%--------------
% interior node
%--------------

x7=(x1+x2+x3)/3.0; y7=(y1+y2+y3)/3.0;

%-----------------------------------------
% interior mid-side nodes and normal vectors
%-----------------------------------------

x8 = 0.5*(x1+x7); y8=0.5*(y1+y7);
d8 = sqrt((x7-x1)^2+(y7-y1)^2);
nx8=-(y7-y1)/d8; ny8=(x7-x1)/d8;

x9 = 0.5*(x2+x7); y9=0.5*(y2+y7);
d9 = sqrt((x7-x2)^2+(y7-y2)^2);
nx9=-(y7-y2)/d9; ny9=(x7-x2)/d9;

x10 = 0.5*(x3+x7); y10=0.5*(y3+y7);
d10 = sqrt((x7-x3)^2+(y7-y3)^2);
nx10=-(y7-y3)/d10; ny10=(x7-x3)/d10;

%----------------------------
% define the coefficient matrix
%----------------------------

D = [ ...
    % first sub-element: (6 eqns)
...
```

Function HCT_sys: Continuing $\longrightarrow$

```
1  x1  y1  x1^2  x1*y1  y1^2  x1^3  x1^2*y1  x1*y1^2  y1^3 ...
0 0 0 0 0 0 0 0 0 0 0 0 0 0 0 0 0 0 0 0 0 ; ...
...
0  1  0  2*x1  y1  0  3*x1^2  2*x1*y1  y1^2  0    ...
0 0 0 0 0 0 0 0 0 0 0 0 0 0 0 0 0 0 0 0 0 ; ...
...
0  0  1  0  x1  2*y1  0  x1^2  2*x1*y1  3*y1^2 ...
0 0 0 0 0 0 0 0 0 0 0 0 0 0 0 0 0 0 0 0 ; ...
...
1  x2  y2  x2^2  x2*y2  y2^2  x2^3  x2^2*y2  x2*y2^2  y2^3 ...
0 0 0 0 0 0 0 0 0 0 0 0 0 0 0 0 0 0 0 0 0 ; ...
...
0  1  0  2*x2  y2  0  3*x2^2  2*x2*y2  y2^2  0    ...
0 0 0 0 0 0 0 0 0 0 0 0 0 0 0 0 0 0 0 0 0 ; ...
...
0  0  1  0  x2  2*y2  0  x2^2  2*x2*y2  3*y2^2 ...
0 0 0 0 0 0 0 0 0 0 0 0 0 0 0 0 0 0 0 0 0 ; ...
...
   % second sub-element: (6 eqns)
...
0 0 0 0 0 0 0 0 0 ...
1  x2  y2  x2^2  x2*y2  y2^2  x2^3  x2^2*y2  x2*y2^2  y2^3 ...
0 0 0 0 0 0 0 0 0 ; ...
0 0 0 0 0 0 0 0 0 ...
0  1  0  2*x2  y2  0  3*x2^2  2*x2*y2  y2^2  0    ...
0 0 0 0 0 0 0 0 0 ; ...
0 0 0 0 0 0 0 0 0 ...
0  0  1  0  x2  2*y2  0  x2^2  2*x2*y2  3*y2^2 ...
0 0 0 0 0 0 0 0 0 ; ...
...
0 0 0 0 0 0 0 0 0 0 ...
1  x3  y3  x3^2  x3*y3  y3^2  x3^3  x3^2*y3  x3*y3^2  y3^3    ...
0 0 0 0 0 0 0 0 0 ; ...
0 0 0 0 0 0 0 0 0 ...
0 1 0 2*x3 y3 0  3*x3^2  2*x3*y3  y3^2    0    ...
0 0 0 0 0 0 0 0 0 ; ...
0 0 0 0 0 0 0 0 0 ...
0  0  1  0  x3  2*y3  0  x3^2  2*x3*y3  3*y3^2 ...
0 0 0 0 0 0 0 0 0 ; ...
...
   % third sub-element: (6 eqns)
...
0 0 0 0 0 0 0 0 0 0 0 0 0 0 0 0 0 0 0 0 ...
1  x3  y3  x3^2  x3*y3  y3^2  x3^3  x3^2*y3  x3*y3^2  y3^3 ; ...
0 0 0 0 0 0 0 0 0 0 0 0 0 0 0 0 0 0 0 0 ...
0  1  0  2*x3  y3    0  3*x3^2  2*x3*y3   y3^2  0 ; ...
0 0 0 0 0 0 0 0 0 0 0 0 0 0 0 0 0 0 0 0 ...
0  0  1  0  x3  2*y3  0  x3^2  2*x3*y3  3*y3^2 ; ...
...
0 0 0 0 0 0 0 0 0 0 0 0 0 0 0 0 0 0 0 0 ...
1  x1  y1  x1^2  x1*y1  y1^2  x1^3  x1^2*y1  x1*y1^2  y1^3 ; ...
0 0 0 0 0 0 0 0 0 0 0 0 0 0 0 0 0 0 0 0 ...
0  1  0  2*x1    y1  0  3*x1^2  2*x1*y1  y1^2    0 ; ...
```

Function HCT_sys: $\longrightarrow$ Continuing $\longrightarrow$

```
0 0 0 0 0 0 0 0 0 0 0 0 0 0 0 0 0 0 0 0 ...
0   0   1   0       x1   2*y1   0   x1^2   2*x1*y1   3*y1^2 ; ...
...
   % mid-side node 4: (1 eqn)
...
0   nx4  ny4  2*x4*nx4  y4*nx4+x4*ny4  2*y4*ny4  ...
  3*x4^2*nx4  x4*(2*y4*nx4+x4*ny4)  y4*(y4*nx4+2*x4*ny4)  3*y4^2*ny4  ...
0 0 0 0 0 0 0 0 0 0 0 0 0 0 0 0 0 0 0 0 ;
...
   % mid-side node 5: (1 eqn)
...
0 0 0 0 0 0 0 0 0 ...
0   nx5  ny5  2*x5*nx5  y5*nx5+x5*ny5  2*y5*ny5  ...
  3*x5^2*nx5  x5*(2*y5*nx5+x5*ny5)  y5*(y5*nx5+2*x5*ny5)  3*y5^2*ny5  ...
0 0 0 0 0 0 0 0 0 ; ...
...
   % mid-side node 6: (1 eqn)
...
0 0 0 0 0 0 0 0 0 0 0 0 0 0 0 0 0 0 0 0 ...
0   nx6  ny6  2*x6*nx6  y6*nx6+x6*ny6  2*y6*ny6  ...
  3*x6^2*nx6  x6*(2*y6*nx6+x6*ny6)  y6*(y6*nx6+2*x6*ny6)  3*y6^2*ny6; ...
...
...
   % continuity at interior node 7 for sub-elements 1 and 2:   (3 eqns)
...
 1   x7   y7   x7^2   x7*y7   y7^2   x7^3   x7^2*y7   x7*y7^2   y7^3  ...
-1  -x7  -y7  -x7^2  -x7*y7  -y7^2  -x7^3  -x7^2*y7  -x7*y7^2  -y7^3  ...
0 0 0 0 0 0 0 0 0 0 ; ...
...
0   1   0   2*x7   y7   0   3*x7^2   2*x7*y7   y7^2   0  ...
0  -1   0  -2*x7  -y7   0  -3*x7^2  -2*x7*y7  -y7^2   0  ...
0 0 0 0 0 0 0 0 0 0 ; ...
...
0   0   1   0   x7   2*y7   0   x7^2   2*x7*y7   3*y7^2 ...
0   0  -1   0  -x7  -2*y7   0  -x7^2  -2*x7*y7  -3*y7^2 ...
0 0 0 0 0 0 0 0 0 0 ; ...
...
   % continuity at interior node 7 for sub-elements 2 and 3: (3 eqns)
...
0 0 0 0 0 0 0 0 0 ...
 1   x7   y7   x7^2   x7*y7   y7^2   x7^3   x7^2*y7   x7*y7^2   y7^3   ...
-1  -x7  -y7  -x7^2  -x7*y7  -y7^2  -x7^3  -x7^2*y7  -x7*y7^2  -y7^3 ;  ...
...
0 0 0 0 0 0 0 0 0 ...
0   1   0   2*x7   y7   0   3*x7^2   2*x7*y7   y7^2   0  ...
0  -1   0  -2*x7  -y7   0  -3*x7^2  -2*x7*y7  -y7^2   0 ;
...
0 0 0 0 0 0 0 0 0 ...
0   0   1   0   x7   2*y7 0   x7^2   2*x7*y7   3*y7^2 ...
0   0  -1   0  -x7  -2*y7 0  -x7^2  -2*x7*y7  -3*y7^2 ;
...
   % continuity of normal derivative at mid-side node 8: (1 eqn)
...
```

Function HCT_sys: $\longrightarrow$ Continuing $\longrightarrow$

```
0   nx8   ny8   2*x8*nx8   y8*nx8+x8*ny8   2*y8*ny8   ...
    3*x8^2*nx8   x8*(2*y8*nx8+x8*ny8)   y8*(y8*nx8+2*x8*ny8) ...
    3*y8^2*ny8   ...
0 0 0 0 0 0 0 0 0 0 ...
0  -nx8  -ny8  -2*x8*nx8  -y8*nx8-x8*ny8  -2*y8*ny8 ...
   -3*x8^2*nx8  -x8*(2*y8*nx8+x8*ny8)  -y8*(y8*nx8+2*x8*ny8) ...
   -3*y8^2*ny8  ;...
...
% Continuity of normal derivative at mid-side node 9: (1 eqn)
...
0   nx9   ny9   2*x9*nx9   y9*nx9+x9*ny9   2*y9*ny9   ...
    3*x9^2*nx9   x9*(2*y9*nx9+x9*ny9)   y9*(y9*nx9+2*x9*ny9) ...
    3*y9^2*ny9 ...
0  -nx9  -ny9  -2*x9*nx9  -y9*nx9-x9*ny9  -2*y9*ny9   ...
   -3*x9^2*nx9  -x9*(2*y9*nx9+x9*ny9)  -y9*(y9*nx9+2*x9*ny9) ...
   -3*y9^2*ny9 ...
0 0 0 0 0 0 0 0 0 0 ; ...
...
% Continuity of normal derivative at mid-side node 10: (1 eqn)
...
0 0 0 0 0 0 0 0 0 0 ...
0  nx10  ny10  2*x10*nx10  y10*nx10+x10*ny10  2*y10*ny10  ...
   3*x10^2*nx10  x10*(2*y10*nx10+x10*ny10)  y10*(y10*nx10+2*x10*ny10) ...
   3*y10^2*ny10 ...
0 -nx10 -ny10 -2*x10*nx10 -y10*nx10-x10*ny10 -2*y10*ny10  ...
  -3*x10^2*nx10 -x10*(2*y10*nx10+x10*ny10) -y10*(y10*nx10+2*x10*ny10) ...
  -3*y10^2*ny10  ;
];

%----------------------------
% define the right-hand side
%----------------------------

b=zeros(1,30);

b(1)  = dof(1);  b(2)  = dof(2);  b(3)  = dof(3);
b(4)  = dof(4);  b(5)  = dof(5);  b(6)  = dof(6);
b(7)  = dof(4);  b(8)  = dof(5);  b(9)  = dof(6);
b(10) = dof(7);  b(11) = dof(8);  b(12) = dof(9);
b(13) = dof(7);  b(14) = dof(8);  b(15) = dof(9);
b(16) = dof(1);  b(17) = dof(2);  b(18) = dof(3);
b(19) = dof(10); b(20) = dof(11); b(21) = dof(12);

%----------------------
% solve the linear system
%----------------------

a = b/D';
b';

return
```

Function HCT_sys: ($\longrightarrow$ Continued.) Computation of the cubic polynomial coefficients for the HCT element.

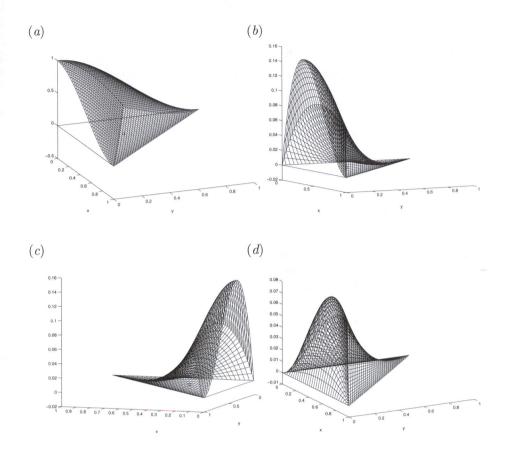

Figure 5.4.8 Element interpolation functions (modes) for the HCT element. Graphs of (a) ψ_1 corresponding to f_1, (b) ψ_2 corresponding to f_1^x, (c) ψ_3 corresponding to f_1^y, and (d) ψ_{10} corresponding to f_4^n.

5.4.5   An echelon of plate bending elements

The four elements discussed in this section are examples taken from an extensive library of triangular and quadrilateral elements designed specifically for plate bending problems. A variety of conforming and non-conforming elements are reviewed and tabulated by Hrabok and Hrudey [28]. While the use of conforming elements guarantees that the finite element solution converges to the exact solution as the element grid becomes finer, some non-conforming elements have also been used with success for reasons discussed by Ciarlet [10].

PROBLEMS

5.4.1 *Element interpolation functions and global modes.*

Complete the definition of the global modes discussed in the text with reference to Figure 5.4.2.

5.4.2 *A Hermite, 4-node, 12-mode rectangle.*

Consider the 4-node quadrilateral element shown in Figure 3.8.1, and introduce the value of the function f and its gradient at each node,

$$f_i, \qquad f_i^x \equiv \left(\frac{\partial w}{\partial x}\right)_i, \qquad f_i^y \equiv \left(\frac{\partial w}{\partial y}\right)_i, \qquad (5.4.45)$$

for $i = 1, 2, 3, 4$, yielding twelve degrees of freedom. In the finite element literature, it is a standard practice to approximate the function $f(x, y)$ over the element with an incomplete cubic, as

$$
\begin{aligned}
f(x, y) \simeq \quad & a_{00} \\
& + a_{10}\, x + a_{01}\, y \\
& + a_{20}\, x^2 + a_{11}\, x\,y + a_{02}\, y^2 \\
& + a_{30}\, x^3 + a_{21}\, x^2\, y + a_{12}\, x\,y^2 + a_{03}\, y^3 \\
+ a_{31}\, x^3\, y \quad & \qquad\qquad\qquad\qquad\quad + a_{13}\, x\,y^3,
\end{aligned}
\qquad (5.4.46)
$$

involving twelve coefficients, a_{ij}.

(*a*) Write a program that computes the interpolation functions (modes) corresponding to the twelve degrees of freedom, and prepare graphs of the three modes associated with a vertex of your choice.

(*b*) Discuss whether or not this element is conforming.

5.5   Finite element methods for plate bending

A variety of finite element formulations are available for computing plate bending deflection. Direct methods target the governing differential equations, while indirect methods are based on functional minimization associated with a variational formulation. In this section, we discuss two direct formulations, one based on the biharmonic equation and the second based on force and moment equilibrium equations. Indirect methods are discussed extensively, and sometimes exclusively, in the literature on finite element methods in structural and solid mechanics (e.g., [21, 33]).

5.5.1   Formulation as a biharmonic equation

The formulation based on the biharmonic equation (5.3.42), repeated here for convenience,

$$\nabla^4 f = -\frac{1}{E_B}\, w(x,y), \qquad (5.5.1)$$

is the counterpart of the beam bending formulation discussed in Section 1.7, culminating in a fourth-order differential equation.

To develop the Galerkin finite element equations, we multiply the governing equation (5.5.1) with each one of the global interpolation functions associated with the available degrees of freedom, and integrate the product over the solution domain, D, to obtain

$$\iint_D \phi_i(x,y)\, \nabla^4 f\, \mathrm{d}x\, \mathrm{d}y = -\frac{1}{E_B} \iint_D \phi_i(x,y)\, w(x,y)\, \mathrm{d}x\, \mathrm{d}y. \qquad (5.5.2)$$

Integrating by parts on the left-hand side to reduce the order of the biharmonic operator, we write

$$\iint_D \phi_i\, \nabla^4 f\, \mathrm{d}x\, \mathrm{d}y = \iint_D \phi_i\, \nabla \cdot \left(\nabla (\nabla^2 f) \right) \mathrm{d}x\, \mathrm{d}y$$

$$= \iint_D \nabla \cdot \left(\phi_i\, \nabla (\nabla^2 f) \right) \mathrm{d}x\, \mathrm{d}y - \iint_D \nabla\phi_i \cdot \nabla(\nabla^2 f)\, \mathrm{d}x\, \mathrm{d}y$$

$$= -\oint_C \phi_i\, \mathbf{n} \cdot \nabla (\nabla^2 f)\, \mathrm{d}l - \iint_D \nabla\phi_i \cdot \nabla (\nabla^2 f)\, \mathrm{d}x\, \mathrm{d}y, \qquad (5.5.3)$$

where C is the boundary of D, $\mathbf{n}$ is the *inward* unit vector normal to the C, and l is the arc length along C. Integrating also by parts the last integral in (5.5.3), we write

$$\iint_D \nabla\phi_i \cdot \nabla (\nabla^2 f)\, \mathrm{d}x\, \mathrm{d}y$$

$$= \iint_D \nabla \cdot \left(\nabla\phi_i\, (\nabla^2 f) \right) \mathrm{d}x\, \mathrm{d}y - \iint_D \nabla^2\phi_i\, \nabla^2 f\, \mathrm{d}x\, \mathrm{d}y$$

$$= -\oint_C (\mathbf{n} \cdot \nabla\phi_i)\, \nabla^2 f\, \mathrm{d}l - \iint_D \nabla^2\phi_i\, \nabla^2 f\, \mathrm{d}x\, \mathrm{d}y. \qquad (5.5.4)$$

Combining (5.5.3) and (5.5.4), we obtain

$$\iint_D \phi_i\, \nabla^4 f\, \mathrm{d}x\, \mathrm{d}y = -\oint_C \phi_i\, \mathbf{n} \cdot \nabla(\nabla^2 f)\, \mathrm{d}l$$

$$+ \oint_C (\mathbf{n} \cdot \nabla\phi_i)\, \nabla^2 f\, \mathrm{d}l + \iint_D \nabla^2\phi_i\, \nabla^2 f\, \mathrm{d}x\, \mathrm{d}y. \qquad (5.5.5)$$

Finally, we substitute this expression in the left-hand side of (5.5.2), and rearrange to obtain

$$
\iint_D \nabla^2 \phi_i \, \nabla^2 f \, dx \, dy = \oint_C \phi_i \, \mathbf{n} \cdot \nabla(\nabla^2 f) \, dl
$$

$$
- \oint_C (\mathbf{n} \cdot \nabla \phi_i) \, \nabla^2 f dl - \frac{1}{E_B} \iint_D \phi_i \, w \, dx \, dy. \tag{5.5.6}
$$

In terms of the trace of the bending moments tensor given in (5.3.37), $M \equiv M_{xx} + M_{yy} = -E_B(1+\nu)\nabla^2 f$, and the transverse shear tension vector given in (5.3.41), $\mathbf{q} = -E_B \, \nabla(\nabla^2 f)$, equation (5.5.6) becomes

$$
E_B \iint_D \nabla^2 \phi_i \, \nabla^2 f \, dx \, dy = -\oint_C \phi_i \, \mathbf{n} \cdot \mathbf{q} \, dl
$$

$$
+ \frac{1}{1+\nu} \oint_C M \, \mathbf{n} \cdot \nabla \phi_i \, dl - \iint_D \phi_i \, w \, dx \, dy, \tag{5.5.7}
$$

where ν is the Poisson ratio. The Galerkin equation (5.5.7) is the counterpart of the beam equation (1.7.33).

The contour integrals on the right-hand sides of (5.5.6) and (5.5.7) are nonzero only if ϕ_i is a mode associated with a boundary node. If ϕ_i is a mode associated with an interior node, these integrals are zero, yielding the simplified equation

$$
\iint_D \nabla^2 \phi_i \, \nabla^2 f \, dx \, dy = -\frac{1}{E_B} \iint_D \phi_i \, w \, dx \, dy. \tag{5.5.8}
$$

For a clamped plate, the transverse displacement and its gradient are specified along the boundary, C. Consequently, the corresponding projections are excluded from the finite element formulation, and the simplified formulation applies.

Formulation of a linear system

Assume that the finite element expansion involves N_G^m modes associated with the available degrees of freedom, and introduce the global expansion

$$
f(x,y) = \sum_{j=1}^{N_G^m} h_j^G \, \phi_j(x,y), \tag{5.5.9}
$$

where the coefficients, h_j^G, represent either the values of the solution, or the spatial derivatives of the solution at the global nodes. Inserting this expression in (5.5.8), we derive the linear system

$$K_{ij} h_j^G = -\frac{1}{E_B} b_i, \qquad (5.5.10)$$

where summation over j is implied on the left-hand side,

$$K_{ij} \equiv \iint_D \nabla^2 \phi_i \, \nabla^2 \phi_j \, \mathrm{d}x \, \mathrm{d}y, \qquad (5.5.11)$$

as the *global bending stiffness matrix,* and

$$b_i \equiv \iint_{E_l} \phi_i \, w \, \mathrm{d}x \, \mathrm{d}y. \qquad (5.5.12)$$

The global stiffness matrix and right-hand side can be compiled in the usual way from corresponding element stiffness matrices,

$$A_{ij}^{(l)} \equiv \iint_{E_l} \nabla^2 \psi_i \, \nabla^2 \psi_j \, \mathrm{d}x \, \mathrm{d}y,$$

$$\qquad (5.5.13)$$

$$b_i^{(l)} \equiv \iint_{E_l} \psi_i \, w \, \mathrm{d}x \, \mathrm{d}y,$$

where E_l denotes the lth element. Like the element diffusion matrix, the element stiffness matrix is singular (Problem 5.5.1).

FSELIB function ebsm_HCT, listed in the text, evaluates the element bending stiffness matrix and corresponding element integral, $b_i^{(l)}$, for the HCT element discussed in Section 5.4.4. To compute the element integrals, each sub-triangle of the HCT triangle is mapped from the physical xy plane to the parametric $\xi\eta$ plane using the mapping functions (3.2.10). The integrals over the sub-triangles are then computed using a Gaussian triangle quadrature.

Code for the bending of a clamped plate

FSELIB code bend_HCT, listed in the text, solves the inhomogeneous biharmonic equation governing the deflection of a plate with an entirely clamped edge, using the HCT element discussed in Section 5.4.4. The general structure of the code is similar to that of of the codes for steady diffusion discussed in previous chapters.

One new feature is that, because the normal derivative at the mid-side nodes of an element points into each element by convention, the sign must be switched in assembling the corresponding entries of the global linear system after the first contribution has been made. This is done by introducing the index *Idid*, which is switched from the value of 1 to the value of -1 after the first contribution has been made.

```
function [ebsm, rhs, arel] = ebsm_HCT (x1,y1, x2,y2, x3,y3, NQ)

%=====================================================
%
% Evaluation of the element bending stiffness matrix
% and right-hand side for the HCT triangle,
% using a Gauss integration quadrature.
%
% NQ:    number of quadrature base points
% arel: element area
%
%=====================================================

%----------------------------------
% compute the area of the triangle
%----------------------------------

d12x = x1-x2; d12y = y1-y2;
d31x = x3-x1; d31y = y3-y1;

arel = 0.5*(d31x*d12y - d31y*d12x);

%--------------
% interior node
%--------------

x7 = (x1+x2+x3)/3.0; y7 = (y1+y2+y3)/3.0;

%---------------------------
% read the triangle quadrature
%---------------------------

[xi, eta, w] = gauss_trgl(NQ);

%----------------------------------------------
% initialize the element bending stiffness matrix
% and right-hand side
%----------------------------------------------

for k=1:12
 for l=1:12
  ebsm(k,l) = 0.0;
 end
 rhs(k) = 0.0;
end

%-----------------------------------
% compute the thirty cubic coefficients
% and put them in the vector: am
%-----------------------------------

for mode=1:12

  dof=zeros(1,12); dof(mode)=1.0;
```

Function ebsm_HCT: Continuing $\longrightarrow$

```
[a] = HCT_sys (x1,y1, x2,y2, x3,y3, dof);

for i=1:30
  am(i,mode) = a(i);
end

end

%----------------------
% perform the quadrature
%----------------------

for pass=1:3  % run over the 3 sub-triangles

 for i=1:NQ    % quadrature

%----
% modes and laplacian:
%---

 for mode=1:12

   for j=1:30
     c(j) = am(j,mode);
   end

   if(pass==1)        % triangle 1-2-7

      x = x1 + (x2-x1)*xi(i) + (x7-x1)*eta(i);
      y = y1 + (y2-y1)*xi(i) + (y7-y1)*eta(i);

      psi(mode) = c(1) + c(2)*x   +c(3)*y ...
                  + c(4)*x^2 +c(5)*x*y + c(6)*y^2 ...
                  + c(7)*x^3 +c(8)*x^2*y+ c(9)*x*y^2+c(10)*y^3;

      lpsi(mode) = 2.0*c(4)+2.0*c(6) ...
                  +6.0*c(7)*x + 2.0*c(8)*y ...
                  +2.0*c(9)*x + 6.0*c(10)*y;

    elseif(pass==2)   % triangle 2-3-7

      x = x2 + (x3-x2)*xi(i) + (x7-x2)*eta(i);
      y = y2 + (y3-y2)*xi(i) + (y7-y2)*eta(i);

      psi(mode) = c(11)+c(12)*x+c(13)*y...
                  +c(14)*x^2+c(15)*x*y+c(16)*y^2 ...
                  +c(17)*x^3+c(18)*x^2*y...
                  +c(19)*x*y^2+c(20)*y^3;

      lpsi(mode) = 2.0*c(14)+2.0*c(16) ...
                  +6.0*c(17)*x+2.0*c(18)*y...
                  +2.0*c(19)*x+6.0*c(20)*y;

      else              % triangle 3-1-7
```

Function ebsm_HCT: $\longrightarrow$ Continuing $\longrightarrow$

```
     x = x3 + (x1-x3)*xi(i) + (x7-x3)*eta(i);
     y = y3 + (y1-y3)*xi(i) + (y7-y3)*eta(i);

     psi(mode) = c(21)+c(22)*x+c(23)*y ...
                    +c(24)*x^2...
                    +c(25)*x*y...
                    +c(26)*y^2 ...
                    +c(27)*x^3+c(28)*x^2*y...
                    +c(29)*x*y^2+c(30)*y^3;

     lpsi(mode) = 2.0*c(24)+2.0*c(26) ...
                      +6.0*c(27)*x+2.0*c(28)*y...
                      +2.0*c(29)*x+6.0*c(30)*y;

    end
   end    % end of loop over modes
%--

   cf = arel*w(i)/3.0;

   for k=1:12

    for l=k:12
      ebsm(k,l) = ebsm(k,l) + lpsi(k)*lpsi(l)*cf;
    end

    Wload = 1.0;                    % uniform load

    rhs(k) = rhs(k) + Wload*psi(k)*cf;

   end

  end    % end of quadrature
 end % end of run over sub-triangles

%-------------------------------
% fill in lower diagonal of ebsm
%-------------------------------

 for k=1:12
  for l=1:k-1
   ebsm(k,l) = ebsm(l,k);
  end
 end

%-----
% done
%-----

return;
```

Function ebsm_HCT: ($\longrightarrow$ Continued.) Evaluation of the element bending stiffness matrix for the biharmonic equation.

```
%=============
% CODE bend_HCT
%
% Solution of the biharmonic equation
% for plate bending using the HCT element.
%
%=============

%-----------
% input data
%-----------

E_B = 1.0; % bending modulus
ndiv = 2;  % discretization level
NQ=4;      % quadrature order

%-----------
% triangulate
%-----------

 [ne,ng,p,c,efl,gfl] = trgl6_sqr(ndiv);
% [ne,ng,p,c,efl,gfl] = trgl6_disk(ndiv);

%-------------------------------------------------
% make sure the mid-nodes are along straight edges
%-------------------------------------------------

for i=1:ne
 for j=1:2
  p(c(i,4),j)=0.5*( p(c(i,1),j)+p(c(i,2),j) );
  p(c(i,5),j)=0.5*( p(c(i,2),j)+p(c(i,3),j) );
  p(c(i,6),j)=0.5*( p(c(i,3),j)+p(c(i,1),j) );
 end
end

%----------------------------------------
% mark the global vertex and edge nodes
%----------------------------------------

for i=1:ne
 gfln(c(i,1)) = 0;        % vertex nodes
    gfln(c(i,2)) = 0;
       gfln(c(i,3)) = 0;
 gfln(c(i,4)) = 1;        % edge nodes
    gfln(c(i,5)) = 1;
       gfln(c(i,6)) = 1;
end

%----------------------------------------
% count the number of global edge nodes
%----------------------------------------

nge=0; % number of edge nodes
```

Code bend_HCT: Continuing ⟶

```
for i=1:ng
 if(gfln(i) == 1)
  nge = nge+1;
 end
end

%----------------------------
% total number of global modes
%----------------------------

ngv = ng-nge;       % vertex nodes

ngm = 3*ngv+nge;  % number of global modes

%-------------------------------------------
% mark the position of the rows and columns
% of the modes in the global system
%-------------------------------------------

for i=1:ng

  s(i) = 1;

  for j=1:i-1
   if(gfln(j) == 1)      % edge node, 1 dof
     s(i) = s(i)+1;
   else                  % vertex node, 3 dofs
     s(i) = s(i)+3;
   end
  end

end

%-----
% initialize the orientation index of the normal derivative
% at the mid-side nodes
%-----

for i=1:ng
 Idid(i)=1.0;
end

%------------------------------------------------
% assemble the global bending stiffness matrix
% and right-hand side
%------------------------------------------------

gbsm = zeros(ngm,ngm); % initialize
b = zeros(1,ngm);      % right-hand side

for l=1:ne              % loop over the elements

% compute the element bending stiffness matrix and rhs
% and right-hand side
```

Code bend_HCT: $\longrightarrow$ Continuing $\longrightarrow$

```
j1=c(1,1); j2=c(1,2); j3=c(1,3);    % global element labels
j4=c(1,4); j5=c(1,5); j6=c(1,6);

x1=p(j1,1); y1=p(j1,2);             % three vertices
x2=p(j2,1); y2=p(j2,2);
x3=p(j3,1); y3=p(j3,2);

[ebsm_elm, rhs, arel] = ebsm_HCT (x1,y1, x2,y2, x3,y3, NQ);

e(1) = s(j1); e(2) = s(j1)+1; e(3) = s(j1)+2;  % position in
e(4) = s(j2); e(5) = s(j2)+1; e(6) = s(j2)+2;  % global matrix
e(7) = s(j3); e(8) = s(j3)+1; e(9) = s(j3)+2;
e(10)= s(j4); e(11)= s(j5);   e(12)= s(j6);

    for i=1:12

        fci = 1.0;
             if(i==10) fci = Idid(j4);
        elseif(i==11) fci = Idid(j5);
        elseif(i==12) fci = Idid(j6);
        end

        for j=1:12
          fcj = 1.0;
               if(j==10) fcj = Idid(j4);
          elseif(j==11) fcj = Idid(j5);
          elseif(j==12) fcj = Idid(j6);
          end
            gbsm(e(i),e(j)) = gbsm(e(i),e(j)) + fci*fcj*ebsm_elm(i,j);
        end

        b(e(i)) = b(e(i)) - fci*rhs(i);

    end

  Idid(j4) = -1; Idid(j5) = -1; Idid(j6) = -1;

end

%-----------------------------------------------------------
% implement the homogeneous Dirichlet boundary condition
% no need to modify the right-hand side
%-----------------------------------------------------------

for j=1:ng

  if(gfl(j,1)==1)    % boundary node

    j1 = s(j);
    j2 = s(j)+1;
    j3 = s(j)+2;
```

Code bend_HCT: $\longrightarrow$ Continuing $\longrightarrow$

```
     for i=1:ngm

       gbsm(i,j1)=0.0; gbsm(j1,i)=0.0;

       if(gfln(j) == 0)     % vertex node
          gbsm(i,j2)=0.0; gbsm(j2,i)=0.0;
          gbsm(i,j3)=0.0; gbsm(j3,i)=0.0;
       end
     end

       gbsm(j1,j1) = 1.0; b(j1) = 0.0;

       if(gfln(j) == 0)     % vertex node
          gbsm(j2,j2) = 1.0; b(j2) = 0.0;
          gbsm(j3,j3) = 1.0; b(j3) = 0.0;
       end

    end
end

%-----------------------
% solve the linear system
%-----------------------

sol = b/gbsm';

%-----------------------------------------
% extract the deflection at the vertices
%-----------------------------------------

for i=1:ng
   if(gfln(i) == 0)
     f(i) = sol(s(i));
   end
end

%---------------------------------
% interpolate the mid-side values
%---------------------------------

for l=1:ne

  j1=c(l,1); j2=c(l,2); j3=c(l,3);    % global element labels
  j4=c(l,4); j5=c(l,5); j6=c(l,6);

  e(1) = s(j1); e(2) = s(j1)+1; e(3) = s(j1)+2; % position in the
  e(4) = s(j2); e(5) = s(j2)+1; e(6) = s(j2)+2; % global matrix
  e(7) = s(j3); e(8) = s(j3)+1; e(9) = s(j3)+2;
  e(10)= s(j4); e(11)= s(j5);   e(12)= s(j6);

  for j=1:12
    dof(j) = sol(e(j));
  end
```

Code bend_HCT: $\longrightarrow$ Continuing $\longrightarrow$

```
x1=p(j1,1); y1=p(j1,2);    % three vertices
x2=p(j2,1); y2=p(j2,2);
x3=p(j3,1); y3=p(j3,2);

[a] = HCT_sys (x1,y1, x2,y2, x3,y3, dof);

x=p(j4,1); y=p(j4,2);

f(j4) = a(1)+a(2)*x+a(3)*y+a(4)*x^2+a(5)*x*y+a(6)*y^2 ...
        +a(7)*x^3+a(8)*x^2*y+a(9)*x*y^2+a(10)*y^3;

x=p(j5,1); y=p(j5,2);

f(j5) = a(11)+a(12)*x+a(13)*y+a(14)*x^2+a(15)*x*y+a(16)*y^2 ...
        +a(17)*x^3+a(18)*x^2*y+a(19)*x*y^2+a(20)*y^3;

x=p(j6,1); y=p(j6,2);

f(j6) = a(21)+a(22)*x+a(23)*y+a(24)*x^2+a(25)*x*y+a(26)*y^2 ...
        +a(27)*x^3+a(28)*x^2*y+a(29)*x*y^2+a(30)*y^3;
end

%-----------------------------
% plot using an fselib function
%-----------------------------

plot_6 (ne,ng,p,c,f);

%-----------------------------
% extended connectivity matrix
% for 3-node sub-triangles
%-----------------------------

Ic=0;

for i=1:ne
  Ic=Ic+1; c3(Ic,1) = c(i,1); c3(Ic,2) = c(i,4); c3(Ic,3) = c(i,6);
  Ic=Ic+1; c3(Ic,1) = c(i,4); c3(Ic,2) = c(i,2); c3(Ic,3) = c(i,5);
  Ic=Ic+1; c3(Ic,1) = c(i,5); c3(Ic,2) = c(i,3); c3(Ic,3) = c(i,6);
  Ic=Ic+1; c3(Ic,1) = c(i,4); c3(Ic,2) = c(i,5); c3(Ic,3) = c(i,6);
end

%-----------------------------
% plot using a matlab function
%-----------------------------

trimesh (c3,p(:,1),p(:,2),f); xlabel('x'); ylabel('y');

%-----
% done
%-----
```

Code bend_HCT: ($\longrightarrow$ Continued.) Finite element solution of the forced biharmonic equation describing the bending of a clamped plate.

Figure 5.5.1 shows the graphics output of the code for a circular and a square plate, subject to a uniformly distributed vertical load corresponding to a constant load density function, w. In the case of the circular plate, the exact analytical solution predicts that the maximum deflection, occurring at the center, is given by

$$f_{Max} = \frac{wa^4}{64E_B} \simeq 0.0156 \, \frac{wa^4}{E_B}, \tag{5.5.14}$$

where a is the plate radius ([65], pp. 54–55). The finite element solution for discretization level $ndiv = 3$ predicts $f_{Max} = 0.0154 \, wa^4/E_B$, which is in good agreement with the exact value. In the case of the square plate, an approximate series solution predicts the maximum center-point deflection

$$f_{Max} \simeq 0.0202 \, \frac{wa^4}{E_B}, \tag{5.5.15}$$

where a is the plate edge half-width ([65], p. 202). The finite element solution for discretization level $ndiv = 3$ is in perfect agreement with this prediction up to the fourth decimal place.

5.5.2 Formulation as a system of Poisson-like equations

In this approach, the constitutive equations (5.3.35) and the equilibrium equation (5.3.27) are restated as a system of four Poisson-like equations,

$$
\begin{aligned}
\frac{\partial^2 f}{\partial x^2} + \nu \frac{\partial^2 f}{\partial y^2} &= -\frac{1}{E_B} M_{xx}, \\
\frac{\partial^2 f}{\partial y^2} + \nu \frac{\partial^2 f}{\partial x^2} &= -\frac{1}{E_B} M_{yy}, \\
\frac{\partial^2 f}{\partial x \, \partial y} &= -\frac{1}{E_B (1-\nu)} M_{xy}, \\
\frac{\partial^2 M_{xx}}{\partial x^2} + 2 \frac{\partial^2 M_{xy}}{\partial x \, \partial y} + \frac{\partial^2 M_{yy}}{\partial y^2} &= w.
\end{aligned}
\tag{5.5.16}
$$

The four unknowns functions include the deflection, f, and the bending and twisting moments, M_{xx}, M_{yy}, and M_{xy}.

In the finite element formulation, each of these functions is expanded in terms of the global modes and associated degrees of freedom, as shown in (5.4.22),

$$
\begin{bmatrix} M_{xx} \\ M_{yy} \\ M_{xy} \\ f \end{bmatrix} = \sum_{j=1}^{N_G^m} \begin{bmatrix} M_{xx_j}^G \\ M_{yy_j}^G \\ M_{xy_j}^G \\ f_j^G \end{bmatrix} \phi_j(x,y),
\tag{5.5.17}
$$

(*a*)

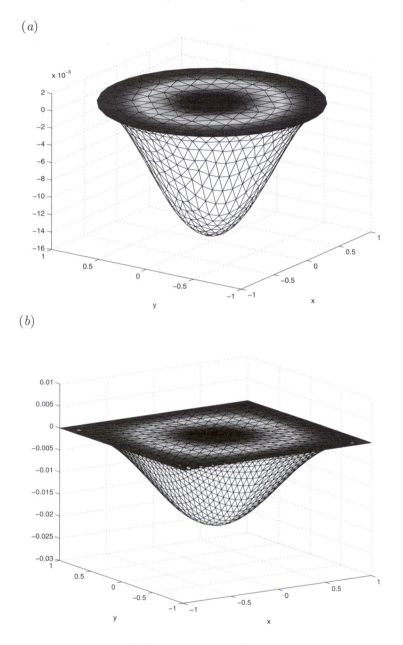

(*b*)

Figure 5.5.1 Bending of (*a*) a circular, and (*b*) square entirely clamped plate, subject to a uniformly distributed vertical load representing, for example, the gravitational force.

where N_G^m is the number of global modes, and the vector on the right-hand side contains the respective coefficients of the global nodes. To simplify the notation, we introduce the vector field

$$\mathbf{F}(x, y) \equiv \begin{bmatrix} M_{xx}(x, y) \\ M_{yy}(x, y) \\ M_{xy}(x, y) \\ f(x, y) \end{bmatrix}, \tag{5.5.18}$$

and express (5.5.17) in the compact form

$$\mathbf{F}(x, y) = \sum_{j=1}^{N_G^m} \mathbf{F}_j^G \, \phi_j(x, y), \tag{5.5.19}$$

where

$$\mathbf{F}_j^G \equiv \begin{bmatrix} M_{xx\,j}^G \\ M_{yy\,j}^G \\ M_{xy\,j}^G \\ f_j^G \end{bmatrix} \tag{5.5.20}$$

is the modal coefficient vector.

Next, we carry out the Galerkin projection by multiplying each equation in (5.5.16) with ϕ_i, and then integrate by parts to eliminate the second derivatives on the left-hand sides. The Galerkin projection of the first equation reads

$$\iint_D \phi_i \left(\frac{\partial^2 f}{\partial x^2} + \nu \frac{\partial^2 f}{\partial y^2} \right) \mathrm{d}x \, \mathrm{d}y = -\frac{1}{E_B} \iint_D \phi_i \, M_{xx} \, \mathrm{d}x \, \mathrm{d}y. \tag{5.5.21}$$

Integrating by parts on the left-hand side, we find

$$\iint_D \phi_i \left(\frac{\partial^2 f}{\partial x^2} + \nu \frac{\partial^2 f}{\partial y^2} \right) \mathrm{d}x \, \mathrm{d}y$$

$$= \iint_D \left[\frac{\partial}{\partial x} \left(\phi_i \frac{\partial f}{\partial x} \right) + \nu \frac{\partial}{\partial y} \left(\phi_i \frac{\partial f}{\partial y} \right) - \frac{\partial \phi_i}{\partial x} \frac{\partial f}{\partial x} - \nu \frac{\partial \phi_i}{\partial y} \frac{\partial f}{\partial y} \right] \mathrm{d}x \, \mathrm{d}y$$

$$= -\oint_C \phi_i \left(\frac{\partial f}{\partial x} n_x + \nu \frac{\partial f}{\partial y} n_y \right) \mathrm{d}l - \iint_D \left(\frac{\partial \phi_i}{\partial x} \frac{\partial f}{\partial x} + \nu \frac{\partial \phi_i}{\partial y} \frac{\partial f}{\partial y} \right) \mathrm{d}x \, \mathrm{d}y, \tag{5.5.22}$$

where $\mathbf{n}$ is the *inward* unit normal vector. Substituting this expression in (5.5.21) and rearranging, we derive the final result

$$\iint_D \left(\frac{\partial \phi_i}{\partial x} \frac{\partial f}{\partial x} + \nu \frac{\partial \phi_i}{\partial y} \frac{\partial f}{\partial y} \right) \mathrm{d}x \, \mathrm{d}y - \frac{1}{E_B} \iint_D \phi_i \, M_{xx} \, \mathrm{d}x \, \mathrm{d}y \tag{5.5.23}$$

$$= -\oint_C \phi_i \left(\frac{\partial f}{\partial x} n_x + \nu \frac{\partial f}{\partial y} n_y \right) \mathrm{d}l.$$

Working in a similar manner with the second and third equations in (5.5.16), we find

$$\iint_D \left(\nu \frac{\partial \phi_i}{\partial x} \frac{\partial f}{\partial x} + \frac{\partial \phi_i}{\partial y} \frac{\partial f}{\partial y} \right) dx \, dy - \frac{1}{E_B} \iint_D \phi_i \, M_{yy} \, dx \, dy$$

$$= - \oint_C \phi_i \left(\nu \frac{\partial f}{\partial x} n_x + \frac{\partial f}{\partial y} n_y \right) dl,$$

(5.5.24)

and

$$\iint_D \left(\frac{\partial \phi_i}{\partial x} \frac{\partial f}{\partial y} + \frac{\partial \phi_i}{\partial y} \frac{\partial f}{\partial x} \right) dx \, dy - \frac{2}{E_B \, (1 - \nu)} \iint_D \phi_i \, M_{xy} \, dx \, dy$$

$$= - \oint_C \phi_i \left(\frac{\partial f}{\partial x} n_y + \frac{\partial f}{\partial y} n_x \right) dl.$$

(5.5.25)

The fourth equation in (5.5.16) in conjunction with the equilibrium equation (5.3.26), $\mathbf{q} = \nabla \cdot \mathbf{M}$, yields

$$\iint_D \left(\frac{\partial \phi_i}{\partial x} \frac{\partial M_{xx}}{\partial x} + \frac{\partial \phi_i}{\partial y} \frac{\partial M_{yy}}{\partial y} + \frac{\partial \phi_i}{\partial y} \frac{\partial M_{xy}}{\partial x} + \frac{\partial \phi_i}{\partial x} \frac{\partial M_{xy}}{\partial y} \right) dx \, dy$$

$$= - \oint_C \phi_i \left[\left(\frac{\partial M_{xx}}{\partial x} + \frac{\partial M_{xy}}{\partial y} \right) n_x + \left(\frac{\partial M_{xy}}{\partial x} + \frac{\partial M_{yy}}{\partial y} \right) n_y \right] dl - \iint_D \phi_i \, w \, dx \, dy$$

$$= - \oint_C \phi_i \, \mathbf{n} \cdot \mathbf{q} \, dl - \iint_D \phi_i \, w \, dx \, dy.$$

(5.5.26)

In the next step, we introduce expansions (5.5.17), and express equations (5.5.23)–(5.5.26) in the form

$$K_{ij}^{xx} f_j^G + \nu \, K_{ij}^{yy} f_j^G - M_{ij} \, M_{xx_j}^G = \mathcal{K}_i,$$

$$\nu \, K_{ij}^{xx} f_j^G + K_{ij}^{yy} f_j^G - M_{ij} \, M_{yy_j}^G = \mathcal{L}_i,$$

$$K_{ij}^{yx} f_j^G + K_{ij}^{xy} f_j^G - \frac{2}{1 - \nu} M_{ij} \, M_{xy_j}^G = \mathcal{M}_i,$$

(5.5.27)

$$K_{ij}^{xx} M_{xx_j}^G + K_{ij}^{yy} M_{yy_j}^G + K_{ij}^{yx} M_{xy_j}^G + K_{ij}^{xy} M_{xy_j}^G = \mathcal{N}_i,$$

where summation of the repeated index j is implied on the left-hand sides in the range $j = 1, 2, \ldots, N_G^m$, subject to the following definitions:

- The mass matrix is defined in the usual way as:

$$M_{ij} = \frac{1}{E_B} \iint_D \phi_i \, \phi_j \, dx \, dy.$$

(5.5.28)

- The directional diffusion-like matrices are defined as:

$$K_{ij}^{xx} = \iint_D \frac{\partial \phi_i}{\partial x} \frac{\partial \phi_j}{\partial x} \, dx \, dy,$$

$$K_{ij}^{xy} = \iint_D \frac{\partial \phi_i}{\partial x} \frac{\partial \phi_j}{\partial y} \, dx \, dy,$$

$$K_{ij}^{yx} = K_{ji}^{xy} = \iint_D \frac{\partial \phi_i}{\partial y} \frac{\partial \phi_j}{\partial x} \, dx \, dy, \qquad (5.5.29)$$

$$K_{ij}^{yy} = \iint_D \frac{\partial \phi_i}{\partial y} \frac{\partial \phi_j}{\partial y} \, dx \, dy.$$

- The right-hand sides are defined as:

$$\mathcal{K}_i = -\oint_C \phi_i \left(\frac{\partial f}{\partial x} n_x + \nu \frac{\partial f}{\partial y} n_y \right) dl,$$

$$\mathcal{L}_i = -\oint_C \phi_i \left(\nu \frac{\partial f}{\partial x} n_x + \frac{\partial f}{\partial y} n_y \right) dl,$$

$$\mathcal{M}_i = -\oint_C \phi_i \left(\frac{\partial f}{\partial x} n_y + \frac{\partial f}{\partial y} n_x \right) dl, \qquad (5.5.30)$$

$$\mathcal{N}_i = -\oint_C \phi_i \, \mathbf{n} \cdot \mathbf{q} \, dl - \iint_D \phi_i \, w \, dx \, dy.$$

Equations (5.5.27) can be collected into the matrix form

$$\sum_{j=1}^{N_G^m} \begin{bmatrix} -M_{ij} & 0 & 0 & K_{ij}^{xx} + \nu \, K_{ij}^{yy} \\ 0 & -M_{ij} & 0 & \nu \, K_{ij}^{xx} + K_{ij}^{yy} \\ 0 & 0 & -\frac{2}{1-\nu} M_{ij} & K_{ij}^{xy} + K_{ij}^{yx} \\ K_{ij}^{xx} & K_{ij}^{yy} & K_{ij}^{xy} + K_{ij}^{yx} & 0 \end{bmatrix} \cdot \mathbf{F}_j^G = \begin{bmatrix} \mathcal{K}_i \\ \mathcal{L}_i \\ \mathcal{M}_i \\ \mathcal{N}_i \end{bmatrix},$$

$$(5.5.31)$$

where the modal coefficient vector $\mathbf{F}_j^G$ was defined in (5.5.19).

The coefficient matrix on the right-hand side can be made symmetric by subtracting from the first equation the second equation multiplied by ν, and from the second equation the first equation multiplied by ν, and then dividing the resulting equations by $1 - \nu^2$, to obtain

$$\sum_{j=1}^{N_G^m} \boldsymbol{\Phi}_{ij} \cdot \mathbf{F}_j^G = \begin{bmatrix} \frac{1}{1-\nu^2} (\mathcal{K}_i - \nu \, \mathcal{L}_i) \\ \frac{1}{1-\nu^2} (\mathcal{L}_i - \nu \, \mathcal{K}_i) \\ \mathcal{M}_i \\ \mathcal{N}_i \end{bmatrix} = \begin{bmatrix} -\oint_C \phi_i \frac{\partial f}{\partial x} \, dl \\ -\oint_C \phi_i \frac{\partial f}{\partial y} \, dl \\ \mathcal{M}_i \\ \mathcal{N}_i \end{bmatrix}, \qquad (5.5.32)$$

where

$$
\mathbf{\Phi}_{ij} =
\begin{bmatrix}
-M'_{ij} & \nu\, M'_{ij} & 0 & K^{xx}_{ij} \\
\nu\, M'_{ij} & -M'_{ij} & 0 & K^{yy}_{ij} \\
0 & 0 & -\frac{2}{1+\nu}\, M_{ij} & K^{xy}_{ij} + K^{yx}_{ij} \\
K^{xx}_{ij} & K^{yy}_{ij} & K^{xy}_{ij} + K^{yx}_{ij} & 0
\end{bmatrix},
\tag{5.5.33}
$$

and

$$
M'_{ij} \equiv \frac{1}{1-\nu^2}\, M_{ij} = \frac{12}{E\,h^3} \iint_D \phi_i\, \phi_j \, \mathrm{d}x\, \mathrm{d}y.
\tag{5.5.34}
$$

In practice, the domain integrals are assembled from corresponding element integrals involving the element interpolation functions.

One-element solution

For illustration, we consider a solution domain consisting of a single 4-node, 10-mode element, as discussed in Section 5.4.1, whereupon the linear system (5.5.32) reduces to

$$
\sum_{j=1}^{10} \mathbf{\Phi}_{ij} \cdot \mathbf{F}^G_j = -
\begin{bmatrix}
\oint_C \phi_i\, \frac{\partial f}{\partial x}\, \mathrm{d}l \\[2mm]
\oint_C \phi_i\, \frac{\partial f}{\partial y}\, \mathrm{d}l \\[2mm]
\oint_C \phi_i \left(\frac{\partial f}{\partial x}\, n_y + \frac{\partial f}{\partial y}\, n_x \right) \mathrm{d}l \\[2mm]
\oint_C \phi_i\, \mathbf{n} \cdot \mathbf{q}\, \mathrm{d}l + \iint_E \phi_i\, w\, \mathrm{d}x\, \mathrm{d}y
\end{bmatrix},
\tag{5.5.35}
$$

for $i = 1, 2, \ldots, 10$, where E denotes the element surface, and C denotes the element contour. For a plate with simply-supported edges, the boundary conditions require that the deflection is restricted, $f = 0$, and the bending moments are zero at the three vertices. Consequently, the nine modal coefficients associated with the vertices are zero, $\mathbf{F}^G_j = \mathbf{0}$, for $j = 1, 2, \ldots, 9$, and the left-hand side of (5.5.35) involves a single term, $\mathbf{\Phi}_{i,10} \cdot \mathbf{F}^G_{10}$.

Applying (5.5.35) for the tenth global mode corresponding to the interior element node, $i = 10$, and recalling that ϕ_{10} is zero around the perimeter of the triangle, we obtain a system of four linear equations for the bending and twisting moments and the displacement at the interior node,

$$
\mathbf{\Phi}_{10,10} \cdot
\begin{bmatrix}
M_{xx} \\
M_{yy} \\
M_{xy} \\
f
\end{bmatrix}^G_{10}
= -
\begin{bmatrix}
0 \\
0 \\
0 \\
\iint_E \phi_{10}\, w\, \mathrm{d}x\, \mathrm{d}y
\end{bmatrix}.
\tag{5.5.36}
$$

PROBLEMS

5.5.1 *Element stiffness matrix for the biharmonic equation.*

Verify by numerical computation that the element stiffness matrix for the HCT triangle is singular.

5.5.2 *Bending of a rectangular plate.*

Run the code `bend_HCT` to compute the bending of a rectangular plate with aspect ratio 2:1 subject to a uniform load, and compare your results with the series solution predicting the maximum deflection

$$f_{Max} \simeq 0.0406 \, \frac{wa^4}{E_B}, \tag{5.5.37}$$

where a is the small plate edge half-width ([65], p. 202).

5.5.3 *One-element bending problem.*

Write a code that formulates and solves system (5.5.36), run the code for a triangle with a shape of your choice, and discuss the results of your computation.

5.6   Viscous flow

The motion of a fluid is governed by Cauchy's equation expressing Newton's second law of motion for a small fluid parcel idealized as an infinitesimal point particle (e.g., [42]). Cauchy's equation relates the acceleration of a point particle, $\mathbf{a}$, to the divergence of the Cauchy stress tensor, $\boldsymbol{\sigma}$, introduced in Section 5.1.1, to the body force, $\mathbf{b}$,

$$\rho \, \mathbf{a} = \nabla \cdot \boldsymbol{\sigma} + \mathbf{b}, \tag{5.6.1}$$

where ρ is the fluid density. In the case of the gravitational force, $\mathbf{b} = \rho \mathbf{g}$, where $\mathbf{g} = (g_x, g_y, g_z)$ is the acceleration of gravity. The Cartesian components of (5.6.1) with the gravitational body force read

$$\rho \, a_x = \frac{\partial \sigma_{xx}}{\partial x} + \frac{\partial \sigma_{yx}}{\partial y} + \frac{\partial \sigma_{zx}}{\partial z} + \rho \, g_x,$$

$$\rho \, a_y = \frac{\partial \sigma_{xy}}{\partial x} + \frac{\partial \sigma_{yy}}{\partial y} + \frac{\partial \sigma_{zy}}{\partial z} + \rho \, g_y, \tag{5.6.2}$$

$$\rho \, a_z = \frac{\partial \sigma_{xz}}{\partial x} + \frac{\partial \sigma_{yz}}{\partial y} + \frac{\partial \sigma_{zz}}{\partial z} + \rho \, g_z.$$

To evaluate the point particle acceleration, we introduce the material derivative, D/Dt, expressing the rate of change of a variable following the motion of a material point particle. By definition, the point particle acceleration is equal to the material derivative of the fluid velocity, $\mathbf{u} = (u_x, u_y, u_z)$,

$$\mathbf{a} = \frac{D\mathbf{u}}{Dt} \equiv \frac{\partial \mathbf{u}}{\partial t} + \mathbf{u} \cdot \nabla \mathbf{u}. \tag{5.6.3}$$

In index notation,

$$a_i = \frac{\partial u_i}{\partial t} + u_x \frac{\partial u_i}{\partial x} + u_y \frac{\partial u_i}{\partial y} + u_z \frac{\partial u_i}{\partial z}, \tag{5.6.4}$$

for $i = x, y, z$.

Mass conservation for an incompressible fluid requires the continuity equation,

$$\nabla \cdot \mathbf{u} = 0. \tag{5.6.5}$$

Explicitly,

$$\frac{\partial u_j}{\partial x_j} = \frac{\partial u_x}{\partial x} + \frac{\partial u_y}{\partial y} + \frac{\partial u_z}{\partial z} = 0, \tag{5.6.6}$$

where summation of the repeated index, j, is implied on the left-hand side over $i = 1, 2, 3$ or x, y, z. To simplify the notation, we have denoted $x_1 = x$, $x_2 = y$, $x_3 = z$, and $u_1 = u_x$, $u_2 = u_y$, $u_3 = u_z$.

In the case of an incompressible *Newtonian* fluid, the stress is related to the pressure and to the rate-of-deformation tensor,

$$\mathbf{E} \equiv \frac{1}{2} \left(\nabla \mathbf{u} + (\nabla \mathbf{u})^T \right), \tag{5.6.7}$$

by the linear constitutive equation

$$\boldsymbol{\sigma} = -p\,\mathbf{I} + \mu\,2\mathbf{E}, \tag{5.6.8}$$

where p is the pressure, $\mathbf{I}$ is the identity matrix, μ is the fluid viscosity, and the superscript T denotes the matrix transpose. In index notation,

$$\sigma_{ij} = -p\,\delta_{ij} + \mu \left(\frac{\partial u_i}{\partial x_j} + \frac{\partial u_j}{\partial x_i} \right), \tag{5.6.9}$$

where δ_{ij} is Kronecker's delta.

Substituting (5.6.3) and (5.6.8) in (5.6.1), and simplifying by use of the continuity equation, we derive the Navier-Stokes equation,

$$\rho \left(\frac{\partial \mathbf{u}}{\partial t} + \mathbf{u} \cdot \nabla \mathbf{u} \right) = -\nabla p + \mu \, \nabla^2 \mathbf{u} + \rho \, \mathbf{g} \tag{5.6.10}$$

(Problem 5.6.1). In index notation, the ith component of the Navier-Stokes equation reads

$$\rho \left(\frac{\partial u_i}{\partial t} + u_j \frac{\partial u_i}{\partial x_j} \right) = -\frac{\partial p}{\partial x_i} + \mu \, \nabla^2 u_i + \rho \, g_i, \tag{5.6.11}$$

for $i = 1, 2, 3$ or x, y, z, where summation of the repeated index, j, is implied on the left-hand side.

When the fluid density is uniform throughout the domain of flow, the effect of gravity can be incorporated into a modified pressure defined as

$$p^{Mod} \equiv p - \rho \left(g_x \, x + g_y \, y + g_z \, z \right) = p - \rho \, \mathbf{g} \cdot \mathbf{x}. \tag{5.6.12}$$

Written in terms of the modified pressure, the gravitational term on the right-hand side of the Navier-Stokes equation does not appear. For simplicity, heretoforth we drop the qualifier *Mod*, bearing in mind that the pressure has been modified to incorporate the body force. The distinction between the two pressures arises only when boundary conditions involving the physical pressure are specified.

In the case of two-dimensional flow in the xy plane, the x and y components of the Navier-Stokes equation read

$$\rho \left(\frac{\partial u_x}{\partial t} + u_x \frac{\partial u_x}{\partial x} + u_y \frac{\partial u_x}{\partial y} \right) = -\frac{\partial p}{\partial x} + \mu \left(\frac{\partial^2 u_x}{\partial x^2} + \frac{\partial^2 u_x}{\partial y^2} \right),$$

$$\rho \left(\frac{\partial u_y}{\partial t} + u_x \frac{\partial u_y}{\partial x} + u_y \frac{\partial u_y}{\partial y} \right) = -\frac{\partial p}{\partial y} + \mu \left(\frac{\partial^2 u_y}{\partial x^2} + \frac{\partial^2 u_y}{\partial y^2} \right),$$

$$\tag{5.6.13}$$

and the continuity equation reads

$$\frac{\partial u_x}{\partial x} + \frac{\partial u_y}{\partial y} = 0. \tag{5.6.14}$$

To solve the Navier-Stokes equation, we require boundary conditions at the physical or computational boundaries of the flow. Typically, we specify the velocity, $\mathbf{u}$, at a solid surface, the traction $\mathbf{f} \equiv \mathbf{n} \cdot \boldsymbol{\sigma}$, at a free surface, and the jump in the traction at a fluid interface, where $\mathbf{n}$ is the unit normal vector. A velocity boundary condition is a Dirichlet boundary condition, and a traction boundary condition is a Neumann boundary condition.

5.6.1   Velocity and pressure interpolation

As a first step toward implementing the finite element method, we discretize the domain of flow into finite elements, as discussed in Chapters 3 and 4, and approximate the velocity and pressure fields using two *different* sets of global interpolation functions, as

$$\mathbf{u} = \sum_{j=1}^{N_G^u} \mathbf{u}_j(t) \, \phi_j^u, \qquad p = \sum_{j=1}^{N_G^p} p_j(t) \, \phi_j^p, \tag{5.6.15}$$

where:

- N_G^u is the number of velocity global interpolation nodes, and $\mathbf{u}_j(t)$ are the corresponding nodal values of the velocity vector, for $j = 1, 2, \ldots, N_G^u$.

- N_G^p is the number of pressure global interpolation nodes, and $p_j(t)$ are the corresponding nodal values of the pressure, for $j = 1, 2, \ldots, N_G^p$.

Specific examples will be given in Section 5.7 in the context of Stokes flow.

5.6.2 Galerkin projections

Consider a flow in a two-dimensional domain, D, that is bounded by a single contour or a collection of contours, denoted by C. To derive the Galerkin finite element equations, we project each component of the Navier-Stokes equation (5.6.10) on the velocity interpolation functions, ϕ_j^u, to obtain the vectorial equation

$$\iint_D \phi_i^u \, \rho \left(\frac{\partial \mathbf{u}}{\partial t} + \mathbf{u} \cdot \nabla \mathbf{u} \right) \mathrm{d}x \, \mathrm{d}y \tag{5.6.16}$$

$$= \iint_D \phi_i^u \left(-\nabla p + \mu \, \nabla^2 \mathbf{u} \right) \mathrm{d}x \, \mathrm{d}y,$$

for $i = 1, 2, \ldots, N_G^u$. Moreover, we project the continuity equation (5.6.5) on the pressure interpolation functions, ϕ_j^p, to obtain

$$\iint_D \phi_i^p \, (\nabla \cdot \mathbf{u}) \, \mathrm{d}x \, \mathrm{d}y = 0, \tag{5.6.17}$$

for $i = 1, 2, \ldots, N_G^p$.

Next, we reduce the order of differentiation defined by the Laplacian of the velocity on the right-hand side of (5.6.16). For this purpose, we restate the Laplacian as the divergence of the gradient, and write

$$\iint_D \phi_i^u \, (\nabla^2 \mathbf{u}) \, \mathrm{d}x \, \mathrm{d}y = \iint_D \phi_i^u \, (\nabla \cdot \nabla \mathbf{u}) \, \mathrm{d}x \, \mathrm{d}y$$

$$\tag{5.6.18}$$

$$= \iint_D \nabla \cdot (\phi_i^u \, \nabla \mathbf{u}) \, \mathrm{d}x \, \mathrm{d}y - \iint_D \nabla \phi_i^u \cdot \nabla \mathbf{u} \, \mathrm{d}x \, \mathrm{d}y.$$

Using the Gauss divergence theorem to convert the first integral on the right-hand side to a contour integral along the boundary, C, we obtain

$$\iint_D \phi_i^u \, (\nabla^2 \mathbf{u}) \, \mathrm{d}x \, \mathrm{d}y = -\oint_C \phi_i^u \, \mathbf{n} \cdot \nabla \mathbf{u} \, \mathrm{d}l - \iint_D \nabla \phi_i^u \cdot \nabla \mathbf{u} \, \mathrm{d}x \, \mathrm{d}y, \tag{5.6.19}$$

where $\mathbf{n}$ is the unit vector normal to C pointing *into* the flow, and l is the arc length along C.

Working similarly with the pressure term on the right-hand side of (5.6.16), we find

$$\iint_D \phi_i^u \, \nabla p \, \mathrm{d}x \, \mathrm{d}y = \iint_D \nabla(\phi_i^u \, p) \, \mathrm{d}x \, \mathrm{d}y - \iint_D \nabla \phi_i^u \, p \, \mathrm{d}x \, \mathrm{d}y$$

$$\tag{5.6.20}$$

$$= -\oint_C \phi_i^u \, p \, \mathbf{n} \, \mathrm{d}l - \iint_D \nabla \phi_i^u \, p \, \mathrm{d}x \, \mathrm{d}y.$$

Substituting (5.6.19) and (5.6.20) in (5.6.16), we derive the final result

$$
\iint_D \phi_i^u \, \rho \left(\frac{\partial \mathbf{u}}{\partial t} + \mathbf{u} \cdot \nabla \mathbf{u} \right) \mathrm{d}x \, \mathrm{d}y
$$

$$
= \oint_C \phi_i^u \, p \, \mathbf{n} \, \mathrm{d}l - \mu \oint_C \phi_i^u \, \mathbf{n} \cdot \nabla \mathbf{u} \, \mathrm{d}l \tag{5.6.21}
$$

$$
+ \iint_D \nabla \phi_i^u \, p \, \mathrm{d}x \, \mathrm{d}y - \mu \iint_D \nabla \phi_i^u \cdot \nabla \mathbf{u} \, \mathrm{d}x \, \mathrm{d}y,
$$

for $i = 1, 2, \ldots N_G^u$. The x and y components of this equation read

$$
\iint_D \phi_i^u \, \rho \left(\frac{\partial \mathbf{u}}{\partial t} + \mathbf{u} \cdot \nabla \mathbf{u} \right)_x \mathrm{d}x \, \mathrm{d}y
$$

$$
= \oint_C \phi_i^u \, p \, n_x \, \mathrm{d}l - \mu \oint_C \phi_i^u \, \mathbf{n} \cdot \nabla u_x \, \mathrm{d}l \tag{5.6.22}
$$

$$
+ \iint_D \frac{\partial \phi_i^u}{\partial x} \, p \, \mathrm{d}x \, \mathrm{d}y - \mu \iint_D \nabla \phi_i^u \cdot \nabla u_x \, \mathrm{d}x \, \mathrm{d}y,
$$

and

$$
\iint_D \phi_i^u \, \rho \left(\frac{\partial \mathbf{u}}{\partial t} + \mathbf{u} \cdot \nabla \mathbf{u} \right)_y \mathrm{d}x \, \mathrm{d}y
$$

$$
= \oint_C \phi_i^u \, p \, n_y \, \mathrm{d}l - \mu \oint_C \phi_i^u \, \mathbf{n} \cdot \nabla u_y \, \mathrm{d}l \tag{5.6.23}
$$

$$
+ \iint_D \frac{\partial \phi_i^u}{\partial y} \, p \, \mathrm{d}x \, \mathrm{d}y - \mu \iint_D \nabla \phi_i^u \cdot \nabla u_y \, \mathrm{d}x \, \mathrm{d}y,
$$

where the x and y subscripts on the left-hand sides denote the corresponding Cartesian components of the enclosed quantity.

To simplify the notation, we can multiply the x component (5.6.22) by the arbitrary constant a, and the y component (5.6.23) by the arbitrary constant b, and add the resulting equations to obtain

$$
\iint_D \rho \, \mathbf{h} \cdot \left(\frac{\partial \mathbf{u}}{\partial t} + \mathbf{u} \cdot \nabla \mathbf{u} \right) \mathrm{d}x \, \mathrm{d}y
$$

$$
= \oint_C \mathbf{n} \cdot \mathbf{h} \, p \, \mathrm{d}l - \mu \oint_C \mathbf{n} \cdot (\nabla \mathbf{u}) \cdot \mathbf{h} \, \mathrm{d}l \tag{5.6.24}
$$

$$
+ \iint_D p \, \nabla \cdot \mathbf{h} \, \mathrm{d}x \, \mathrm{d}y - \mu \iint_D \nabla \mathbf{h} : \nabla \mathbf{u} \, \mathrm{d}x \, \mathrm{d}y,
$$

where

$$
\mathbf{h} \equiv \phi_i^u \begin{bmatrix} a \\ b \end{bmatrix} \tag{5.6.25}
$$

is a modulated interpolation function. The double dot product in the last integral of (5.6.24) signifies the sum of the products of corresponding elements of the matrices on either side. For every global interpolation function, we introduce two independent vectors **h** corresponding to different choices of the parameters a and b, which can be combined to yield the x and y components shown in (5.6.22) and (5.6.23).

We conclude the finite element formulation by substituting the velocity and pressure finite element expansions (5.6.15) in (5.6.21) and (5.6.17), and thus derive:

- A system of ordinary differential equations (ODEs) for the nodal velocities, $\mathbf{u}_i(t)$, also involving the nodal pressures, $p_i(t)$. If the flow is steady, we obtain a system of algebraic equations.

- A complementary linear system of linear algebraic equations for the nodal velocities associated with the continuity equation, effectively acting as constraints.

The structure and properties of this system will be discussed in Section 5.7 in the context of Stokes flow.

PROBLEMS

5.6.1 *Navier-Stokes equation from the Cauchy equation of motion.*

Derive the Navier-Stokes equation from the Cauchy equation of motion, as discussed in the text.

5.6.2 *Galerkin projection of Cauchy's equation of motion.*

Perform the Galerkin projection of the Cauchy equation of motion with the velocity interpolation function, ϕ_i^u, to derive the Galerkin equation

$$\iint_D \phi_i^u \, \rho \left(\frac{\partial \mathbf{u}}{\partial t} + \mathbf{u} \cdot \nabla \mathbf{u} \right) \mathrm{d}x \, \mathrm{d}y$$

$$= - \oint_C \phi_i^u \, \mathbf{f} \, \mathrm{d}l - \iint_D \nabla \phi_i^u \cdot \boldsymbol{\sigma} \, \mathrm{d}x \, \mathrm{d}y, \qquad (5.6.26)$$

for $i = 1, 2, \ldots N_G^u$, where $\mathbf{f} \equiv \mathbf{n} \cdot \boldsymbol{\sigma}$ is the boundary traction, and $\mathbf{n}$ is the unit normal vector pointing into the flow. This equation is a point of departure for computing free-surface flows where a condition for the boundary traction is specified.

5.7 Stokes flow

Consider a fluid flow in a domain of overall size L, where the magnitude of the velocity field is comparable to the characteristic scale U. In the case of flow through a circular tube, the length L can be identified with the tube diameter, and the velocity U can be identified with the fluid velocity at the tube centerline.

In the case of flow due to a settling particle, the length L can be identified with the particle diameter, and the velocity U can be identified with the particle settling velocity.

Analysis shows that, when the Reynolds number, $Re = \rho U L/\mu$, is sufficiently small, typically less than unity, the left-hand side of the Navier-Stokes equation expressing the effect of fluid inertia is small compared to the right-hand side expressing the pressure, viscous, and body forces, and may be neglected without significant error (e.g., [42]). In the case of unsteady flow, one consequence of dropping the time derivative $\partial \mathbf{u}/\partial t$ is that the evolution becomes *quasi-steady*, which means that time-dependence is due exclusively to the changing boundary geometry. In the absence of inertia, if the boundaries are stationary, the flow is steady.

Setting the left-hand side of the Navier-Stokes equation equal to zero, we obtain the Stokes equation describing creeping flow. In terms of the modified pressure incorporating the effect of gravity, the Stokes equation reads

$$-\nabla p + \mu \nabla^2 \mathbf{u} = \mathbf{0}. \tag{5.7.1}$$

The numerical task is to solve this equation together with the continuity equation, $\nabla \cdot \mathbf{u} = 0$, subject to appropriate conditions on the physical or computational boundaries of the flow.

5.7.1 Galerkin finite element equations

In the case of Stokes flow, the Galerkin finite element projection of the equation of motion originating from (5.6.21) reduces to

$$\oint_C \phi_i^u \, p \, \mathbf{n} \, dl - \mu \oint_C \phi_i^u \, \mathbf{n} \cdot \nabla \mathbf{u} \, dl$$
$$+ \iint_D \nabla \phi_i^u \, p \, dx \, dy - \mu \iint_D \nabla \phi_i^u \cdot \nabla \mathbf{u} \, dx \, dy = \mathbf{0}, \tag{5.7.2}$$

for $i = 1, 2, \ldots, N_G^u$, where $\mathbf{n}$ is the unit vector normal to the boundary, C, pointing into the flow. The x and y components of this vectorial equation read

$$\oint_C \phi_i^u \, p \, n_x \, dl - \mu \oint_C \phi_i^u \, \mathbf{n} \cdot \nabla u_x \, dl$$
$$+ \iint_D \frac{\partial \phi_i^u}{\partial x} \, p \, dx \, dy - \mu \iint_D \nabla \phi_i^u \cdot \nabla u_x \, dx \, dy = 0, \tag{5.7.3}$$

and

$$\oint_C \phi_i^u \, p \, n_y \, dl - \mu \oint_C \phi_i^u \, \mathbf{n} \cdot \nabla u_y \, dl$$
$$+ \iint_D \frac{\partial \phi_i^u}{\partial y} \, p \, dx \, dy - \mu \iint_D \nabla \phi_i^u \cdot \nabla u_y \, dx \, dy = 0. \tag{5.7.4}$$

Substituting the velocity and pressure finite element expansions (5.6.15) in (5.7.2), and noting that

$$\nabla u_x = \sum_{j=1}^{N_G^u} (\nabla \phi_j^u)\, u_{x_j}, \qquad \nabla u_y = \sum_{j=1}^{N_G^u} (\nabla \phi_j^u)\, u_{y_j}, \qquad (5.7.5)$$

and thus

$$\nabla \mathbf{u} = \sum_{j=1}^{N_G^u} (\nabla \phi_j^u)\, \mathbf{u}_j, \qquad (5.7.6)$$

we obtain

$$\sum_{j=1}^{N_G^p} p_j \oint_C \phi_i^u\, \phi_j^p\, \mathbf{n}\, \mathrm{d}l - \mu \sum_{j=1}^{N_G^u} \mathbf{u}_j \oint_C \phi_i^u\, \mathbf{n} \cdot \nabla \phi_j^u\, \mathrm{d}l$$

$$\hspace{8cm} (5.7.7)$$

$$+ \sum_{j=1}^{N_G^p} p_j \iint_D (\nabla \phi_i^u)\, \phi_j^p\, \mathrm{d}x\, \mathrm{d}y - \mu \sum_{j=1}^{N_G^u} \mathbf{u}_j \iint_D \nabla \phi_i^u \cdot \nabla \phi_j^u\, \mathrm{d}x\, \mathrm{d}y = \mathbf{0}.$$

In terms of the x and y components of the velocity at the jth node encapsulated in the nodal vector, $\mathbf{u}_j = (u_{x_j}, u_{y_j})$, the x and y components of this vectorial equation read

$$\sum_{j=1}^{N_G^p} p_j \oint_C \phi_i^u\, \phi_j^p\, n_x\, \mathrm{d}l - \mu \sum_{j=1}^{N_G^u} u_{x_j} \oint_C \phi_i^u\, \mathbf{n} \cdot \nabla \phi_j^u\, \mathrm{d}l$$

$$\hspace{8cm} (5.7.8)$$

$$+ \sum_{j=1}^{N_G^p} p_j \iint_D \frac{\partial \phi_i^u}{\partial x}\, \phi_j^p\, \mathrm{d}x\, \mathrm{d}y - \mu \sum_{j=1}^{N_G^u} u_{x_j} \iint_D \nabla \phi_i^u \cdot \nabla \phi_j^u\, \mathrm{d}x\, \mathrm{d}y = 0.$$

and

$$\sum_{j=1}^{N_G^p} p_j \oint_C \phi_i^u\, \phi_j^p\, n_y\, \mathrm{d}l - \mu \sum_{j=1}^{N_G^u} u_{y_j} \oint_C \phi_i^u\, \mathbf{n} \cdot \nabla \phi_j^u\, \mathrm{d}l$$

$$\hspace{8cm} (5.7.9)$$

$$+ \sum_{j=1}^{N_G^p} p_j \iint_D \frac{\partial \phi_i^u}{\partial y}\, \phi_j^p\, \mathrm{d}x\, \mathrm{d}y - \mu \sum_{j=1}^{N_G^u} u_{y_j} \iint_D \nabla \phi_i^u \cdot \nabla \phi_j^u\, \mathrm{d}x\, \mathrm{d}y = 0.$$

The last integral on the right-hand sides is recognized as the global diffusion matrix for the velocity interpolation functions, denoted by

$$D_{ij}^u \equiv \iint_D \nabla \phi_i^u \cdot \nabla \phi_j^u\, \mathrm{d}x\, \mathrm{d}y. \qquad (5.7.10)$$

Substituting further the velocity finite element expansion in the Galerkin projection of the continuity equation (5.6.17), we find

$$\sum_{j=1}^{N_G^u} \mathbf{u}_j \cdot \iint_D \phi_i^p \, \nabla \phi_j^u \, \mathrm{d}x \, \mathrm{d}y = 0, \tag{5.7.11}$$

for $i = 1, 2, \ldots, N_G^p$.

In compact notation, equations (5.7.7) and (5.7.11) take the form

$$\sum_{j=1}^{N_G^p} \mathbf{Q}_{ij} \, p_j - \mu \sum_{j=1}^{N_G^u} R_{ij} \, \mathbf{u}_j + \sum_{j=1}^{N_G^p} \mathbf{D}_{ji} \, p_j - \mu \sum_{j=1}^{N_G^u} D_{ij}^u \, \mathbf{u}_j = \mathbf{0}, \tag{5.7.12}$$

and

$$\sum_{j=1}^{N_G^u} \mathbf{D}_{ij} \cdot \mathbf{u}_j = 0, \tag{5.7.13}$$

where

$$\mathbf{Q}_{ij} \equiv \oint_C \phi_i^u \, \phi_j^p \, \mathbf{n} \, \mathrm{d}l, \qquad R_{ij} \equiv \oint_C \phi_i^u \, \mathbf{n} \cdot \nabla \phi_j^u \, \mathrm{d}l, \tag{5.7.14}$$

and

$$\mathbf{D}_{ij} \equiv \iint_D \phi_i^p \, \nabla \phi_j^u \, \mathrm{d}x \, \mathrm{d}y. \tag{5.7.15}$$

Specifically, $\mathbf{Q}_{ij} = (Q_{x_{ij}}, Q_{y_{ij}})$, where

$$Q_{x_{ij}} = \oint_C \phi_i^u \, \phi_j^p \, n_x \, \mathrm{d}l, \qquad Q_{y_{ij}} = \oint_C \phi_i^u \, \phi_j^p \, n_y \, \mathrm{d}l, \tag{5.7.16}$$

and $\mathbf{D}_{ij} = (D_{x_{ij}}, D_{y_{ij}})$, where

$$D_{x_{ij}} = \iint_D \phi_i^p \, \frac{\partial \phi_j^u}{\partial x} \, \mathrm{d}l, \qquad D_{y_{ij}} = \iint_D \phi_i^p \, \frac{\partial \phi_j^u}{\partial y} \, \mathrm{d}l. \tag{5.7.17}$$

Subject to these definitions, the x and y components of (5.7.12) read

$$\sum_{j=1}^{N_G^p} Q_{x_{ij}} \, p_j - \mu \sum_{j=1}^{N_G^u} R_{ij} \, u_{x_j} + \sum_{j=1}^{N_G^p} D_{x_{ji}} \, p_j - \mu \sum_{j=1}^{N_G^u} D_{ij}^u \, u_{x_j} = 0,$$

$$\tag{5.7.18}$$

$$\sum_{j=1}^{N_G^p} Q_{y_{ij}} \, p_j - \mu \sum_{j=1}^{N_G^u} R_{ij} \, u_{y_j} + \sum_{j=1}^{N_G^p} D_{y_{ji}} \, p_j - \mu \sum_{j=1}^{N_G^u} D_{ij}^u \, u_{y_j} = 0.$$

If the velocity is provided as a boundary condition all around the flow boundary, C, the projection of the Stokes equation on ϕ_i^u is excluded for those nodes that lie on the boundary. For all other nodes, because $\phi_i^u = 0$ on C, the matrices R_{ij} and $\mathbf{Q}_{ij}$ are identically zero. Thus, the Galerkin projection for the interior velocity nodes simplifies to

$$\mu \sum_{j=1}^{N_G^u} D_{ij}^u\, \mathbf{u}_j - \sum_{j=1}^{N_G^p} \mathbf{D}_{ji}\, p_j = \mathbf{0}, \tag{5.7.19}$$

whose x and y components are

$$\mu \sum_{j=1}^{N_G^u} D_{ij}^u\, u_{x_j} - \sum_{j=1}^{N_G^p} D_{x_{ji}}\, p_j = 0,$$

$$\mu \sum_{j=1}^{N_G^u} D_{ij}^u\, u_{y_j} - \sum_{j=1}^{N_G^p} D_{y_{ji}}\, p_j = 0. \tag{5.7.20}$$

Writing these equations for all velocity nodes, bearing in mind that additional terms should have been included for the boundary nodes, and appending to the system thus obtained the continuity equation (5.7.13), we derive the grand linear system

$$\begin{bmatrix} \mu\,\mathbf{D}^u & \mathbf{0} & -\mathbf{D}_x^T \\ \mathbf{0} & \mu\,\mathbf{D}^u & -\mathbf{D}_y^T \\ -\mathbf{D}_x & -\mathbf{D}_y & \mathbf{0} \end{bmatrix} \cdot \begin{bmatrix} u_{x_1} \\ u_{x_2} \\ \vdots \\ u_{x_{N_G^u}} \\ - \\ u_{y_1} \\ u_{y_2} \\ \vdots \\ u_{y_{N_G^u}} \\ - \\ p_1 \\ p_2 \\ \vdots \\ p_{N_G^p} \end{bmatrix} = \mathbf{0}, \tag{5.7.21}$$

where $\mathbf{D}^u$ is a square $N_G^u \times N_G^u$ matrix, $\mathbf{D}_x$ and $\mathbf{D}_y$ are rectangular $N_G^p \times N_G^u$ matrices, and the superscript T denotes the matrix transpose. Note that the block coefficient matrix on the left-hand side of the grand linear system (5.7.21) is symmetric.

In the simplest implementation, the pressure is approximated with a constant function over each element, which amounts to placing pressure interpolation nodes at designated element centers and setting $\phi_j^p = 1$ over the jth element, and $\phi_j^p = 0$ outside the element. Accordingly,

$$D_{x_{ij}} = \iint_{E_i} \frac{\partial \phi_j^u}{\partial x} \, dx \, dy, \qquad D_{y_{ij}} = \iint_{E_i} \frac{\partial \phi_j^u}{\partial y} \, dx \, dy, \qquad (5.7.22)$$

where E_j denotes the jth element. The integrals in (5.7.22) are non-zero only if the jth global velocity node is an ith-element node.

5.7.2　Code for Stokes flow in a rectangular cavity

Consider flow in a rectangular cavity that is filled with a liquid, where the motion is driven by the translation of the cavity lid parallel to itself with constant velocity, V, as illustrated in Figure 5.7.1. The non-slip and no-penetration boundary conditions require that the x and y velocity components, u_x and u_y, are zero over the bottom and side walls. At the cavity lid, $u_x = V$ and $u_y = 0$.

FSELIB code `cvt6`, listed in the text, assembles and solves the linear system (5.7.21) for the nodal velocities and pressures. The structure of the code is similar to that of code `lap16_d` for Laplace's equation in a disk-like domain subject to the Dirichlet boundary condition, as discussed in Section 4.2, and some parts are virtually identical. Noteworthy new features include the following:

- The domain of flow is discretized into 6-node triangular elements using the FSELIB function `trgl6_sqr`, which is similar to the function `trgl6_disk` for a disk-like domain discussed in Section 4.2. The union of the elements is defined by a group of N_G geometrical global nodes.

- The x and y velocity components are interpolated isoparametrically, that is, they are approximated with quadratic functions, whereas the pressure is approximated with a constant function over each element, resulting in a *mixed* finite element expansion. Accordingly, the number of velocity and pressure global nodes are

$$N_G^u = N_G, \qquad N_G^p = N_E, \qquad (5.7.23)$$

  where N_E is the number of elements, and N_G is the number of global nodes.

- The computation of the global diffusion matrix, $\mathbf{D}^u$, named *gdm* in the code, follows the procedure described in Section 4.2 for code `lap16_d`.

- To compile the matrices $\mathbf{D}_x$ and $\mathbf{D}_y$ defined in (5.7.17), we run over the elements regarded as pressure nodes, compute the matrix entries with the help of the function `cvt6_D`, listed in the text, and make appropriate contributions to the finite element equations.

(a)

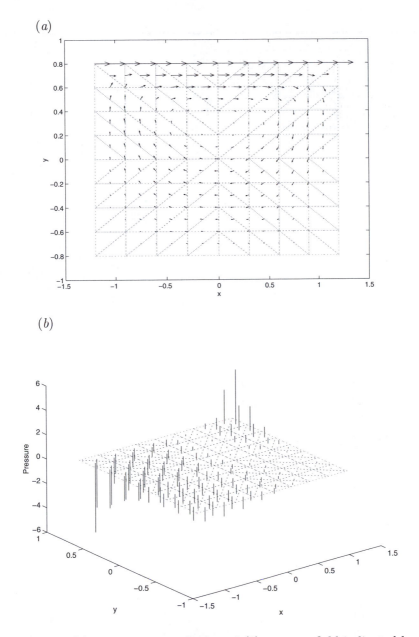

(b)

Figure 5.7.1 (a) Velocity vector field, and (b) pressure field indicated by vertical bars passing through the element centroids, for Stokes flow in a rectangular cavity driven by a translating lid, produced by the FSELIB code cvt6.

```
%==================================
%
% CODE cvt6
%
% Stokes flow in a cavity computed
% with 6-node triangular elements
%
%==================================

%-----------
% input data
%-----------

visc = 1.0; % viscosity
V = 1.0;    % lid velocity
NQ = 6;     % gauss-triangle quadrature
ndiv = 1;   % discretization level

%------------
% triangulate
%------------

[ne,ng,p,c,efl,gfl] = trgl6_sqr (ndiv);

%-----------------------------
% specify the boundary velocity
%-----------------------------

for i=1:ng

  if(gfl(i,1)==1)  % boundary node

    gfl(i,2) = 0.0; % x velocity
    gfl(i,3) = 0.0; % y velocity

    if(p(i,2) > 0.99)  % point is on the lid
     gfl(i,2) = V;     % x velocity of the lid
    end

  end

end

%----------------------------
% deform square to a rectangle
%----------------------------

def = 0.2; % example

for i=1:ng
 p(i,1)=p(i,1)*(1.0 + def);
 p(i,2)=p(i,2)*(1.0 - def);
end
```

Code cvt6: Continuing →

```
%----------------------------------------
% assemble: the global diffusion matrix
%           the Dx and Dy matrices
%----------------------------------------

gdm = zeros(ng,ng); % initialize

gDx = zeros(ne,ng); % initialize

gDy = zeros(ne,ng); % initialize

for l=1:ne          % loop over the elements

% compute the element diffusion matrix

 j=c(l,1); x1=p(j,1); y1=p(j,2);
 j=c(l,2); x2=p(j,1); y2=p(j,2);
 j=c(l,3); x3=p(j,1); y3=p(j,2);
 j=c(l,4); x4=p(j,1); y4=p(j,2);
 j=c(l,5); x5=p(j,1); y5=p(j,2);
 j=c(l,6); x6=p(j,1); y6=p(j,2);

 [edm_elm, arel] = edm6 ...
 ...
      (x1,y1, x2,y2, x3,y3, x4,y4, x5,y5, x6,y6, NQ);

% compute the element matrices Dx, Dy

 [Dx, Dy] = cvt6_D ...
 ...
      (x1,y1, x2,y2, x3,y3, x4,y4, x5,y5, x6,y6, NQ);

   for i=1:6

     i1 = c(l,i);
     for j=1:6
       j1 = c(l,j);
       gdm(i1,j1) = gdm(i1,j1) + edm_elm(i,j);
     end
     gDx(l,i1) = Dx(i);
     gDy(l,i1) = Dy(i);

   end

end   % end of loop over the elements

%------------------------
% compile the grand matrix
%------------------------

for i=1:ng     % first block

 ngi = ng+i;
```

Code cvt6: $\longrightarrow$ Continuing $\longrightarrow$

```
  for j=1:ng
   ngj = ng+j;
   Gm(i,j)   = visc*gdm(i,j);   Gm(i,ngj) = 0.0;
   Gm(ngi,j) = 0.0;              Gm(ngi,ngj) = Gm(i,j);
  end

  for j=1:ne
   Gm(i,  2*ng+j) = -gDx(j,i);
   Gm(ngi,2*ng+j) = -gDy(j,i);
  end

 end

 for i=1:ne      % second block

  for j=1:ng
   Gm(2*ng+i,j)    = -gDx(i,j);
   Gm(2*ng+i,ng+j) = -gDy(i,j);
  end

  for j=1:ne
   Gm(2*ng+i,2*ng+j) = 0;
  end

 end

%------------
% system size
%------------

nsys = 2*ng+ne;

%-----------------------------------------------
% compute the right-hand side and implement the
% velocity (Dirichlet) boundary condition
%-----------------------------------------------

for i=1:nsys
 b(i) = 0.0;   % initialize
end

for j=1:ng
 if(gfl(j,1)==1)
    for i=1:nsys
     b(i) = b(i) - Gm(i,j) * gfl(j,2) - Gm(i,ng+j) * gfl(j,3);
     Gm(i,j) = 0; Gm(i,ng+j) = 0;
     Gm(j,i) = 0; Gm(ng+j,i) = 0;
    end

    Gm(j,j) = 1.0; Gm(ng+j,ng+j) = 1.0;
    b(j) = gfl(j,2); b(ng+j) = gfl(j,3);
 end
end
```

Code cvt6: $\longrightarrow$ Continuing $\longrightarrow$

```
%----------------------
% solve the linear system
%----------------------

Gm(:,nsys) = [];  % remove the last (nsys) column

Gm(nsys,:) = [];  % remove the last (nsys) row

b(:,nsys)  = [];  % remove the last (nsys) element

f = b/Gm';

%---------------------
% recover the velocity
%---------------------

for i=1:ng
  ux(i) = f(i);
  uy(i) = f(ng+i);
end

%---------------------
% recover the pressure
%---------------------

for i=1:ne-1
  press(i) = f(2*ng+i);
end

press(ne) = 0.0;   % arbitrary reference level

%----------------------
% plot the element edges
%----------------------

for i=1:ne

  i1=c(i,1);i2=c(i,2);i3=c(i,3);
  i4=c(i,4);i5=c(i,5);i6=c(i,6);

  xp(1)=p(i1,1); xp(2)=p(i4,1); xp(3)=p(i2,1); xp(4)=p(i5,1);
                 xp(5)=p(i3,1); xp(6)=p(i6,1); xp(7)=p(i1,1);

  yp(1)=p(i1,2); yp(2)=p(i4,2); yp(3)=p(i2,2); yp(4)=p(i5,2);
                 yp(5)=p(i3,2); yp(6)=p(i6,2); yp(7)=p(i1,2);

  plot(xp, yp,'-'); hold on;  % plot(xp, yp,'-o');

end

%----------------------------
% plot the velocity vector field
%----------------------------
```

Code cvt6: $\longrightarrow$ Continuing $\longrightarrow$

```
plot_6 (ne,ng,p,c,ux);

quiver (p(:,1)',p(:,2)',ux,uy);

xlabel('x'); ylabel('y');

%-----------------------
% plot the pressure field
%-----------------------

figure;    % new graphic window

for i=1:ne

 i1=c(i,1); i2=c(i,2); i3=c(i,3);

 % compute the element centroid:

 xplt(1) = (p(i1,1)+p(i2,1)+p(i3,1))/3.0;
 yplt(1) = (p(i1,2)+p(i2,2)+p(i3,2))/3.0;
 zplt(1) = press(i);

 xplt(2) = xplt(1); yplt(2) = yplt(1);
 zplt(2) = 0.0;

 plot3(xplt,yplt,zplt,'-'); hold on;

end

xlabel('x'); ylabel('y'); zlabel('Pressure');

%-----------------------
% plot the element edges
%-----------------------

for i=1:ne

 i1=c(i,1);i2=c(i,2);i3=c(i,3);
 i4=c(i,4);i5=c(i,5);i6=c(i,6);

 xp(1)=p(i1,1); xp(2)=p(i4,1); xp(3)=p(i2,1); xp(4)=p(i5,1);
               xp(5)=p(i3,1); xp(6)=p(i6,1); xp(7)=p(i1,1);

 yp(1)=p(i1,2); yp(2)=p(i4,2); yp(3)=p(i2,2); yp(4)=p(i5,2);
               yp(5)=p(i3,2); yp(6)=p(i6,2); yp(7)=p(i1,2);

 plot(xp, yp, ':'); hold on;

end

%-----
% done
%-----
```

Code cvt6: ($\longrightarrow$ Continued.) Code for Stokes flow in a rectangular cavity.

- To implement the velocity boundary condition, we apply a straightforward generalization of Algorithm 2.1.2.

- Because the pressure gradient, but not the pressure itself, appears in the Stokes or Navier-Stokes equation, the solution for the pressure is defined only up to an arbitrary constant. Accordingly, the pressure of an arbitrary element can be assigned an arbitrary, inconsequential value.

  In the code, the pressure at the last pressure node is set to zero, $p_{N_G^p} = 0$. This is implemented by discarding the last equation of the grand system and the last column of the grand matrix.

- The velocity vector field is visualized using the MATLAB function `quiver`, whereas the pressure field is visualized using low-level graphics functions by drawing vertical bars passing through the element centroids.

Results of a computation are shown in Figure 5.7.1. It is important to note that, after the boundary conditions have been implemented, the number of equations involving the nodal values of the pressure is reduced from $2N_G^u$, corresponding to the first two row blocks of (5.7.21), to $2(N_G^u)_i$, where $(N_G^u)_i$ is the number of *interior* global nodes. For a solution to exist, the criterion

$$\frac{2(N_G^u)_i}{N_G^p} > 1 \qquad (5.7.24)$$

must be fulfilled, otherwise the computation will fail. In practice, the fraction in (5.7.24) must be greater than two or three to ensure numerical stability. In the case of the driven-cavity flow, if we had used 3-node triangular elements with an isoparametric velocity interpolation and constant element pressures, criterion (5.7.24) would fail for any discretization level.

5.7.3 Triangularization

Let us return to the grand finite element equation (5.7.21) and introduce the square, $2N_G^u \times 2N_G^u$ Laplacian matrix

$$\mathbf{L} \equiv \begin{bmatrix} \mathbf{D}^u & \mathbf{0} \\ \mathbf{0} & \mathbf{D}^u \end{bmatrix}, \qquad (5.7.25)$$

and the $N_G^p \times 2N_G^u$ rectangular gradient matrix

$$\mathbf{D} \equiv \begin{bmatrix} \mathbf{D}_x & \mathbf{D}_y \end{bmatrix}. \qquad (5.7.26)$$

Subject to these definitions, equation (5.7.21) takes the form

$$\begin{bmatrix} \mu\,\mathbf{L} & -\mathbf{D}^T \\ -\mathbf{D} & \mathbf{0} \end{bmatrix} \cdot \begin{bmatrix} \mathbf{U} \\ \mathbf{P} \end{bmatrix} = \mathbf{0}, \qquad (5.7.27)$$

```
function [Dx, Dy] = cvt6_D ...
   ...
   (x1,y1, x2,y2, x3,y3, x4,y4, x5,y5, x6,y6, NQ)

%=================================================
% Computation of the gradient matrices Dx and Dy
% required by cvt6 for viscous flow
%=================================================

%---------------------------------
% compute the mapping coefficients:
%---------------------------------

[al, be, ga] = elm6_abc ...
   ...
   (x1,y1, x2,y2, x3,y3, x4,y4, x5,y5, x6,y6);

%----------------------------
% read the triangle quadrature
%----------------------------

[xi, eta, w] = gauss_trgl(NQ);

%---------------------------------
% initialize the matrices Dx and Dy
%---------------------------------

for k=1:6
   Dx(k) = 0.0;
   Dy(k) = 0.0;
end

%----------------------
% perform the quadrature
%----------------------
```

Function cvt6_D: $\longrightarrow$ Continuing $\longrightarrow$

where

$$\mathbf{U} \equiv \begin{bmatrix} u_{x_1} \\ u_{x_2} \\ \vdots \\ u_{x_{N_G^u}} \\ - \\ u_{y_1} \\ u_{y_2} \\ \vdots \\ u_{y_{N_G^u}} \end{bmatrix}, \qquad \mathbf{P} \equiv \begin{bmatrix} p_1 \\ p_2 \\ \vdots \\ p_{N_G^p} \end{bmatrix}, \qquad (5.7.28)$$

```
for i=1:NQ

  [psi, gpsi, hs] = elm6_interp ...
  ...
      (x1,y1, x2,y2, x3,y3, x4,y4, x5,y5, x6,y6 ...
      ,al,be,ga,
      ,xi(i),eta(i));

  cf = 0.5*hs*w(i)

  for k=1:6
    Dx(k) = Dx(k) + gpsi(k,1)*cf;
    Dy(k) = Dy(k) + gpsi(k,2)*cf;
  end

end

%-----
% done
%-----

return;
```

Function cvt6_D: Computation of the matrices $\mathbf{D}^x$ and $\mathbf{D}^y$ required by the finite element code cvt6 for viscous flow.

are the velocity and pressure vectors. After implementing the boundary conditions, the homogeneous system (5.7.27) reduces to the inhomogeneous system

$$
\begin{bmatrix} \mu\,\widehat{\mathbf{L}} & -\widehat{\mathbf{D}}^T \\ -\widehat{\mathbf{D}} & \mathbf{0} \end{bmatrix} \cdot \begin{bmatrix} \mathbf{U} \\ \mathbf{P} \end{bmatrix} = \begin{bmatrix} \mathbf{b}_U \\ \mathbf{b}_P \end{bmatrix},
\tag{5.7.29}
$$

where the hat designates modified matrices, and $\mathbf{b}_U$, $\mathbf{b}_P$ are properly constructed right-hand sides.

To compute the solution of the modified linear system, we may apply the method of Gauss elimination discussed in Section C.1 of Appendix C, and solve the first block equation in (5.7.29) for the velocity vector $\mathbf{U}$, finding

$$
\mathbf{U} = \frac{1}{\mu}\,\widehat{\mathbf{L}}^{-1} \cdot (\widehat{\mathbf{D}}^T \cdot \mathbf{P} + \mathbf{b}_U).
\tag{5.7.30}
$$

The inverse matrix, $\widehat{\mathbf{L}}^{-1}$, is composed of two diagonal blocks, each containing the inverse of the modified diffusion matrix, $\widehat{\mathbf{D}}^u$. Substituting (5.7.30) in the second block equation of (5.7.29), we find

$$
-\widehat{\mathbf{D}} \cdot \widehat{\mathbf{L}}^{-1} \cdot (\widehat{\mathbf{D}}^T \cdot \mathbf{P} + \mathbf{b}_U) = \mu\,\mathbf{b}_P.
\tag{5.7.31}
$$

Replacing the second block equation in (5.7.29) with (5.7.31) and rearranging, we derive the upper-triangular system

$$\begin{bmatrix} \mu \widehat{\mathbf{L}} & -\widehat{\mathbf{D}}^T \\ \mathbf{0} & \widehat{\mathbf{D}} \cdot \widehat{\mathbf{L}}^{-1} \cdot \widehat{\mathbf{D}}^T \end{bmatrix} \cdot \begin{bmatrix} \mathbf{U} \\ \mathbf{P} \end{bmatrix} = \begin{bmatrix} \mathbf{b}_U \\ -\mu\, \mathbf{b}_P - \widehat{\mathbf{D}} \cdot \widehat{\mathbf{L}}^{-1} \cdot \mathbf{b}_U \end{bmatrix}, \quad (5.7.32)$$

which can be solved by backward substitution according to the Uzawa's algorithm: solve (5.7.31) for $\mathbf{P}$, and then recover $\mathbf{U}$ from (5.7.30). The matrix

$$\mathbf{S} \equiv \widehat{\mathbf{D}} \cdot \widehat{\mathbf{L}}^{-1} \cdot \widehat{\mathbf{D}}^T, \qquad (5.7.33)$$

occupying the southeastern block of (5.7.32), is known as the pressure matrix. Since $\widehat{\mathbf{L}}$ is symmetric, $\mathbf{S}$ is also symmetric. In practice, the pressure equation (5.7.31) is solved using iterative methods, such as the method of conjugate gradients discussed in Appendix C.

PROBLEM

5.7.1 *Stokes flow in a cavity.*

Run code `cvt6` for flow in a slender cavity with width to depth ratio equal to 1:5, and discuss the pressure distribution and structure of the velocity field far from the side-walls.

5.8 Navier-Stokes flow

To develop the finite element formulation of flow at non-zero Reynolds numbers, we substitute the finite element expansion for the velocity in the left-hand side of the Navier-Stokes equation (5.6.10), denoted by **LHS**, and obtain

$$\mathbf{LHS} = \rho \sum_{j=1}^{N_G^u} \frac{d\mathbf{u}_j}{dt}\, \phi_j^u + \rho \sum_{j=1}^{N_G^u} \phi_j^u\, \mathbf{u}_j \cdot \sum_{k=1}^{N_G^u} \nabla \phi_k^u\, \mathbf{u}_k$$

$$= \rho \sum_{j=1}^{N_G^u} \frac{d\mathbf{u}_j}{dt}\, \phi_j^u + \rho \sum_{j=1}^{N_G^u} \sum_{k=1}^{N_G^u} \phi_j^u\, (\mathbf{u}_j \cdot \nabla \phi_k^u)\, \mathbf{u}_k. \qquad (5.8.1)$$

The x and y components of this expression are

$$LHS_x = \rho \sum_{j=1}^{N_G^u} \frac{du_{x_j}}{dt}\, \phi_j^u + \rho \sum_{j=1}^{N_G^u} \sum_{k=1}^{N_G^u} \phi_j^u\, (\mathbf{u}_j \cdot \nabla \phi_k^u)\, u_{x_k},$$

$$(5.8.2)$$

$$LHS_y = \rho \sum_{j=1}^{N_G^u} \frac{du_{x_j}}{dt}\, \phi_j^u + \rho \sum_{j=1}^{N_G^u} \sum_{k=1}^{N_G^u} \phi_j^u\, (\mathbf{u}_j \cdot \nabla \phi_k^u)\, u_{y_k}.$$

To carry out the Galerkin projection, we multiply each of these expressions by ϕ_i^u, and integrate over the solution domain to obtain

$$\mathbf{LHS} = \rho \sum_{j=1}^{N_G^u} M_{ij}^u \frac{d\mathbf{u}_j}{dt} + \rho \sum_{j=1}^{N_G^u} \sum_{k=1}^{N_G^u} (\mathbf{u}_j \cdot \mathbf{N}_{jik}) \, \mathbf{u}_k, \tag{5.8.3}$$

where

$$M_{ij}^u = \iint_D \phi_i^u \, \phi_j^u \, dx \, dy \tag{5.8.4}$$

is the mass matrix, and

$$\mathbf{N}_{jik} = \iint_D \phi_j^u \, \phi_i^u \, (\nabla \phi_k^u) \, dx \, dy \tag{5.8.5}$$

is the advection matrix.

Setting the left-hand side of (5.8.3) equal to the right-hand side of the Navier-Stokes equation as given in (5.7.12), we find

$$\rho \sum_{j=1}^{N_G^u} M_{ij}^u \frac{d\mathbf{u}_j}{dt} + \rho \sum_{j=1}^{N_G^u} \sum_{k=1}^{N_G^u} (\mathbf{u}_j \cdot \mathbf{N}_{jik}) \, \mathbf{u}_k \tag{5.8.6}$$

$$= \sum_{j=1}^{N_G^p} \mathbf{Q}_{ij} \, p_j - \mu \sum_{j=1}^{N_G^u} R_{ij} \, \mathbf{u}_j + \sum_{j=1}^{N_G^p} \mathbf{D}_{ji} \, p_j - \mu \sum_{j=1}^{N_G^u} D_{ij}^u \, \mathbf{u}_j,$$

which is to be solved together with the Galerkin projection of the continuity equation (5.7.13), repeated here for convenience,

$$\sum_{j=1}^{N_G^u} \mathbf{D}_{ij} \cdot \mathbf{u}_j = 0. \tag{5.8.7}$$

Explicitly, the x and y components of (5.8.6) read

$$\rho \sum_{j=1}^{N_G^u} M_{ij}^u \frac{du_{x_j}}{dt} + \rho \sum_{j=1}^{N_G^u} \sum_{k=1}^{N_G^u} (\mathbf{u}_j \cdot \mathbf{N}_{jik}) \, u_{x_k} \tag{5.8.8}$$

$$= \sum_{j=1}^{N_G^p} Q_{x_{ij}} \, p_j - \mu \sum_{j=1}^{N_G^u} R_{ij} \, u_{x_j} + \sum_{j=1}^{N_G^p} D_{x_{ji}} \, p_j - \mu \sum_{j=1}^{N_G^u} D_{ij}^u \, u_{x_j} = 0,$$

and

$$\rho \sum_{j=1}^{N_G^u} M_{ij}^u \frac{du_{y_j}}{dt} + \rho \sum_{j=1}^{N_G^u} \sum_{k=1}^{N_G^u} (\mathbf{u}_j \cdot \mathbf{N}_{jik}) \, u_{y_k} \tag{5.8.9}$$

$$= \sum_{j=1}^{N_G^p} Q_{y_{ij}} \, p_j - \mu \sum_{j=1}^{N_G^u} R_{ij} \, u_{y_j} + \sum_{j=1}^{N_G^p} D_{y_{ji}} \, p_j - \mu \sum_{j=1}^{N_G^u} D_{ij}^u \, u_{y_j}.$$

5.8.1 Steady state

At steady steady, $\mathrm{d}\mathbf{u}_j/\mathrm{d}t = \mathbf{0}$, we obtain a nonlinear quadratic system of algebraic equations, which must be solved by iterative methods. Two options are Newton and quasi-Newton methods (e.g., [43]), and iterative methods based on functional minimization involving the solution of a sequence of discrete Stokes flow problems [22].

5.8.2 Time integration

One way of integrating (5.8.6) forward in time from a specified initial state involves the following steps:

1. Select a time step, Δt.

2. Apply the differential equation at time $t + \frac{1}{2}\Delta t$.

3. Approximate the time derivative on the left-hand side with a centered difference, and implement the Crank-Nicolson discretization for the viscous term to obtain

$$
\frac{\rho}{\Delta t} \sum_{j=1}^{N_G^u} M_{ij}^u \left(\mathbf{u}_j^{n+1} - \mathbf{u}_j^n\right) + \rho \sum_{j=1}^{N_G^u} \sum_{k=1}^{N_G^u} \left[(\mathbf{u}_j \cdot \mathbf{N}_{jik}) \, \mathbf{u}_k \right]^{n+1/2}
$$

$$
= \sum_{j=1}^{N_G^p} \mathbf{Q}_{ij} \, p_j^{n+1/2} - \mu \, \frac{1}{2} \sum_{j=1}^{N_G^u} R_{ij} \left(\mathbf{u}_j^n + \mathbf{u}_j^{n+1}\right) \qquad (5.8.10)
$$

$$
+ \sum_{j=1}^{N_G^p} \mathbf{D}_{ji} \, p_j^{n+1/2} - \mu \, \frac{1}{2} \sum_{j=1}^{N_G^u} D_{ij}^u \left(\mathbf{u}_j^n + \mathbf{u}_j^{n+1}\right),
$$

where the superscripts $n, n+1/2$, and $n+1$ denote evaluation, respectively, at times $t, t + \frac{1}{2}\Delta t$, and $t + \Delta t$.

The second term on the left-hand side of (5.8.10), expressing the nonlinear effect of convection, is assumed to be available by extrapolation from previous steps. To simplify the algorithm, at the first step, this term can be evaluated at the initial instant.

The continuity equation requires

$$
\sum_{j=1}^{N_G^u} \mathbf{D}_{ij} \cdot \mathbf{u}_j^{n+1} = 0. \qquad (5.8.11)
$$

The unknowns in (5.8.10) and (5.8.11) are the nodal velocities, $\mathbf{u}_j^{n+1}$, and the nodal pressures, $p_j^{n+1/2}$. Rearranging (5.8.10) to bring the unknowns to the left-hand side, we obtain

$$
\mu \sum_{j=1}^{N_G^u} (\frac{1}{\nu \Delta t} M_{ij}^u + \frac{1}{2} D_{ij}^u) \, \mathbf{u}_j^{n+1} - \sum_{j=1}^{N_G^p} Q_{ij} \, p_j^{n+1/2} + \mu \, \frac{1}{2} \sum_{j=1}^{N_G^u} R_{ij} \, \mathbf{u}_j^{n+1}
$$

$$
- \sum_{j=1}^{N_G^p} \mathbf{D}_{ji} \, p_j^{n+1/2} = \mu \sum_{j=1}^{N_G^u} (\frac{1}{\nu \Delta t} M_{ij}^u - \frac{1}{2} D_{ij}^u) \, \mathbf{u}_j^n - \mu \, \frac{1}{2} \sum_{j=1}^{N_G^u} R_{ij} \, \mathbf{u}_j^n + \mathbf{F}_i,
$$

$$(5.8.12)$$

where $\nu \equiv \mu/\rho$ is the kinematic viscosity, and the vector $\mathbf{F}_i$ on the right-hand side encapsulates the nonlinear convection term,

$$
\mathbf{F}_i \equiv -\rho \sum_{j=1}^{N_G^u} \sum_{k=1}^{N_G^u} \Big[(\mathbf{u}_j \cdot \mathbf{N}_{jik}) \, \mathbf{u}_k \Big]^{n+1/2}. \tag{5.8.13}
$$

If the velocity is specified as a boundary condition all around the boundary C, the projection of the Navier-Stokes equation on ϕ_i^u is excluded for those nodes lying on the boundary. For all other nodes, because $\phi_i^u = 0$ on C, the matrices R_{ij} and $\mathbf{Q}_{ij}$ are identically zero. Writing equation (5.8.13) for all velocity nodes, bearing in mind that additional terms should have been included for the boundary nodes, and appending to the system thus derived the continuity equation (5.8.11), we obtain the grand system

$$
\begin{bmatrix}
\mu \, (\frac{1}{\nu \Delta t} \mathbf{M}^u + \frac{1}{2} \mathbf{D}^u) & \mathbf{0} & -\mathbf{D}_x^T \\[2mm]
\mathbf{0} & \mu \, (\frac{1}{\nu \Delta t} \mathbf{M}^u + \frac{1}{2} \mathbf{D}^u) & -\mathbf{D}_y^T \\[2mm]
-\mathbf{D}_x & -\mathbf{D}_y & \mathbf{0}
\end{bmatrix} \cdot \mathbf{X}^{n+1} = \mathbf{RHS},
$$

$$(5.8.14)$$

involving a symmetric coefficient matrix, where the right-hand side is given by

$$
\mathbf{RHS} = \begin{bmatrix}
\mu \, (\frac{1}{\nu \Delta t} \mathbf{M}^u - \frac{1}{2} \mathbf{D}^u) & \mathbf{0} \\[2mm]
\mathbf{0} & \mu \, (\frac{1}{\nu \Delta t} \mathbf{M}^u - \frac{1}{2} \mathbf{D}^u) \\[2mm]
\mathbf{0} & \mathbf{0}
\end{bmatrix} \cdot \mathbf{U}^n - \mathbf{b},
$$

The unknown vector, $\mathbf{X}^{n+1}$, velocity vector, $\mathbf{U}^n$, and nonlinear term, $\mathbf{b}$, are displayed in Table 5.8.1. In the case of unsteady Stokes flow (Problem 5.8.1), the nonlinear term, $\mathbf{b}$, is zero. The solution of the linear system can be found as discussed in Section 5.5.1 for Stokes flow.

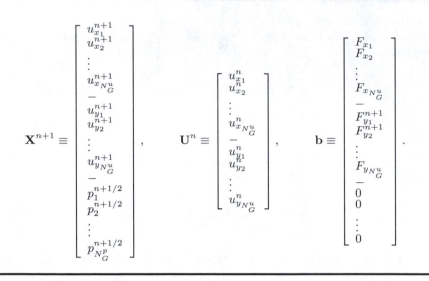

Table 5.8.1 Unknown solution vector, velocity vector, and nonlinear term on the right-hand side of system (5.8.14).

5.8.3   Formulation based on the pressure Poisson equation

In the formulations discussed previously in this section, the equation of motion was solved simultaneously with the continuity equation by the finite element method. In an alternative approach, the continuity equation is replaced by a Poisson equation that emerges by taking the divergence of the Navier-Stokes equation, and then enforcing the incompressibility condition, finding

$$\nabla^2 p = -\rho \, \nabla \cdot \left(\mathbf{u} \cdot \nabla \mathbf{u} \right). \tag{5.8.15}$$

A variety of methods are available for integrating the equation of motion in space or time together with the pressure Poisson equation, as discussed in texts of computational fluid dynamics (CFD) (e.g., [30, 42]).

PROBLEM

5.8.1 *Unsteady Stokes flow.*

Unsteady Stokes flow is governed by the unsteady Stokes equation

$$\rho \, \frac{\partial \mathbf{u}}{\partial t} = -\nabla p + \mu \, \nabla^2 \mathbf{u}, \tag{5.8.16}$$

subject to the continuity equation expressed by (5.6.5).

Modify the FSELIB code cvt6 into a code named cvt6_ucn that computes the evolution of a flow from a specified initial state using the Crank-Nicolson method. Run the code to describe the onset of steady lid-driven cavity flow from a quiescent initial condition, subject to impulsively started lid translation where the lid velocity suddenly jumps to a constant value at the origin of computational time.

Finite and spectral element methods in three dimensions

6

The formulation of the Galerkin finite and spectral element methods in three dimensions is a straightforward generalization of that in two dimensions discussed in Chapters 3–5. However, not surprisingly, the implementation presents certain new pragmatic challenges associated with a more demanding bookkeeping, the need for the efficient computation of the various quantities involved, and the affordable solution of the final system of algebraic or ordinary differential equations. Highlights of the fundmental concepts and basic procedures will be discussed in this chapter with reference to the convection–diffusion equation.

6.1 Convection–diffusion in three dimensions

Consider unsteady heat transport in a three-dimensional domain, in the presence of a distributed source as illustrated in Figure 6.1.1(a). The evolution of the temperature field, $f(x, y, z, t)$, is governed by the convection–diffusion equation

$$\rho\, c_p \left(\frac{\partial f}{\partial t} + u_x\, \frac{\partial f}{\partial x} + u_y\, \frac{\partial f}{\partial y} + u_z\, \frac{\partial f}{\partial z} \right)$$
$$= k \left(\frac{\partial^2 f}{\partial x^2} + \frac{\partial^2 f}{\partial y^2} + \frac{\partial^2 f}{\partial z^2} \right) + s, \qquad (6.1.1)$$

where ρ, c_p, and k are the medium density, heat capacity, and thermal conductivity, $u_x(x, y, z, f, t)$, $u_y(x, y, z, f, t)$, and $u_z(x, y, z, f, t)$ are the x, y, and z components of the convection velocity, and $s(x, y, z, f, t)$ is the distributed source. Note that (6.1.1) is the three-dimensional counterpart of the one-dimensional rod equation (1.6.1) and of the two-dimensional plate equation (3.1.1). For simplicity, we shall assume that the convection velocity and source term depend on x, y, z and t explicitly, but not implicitly through f, that is, $u_x(x, y, z, t)$, $u_y(x, y, z, t)$, $u_z(x, y, z, t)$, and $s(x, y, z, t)$.

In vector notation, the convection–diffusion equation (6.1.1) takes the compact form

$$\frac{\partial f}{\partial t} + \mathbf{u} \cdot \nabla f = \kappa\, \nabla^2 f + \frac{s}{\rho\, c_p}, \qquad (6.1.2)$$

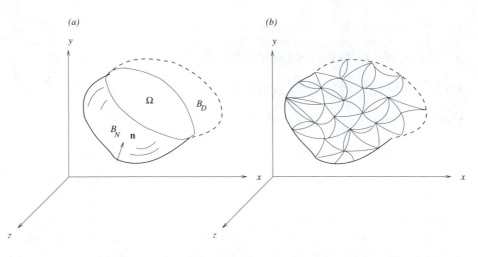

Figure 6.1.1 (*a*) Illustration of heat conduction in a three-dimensional domain
with arbitrary geometry. (*b*) Finite element discretization of the solution
domain into three-dimensional elements.

where $\kappa \equiv k/(\rho \, c_p)$ is the thermal diffusivity,

$$\nabla f = \left(\frac{\partial f}{\partial x}, \frac{\partial f}{\partial y}, \frac{\partial f}{\partial z} \right) \qquad (6.1.3)$$

is the three-dimensional gradient, and

$$\nabla^2 f \equiv \nabla \cdot \nabla f = \frac{\partial^2 f}{\partial x^2} + \frac{\partial^2 f}{\partial y^2} + \frac{\partial^2 f}{\partial z^2} \qquad (6.1.4)$$

is the three-dimensional Laplacian.

Equation (6.1.2) is to be solved in a domain, Ω, that is enclosed by the
surface B, subject to two complementary boundary conditions: the *Neumann
boundary condition* specifying the normal derivative of the unknown function,
and the *Dirichlet boundary condition* specifying the boundary distribution of the
unknown function. In the present problem, the boundary conditions prescribe:

- The heat flux along the Neumann portion of B, denoted by B_N and drawn
  with the solid line in Figure 6.1.1(*a*),

$$q(\mathbf{x}) \equiv -k \, \mathbf{n} \cdot \nabla f \equiv -k \, \frac{\partial f}{\partial l_n}, \qquad (6.1.5)$$

where $\mathbf{n}$ is the unit vector normal to B_N pointing into the solution domain,
as shown in Figure 6.1.1(*a*), l_n is the arc length in the normal direction,
and $q(\mathbf{x})$ is a given function of position over B_N.

If $q(\mathbf{x}) > 0$, in which case $\mathbf{n} \cdot \nabla f < 0$, heat enters the solution domain Ω, whereas if $q(\mathbf{x}) < 0$, in which case $\mathbf{n} \cdot \nabla f > 0$, heat escapes from the solution domain across the Neumann boundary, B_N.

- The temperature distribution along the complementary Dirichlet portion of the boundary, denoted by B_D and drawn with the broken line in Figure 6.1.1(a),

$$f(\mathbf{x}) = g(\mathbf{x}), \qquad (6.1.6)$$

where $B_D = B - B_N$ is the complement of B_N, and $g(\mathbf{x})$ is a given function of position over B_D.

As a first step toward applying the finite or spectral element method, the solution domain is discretized into three-dimensional elements defined by a small group of geometrical nodes whose collection comprises a larger set of unique global nodes. For simplicity, in the remainder of this section we discuss the isoparametric representation where the interpolation nodes coincide with the geometrical element nodes.

6.1.1 Galerkin projection

To develop the finite element equations, we carry out the Galerkin projection of the governing differential equation (6.1.1), using as weighting functions the global cardinal interpolation functions associated with the unique global nodes, $\phi_i(x, y, z)$. By definition, the function $\phi_i(x, y, z)$ takes the value of unity at the ith global interpolation node and the value of zero at all other global interpolation nodes.

Multiplying first the right-hand side of (6.1.1) by $\phi_i(x, y, z)$, integrating the product over the solution domain, Ω, and using the Gauss divergence theorem under the assumption that ϕ_i is continuous throughout Ω, we find

$$\iiint_\Omega \phi_i \left[k \nabla^2 f + s \right] \mathrm{d}x \, \mathrm{d}y \, \mathrm{d}z$$

$$= \iiint_\Omega \left[\phi_i \, k \, (\nabla \cdot \nabla f) + \phi_i \, s \right] \mathrm{d}x \, \mathrm{d}y \, \mathrm{d}z$$

$$= k \iiint_\Omega \nabla \cdot (\phi_i \, \nabla f) \, \mathrm{d}x \, \mathrm{d}y \, \mathrm{d}z + \iiint_\Omega \left[-k \, \nabla \phi_i \cdot \nabla f + \phi_i \, s \right] \mathrm{d}x \, \mathrm{d}y \, \mathrm{d}z$$

$$= -k \iint_B \phi_i \, \mathbf{n} \cdot \nabla f \, \mathrm{d}S + \iiint_\Omega \left[-k \, \nabla \phi_i \cdot \nabla f + \phi_i \, s \right] \mathrm{d}x \, \mathrm{d}y \, \mathrm{d}z.$$

$$(6.1.7)$$

Substituting the boundary flux definition in the boundary integral shown in the last line, $q \equiv -k \, \mathbf{n} \cdot \nabla f$, and rearranging, we find

$$\iiint_\Omega \phi_i \left[k \, \nabla^2 f + s \right] \mathrm{d}x \, \mathrm{d}y \, \mathrm{d}z = -k \iiint_\Omega \nabla f \cdot \nabla \phi_i \, \mathrm{d}x \, \mathrm{d}y \, \mathrm{d}z + Q_i + S_i,$$

(6.1.8)

where

$$Q_i \equiv \iint_B \phi_i \, q \, \mathrm{d}S, \qquad S_i \equiv \iiint_\Omega \phi_i \, s \, \mathrm{d}x \, \mathrm{d}y \, \mathrm{d}z, \qquad (6.1.9)$$

are the boundary and volume integrals of the surface flux and distributed source, both weighted by the ith global interpolation functions. Note that, if the ith node is an interior node, ϕ_i is zero along the whole of the boundary, B, and $Q_i = 0$.

Next, we multiply the left-hand side of (6.1.1) by ϕ_i, integrate the product over the solution domain, Ω, set the result equal to the right-hand side of (6.1.8), divide each term by $\rho \, c_p$, and thus derive the targeted Galerkin equation

$$\iiint_\Omega \phi_i \, \frac{\partial f}{\partial t} \, \mathrm{d}x \, \mathrm{d}y \, \mathrm{d}z + \iiint_\Omega \phi_i \, \mathbf{u} \cdot \nabla f \, \mathrm{d}x \, \mathrm{d}y \, \mathrm{d}z$$

$$= -\kappa \iiint_\Omega \nabla \phi_i \cdot \nabla f \, \mathrm{d}x \, \mathrm{d}y \, \mathrm{d}z + \frac{1}{\rho \, c_p} \left(Q_i + S_i \right), \qquad (6.1.10)$$

which provides us with a basis for the Galerkin finite element method.

6.1.2 Galerkin finite element equations

In the isoparametric interpolation, the requisite solution f is expressed in terms of the *a priori* unknown values at the global interpolation nodes, $f_j(t)$, and associated global cardinal interpolation functions, ϕ_j, as

$$f(x, y, z, t) = \sum_{j=1}^{N_G} f_j(t) \, \phi_j(x, y, z). \qquad (6.1.11)$$

Inserting this and a similar expansion for the source term, s, in (6.1.10), and recalling the definition of the boundary and source integrals Q_i and S_i, we obtain

$$\sum_{j=1}^{N_G} M_{ij} \, \frac{\mathrm{d}f_j}{\mathrm{d}t} + \sum_{j=1}^{N_G} N_{ij} \, f_j$$

$$= -\kappa \sum_{j=1}^{N_G} D_{ij} \, f_j + \frac{1}{\rho \, c_p} \iint_B \phi_i \, q \, \mathrm{d}l + \frac{1}{\rho \, c_p} \sum_{j=1}^{N_G} M_{ij} \, s_j, \qquad (6.1.12)$$

where

$$D_{ij} \equiv \iiint_{\Omega} \nabla \phi_i \cdot \nabla \phi_j \, dx \, dy \, dz \qquad (6.1.13)$$

is the global diffusion matrix,

$$M_{ij} \equiv \iiint_{\Omega} \phi_i \, \phi_j \, dx \, dy \, dz \qquad (6.1.14)$$

is the global mass matrix, and

$$N_{ij} \equiv \iiint_{\Omega} \phi_i \, \mathbf{u} \cdot \nabla \phi_j \, dx \, dy \, dz \qquad (6.1.15)$$

is the global advection matrix.

Applying (6.1.12) for the global interpolation functions associated with the interpolation nodes where the Dirichlet boundary condition is *not* prescribed, we obtain a system of ordinary differential equations for the unknown nodal values,

$$\mathbf{M} \cdot \frac{d\mathbf{f}}{dt} + \mathbf{N} \cdot \mathbf{f} = \kappa \left(-\mathbf{D} \cdot \mathbf{f} + \mathbf{b} \right), \qquad (6.1.16)$$

where the vector $\mathbf{b}$ on the right-hand side,

$$b_i \equiv \frac{1}{k} \left(\iint_{B} \phi_i \, q \, dS + \sum_{j=1}^{N_G} M_{ij} \, s_j \right), \qquad (6.1.17)$$

incorporates the given source term and the prescribed Neumann boundary conditions. The first integral on the right-hand side of (6.1.17) is nonzero only if the ith node is a boundary node.

6.1.3 Element matrices

In practice, the global matrices are compiled from corresponding element matrices which are defined in terms of the local element interpolation functions, $\psi_i^{(l)}$, using a connectivity table, as discussed in previous chapters. The lth-element diffusivity matrix is given by

$$A_{ij}^{(l)} \equiv \iiint_{E_l} \nabla \psi_i^{(l)} \cdot \nabla \psi_j^{(l)} \, dx \, dy \, dz, \qquad (6.1.18)$$

the corresponding mass matrix is given by

$$B_{ij}^{(l)} \equiv \iiint_{E_l} \psi_i^{(l)} \, \psi_j^{(l)} \, dx \, dy \, dz, \qquad (6.1.19)$$

and the corresponding advection matrix is given by

$$C_{ij}^{(l)} \equiv \iiint_{E_l} \psi_i^{(l)} \, \mathbf{u} \cdot \nabla \psi_j^{(l)} \, dx \, dy \, dz, \qquad (6.1.20)$$

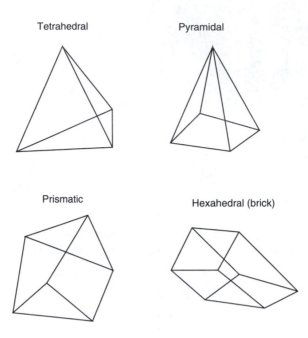

Figure 6.1.2 Element types for discretizing a three-dimensional domain.

where E_l stands for the lth element, and the integration is performed over the element volume.

The assembly methodology hinges on the observation that these integrals are nonzero only if nodes i and j lie inside, in the faces, or at the vertices of the lth element. The assembly algorithm is identical to Algorithm 3.1.2 for the corresponding problem in two dimensions, subject to the convention that the entry $gfl(m, 2)$ contains the prescribed boundary value (Dirichlet boundary condition) at the position of the mth global node. In some finite element codes, these values are hosted by a different vector.

6.1.4   Element types

Several element types are available for discretizing a three-dimensional domain, including tetrahedral and hexahedral elements discussed in the remainder of this chapter, and five-faced pyramidal and prismatic elements depicted in Figure 6.1.2. Being the counterparts of the two-dimensional triangular and quadrilateral elements, the tetrahedral and hexahedral elements are favored in general-purpose finite and spectral element applications.

PROBLEM

6.1.1 *Flux across a spherical surface.*

Departing from (6.1.5), show that the inward flux across a spherical boundary is given by

$$q(\mathbf{x}) = k \, \frac{\partial f}{\partial r}, \qquad (6.1.21)$$

where r is the distance from the center. Then explain on physical grounds why the flux is positive when $\partial f / \partial r > 0$, and negative otherwise.

6.2   4-node tetrahedral elements

Tetrahedral elements have four planar or curved *faces* intersecting at six straight or curved *edges* subtended between the four vertices. The word "tetrahedron" derives from the Greek word "$\tau\epsilon\tau\rho\alpha\epsilon\delta\rho o\nu$," which consists of the words "$\tau\epsilon\sigma\sigma\epsilon\rho\alpha$," meaning "four," and "$\epsilon\delta\rho\alpha$," meaning "base" or "side." Tetrahedral elements are the counterparts of the triangular elements with straight or curved edges in two dimensions previously discussed in Chapters 3–5.

Another composite Greek word is "$\varphi\iota\lambda o\tau\iota\mu o$," which consists of the words "$\varphi\iota\lambda o\varsigma$," meaning "friend," and "$\tau\iota\mu\eta$," meaning "honor." Linguists have argued that "filotimo" is impossible to translate, and can be understood only if honor is interpreted as a psychologically internalized yardstick of generosity, goodness, and integrity. The best way to realize the meaning of the word is to observe how ordinary Greeks conduct their lives at home and abroad.

The simplest tetrahedron has four planar faces and six straight edges defined by four geometrical vertex nodes,

$$\mathbf{x}_i^E = (x_i^E, y_i^E, z_i^E), \qquad (6.2.1)$$

where $i = 1, 2, 3, 4$, as shown in Figure 6.2.1(a). The superscript E emphasizes that these are local or "element" nodes, which can be mapped to the unique global nodes through a connectivity table. Face 1 is defined by nodes 2, 3, and 4, face 2 is defined by nodes 3, 1, and 4, face 3 is defined by nodes 1, 2, and 4, and face 4 is defined by nodes 1, 2, and 3. The vertices 1, 2, and 3 are labeled such that the perimeter of face 4 defined by these vertices is traced in the counter-clockwise direction when the tetrahedron is observed from the inside.

The first *altitude* of the tetrahedron is the minimum distance of vertex 4 from the face defined by vertices 1, 2, and 3, drawn with the dashed line in Figure 6.2.1(a). The corresponding *face altitudes* are the minimum distances of vertex 4 from the three edges defined by vertices 1, 2, and 3, drawn with the dotted lines in Figure 6.2.1(a). Similar definitions apply to the other vertices. If the tetrahedron is regular, all altitudes and face altitudes are the same.

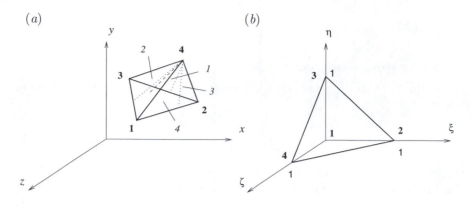

Figure 6.2.1 A 4-node tetrahedral element with six straight edges in the (*a*) physical, and (*b*) parametric space. The node labels are printed in bold, and the face labels are printed in italic.

FSELIB script `tetra4`, listed in the text, visualizes the element with the faces painted in red, blue, yellow, and green. The graphics display generated by a modification of this script where the commented out MATLAB function `plot3` is used in place of the MATLAB function `patch` to produce a wire-frame image, is shown in Figure 6.2.2(*a*).

6.2.1 Parametric representation

To describe the element in parametric form, we map it from the physical *xyz* space to an orthogonal tetrahedron with three perpendicular faces of equal size in the $\xi\eta\zeta$ space, as shown in Figure 6.2.1(*b*). The volume of the parametric tetrahedron is equal to $1/6$. The mapping from the physical to the parametric space is mediated by the function

$$\mathbf{x} = \mathbf{x}_1^E\, \psi_1(\xi, \eta, \zeta) + \mathbf{x}_2^E\, \psi_2(\xi, \eta, \zeta) \qquad (6.2.2)$$
$$+\mathbf{x}_3^E\, \psi_3(\xi, \eta, \zeta) + \mathbf{x}_4^E\, \psi_4(\xi, \eta, \zeta),$$

where $\psi_i(\xi, \eta, \zeta)$, for $i = 1$–4, are element node interpolation functions satisfying familiar cardinal properties: $\psi_i = 1$ at the ith element node, and $\psi_i = 0$ at the other three element nodes.

Working as in Section 3.2 for the 3-node triangular element, we find that the tetrahedral element interpolation functions are given by

$$\psi_1(\xi, \eta, \zeta) = \omega, \qquad \psi_2(\xi, \eta, \zeta) = \xi,$$

$$\qquad (6.2.3)$$

$$\psi_3(\xi, \eta, \zeta) = \eta, \qquad \psi_4(\xi, \eta, \zeta) = \zeta,$$

where

$$\omega \equiv 1 - \xi - \eta - \zeta. \qquad (6.2.4)$$

```
% vertex coordinates

x(1)= 0.0;  y(1)= 0.0;  z(1)= 0.0;
x(2)= 0.9;  y(2)= 0.2;  z(2)= 0.02;
x(3)= 0.1;  y(3)= 1.1;  z(3)=-0.05;
x(4)=-0.1;  y(4)=-0.1;  z(4)= 1.2;

% face node labels

s(1,1) = 2;  s(1,2) = 3;  s(1,3) = 4;   % first  face
s(2,1) = 3;  s(2,2) = 1;  s(2,3) = 4;   % second face
s(3,1) = 1;  s(3,2) = 2;  s(3,3) = 4;   % third  face
s(4,1) = 1;  s(4,2) = 2;  s(4,3) = 3;   % fourth face

% paint the faces

for i=1:4
 for j=1:3
  xp(j)=x(s(i,j)); yp(j)=y(s(i,j)); zp(j)=z(s(i,j));
 end
  if(i==1) patch(xp,yp,zp,'r'); end
  if(i==2) patch(xp,yp,zp,'b'); end
  if(i==3) patch(xp,yp,zp,'y'); end
  if(i==4) patch(xp,yp,zp,'g'); end
% plot3(xp,yp,zp); hold on;
end

xlabel('x'); ylabel('y'); zlabel('z');
```

Script tetra4: Abbreviation of the FSELIB script tetra4 for visualizing a
tetrahedron.

(a) (b)

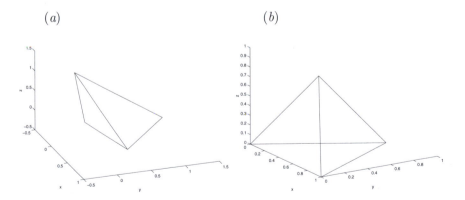

Figure 6.2.2 (a) Visualization of a tetrahedral element using MATLAB graphics functions, and (b) illustration of the standard regular tetrahedron.

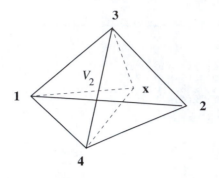

Figure 6.2.3 The barycentric coordinates are defined with respect to the volumes of sub-tetrahedra defined by a field point and each face.

Physically, the coordinate ξ is the ratio between (a) the volume of a tetrahedron having one vertex at the field point $\mathbf{x}$ and three vertices at the tetrahedral vertices 4, 3, 1, and (b) the volume of the whole tetrahedron, $\xi = V_2/V$, as shown in Figure 6.2.3. Similar interpretations apply to the rest of the tetrahedral barycentric coordinates, η, ζ, and ω.

Substituting the element interpolation functions in (6.2.2), we derive a mapping function that is a *complete* linear function of ξ, η, and ζ, consisting of a constant term, a term that is linear in ξ, a term that is linear in η, and a term that is linear in ζ,

$$\mathbf{x} = \mathbf{x}_1^E + (\mathbf{x}_2^E - \mathbf{x}_1^E)\,\xi + (\mathbf{x}_3^E - \mathbf{x}_1^E)\,\eta + (\mathbf{x}_4^E - \mathbf{x}_1^E)\,\zeta. \tag{6.2.5}$$

Explicitly, the x, y, and z components of the mapping are given by

$$\begin{aligned}
x &= x_1^E + (x_2^E - x_1^E)\,\xi + (x_3^E - x_1^E)\,\eta + (x_4^E - x_1^E)\,\zeta, \\
y &= y_1^E + (y_2^E - y_1^E)\,\xi + (y_3^E - y_1^E)\,\eta + (y_4^E - y_1^E)\,\zeta, \\
z &= z_1^E + (z_2^E - z_1^E)\,\xi + (z_3^E - z_1^E)\,\eta + (z_4^E - z_1^E)\,\zeta.
\end{aligned} \tag{6.2.6}$$

For a specified location in the physical space, (x, y, z), the corresponding barycentric coordinates, (ξ, η, ζ), can be found by solving a system of three linear equations originating from (6.2.6).

The standard orthogonal tetrahedron in the $\xi\eta\zeta$ space can be further mapped to a regular tetrahedron with edges of unit length in the $\hat{\xi}\hat{\eta}\hat{\zeta}$ space, as shown in Figure 6.2.2(b), using the transformation rules

$$\hat{\xi} = \xi + \frac{1}{2}\,(\xi + \zeta), \qquad \hat{\eta} = \frac{\sqrt{3}}{2}\,(\eta + \frac{1}{3}\,\zeta), \qquad \hat{\zeta} = \sqrt{\frac{2}{3}}\,\zeta. \tag{6.2.7}$$

The third rule shows that all altitudes of the regular octahedron are equal to $\sqrt{2/3}$.

6.2.2 Integral over the volume of the tetrahedron

The Jacobian matrix of the mapping from the physical xyz space to the parametric $\xi\eta\zeta$ space is defined as

$$
\mathbf{J} \equiv
\begin{bmatrix}
\frac{\partial x}{\partial \xi} & \frac{\partial x}{\partial \eta} & \frac{\partial x}{\partial \zeta} \\[6pt]
\frac{\partial y}{\partial \xi} & \frac{\partial y}{\partial \eta} & \frac{\partial y}{\partial \zeta} \\[6pt]
\frac{\partial z}{\partial \xi} & \frac{\partial z}{\partial \eta} & \frac{\partial z}{\partial \zeta}
\end{bmatrix}.
\tag{6.2.8}
$$

The determinant of the Jacobian matrix is the *volume metric coefficient*,

$$
h_V \equiv \mathrm{Det}(\mathbf{J}) = \left(\frac{\partial \mathbf{x}}{\partial \xi} \times \frac{\partial \mathbf{x}}{\partial \eta} \right) \cdot \frac{\partial \mathbf{x}}{\partial \zeta}.
\tag{6.2.9}
$$

Substituting the linear mapping functions shown in (6.2.6), we find

$$
\frac{\partial \mathbf{x}}{\partial \xi} = \mathbf{x}_2^E - \mathbf{x}_1^E, \qquad
\frac{\partial \mathbf{x}}{\partial \eta} = \mathbf{x}_3^E - \mathbf{x}_1^E, \qquad
\frac{\partial \mathbf{x}}{\partial \zeta} = \mathbf{x}_4^E - \mathbf{x}_1^E,
\tag{6.2.10}
$$

and calculate

$$
\mathbf{J} =
\begin{bmatrix}
x_2^E - x_1^E & x_3^E - x_1^E & x_4^E - x_1^E \\[4pt]
y_2^E - y_1^E & y_3^E - y_1^E & y_4^E - y_1^E \\[4pt]
z_2^E - z_1^E & z_3^E - z_1^E & z_4^E - z_1^E
\end{bmatrix}.
\tag{6.2.11}
$$

Invoking the geometrical interpretation of the outer vector product in (6.2.9) (see Appendix D), we find that h_V is equal to the volume of a parallelepiped with three sides along the vectors $\mathbf{x}_2^E - \mathbf{x}_1^E$, $\mathbf{x}_3^E - \mathbf{x}_1^E$, and $\mathbf{x}_4^E - \mathbf{x}_1^E$, which is equal to six times the volume of the tetrahedron in the physical xyz space. Thus, the volume metric coefficient for the tetrahedron is

$$
h_V = 6\,V,
\tag{6.2.12}
$$

where V is the volume of the tetrahedron in the physical xyz space.

Using elementary calculus, we find that the integral of a function $f(x, y, z)$ over the volume of the physical tetrahedron in the xyz space can be expressed as an integral over the parametric tetrahedron in the $\xi\eta\zeta$ space, as

$$
\iiint f(x, y, z)\, \mathrm{d}x\, \mathrm{d}y \mathrm{d}z = \iiint f(\xi, \eta, \zeta)\, h_V\, \mathrm{d}\xi\, \mathrm{d}\eta\, \mathrm{d}\zeta
$$

$$
= 6\,V \iiint f(\xi, \eta, \zeta)\, \mathrm{d}\xi\, \mathrm{d}\eta\, \mathrm{d}\zeta.
\tag{6.2.13}
$$

$$
= 6\,V \int_0^1 \left[\int_0^{1-\zeta} \left(\int_0^{1-\zeta-\xi} f(\xi, \eta, \zeta)\, \mathrm{d}\eta \right) \mathrm{d}\xi \right] \mathrm{d}\zeta
$$

The integral in the parametric space can be computed analytically when the function f is a polynomial, and numerically under more general circumstances, as will be discussed later in this section.

6.2.3 Element matrices

In the *isoparametric interpolation*, a function of interest defined over the element is expressed in a form that is analogous to that shown in (6.2.2),

$$f(x, y, z, t) = \sum_{i=1}^{4} f_i^E(t)\, \psi_i(\xi, \eta, \zeta), \qquad (6.2.14)$$

with the understanding that the point $\mathbf{x} = (x, y, z)$ is mapped to the parametric point (ξ, η, ζ), and *vice versa*, through (6.2.2). Accordingly, the lth-element diffusion matrix is given by

$$A_{ij}^{(l)} \equiv \iiint_{E_l} \nabla\psi_i \cdot \nabla\psi_j \, \mathrm{d}x\, \mathrm{d}y\, \mathrm{d}z = 6\, V \iiint \nabla\psi_i \cdot \nabla\psi_j \, \mathrm{d}\xi\, \mathrm{d}\eta\, \mathrm{d}\zeta, \quad (6.2.15)$$

for $i, j = 1, 2, 3, 4$, where

$$\nabla\psi_i = \left(\frac{\partial\psi_i}{\partial x}, \quad \frac{\partial\psi_i}{\partial y}, \quad \frac{\partial\psi_i}{\partial z} \right) \qquad (6.2.16)$$

is the three-dimensional gradient. The corresponding mass matrix is given by

$$B_{ij}^{(l)} = \iiint_{E_l} \psi_i\, \psi_j \, \mathrm{d}x\, \mathrm{d}y\, \mathrm{d}z = 6\, V \iiint \psi_i\, \psi_j \, \mathrm{d}\xi\, \mathrm{d}\eta\, \mathrm{d}\zeta, \qquad (6.2.17)$$

and the corresponding advection matrix is given by

$$C_{ij}^{(l)} = \iiint_{E_l} \psi_i\, \mathbf{u} \cdot \nabla\psi_j \, \mathrm{d}x\, \mathrm{d}y\, \mathrm{d}z = 6\, V \iiint \psi_i\, \mathbf{u} \cdot \nabla\psi_j \, \mathrm{d}\xi\, \mathrm{d}\eta\, \mathrm{d}\zeta, \quad (6.2.18)$$

where $\mathbf{u} = (u_x, u_y, u_z)$ is the advection velocity.

Computation of the gradient

To compute the element diffusion and advection matrices, we require the gradients of the element interpolation functions, $\nabla\psi_i$, for $i = 1, 2, 3, 4$. These can be found readily using the relations

$$\frac{\partial\mathbf{x}}{\partial\xi} \cdot \nabla\psi_i = \frac{\partial\psi_i}{\partial\xi}, \qquad \frac{\partial\mathbf{x}}{\partial\eta} \cdot \nabla\psi_i = \frac{\partial\psi_i}{\partial\eta}, \qquad \frac{\partial\mathbf{x}}{\partial\zeta} \cdot \nabla\psi_i = \frac{\partial\psi_i}{\partial\zeta}, \qquad (6.2.19)$$

which state that the projection of the gradient vector on the ξ-line vector, $\partial\mathbf{x}/\partial\xi$, is the partial derivative with respect to ξ; similar statements are made for η and ζ. Explicitly, equations (6.2.19) read

$$\frac{\partial x}{\partial\xi}\frac{\partial\psi_i}{\partial x} + \frac{\partial y}{\partial\xi}\frac{\partial\psi_i}{\partial y} + \frac{\partial z}{\partial\xi}\frac{\partial\psi_i}{\partial z} = \frac{\partial\psi_i}{\partial\xi},$$

$$\frac{\partial x}{\partial\eta}\frac{\partial\psi_i}{\partial x} + \frac{\partial y}{\partial\eta}\frac{\partial\psi_i}{\partial y} + \frac{\partial z}{\partial\eta}\frac{\partial\psi_i}{\partial z} = \frac{\partial\psi_i}{\partial\eta}, \qquad (6.2.20)$$

$$\frac{\partial x}{\partial\zeta}\frac{\partial\psi_i}{\partial x} + \frac{\partial y}{\partial\zeta}\frac{\partial\psi_i}{\partial y} + \frac{\partial z}{\partial\zeta}\frac{\partial\psi_i}{\partial z} = \frac{\partial\psi_i}{\partial\zeta}.$$

Collecting these equations in a matrix form, we write

$$
\mathbf{J}^T \cdot \nabla \psi_i =
\begin{bmatrix}
\frac{\partial \psi_i}{\partial \xi} \\[2mm]
\frac{\partial \psi_i}{\partial \eta} \\[2mm]
\frac{\partial \psi_i}{\partial \zeta}
\end{bmatrix},
\tag{6.2.21}
$$

where $\mathbf{J}^T$ is the transpose of the Jacobian matrix defined in (6.2.8),

$$
\mathbf{J}^T \equiv
\begin{bmatrix}
\frac{\partial x}{\partial \xi} & \frac{\partial y}{\partial \xi} & \frac{\partial z}{\partial \xi} \\[2mm]
\frac{\partial x}{\partial \eta} & \frac{\partial y}{\partial \eta} & \frac{\partial z}{\partial \eta} \\[2mm]
\frac{\partial x}{\partial \zeta} & \frac{\partial y}{\partial \zeta} & \frac{\partial z}{\partial \zeta}
\end{bmatrix}.
\tag{6.2.22}
$$

The determinant of $\mathbf{J}^T$ is equal to the determinant of $\mathbf{J}$, which is equal to the volume metric coefficient, $h_V = 6\,V$.

Using (6.2.10), we find that equations (6.2.19) take the specific form,

$$
\begin{aligned}
(\mathbf{x}_2^E - \mathbf{x}_1^E) \cdot \nabla \psi_i &= \frac{\partial \psi_i}{\partial \xi}, \\[2mm]
(\mathbf{x}_3^E - \mathbf{x}_1^E) \cdot \nabla \psi_i &= \frac{\partial \psi_i}{\partial \eta}, \\[2mm]
(\mathbf{x}_4^E - \mathbf{x}_1^E) \cdot \nabla \psi_i &= \frac{\partial \psi_i}{\partial \zeta},
\end{aligned}
\tag{6.2.23}
$$

and accordingly,

$$
\mathbf{J}^T =
\begin{bmatrix}
x_2^E - x_1^E & y_2^E - y_1^E & z_2^E - z_1^E \\[2mm]
x_3^E - x_1^E & y_3^E - y_1^E & z_3^E - z_1^E \\[2mm]
x_4^E - x_1^E & y_4^E - y_1^E & z_4^E - z_1^E
\end{bmatrix},
\tag{6.2.24}
$$

which is in agreement with (6.2.8). Substituting the specific expressions of the cardinal interpolation functions given in (6.2.3), we derive the linear systems

$$
\mathbf{J}^T \cdot \nabla \psi_1 = -
\begin{bmatrix}
1 \\ 1 \\ 1
\end{bmatrix},
\qquad
\mathbf{J}^T \cdot \nabla \psi_2 =
\begin{bmatrix}
1 \\ 0 \\ 0
\end{bmatrix},
$$

$$
\tag{6.2.25}
$$

$$
\mathbf{J}^T \cdot \nabla \psi_3 =
\begin{bmatrix}
0 \\ 1 \\ 0
\end{bmatrix},
\qquad
\mathbf{J}^T \cdot \nabla \psi_4 =
\begin{bmatrix}
0 \\ 0 \\ 1
\end{bmatrix}.
$$

Applying Cramer's rule, we find the solutions

$$\nabla\psi_1 = \frac{1}{6V} \begin{bmatrix} -y_{32}\,z_{42} + z_{32}\,y_{42} \\ x_{32}\,z_{42} - z_{32}\,x_{42} \\ -x_{32}\,y_{42} + y_{32}\,x_{42} \end{bmatrix},$$

$$\nabla\psi_2 = \frac{1}{6V} \begin{bmatrix} y_{31}\,z_{41} - z_{31}\,y_{41} \\ -x_{31}\,z_{41} + z_{31}\,x_{41} \\ x_{31}\,y_{41} - y_{31}\,x_{41} \end{bmatrix},$$

$$\nabla\psi_3 = \frac{1}{6V} \begin{bmatrix} -y_{21}\,z_{41} + z_{21}\,y_{41} \\ x_{21}\,z_{41} - z_{21}\,x_{41} \\ -x_{21}\,y_{41} + z_{21}\,y_{41} \end{bmatrix}, \qquad (6.2.26)$$

$$\nabla\psi_4 = \frac{1}{6V} \begin{bmatrix} y_{21}\,z_{31} - z_{21}\,y_{31} \\ -x_{21}\,z_{31} + z_{21}\,x_{31} \\ x_{21}\,y_{31} - y_{21}\,x_{31} \end{bmatrix},$$

where $\mathbf{x}_{21} = \mathbf{x}_2^E - \mathbf{x}_1^E$, $\mathbf{x}_{31} = \mathbf{x}_3^E - \mathbf{x}_1^E$, $\mathbf{x}_{41} = \mathbf{x}_4^E - \mathbf{x}_1^E$, $\mathbf{x}_{32} = \mathbf{x}_3^E - \mathbf{x}_2^E$, and $\mathbf{x}_{42} = \mathbf{x}_4^E - \mathbf{x}_2^E$. The gradient $\nabla\psi_1$ is perpendicular to the face 234, the gradient $\nabla\psi_2$ is perpendicular to the face 314, the gradient $\nabla\psi_3$ is perpendicular to the face 124, and the gradient $\nabla\psi_4$ is perpendicular to the face 132. Expressions (6.2.26) are used to evaluate the element diffusion and advection matrices.

Computation of the element diffusion matrix

Because the gradient of the element interpolation functions is a vectorial constant independent of position inside the element, the element diffusion matrix defined in (6.2.15) simplifies to

$$A_{ij}^{(l)} = V\,\nabla\psi_i \cdot \nabla\psi_j, \qquad (6.2.27)$$

for $i, j = 1, 2, 3, 4$, and may thus be computed by direct evaluation using expressions (6.2.26).

Computation of the element mass matrix

Substituting expressions (6.2.3) in (6.2.17), we find that the element mass matrix is given by

$$\mathbf{B}^{(l)} = 6\,V \iiint \begin{bmatrix} \omega^2 & \omega\,\xi & \omega\,\eta & \omega\,\zeta \\ \xi\,\omega & \xi^2 & \xi\,\eta & \xi\,\zeta \\ \eta\,\omega & \eta\,\xi & \eta^2 & \eta\,\zeta \\ \zeta\,\omega & \zeta\,\xi & \zeta\,\eta & \zeta^2 \end{bmatrix} d\xi\,d\eta\,d\zeta. \qquad (6.2.28)$$

To compute these integrals, we use the integration formula

$$\iiint \omega^p\,\xi^q\,\eta^r\,\zeta^s\,d\xi\,d\eta\,d\zeta = \frac{p!\,q!\,r!\,s!}{(p+q+r+s+3)!}, \qquad (6.2.29)$$

to be proved later in this section, where p, q, r and s are non-negative integers, and an exclamation mark denotes the factorial, $m! = 1 \cdot 2 \dots \cdot m$. A straightforward calculation yields

$$
\mathbf{B}^{(l)} = \frac{V}{20}
\begin{bmatrix}
2 & 1 & 1 & 1 \\
1 & 2 & 1 & 1 \\
1 & 1 & 2 & 1 \\
1 & 1 & 1 & 2
\end{bmatrix}.
\tag{6.2.30}
$$

In the case of the 4-node tetrahedron presently discussed, but not more generally, the element mass matrix depends only on the element volume, V, and is independent of the element shape. The sum of all entries of the element mass matrix is equal to the volume of the tetrahedron, V.

The mass-lumped, diagonal element mass matrix, denoted by a hat, is given by

$$
\widehat{\mathbf{B}}^{(l)} = \frac{V}{4}
\begin{bmatrix}
1 & 0 & 0 & 0 \\
0 & 1 & 0 & 0 \\
0 & 0 & 1 & 0 \\
0 & 0 & 0 & 1
\end{bmatrix}.
\tag{6.2.31}
$$

The trace of the lumped element mass matrix is equal to the volume of the tetrahedron.

To prove the integration formula (6.2.29), we use (6.2.13) to write

$$
\iiint \omega^p \, \xi^q \, \eta^r \, \zeta^s \, \mathrm{d}\xi \, \mathrm{d}\eta \, \mathrm{d}\zeta
$$

$$
= \int_0^1 \left[\int_0^{1-\zeta} \left(\int_0^{1-\zeta-\xi} (1 - \zeta - \xi - \eta)^p \, \xi^q \, \eta^r \, \mathrm{d}\eta \right) \mathrm{d}\xi \right] \zeta^s \, \mathrm{d}\zeta, \tag{6.2.32}
$$

$$
= \int_0^1 \left[\int_0^1 \left(\int_0^{1-\hat{\xi}} (1 - \hat{\xi} - \hat{\eta})^p \, \hat{\xi}^q \, \hat{\eta}^r \, \mathrm{d}\hat{\eta} \right) \mathrm{d}\hat{\xi} \right] \zeta^s \, (1 - \zeta)^{p+q+r+2} \, \mathrm{d}\zeta,
$$

where $\hat{\xi} \equiv \xi/(1 - \zeta)$, and $\hat{\eta} \equiv \eta/(1 - \zeta)$. Next, we compute the double integral enclosed by the square brackets in the last expression using the integration formula (3.2.34), finding

$$
\iiint \omega^p \, \xi^q \, \eta^r \, \zeta^s \, \mathrm{d}\xi \, \mathrm{d}\eta \, \mathrm{d}\zeta = \frac{p! \, q! \, r!}{(p + q + r + 2)!} \int_0^1 \zeta^s \, (1 - \zeta)^{p+q+r} \, \mathrm{d}\zeta,
$$

$$
= \frac{p! \, q! \, r!}{(p + q + r + 2)!} \, B(s + 1, p + q + r + 3)
$$

$$
= \frac{p! \, q! \, r!}{(p + q + r + 2)!} \, \frac{s! \, (p + q + r + 2)!}{(p + q + r + s + 3)!}, \tag{6.2.33}
$$

where $B(k, l)$ is the beta function. Simplifying, we derive the integration formula (6.2.29).

Computation of the element advection matrix

When the advection velocity, $\mathbf{u}$, is constant and equal to $\mathbf{U} = (U_x, U_y, U_z)$, the element advection matrix simplifies to

$$C_{ij}^{(l)} = 6\,V\,\mathbf{U} \cdot \nabla \psi_j \iiint \psi_i \; \mathrm{d}\xi \, \mathrm{d}\eta \, \mathrm{d}\zeta \equiv 6\,V\,\alpha\,\mathbf{U} \cdot \nabla \psi_j, \qquad (6.2.34)$$

where α is a numerical coefficient given by

$$\alpha = \iiint \psi_i \; \mathrm{d}\xi \, \mathrm{d}\eta \, \mathrm{d}\zeta = \frac{1}{24}. \qquad (6.2.35)$$

The element advection matrix may thus be computed by direct evaluation using expressions (6.2.26).

6.2.4   Mapping the tetrahedron to a cube

Just like the triangle can be mapped to a square using Duffy's transformation, as illustrated in Figure 3.2.3, the standard tetrahedron in the $\xi\eta\zeta$ space can be mapped to a cube in the $\xi'\eta'\zeta'$ space by a similar transformation, as illustrated in Figure 6.2.4. The mapping is mediated by the functions

$$\xi = \frac{1 + \xi'}{2} \frac{1 - \eta'}{2} \frac{1 - \zeta'}{2},$$

$$(6.2.36)$$

$$\eta = \frac{1 + \eta'}{2} \frac{1 - \zeta'}{2}, \qquad \zeta = \frac{1 + \zeta'}{2},$$

where $-1 \le \xi', \eta', \zeta' \le 1$. The third barycentric coordinate is mapped according to the transformation rule

$$\omega \equiv 1 - \xi - \eta - \zeta = \frac{1 - \xi'}{2} \frac{1 - \eta'}{2} \frac{1 - \zeta'}{2}. \qquad (6.2.37)$$

When $\zeta' = -1$, corresponding to $\zeta = 0$, we recover the two-dimensional triangle transformations (3.2.37) and (3.2.38). For future reference, we note the relations

$$1 - \zeta = \frac{1 - \zeta'}{2}, \qquad 1 - \eta - \zeta = \frac{1 - \eta'}{2} \frac{1 - \zeta'}{2}. \qquad (6.2.38)$$

The inverse mapping functions are given by

$$\xi' = \frac{2\,\xi}{1 - \eta - \zeta} - 1,$$

$$(6.2.39)$$

$$\eta' = 2\,\frac{\eta}{1 - \zeta} - 1, \qquad \zeta' = 2\,\zeta - 1.$$

When $\zeta = 0$ corresponding to $\zeta' = -1$, we recover the inverse two-dimensional triangle transformations (3.2.40).

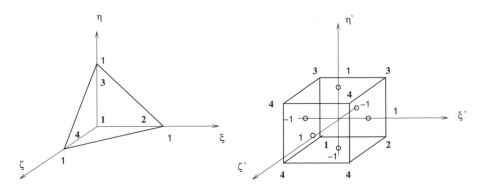

Figure 6.2.4 Mapping of the standard tetrahedron to the standard square in the $\xi'\eta'\zeta'$ space. The node labels are printed in bold.

The integral of a function $f(\xi, \eta, \zeta)$ over the volume of the tetrahedron in the $\xi\eta\zeta$ space can be expressed as an integral over the standard cube, as

$$\iiint f(\xi, \eta, \zeta) \, \mathrm{d}\xi \, \mathrm{d}\eta \, \mathrm{d}\zeta = \int_{-1}^{1} \int_{-1}^{1} \int_{-1}^{1} f(\xi, \eta, \zeta) \, h'_V \, \mathrm{d}\xi' \, \mathrm{d}\eta' \, \mathrm{d}\zeta'. \qquad (6.2.40)$$

The metric coefficient of the transformation, h'_V, is equal to the determinant of the Jacobian matrix of the transformation,

$$h'_V = \mathrm{Det}\left(\begin{bmatrix} \frac{\partial \xi}{\partial \xi'} & \frac{\partial \eta}{\partial \xi'} & \frac{\partial \zeta}{\partial \xi'} \\[6pt] \frac{\partial \xi}{\partial \eta'} & \frac{\partial \eta}{\partial \eta'} & \frac{\partial \zeta}{\partial \eta'} \\[6pt] \frac{\partial \xi}{\partial \zeta'} & \frac{\partial \eta}{\partial \zeta'} & \frac{\partial \zeta}{\partial \zeta'} \end{bmatrix} \right)$$

$$\qquad (6.2.41)$$

$$= \mathrm{Det}\left(\begin{bmatrix} \frac{1}{8}(1-\eta')(1-\zeta') & 0 & 0 \\[6pt] -\frac{1}{8}(1+\xi')(1-\zeta') & \frac{1}{4}(1-\zeta') & 0 \\[6pt] -\frac{1}{8}(1+\xi')(1-\eta') & -\frac{1}{4}(1+\eta') & \frac{1}{2} \end{bmatrix} \right) = \frac{(1-\eta')(1-\zeta')^2}{64}.$$

The integral over the cube on the right-hand side of (6.2.40) can be computed by applying an integration rule or Gaussian quadrature for the individual one-dimensional integrals.

6.2.5 Orthogonal polynomials over the tetrahedron

The orthogonal Proriol polynomials over the triangle discussed in Section 4.3.3 can be extended into a broader family of polynomials that are orthogonal over the volume of the tetrahedron with a flat weighting function, as discussed by Sherwin and Karniadakis [30, 55, 56, 57]. The tetrahedral orthogonal polynomials are given by

$$
\mathcal{P}_{klp} = L_k(\xi') \left(\frac{1 - \eta'}{2} \right)^k \left(\frac{1 - \zeta'}{2} \right)^k J_l^{(2k+1,0)}(\eta')
$$

$$
\times \left(\frac{1 - \zeta'}{2} \right)^l J_p^{(2k+2l+2,0)}(\zeta')
$$

$$
= \left(\frac{1 - \eta'}{2} \right)^k \left(\frac{1 - \zeta'}{2} \right)^{k+l} L_k(\xi') J_l^{(2k+1,0)}(\eta') J_p^{(2k+2l+2,0)}(\zeta'),
$$

$$(6.2.42)$$

where L_q are the Legendre polynomials, and $J_q^{(r,s)}$ are the Jacobi polynomials discussed in Appendix B. Recalling the transformation rules (6.2.38) and noting the presence of the denominators on the right-hand sides of the transformation rule for ξ' and η', as shown in (6.2.39), we find that $\mathcal{P}_{klp}$ is, in fact, a kth-degree polynomial in ξ, a $(k + l)$-degree polynomial in η, and a $(k + l + p)$-degree polynomial in ζ. Explicitly, the first few tetrahedral orthogonal polynomials are found to be

$$
\begin{aligned}
\mathcal{P}_{000} &= 1, \\
\mathcal{P}_{100} &= 2\xi + \eta + \zeta - 1, \\
\mathcal{P}_{010} &= 3\eta + \zeta - 1, \\
\mathcal{P}_{001} &= 4\zeta - 1, \\
\mathcal{P}_{200} &= 6\xi^2 + \eta^2 + \zeta^2 + 6\xi\eta + 6\xi\zeta + 2\eta\zeta - 6\xi - 2\eta - 2\zeta + 1, \\
\mathcal{P}_{110} &= 5\eta^2 + \zeta^2 + 10\xi\eta + 2\xi\zeta + 6\eta\zeta - 2\xi - 6\eta - 2\zeta + 1, \\
\mathcal{P}_{101} &= 6\zeta^2 + 12\xi\zeta + 6\eta\zeta - 2\xi - \eta - 7\zeta + 1, \\
\mathcal{P}_{020} &= 10\eta^2 + \zeta^2 + 8\eta\zeta - 8\eta - 2\zeta + 1, \\
\mathcal{P}_{011} &= 6\zeta^2 + 18\eta\zeta - 3\eta - 7\zeta + 1, \\
\mathcal{P}_{002} &= 15\zeta^2 - 10\zeta + 1.
\end{aligned}
\tag{6.2.43}
$$

To prove orthogonality, we use the integration formula (6.2.40), and find that the integral of the product of the klp and qrs polynomials over the volume of the tetrahedron is given by

$$
\iiint \mathcal{P}_{klp} \, \mathcal{P}_{qrs} \, d\xi \, d\eta \, d\zeta
$$

$$
= \int_{-1}^{1} \int_{-1}^{1} \int_{-1}^{1} \mathcal{P}_{klp} \, \mathcal{P}_{qrs} \frac{(1 - \eta')(1 - \zeta')^2}{64} \, d\xi' \, d\eta' \, d\zeta'
$$

$$
= \frac{1}{8} \int_{-1}^{1} L_k(\xi') \, L_q(\xi') \, d\xi' \times \int_{-1}^{1} J_l^{2k+1,0}(\eta') \, J_r^{2q+1,0}(\eta') \left(\frac{1 - \eta'}{2} \right)^{k+q+1} d\eta'
$$

$$
\times \int_{-1}^{1} J_p^{2k+2l+2,0}(\zeta') \, J_s^{2q+2r+2,0}(\zeta') \left(\frac{1 - \zeta'}{2} \right)^{k+l+q+r+2} d\eta'.
\tag{6.2.44}
$$

The integral with respect to ξ' is non-zero only if $k = q$. When $k = q$, the integral with respect to η' is non-zero only if $l = r$. When $k = q$ and $l = r$, the integral with respect to ζ' is non-zero only if $p = s$. Overall, the integral is non-zero only when $k = q$, $l = r$, and $p = s$, which shows that the polynomials are orthogonal. The self-projection integral is given by

$$\mathcal{G}_{klp} \equiv \iiint \mathcal{P}_{klp}^2(\xi, \eta, \zeta) \, d\xi \, d\eta \, d\zeta = \frac{1}{(2k+1)(2k+2l+2)(2k+2l+2p+3)}. \tag{6.2.45}$$

For example, $\mathcal{G}_{000} = 1/6$, which is equal to the volume of the standard tetrahedron in the $\xi\eta\zeta$ space.

A polynomial in ξ, η, and ζ can be recast in terms of the orthogonal tetrahedral polynomials using the relations

$$1 = \mathcal{P}_{000},$$

$$\xi = \frac{1}{12}(3\mathcal{P}_{000} + 6\mathcal{P}_{100} - 2\mathcal{P}_{010} - \mathcal{P}_{001}),$$

$$\eta = \frac{1}{12}(3\mathcal{P}_{000} + 4\mathcal{P}_{010} - \mathcal{P}_{001}),$$

$$\zeta = \frac{1}{4}(\mathcal{P}_{000} + \mathcal{P}_{001}),$$

$$\xi^2 = \frac{1}{90}(9\mathcal{P}_{000} + 30\mathcal{P}_{100} - 10\mathcal{P}_{010} - 5\mathcal{P}_{001}$$
$$+ 15\mathcal{P}_{200} - 9\mathcal{P}_{110} - 6\mathcal{P}_{101} + 3\mathcal{P}_{020} + 2\mathcal{P}_{011} + \mathcal{P}_{002}),$$

$$\eta^2 = \frac{1}{90}(9\mathcal{P}_{000} + 20\mathcal{P}_{010} - 5\mathcal{P}_{001} + 9\mathcal{P}_{020} - 4\mathcal{P}_{011} + \mathcal{P}_{002}),$$

$$\zeta^2 = \frac{1}{30}(3\mathcal{P}_{000} + 5\mathcal{P}_{001} + 2\mathcal{P}_{002}),$$

$$\xi\eta = \frac{1}{180}(9\mathcal{P}_{000} + 15\mathcal{P}_{100} + 5\mathcal{P}_{010} - 5\mathcal{P}_{001}$$
$$+ 18\mathcal{P}_{110} - 3\mathcal{P}_{101} - 9\mathcal{P}_{020} - \mathcal{P}_{011} + \mathcal{P}_{002}),$$

$$\xi\zeta = \frac{1}{180}(9\mathcal{P}_{000} + 15\mathcal{P}_{100} - 5\mathcal{P}_{010} + 5\mathcal{P}_{001} + 15\mathcal{P}_{101} - 5\mathcal{P}_{011} - 4\mathcal{P}_{002}),$$

$$\eta\zeta = \frac{1}{180}(9\mathcal{P}_{000} + 10\mathcal{P}_{010} + 5\mathcal{P}_{001} + 10\mathcal{P}_{011} - 4\mathcal{P}_{002}),$$

$$\dots \tag{6.2.46}$$

Orthogonal expansion

The tetrahedral orthogonal polynomials provide us with a complete orthogonal basis. Any non-singular function, $f(\xi, \eta, \zeta)$, defined over the standard tetrahedron in the $\xi\eta\zeta$ space, can be approximated with a complete mth-degree polynomial in ξ, η, and ζ, expressed in the form

$$f(\xi, \eta, \zeta) \simeq \sum_{k=0}^{m} \left[\sum_{l=0}^{m-k} \left(\sum_{p=0}^{m-k-l} a_{klp} \, \mathcal{P}_{klp}(\xi, \eta, \zeta) \right) \right], \tag{6.2.47}$$

where the triple sum is designed so that $k+l+p \le m$, and a_{klp} are appropriate coefficients.

The triple sum on the right-hand side of (6.2.47) can be arranged into a union of Pascal triangles with increasing dimensions. The first triangle is a point representing the constant term

$$\mathcal{P}_{000},$$

the second triangle encapsulates the linear functions

$$\mathcal{P}_{100}$$
$$\mathcal{P}_{010} \quad \mathcal{P}_{001},$$

the third triangle encapsulates the quadratic functions

$$\mathcal{P}_{200}$$
$$\mathcal{P}_{110} \quad \mathcal{P}_{101}$$
$$\mathcal{P}_{020} \quad \mathcal{P}_{011} \quad \mathcal{P}_{002},$$

and the $m+1$ triangle encapsulates the mth-order functions. The total number of coefficients in the mth-order expansion is

$$N = 1 + 3 + 6 + \ldots + \frac{1}{2}(m+1)(m+2) = \frac{1}{2}\sum_{i=0}^{m}(i+1)(i+2)$$

$$= \frac{1}{2}\left[\sum_{i=0}^{m}i^2 + 3\sum_{i=0}^{m}i + 2(m+1)\right]$$

$$= \frac{1}{2}\left[\frac{m(m+1)(2m+1)}{6} + 3\frac{m(m+1)}{2} + 2(m+1)\right]$$

$$= \frac{1}{6}(m+1)(m^2+5m+6)$$

$$= \frac{1}{6}(m+1)(m+2)(m+3) = \binom{m+3}{3}, \tag{6.2.48}$$

where (:) denotes the combinatorial. The triangles can be stacked downward to yield Pascal's pyramid, where the constant function $\mathcal{P}_{000}$ is located at the apex.

Multiplying (6.2.47) by $\mathcal{P}_{qrs}$, integrating the product over the volume of the standard tetrahedron, and using the orthogonality property, we find that the expansion coefficients are given by

$$a_{klp} = \frac{1}{\mathcal{G}_{klp}} \iiint f(\xi, \eta, \zeta)\, \mathcal{P}_{klp}(\xi, \eta, \zeta)\, \mathrm{d}\xi\, \mathrm{d}\eta\, \mathrm{d}\zeta, \tag{6.2.49}$$

where $\mathcal{G}_{klp}$ is defined in (6.2.45).

In practice, the polynomials coefficients introduced in (6.2.47) are found by enforcing N interpolation conditions to generate a Vandermonde system of linear equations, as discussed in Section 4.3. The orthogonality of the basis functions guarantees the well-conditioning of the coefficient matrix.

PROBLEMS

6.2.1 *Volume of a tetrahedron.*

(*a*) Show that the volume of a tetrahedron is given by

$$V = \frac{1}{6} \text{Det} \left(\begin{bmatrix} 1 & 1 & 1 & 1 \\ x_1^E & x_2^E & x_3^E & x_4^E \\ y_1^E & y_2^E & y_3^E & y_4^E \\ z_1^E & z_2^E & z_3^E & z_4^E \end{bmatrix} \right). \tag{6.2.50}$$

Hint: The determinant of a matrix remains unchanged when you subtract one column from another.

(*b*) Apply the integration formula (6.2.40) for $f = 1$ to show that the volume of the standard tetrahedron is equal to $1/6$.

6.2.2 *Gradient of the element interpolation functions.*

Prove the geometrical interpretation of the gradients of the element interpolation functions shown in (6.2.26), as discussed in the text.

6.2.4 *Mapping of the prism.*

Derive the mapping of the 6-node prismatic element shown in Figure 6.1.2 to the standard cube.

6.3 High-order and spectral tetrahedral elements

In Section 6.2, we discussed 4-node tetrahedral elements and derived node interpolation functions that are linear in the physical coordinates x, y, and z, as well as in the parametric coordinates ξ, η, and ζ. A complete mth-order polynomial expansion of any suitable function, $f(\xi, \eta, \zeta, t)$, over the volume of the tetrahedron takes the form

$$\begin{aligned} f(\xi, \eta, \zeta, t) = \quad & a_{000} \\ & + a_{100}\, \xi + a_{010}\, \eta + a_{001}\, \zeta \\ & + a_{200}\, \xi^2 + a_{110}\, \xi\eta + a_{101}\, \xi\zeta + a_{020}\, \eta^2 + a_{011}\, \eta\zeta + a_{002}\, \zeta^2 \\ & \cdots \quad \cdots \quad \cdots \quad \cdots \quad \cdots \quad \cdots \quad \cdots \\ & + a_{m,0,0}\, \xi^m + a_{m-1,1,0}\, \xi^{m-1}\eta + \ldots + a_{0,1,m-1}\, \eta\zeta^m + a_{0,0,m}\, \zeta^m, \end{aligned}$$

$$\tag{6.3.1}$$

where a_{ijk} are generally time-dependent coefficients. Note that the sum of the indices $i + j + k$ across each row of (6.3.1) is constant. The first row contains only one coefficient, the second row contains three coefficients, and the ith row contains $\frac{1}{2}(i+1)(i+2)$ coefficients. The total number of coefficients up to the mth row was counted in (6.2.48), and was found to be

$$N = \frac{1}{6}(m+1)(m+2)(m+3).\qquad(6.3.2)$$

When $m = 2$, corresponding to the complete quadratic polynomial, the expansion contains ten terms.

The terms in each row of (6.3.1) can be arranged on a Pascal triangle with increasing dimensions. The first triangle is a point representing the constant term,

$$1$$

the second triangle encapsulates the linear terms

$$\xi$$
$$\eta \quad \zeta,$$

the third triangle encapsulates the quadratic terms

$$\xi^2$$
$$\xi\eta \quad \zeta\xi$$
$$\eta^2 \quad \eta\zeta \quad \zeta^2,$$

and the $m+1$ triangle encapsulates the mth-order terms. The triangles can be stacked downward to yield Pascal's pyramid, where the unit monomial is located at the apex. To compute the expansion coefficients, we require an equal number of conditions or constraints associated either with interpolation nodes or with expansion modes.

6.3.1 Uniform node distribution

One way to ensure that the number of interpolation nodes is equal to the number of coefficients in the complete mth-order expansion, is to deploy the nodes as shown in Figure 6.3.1. In this method, nodal points along the ξ, η, and ζ axes are distributed on a one-dimensional uniform grid defined by a set of $m + 1$ points, $v_i = (i - 1)/m$, where $i = 1, 2, \ldots, m + 1$. The face and interior nodes are identified by the intersection of planes that are parallel to the four faces of the tetrahedron at evenly-spaced intervals, yielding Pascal's pyramid.

The nodes can be labeled by the trio of indices, (i, j, k), corresponding to the ξ, η, and ζ axes, as shown in Figure 6.3.1. For each value of the index i in the range $i = 1, 2, \ldots, m + 1$, the index j takes values in the range $j = 1, 2, \ldots, m + 2 - i$; and for each doublet (i, j), the index k takes values in the range $k = 1, 2, \ldots, m+3-i-j$. The labeling protocol can be permuted without prejudice.

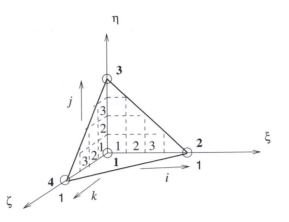

Figure 6.3.1 Uniform node distribution over the standard tetrahedron in parametric space corresponding to a complete polynomial expansion. The nodes are parametrized by the three indices, $i, j,$ and k.

A plane that is parallel to the slanted face of the tetrahedron in the $\xi\eta\zeta$ space corresponds to a constant value of the fourth barycentric coordinate, $\omega \equiv 1 - \xi - \eta - \zeta$, ranging from $\omega = 0$ at the slanted face, to $\omega = 1$ at the origin, $\xi = 0, \eta = 0, \zeta = 0$. A slanted plane hosting nodes is identified by the index

$$l = m + 4 - i - j - k, \tag{6.3.3}$$

which decreases from the value of 1 along the slanted face to the value of $m+1$ at the origin.

FSELIB script `nodes_tetra`, listed in abbreviated form in the text, generates and displays the nodes for a specified polynomial order, m. The graphics display produced by the script for $m = 6$ corresponding to eighty-four nodes, $N = 84$, is shown in Figure 6.3.2. Nodes marked with like symbols lie in slanted planes corresponding to constant values of ω and associated label k.

Node interpolation functions

The node interpolation functions can be deduced by generalizing the rule discussed in Section 4.3 for triangular elements. Consider the ith node of a tetrahedron, and assume that m planes can be found passing through all nodes, except for the ith node. If these planes are described by the linear functions

$$\mathcal{F}_j(\xi, \eta, \zeta) = A_j \xi + B_j \eta + C_j \zeta + D_j = 0, \tag{6.3.4}$$

for $j = 1, 2, \ldots, m$, where $A_j, B_j, C_j,$ and D_j are constant coefficients, then the ith node interpolation function is given by

$$\psi_i(\xi, \eta, \zeta) = \prod_{j=1}^{m} \frac{\mathcal{F}_j(\xi, \eta, \zeta)}{\mathcal{F}_j(\xi_i, \eta_i, \zeta_i)}. \tag{6.3.5}$$

```
%=========================================================
% nodes_tetra
%
% Uniform node distribution in the tetrahedron
% corresponding to a complete mth-order polynomial
%=========================================================

m = 5     % example

%--------------------
% uniform master grid
%--------------------

for i=1:m+1
   v(i) = (i-1.0)/m;
end

%---------------------------
% deploy and count the nodes
%---------------------------

count = 0;

for i=1:m+1;
 xp = v(i);

 for j=1:m+2-i
  yp = v(j);

   for k=1:m+3-i-j
    zp = v(j);

    count = count+1;
    l = m+4-i-j-k;
    if(l==1) plot3(xp,yp,zp,'+'); hold on; end;
    if(l==2) plot3(xp,yp,zp,'o'); hold on; end;
    if(l==3) plot3(xp,yp,zp,'x'); hold on; end;
    if(l==4) plot3(xp,yp,zp,'*'); hold on; end;
    if(l==5) plot3(xp,yp,zp,'+'); hold on; end;
    if(l==6) plot3(xp,yp,zp,'o'); hold on; end;
    if(l==7) plot3(xp,yp,zp,'x'); hold on; end;
    if(l==8) plot3(xp,yp,zp,'*'); hold on; end;
    if(l==9) plot3(xp,yp,zp,'+'); hold on; end;

   end
  end
end

......
```

Script nodes_tetra: Abbreviation of the FSELIB script nodes_tetra for generating and displaying a uniform node distribution over the tetrahedron, for a specified polynomial order, m. The six dots at the end of the listing denote unprinted code that draws the edges.

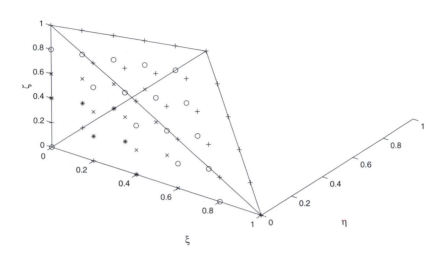

Figure 6.3.2 Uniform grid for polynomial order $m = 6$ involving eighty-four nodes, produced by the FSELIB script `nodes_tetra`. Nodes plotted with like symbols lie in slanted planes corresponding to a constant value of the barycentric coordinate, w and index l.

The cardinal node interpolation functions for the uniform grid can be constructed using this formula, on the observation that, for each node, all other nodes lie in preceding horizontal and vertical planes, and in subsequent slanted planes.

Specifically, the interpolation function corresponding to the (i, j, k) node can be expressed as the product of four functions,

$$\psi_{ijk}(\xi, \eta, \zeta) = \Xi_i^{(i-1)}(\xi) \times H_j^{(j-1)}(\eta) \times Z_k^{(k-1)}(\zeta) \times Y_l^{(l-1)}(w), \qquad (6.3.6)$$

where:

- $\Xi_i^{(i-1)}(\xi)$ is an $(i-1)$-degree polynomial, defined such that:

$$\Xi_1^{(0)}(\xi) = 1,$$

$$(6.3.7)$$

$$\Xi_i^{(i-1)}(\xi) = \frac{(\xi - v_1)(\xi - v_2) \dots (\xi - v_{i-2})(\xi - v_{i-1})}{(v_i - v_1)(v_i - v_2) \dots (v_i - v_{i-2})(v_i - v_{i-1})},$$

for $i = 2, 3, \dots, m + 1$.

- $H_j^{(j-1)}(\eta)$ is a $(j-1)$-degree polynomial, defined such that:

$$H_1^{(0)}(\eta) = 1,$$

(6.3.8)

$$H_j^{(j-1)}(\eta) = \frac{(\eta - v_1)(\eta - v_2)\ldots(\eta - v_{j-2})(\eta - v_{j-1})}{(v_j - v_1)(v_j - v_2)\ldots(v_j - v_{j-2})(v_j - v_{j-1})},$$

for $j = 2, 3, \ldots, m + 1$.

- $Z_k^{(k-1)}(\eta)$ is a $(k-1)$-degree polynomial, defined such that:

$$Z_1^{(0)}(\eta) = 1,$$

(6.3.9)

$$Z_k^{(k-1)}(\eta) = \frac{(\zeta - v_1)(\zeta - v_2)\ldots(\zeta - v_{k-2})(\zeta - v_{k-1})}{(v_k - v_1)(v_k - v_2)\ldots(v_k - v_{k-2})(v_k - v_{k-1})},$$

for $k = 2, 3, \ldots, m + 1$.

- $Y_l^{(l-1)}(\omega)$ is an $(l-1)$-degree polynomial, defined such that:

$$Y_1^{(0)}(\omega) = 1,$$

(6.3.10)

$$Y_l^{(l-1)}(\omega) = \frac{(\omega - v_1)(\omega - v_2)\ldots(\omega - v_{l-2})(\omega - v_{l-1})}{(v_l - v_1)(v_l - v_2)\ldots(v_l - v_{l-2})(v_l - v_{l-1})},$$

for $l = 2, 3, \ldots, m + 1$.

We may now confirm that $\psi_{ijk}(\xi, \eta, \zeta)$ is a polynomial of degree

$$(i - 1) + (j - 1) + (k - 1) + (l - 1) = i + j + k + l - 4 = m \qquad (6.3.11)$$

with respect to ξ, η, and ζ, as required, and that all cardinal interpolation conditions are met.

10-node quadratic element

As an application, we consider the complete quadratic expansion corresponding to $m = 2$, defined by ten interpolation nodes whose labels are printed in bold in Figure 6.3.3. Applying the preceding expressions with $v_1 = 0$, $v_2 = 0.5$, and $v_3 = 1.0$, and setting $l = 6 - i - j - k$, we find that the interpolation function of node $(i, j, k) = (1, 1, 1)$ corresponding to $l = 3$, labeled as node 1 in Figure 6.3.3, is given by

$$\psi_1 \equiv \psi_{1,1,1} = \Xi_1^{(0)}(\xi) \cdot H_1^{(0)}(\eta) \cdot Z_1^{(0)}(\zeta) \cdot Y_3^{(2)}(\omega), \qquad (6.3.12)$$

or

$$\psi_1 = 1 \cdot 1 \cdot 1 \cdot \frac{(\omega - 0.0)(\omega - 0.5)}{(1 - 0.0)(1 - 0.5)} = \omega\,(2\omega - 1). \qquad (6.3.13)$$

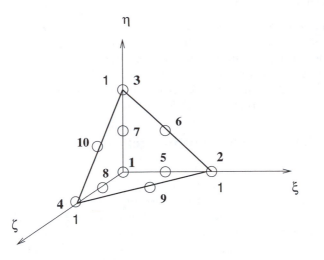

Figure 6.3.3 A 10-node tetrahedron supporting a complete quadratic expansion, $m = 2$. The node labels are printed in bold.

Working similarly for the second node, we find

$$\psi_2 \equiv \psi_{3,1,1} = \Xi_3^{(2)}(\xi) \cdot H_1^{(0)}(\eta) \cdot Z_1^{(0)}(\zeta) \cdot Y_1^{(0)}(\omega), \qquad (6.3.14)$$

or

$$\psi_2 = \frac{(\xi - 0.0)(\xi - 0.5)}{(1 - 0.0)(1 - 0.5)} \cdot 1 \cdot 1 \cdot 1 = \xi\,(2\xi - 1). \qquad (6.3.15)$$

The rest of the node interpolation functions can be computed in a similar fashion. The complete set reads

$$
\begin{aligned}
\psi_1 &\equiv \psi_{1,1,1} = \omega\,(2\omega - 1),\\
\psi_2 &\equiv \psi_{3,1,1} = \xi\,(2\xi - 1),\\
\psi_3 &\equiv \psi_{1,3,1} = \eta\,(2\eta - 1),\\
\psi_4 &\equiv \psi_{1,1,3} = \zeta\,(2\zeta - 1),\\
\psi_5 &\equiv \psi_{2,1,1} = 4\,\xi\,\omega,\\
\psi_6 &\equiv \psi_{2,2,1} = 4\,\xi\,\eta,\\
\psi_7 &\equiv \psi_{1,2,1} = 4\,\eta\,\omega,\\
\psi_8 &\equiv \psi_{1,1,2} = 4\,\zeta\,\omega,\\
\psi_9 &\equiv \psi_{2,1,2} = 4\,\xi\,\zeta,\\
\psi_{10} &\equiv \psi_{1,2,2} = 4\,\eta\,\zeta.
\end{aligned}
\qquad (6.3.16)
$$

Gradient of the node interpolation functions

The gradients of the element node interpolation functions, $\nabla \psi_i$, can be computed as discussed in Section 6.2, using relations (6.2.19). If the position vector, $\mathbf{x}$, is mapped using the linear functions discussed in Section 6.2 based on the vertices of the tetrahedron alone, whereas a function of interest is interpolated using the quadratic functions (6.3.16), then $\nabla \psi_i$ is found by solving system (6.2.21). In this case, the interpolation is super-parametric.

6.3.2 Arbitrary node distributions

To compute the mth-degree node interpolation functions for an arbitrary node distribution over the tetrahedron, we may expand each one of them in a series of monomial products or tetrahedral orthogonal polynomials, as discussed in Section 6.2, and compute the N expansion coefficients by solving systems of linear equations that arise by enforcing the cardinal interpolation conditions. Compiling these systems, we derive an expression for the vector of nodal interpolation functions in terms of the Vandermonde matrix, as discussed in Section 4.3.

Applying Cramer's rule, we find that each node interpolation function can be expressed as the ratio of the determinants of two generalized Vandermonde matrices, as shown in (4.3.27). The node distribution that maximizes the magnitude of the determinant of the Vandermonde matrix in the denominator, subject to the restriction that all nodes lie inside over the faces, or at the vertices of the tetrahedron, comprises the Fekete set.

6.3.3 Spectral node distributions

To guarantee that the interpolating function is a complete mth-degree polynomial of two barycentric coordinates over each face, and thus facilitate enforcing the C^0 continuity condition of the finite element expansion, we deploy a group of interpolation nodes over the faces of the tetrahedron by one of the methods discussed in Section 4.4 for the triangle. The remaining nodes are deployed inside the element according to a new set of criteria.

The nodal set includes $N_v = 4$ vertex nodes, $N_e = 6 \times (m-1)$ non-vertex edge nodes, and $N_f = 4 \times \frac{1}{2}(m-1)(m-2)$ interior face nodes, adding up to

$$N_s = 2\left(m^2 + 1\right) \tag{6.3.17}$$

surface nodes. The number of interior nodes is

$$N_i = N - N_s = \frac{1}{6}(m-3)(m-2)(m-1) = \binom{m-1}{3}, \tag{6.3.18}$$

which is precisely equal to the number of nodes corresponding to the $(m-4)$-order complete polynomial expansion. Note that interior nodes arise only for polynomial orders $m \geq 4$.

(a)

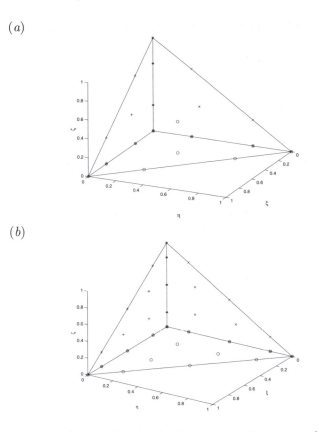

(b)

Figure 6.3.4 (a) A 20-node tetrahedron supporting a complete cubic expansion, $m = 3$, and (b) a 35-node tetrahedron supporting a complete quartic expansion, $m = 4$. For clarity, only the face nodes in the three orthogonal planes are shown in (b).

In the case of the linear expansion, $m = 1$, we assign the four nodes at the four element vertices, as discussed in Section 6.2. In the case of the quadratic expansion, $m = 2$, described by ten nodes, we adopt the uniform node distribution shown in Figure 6.3.3. In the case of the cubic expansion, $m = 3$, we deploy the twenty interpolation nodes at the element faces using one of the methods discussed in Section 4.4 for the triangle. For example, in the distribution illustrated in Figure 6.3.4(a), the edge nodes are distributed at positions corresponding to the zeros of the completed Lobatto polynomial, $Lo_4^c(t) = (1 - t^2) \, Lo_2(t)$, and the face nodes are positioned at the face centroids. In the nodal set illustrated in Figure 6.3.4(b) for $m = 4$, consisting of thirty-five nodes, thirty-four surface nodes are distributed at positions corresponding to the Lobatto triangle grid. Common sense and symmetry arguments suggest placing the remaining solitary interior node at the element centroid located at $\xi = \eta = \zeta = 1/4$.

Chen & Babuška [8] computed node distributions by maximizing the magnitude of the determinant of the generalized Vandermonde matrix in the spirit of the Fekete set, as well as by minimizing the magnitude of the modified Lebesgue function

$$\left(\iiint \sum_{i=1}^{N} |\psi_i(\xi, \eta, \zeta)|^2 \, d\xi \, d\eta \, d\zeta \right)^{1/2}, \tag{6.3.19}$$

where the integral is computed over the volume of the standard tetrahedron, subject to the condition that the nodal sets observe the geometrical symmetries of the tetrahedron and the face nodes are distributed as in the case of two-dimensional interpolation discussed in Section 4.4. Hesthaven & Teng [27] performed a similar computation by minimizing instead an electrostatic potential. More recently, Luo & Pozrikidis [34] proposed a tetrahedral Lobatto grid as an extension of the Lobatto triangle grid discussed in Section 4.4.

6.3.4 Modal expansion

An alternative approach is based on a modal expansion that is similar to that discussed in Section 4.5 for the triangle. Sherwin and Karniadakis [30, 55, 56, 57] proposed approximating a function of interest, $f(\xi, \eta, \zeta)$, defined over the volume of the standard tetrahedron, with an mth-degree polynomial, as

$$f(\xi, \eta, \zeta) \simeq F_v + F_e + F_f + F_i, \tag{6.3.20}$$

where

$$\begin{aligned}
F_v(\xi, \eta, \zeta) = f_1 \, \zeta_1^v(\xi, \eta, \zeta) + f_2 \, \zeta_2^v(\xi, \eta, \zeta) \\
+ f_3 \, \zeta_3^v(\xi, \eta, \zeta) + f_4 \, \zeta_4^v(\xi, \eta, \zeta)
\end{aligned} \tag{6.3.21}$$

is the *vertex part*,

$$F_e(\xi, \eta, \zeta) = \sum_{i=1}^{m-1} c_i^{12} \zeta_i^{12}(\xi, \eta, \zeta) + \sum_{i=1}^{m-1} c_i^{13} \zeta_i^{13}(\xi, \eta, \zeta) + \sum_{i=1}^{m-1} c_i^{23} \zeta_i^{23}(\xi, \eta, \zeta)$$

$$+ \sum_{i=1}^{m-1} c_i^{14} \zeta_i^{14}(\xi, \eta, \zeta) + \sum_{i=1}^{m-1} c_i^{24} \zeta_i^{24}(\xi, \eta, \zeta) + \sum_{i=1}^{m-1} c_i^{34} \zeta_i^{34}(\xi, \eta, \zeta) \tag{6.3.22}$$

is the *edge part* arising when $m \geq 2$,

$$F_f(\xi, \eta, \zeta) = \sum_{i=1}^{m-2} \left(\sum_{j=1}^{m-i-1} c_{ij}^{123} \zeta_{ij}^{123}(\xi, \eta, \zeta) \right) + \sum_{i=1}^{m-2} \left(\sum_{j=1}^{m-i-1} c_{ij}^{134} \zeta_{ij}^{134}(\xi, \eta, \zeta) \right)$$

$$+ \sum_{i=1}^{m-2} \left(\sum_{j=1}^{m-i-1} c_{ij}^{142} \zeta_{ij}^{142}(\xi, \eta, \zeta) \right) + \sum_{i=1}^{m-2} \left(\sum_{j=1}^{m-i-1} c_{ij}^{234} \zeta_{ij}^{234}(\xi, \eta, \zeta) \right) \tag{6.3.23}$$

is the *face part* arising when $m \geq 3$, and

$$F_i(\xi, \eta) = \sum_{i=1}^{m-3} \left[\sum_{j=1}^{m-i-2} \left(\sum_{k=1}^{m-i-j-1} c_{ijk}\, \zeta_{ijk}(\xi, \eta, \zeta) \right) \right] \tag{6.3.24}$$

is the *interior part* arising when $m \geq 4$. Each part is comprised of corresponding modes represented by interpolation functions multiplied by generally time-depended coefficients, sometimes called degrees of freedom, as will be discussed in detail later in this section.

The modal expansion includes $N_v = 4$ vertex modes, $N_e = 6 \times (m-1)$ edge modes, $N_f = 4 \times \frac{1}{2}(m-1)(m-2)$ face modes, and $N_i = \frac{1}{6}(m-3)(m-2)(m-1)$ interior modes. The total number of modes is

$$\begin{aligned} N &= N_v + N_e + N_f + N_i \\ &= 4 + 6\,(m-1) + 2\,(m-1)(m-2) + \frac{1}{6}\,(m-3)(m-2)(m-1) \\ &= \frac{(m+1)(m+2)(m+3)}{6}, \end{aligned} \tag{6.3.25}$$

which is precisely equal to the number of terms in the complete mth-order polynomial expansion in ξ, η, and ζ. In fact, the modes are designed so that the modal expansion expressed by (4.5.1) is a complete mth-degree polynomial in ξ, η, and η. The four-component modal decomposition is motivated by the ease of enforcing continuity of the finite element expansion at nodes, edges, and faces shared by adjacent elements, as discussed in Section 4.5 for the triangle.

Vertex modes

The coefficients f_1, f_2, f_3, and f_4 in (6.3.21) are the values of the function $f(\xi, \eta, \zeta, t)$ at the four vertex nodes labeled 1, 2, 3, and 4, as shown in Figure 6.2.3. To ensure C^0 continuity of the finite element expansion, these values must be shared by neighboring elements at the common nodes. The corresponding cardinal interpolation functions, ζ_1^v, ζ_2^v, ζ_3^v, and ζ_4^v, are identical to the linear nodal interpolation functions given in (6.2.3),

$$\zeta_1^v = 1 - \xi - \eta - \zeta, \qquad \zeta_2^v = \xi,$$

$$\zeta_3^v = \eta, \qquad \zeta_4^v = \zeta. \tag{6.3.26}$$

Edge modes

The coefficients c_i^{12}, c_i^{13}, c_i^{23} c_i^{14}, c_i^{24}, and c_i^{34} in (6.3.22) are associated with six families of the edge interpolation functions, ζ_i^{12}, ζ_i^{13}, ζ_i^{23}, ζ_i^{14}, ζ_i^{24}, and ζ_i^{34}, where $i = 1, 2, \ldots, m-1$. To ensure C^0 continuity of the finite element expansion, these coefficients must be shared by neighboring elements at the common edges. The edge modes take the value of zero along all edges, except for the superscripted edge. Working as in Section 4.5, we find the specific expressions

$$\zeta_i^{12} = \frac{1-\xi'}{2} \frac{1+\xi'}{2} \left(\frac{1-\eta'}{2}\right)^{i+1} \left(\frac{1-\zeta'}{2}\right)^{i+1} Lo_{i-1}(\xi'),$$

$$\zeta_i^{13} = \frac{1+\eta'}{2} \frac{1-\xi'}{2} \frac{1-\eta'}{2} \left(\frac{1-\zeta'}{2}\right)^{i+1} Lo_{i-1}(\eta'),$$

$$\zeta_i^{23} = \frac{1+\xi'}{2} \frac{1+\eta'}{2} \frac{1-\eta'}{2} \left(\frac{1-\zeta'}{2}\right)^{i+1} Lo_{i-1}(\eta'),$$

$$\zeta_i^{14} = \frac{1-\xi'}{2} \frac{1-\eta'}{2} \frac{1-\zeta'}{2} \frac{1+\zeta'}{2} Lo_{i-1}(\zeta'), \qquad (6.3.27)$$

$$\zeta_i^{24} = \frac{1+\xi'}{2} \frac{1-\eta'}{2} \frac{1-\zeta'}{2} \frac{1+\zeta'}{2} Lo_{i-1}(\zeta'),$$

$$\zeta_i^{34} = \frac{1+\eta'}{2} \frac{1-\zeta'}{2} \frac{1+\zeta'}{2} Lo_{i-1}(\zeta'),$$

for $i = 1, 2, \ldots, m-1$, where Lo_k is a Lobatto polynomial.

Face modes

The coefficients c_{ij}^{pqr} in (6.3.23) are associated with four families of face modes represented by the functions ζ_{ij}^{pqr}, where

$$i = 1, 2, \ldots, m-1, \qquad j = 1, 2, \ldots, m-i-1. \qquad (6.3.28)$$

To ensure C^0 continuity of the finite element expansion, these coefficients must be shared by neighboring elements at the common faces. The face modes take the value of zero over all faces, except for the superscripted face. Karniadakis and Sherwin [30, 55, 56, 57] define the face modes as

$$\zeta_{ij}^{123} = \frac{1-\xi'}{2} \frac{1+\xi'}{2} Lo_{i-1}(\xi') \left(\frac{1-\eta'}{2}\right)^{i+1} \frac{1+\eta'}{2} J_{j-1}^{(2i+1,1)}(\eta') \left(\frac{1-\zeta'}{2}\right)^{i+j+1},$$

$$\zeta_{ij}^{124} = \frac{1-\xi'}{2} \frac{1+\xi'}{2} Lo_{i-1}(\xi') \left(\frac{1-\eta'}{2}\right)^{i+1} J_{j-1}^{(2i+1,1)}(\eta') \left(\frac{1-\zeta'}{2}\right)^{i1} \frac{1+\zeta'}{2},$$

$$\zeta_{ij}^{134} = \frac{1-\xi'}{2} \frac{1-\eta'}{2} \frac{1+\eta'}{2} Lo_{i-1}(\eta') \left(\frac{1-\zeta'}{2}\right)^{i+1} \frac{1+\zeta'}{2} J_{j-1}^{(2i+1,1)}(\eta'),$$

$$\zeta_{ij}^{234} = \frac{1+\xi'}{2} \frac{1-\eta'}{2} \frac{1+\eta'}{2} Lo_{i-1}(\eta') \left(\frac{1-\zeta'}{2}\right)^{i+1} \frac{1+\zeta'}{2} J_{j-1}^{(2i+1,1)}(\eta'),$$

$$(6.3.29)$$

where $J_\alpha^{(\beta,\gamma)}$ are the Jacobi polynomials discussed in Section B.8 of Appendix B.

Interior modes

The interior modes, also called bubble modes, are zero over all faces of the tetrahedron and non-zero only inside the element. Requiring partial orthogonality, we find

$$\zeta_{ijk} = \frac{1+\xi'}{2} \frac{1-\xi'}{2} Lo_{i-1}(\xi') \frac{1+\eta'}{2} \left(\frac{1-\eta'}{2}\right)^{i+1}$$

$$\times J_{j-1}^{(2i+1,1)}(\eta') \left(\frac{1-\zeta'}{2}\right)^{i+j+1} \frac{1+\zeta'}{2} J_{k-1}^{(2i+2j+1,1)}(\zeta'). \quad (6.3.30)$$

For example, when $m = 4$, we obtain a single bubble mode corresponding to $i = 1, j = 1$, and $k = 1$, given by

$$\zeta_{111} = \xi \, \eta \, \zeta \, \omega. \quad (6.3.31)$$

Implementation of the modal expansion

The Galerkin finite element method for the modal expansion is implemented as discussed earlier in this chapter for the nodal expansion, the main difference being that the element node interpolation functions are replaced by the non-cardinal, modal interpolation functions, and the interior nodal values are replaced by the corresponding modal expansion coefficients. The numerical computation of the mass and diffusion matrices, consisting of integrals of pairwise products of the modal functions and their gradients, is discussed in detail by Sherwin and Karniadakis [30, 56, 57].

PROBLEMS

6.3.1 *Pascal pyramid.*

Display the structure of the fourth and fifth triangle in the Pascal pyramid involving the cubic and quartic monomial products.

6.3.2 *Bubble modes.*

Explain how the bubble mode (6.3.31) follows from (6.3.30).

6.4   Hexahedral elements

Hexahedral brick-like elements have six planar or curved *faces* intersecting at twelve straight or curved *edges*, The word "hexahedron" derives from the Greek work "εξαεδρον," which is composed of the word "εξι," meaning "six," and the word "εδρα," meaning "base" or "side." Hexahedral elements are the counterparts of the quadrilateral elements in two dimensions discussed in Sections 3.8 and 4.7.

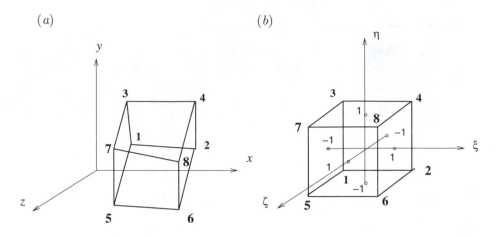

Figure 6.4.1 An 8-node hexahedral element with twelve straight edges in the (a) physical, and (b) parametric space. The element node labels are printed in bold.

The simplest hexahedral element has twelve straight edges defined by eight geometrical vertex nodes,

$$\mathbf{x}_i^E = (x_i^E, y_i^E, z_i^E), \tag{6.4.1}$$

where $i = 1, 2, \ldots, 8$, as shown in Figure 6.4.1(a). The superscript E emphasizes that these are local or "element" nodes, which can be mapped to the unique global nodes through a connectivity table.

FSELIB script hexa8, listed in the text, visualizes the element with the faces painted in red, blue, yellow, and green. The graphics display generated by a modification of this script where the commented out MATLAB function plot3 is used in place of the MATLAB function patch to produce a wire-frame image, is shown in Figure 6.4.2.

6.4.1   Parametric representation

To describe the 8-node element in parametric form, we map it from the physical xyz space to the standard cube whose vertices are located at the points $(\pm1, \pm1, \pm1)$ in the $\xi\eta\zeta$ space, as shown in Figure 6.4.1(b). The volume of the parametric cube is equal to 8. The mapping from the physical to the parametric space is mediated by the function

$$\mathbf{x} = \sum_{i=1}^{8} \mathbf{x}_i^E \, \psi_i(\xi, \eta, \zeta), \tag{6.4.2}$$

where $\psi_i(\xi, \eta, \zeta)$ are element interpolation functions satisfying familiar cardinal properties: $\psi_i = 1$ at the ith element node, and $\psi_i = 0$ at the other seven element nodes, for $i = 1 - 8$.

```
% vertex coordinates

x(1)= 0.0; y(1)= 0.0; z(1)= 0.00;
x(2)= 0.9; y(2)= 0.2; z(2)= 0.02;
x(3)= 0.1; y(3)= 1.1; z(3)=-0.05;
x(4)= 0.8; y(4)= 1.0; z(4)= 0.02;
x(5)= 0.1; y(5)= 0.1; z(5)= 1.01;
x(6)= 0.8; y(6)= 0.6; z(6)= 1.10;
x(7)= 0.2; y(7)= 0.9; z(7)= 0.90;
x(8)= 0.7; y(8)= 0.8; z(8)= 0.80;

% face node labels

s(1,1)=1; s(1,2)=2; s(1,3)=4; s(1,4)=3;
s(2,1)=1; s(2,2)=3; s(2,3)=7; s(2,4)=5;
s(3,1)=6; s(3,2)=5; s(3,3)=7; s(3,4)=8;
s(4,1)=2; s(4,2)=6; s(4,3)=8; s(4,4)=4;
s(5,1)=3; s(5,2)=4; s(5,3)=8; s(5,4)=7;
s(6,1)=2; s(6,2)=6; s(6,3)=5; s(6,4)=1;

% paint the faces

for i=1:6
  for j=1:4
  xp(j)=x(s(i,j)); yp(j)=y(s(i,j)); zp(j)=z(s(i,j));
  end
  if(i==1) patch(xp,yp,zp,'r'); end
  if(i==2) patch(xp,yp,zp,'b'); end
  if(i==3) patch(xp,yp,zp,'y'); end
  if(i==4) patch(xp,yp,zp,'g'); end
  if(i==5) patch(xp,yp,zp,'r'); end
  if(i==6) patch(xp,yp,zp,'b'); end
% plot3(xp,yp,zp); hold on
end

xlabel('x'); ylabel('y'); zlabel('z');
```

Script hexa8: Abbreviation of the FSELIB script hexa8 for generating and displaying a hexahedral element.

Working as in Section 3.8 for the 4-node quadrilateral element, we derive the element interpolation functions

$$\psi_1 = \frac{1-\xi}{2}\frac{1-\eta}{2}\frac{1-\zeta}{2}, \qquad \psi_2 = \frac{1+\xi}{2}\frac{1-\eta}{2}\frac{1-\zeta}{2},$$

$$\psi_3 = \frac{1-\xi}{2}\frac{1+\eta}{2}\frac{1-\zeta}{2}, \qquad \psi_4 = \frac{1+\xi}{2}\frac{1+\eta}{2}\frac{1-\zeta}{2},$$

$$\psi_5 = \frac{1-\xi}{2}\frac{1-\eta}{2}\frac{1+\zeta}{2}, \qquad \psi_6 = \frac{1+\xi}{2}\frac{1-\eta}{2}\frac{1+\zeta}{2}, \qquad (6.4.3)$$

$$\psi_7 = \frac{1-\xi}{2}\frac{1+\eta}{2}\frac{1+\zeta}{2}, \qquad \psi_8 = \frac{1+\xi}{2}\frac{1+\eta}{2}\frac{1+\zeta}{2}.$$

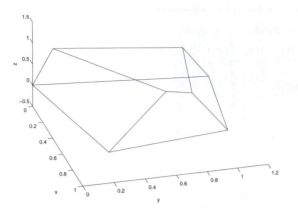

Figure 6.4.2 Visualization of a hexahedral element generated by the FSELIB
   script `hexa8`.

6.4.2   Integral over the volume of the hexahedron

The Jacobian matrix of the mapping from the physical xyz space to the parametric $\xi\eta\zeta$ space is defined in (6.2.8). The determinant of the Jacobian matrix is
the *volume metric coefficient,* h_V defined in (6.2.9). Substituting the quadratic
mapping shown in (6.4.3), we derive a nonlinear expression for h_V in ξ, η, and
ζ. The metric coefficient is constant only if the hexahedral element in physical
space is orthogonal. In contrast, the corresponding metric coefficient of the
4-node hexahedron is constant, independent of the element shape.

   Using elementary calculus, we find that the integral of a function $f(x, y, z)$
over the physical hexahedron in the xyz space can be expressed as an integral
over the parametric cube in the $\xi\eta\zeta$ space, as

$$\iiint f(x, y, z) \, \mathrm{d}x \, \mathrm{d}y \, \mathrm{d}z = \iiint f(\xi, \eta, \zeta) \, h_V \, \mathrm{d}\xi \, \mathrm{d}\eta \, \mathrm{d}\zeta. \qquad (6.4.4)$$

The volume integral over the cube can be computed by the triple application
of a one-dimensional integration quadrature.

   For example, adopting the Lobatto quadrature shown in (2.3.13), we obtain
the approximation

$$\iiint f(\xi, \eta, \zeta) \, h_V \, \mathrm{d}\xi \, \mathrm{d}\eta \, \mathrm{d}\zeta \qquad (6.4.5)$$

$$\simeq \sum_{p_1=1}^{k_1+1} \sum_{p_2=1}^{k_2+1} \sum_{p_3=1}^{k_3+1} f(t_{p_1}, t_{p_2}, t_{p_3}) \, h_V(t_{p_1}, t_{p_2}, t_{p_3}) \, w_{p_1} \, w_{p_2} \, w_{p_3},$$

where k_1, k_2, and k_3 are specified quadrature orders.

6.4.3 High-order and spectral hexahedral elements

Hexahedral elements with a higher number of edge, face, and interior nodes can be developed, as discussed in Section 4.7 for quadrilateral elements.

Grid nodes via tensor-product expansions

In this class of hexahedral elements, the interpolation nodes are deployed on a $(m_1 + 1) \times (m_2 + 1) \times (m_3 + 1)$ Cartesian grid covering the standard cube in the $\xi\eta\zeta$ space. The nodes are identified by the coordinate triplet (ξ_i, η_j, ζ_k), where $i = 1, 2, \ldots, m_1 + 1$, $j = 1, 2, \ldots, m_2 + 1$, and $k = 1, 2, \ldots, m_3 + 1$. To ensure that common nodes are introduced at the element vertices and along the edges, and thereby facilitate enforcing the C^0 continuity condition of the finite element expansion, we require

$$\xi_1 = -1, \quad \xi_{m_1+1} = 1,$$
$$\eta_1 = -1, \quad \eta_{m_2+1} = 1, \tag{6.4.6}$$
$$\zeta_1 = -1, \quad \zeta_{m_3+1} = 1.$$

The cardinal interpolation function of the (i, j, k) node is given by the tensor product

$$\psi_{ijk}(\xi, \eta, \zeta) = \mathcal{L}_i(\xi) \, \mathcal{M}_j(\eta) \, \mathcal{N}_k(\zeta), \tag{6.4.7}$$

where $\mathcal{L}_i(\xi)$ is an m_1-degree Lagrange interpolating polynomial defined with respect to the ξ grid lines,

$$\mathcal{L}_i(\xi) = \frac{(\xi - \xi_1)(\xi - \xi_2) \cdots (\xi - \xi_{i-1})(\xi - \xi_{i+1}) \cdots (\xi - \xi_{m_1+1})}{(\xi_i - \xi_1)(\xi_i - \xi_2) \cdots (\xi_i - \xi_{i-1})(\xi_i - \xi_{i+1}) \cdots (\xi_i - \xi_{m_1+1})}, \tag{6.4.8}$$

$\mathcal{M}_j(\eta)$ is an m_2-degree Lagrange interpolating polynomial defined with respect to the η grid lines,

$$\mathcal{M}_j(\eta) = \frac{(\eta - \eta_1)(\eta - \eta_2) \cdots (\eta - \eta_{j-1})(\eta - \eta_{j+1}) \cdots (\eta - \eta_{m_2+1})}{(\eta_j - \eta_1)(\eta_j - \eta_2) \cdots (\eta_j - \eta_{j-1})(\eta_j - \eta_{j+1}) \cdots (\eta_j - \eta_{m_2+1})}, \tag{6.4.9}$$

and $\mathcal{N}_k(\zeta)$ is an m_3-degree Lagrange interpolating polynomial defined with respect to the ζ grid lines,

$$\mathcal{N}_k(\zeta) = \frac{(\zeta - \zeta_1)(\zeta - \zeta_2) \cdots (\zeta - \zeta_{k-1})(\zeta - \zeta_{k+1}) \cdots (\zeta - \zeta_{m_3+1})}{(\zeta_k - \zeta_1)(\zeta_k - \zeta_2) \cdots (\zeta_k - \zeta_{k-1})(\zeta_k - \zeta_{z+1}) \cdots (\zeta_z - \zeta_{m_3+1})}. \tag{6.4.10}$$

Substituting these expressions in (6.4.7), we obtain a polynomial expansion in (ξ, η, ζ) involving $(m_1 + 1) \times (m_2 + 1) \times (m_3 + 1)$ coefficients.

Spectral node distribution

For best interpolation accuracy and numerical stability, the interior grid lines ξ_i, $i = 2, 3, \ldots, m_1$, should be placed at the zeros of the $(m_1 - 1)$-degree Lobatto polynomial, the interior grid lines η_j, $j = 2, 3, \ldots, m_2$, should be placed at the zeros of the $(m_2 - 1)$-degree Lobatto polynomial, and the interior grid lines ζ_k, $k = 2, 3, \ldots, m_3$, should be placed at the zeros of the $(m_3 - 1)$-degree Lobatto polynomial, yielding a spectral node distribution. When this is done, the interpolation error decreases spectrally, that is, faster than algebraically with respect to the polynomial order m_1, m_2, or m_2. The spectral interpolation properties in three dimensions derive from those of its one-dimensional constituents, as discussed in Chapter 2.

6.4.4   Modal expansion

In the modal expansion, a function of interest, $f(\xi, \eta, \zeta, t)$, defined over the standard cube, is approximated with an mth-degree polynomial, as

$$f(\xi, \eta, \zeta) \simeq F_v + F_e + F_f + F_i, \qquad (6.4.11)$$

where

$$
\begin{aligned}
F_v(\xi, \eta, \zeta) = {}& f_1\, \zeta_1^v(\xi, \eta, \zeta) + f_2\, \zeta_2^v(\xi, \eta, \zeta) \\
& + f_3\, \zeta_3^v(\xi, \eta, \zeta) + f_4\, \zeta_4^v(\xi, \eta, \zeta) \\
& + f_5\, \zeta_5^v(\xi, \eta, \zeta) + f_6\, \zeta_6^v(\xi, \eta, \zeta) \\
& + f_7\, \zeta_7^v(\xi, \eta, \zeta) + f_8\, \zeta_8^v(\xi, \eta, \zeta)
\end{aligned}
\qquad (6.4.12)
$$

is the *vertex part*,

$$
\begin{aligned}
F_e(\xi, \eta, \zeta) = {}& \sum_{i=1}^{m_1-1} c_i^{12} \zeta_i^{12}(\xi, \eta, \zeta) + \sum_{i=1}^{m_1-1} c_i^{34} \zeta_i^{34}(\xi, \eta, \zeta) \\
& + \sum_{i=1}^{m_1-1} c_i^{56} \zeta_i^{56}(\xi, \eta, \zeta) + \sum_{i=1}^{m_1-1} c_i^{78} \zeta_i^{78}(\xi, \eta, \zeta) \\
& + \sum_{i=1}^{m_2-1} c_i^{13} \zeta_i^{13}(\xi, \eta, \zeta) + \sum_{i=1}^{m_2-1} c_i^{24} \zeta_i^{24}(\xi, \eta, \zeta) \\
& + \sum_{i=1}^{m_2-1} c_i^{57} \zeta_i^{57}(\xi, \eta, \zeta) + \sum_{i=1}^{m_2-1} c_i^{68} \zeta_i^{68}(\xi, \eta, \zeta) \\
& + \sum_{i=1}^{m_3-1} c_i^{15} \zeta_i^{15}(\xi, \eta, \zeta) + \sum_{i=1}^{m_3-1} c_i^{26} \zeta_i^{26}(\xi, \eta, \zeta) \\
& + \sum_{i=1}^{m_3-1} c_i^{37} \zeta_i^{37}(\xi, \eta, \zeta) + \sum_{i=1}^{m_3-1} c_i^{48} \zeta_i^{48}(\xi, \eta, \zeta)
\end{aligned}
\qquad (6.4.13)
$$

is the *edge part* arising when $m_1 \geq 2$, or $m_2 \geq 2$, or $m_3 \geq 2$,

$$
F_f(\xi, \eta, \zeta) = \sum_{i=1}^{m_1-1} \left(\sum_{j=1}^{m_2-1} c_{ij}^{1234} \zeta_{ij}^{1234}(\xi, \eta, \zeta) \right) + \sum_{i=1}^{m_1-1} \left(\sum_{j=1}^{m_2-1} c_{ij}^{5678} \zeta_{ij}^{134}(\xi, \eta, \zeta) \right)
$$

$$
+ \sum_{i=1}^{m_1-1} \left(\sum_{j=1}^{m_3-1} c_{ij}^{1256} \zeta_{ij}^{1256}(\xi, \eta, \zeta) \right) + \sum_{i=1}^{m_1-1} \left(\sum_{j=1}^{m_3-1} c_{ij}^{3478} \zeta_{ij}^{3478}(\xi, \eta, \zeta) \right)
$$

$$
+ \sum_{i=1}^{m_2-1} \left(\sum_{j=1}^{m_3-1} c_{ij}^{1357} \zeta_{ij}^{1357}(\xi, \eta, \zeta) \right) + \sum_{i=1}^{m_2-1} \left(\sum_{j=1}^{m_3-1} c_{ij}^{2468} \zeta_{ij}^{2468}(\xi, \eta, \zeta) \right)
$$

$$(6.4.14)$$

is the *face part* arising when $m_1 \geq 2$, or $m_2 \geq 2$, or $m_3 \geq 2$, and

$$
F_i(\xi, \eta) = \sum_{i=1}^{m_1-1} \left[\sum_{j=1}^{m_2-1} \left(\sum_{k=1}^{m_3-1} c_{ijk} \zeta_{ijk}(\xi, \eta, \zeta) \right) \right] \tag{6.4.15}
$$

is the *interior part* arising when $m_1 \geq 2$, or $m_2 \geq 2$, or $m_3 \geq 2$. Adding the $N_V = 8$ vertex modes, the

$$
N_e = 4 \times (m_1 - 1) + 4 \times (m_2 - 1) + 4 \times (m_3 - 1) \tag{6.4.16}
$$

edge modes, the

$$
N_f = 2 \times (m_1 - 1)(m_2 - 1)
$$
$$
+ 2 \times (m_1 - 1)(m_2 - 1) + 2 \times (m_2 - 1)(m_3 - 1) \tag{6.4.17}
$$

face modes, and the

$$
N_i = (m_1 - 1)(m_2 - 1)(m_3 - 1) \tag{6.4.18}
$$

interior nodes, we obtain the total number of modes,

$$
N = N_v + N_e + N_f + N_i
$$
$$
= (m_1 + 1)(m_2 + 1)(m_3 + 1), \tag{6.4.19}
$$

which is precisely equal to that involved in the tensor-product expansion discussed in Section 6.4.3.

Vertex modes

The coefficients $f_1 - f_8$ in (6.4.12) are the values of the function $f(\xi, \eta, \zeta)$ at the eight vertex nodes shown in Figure 6.4.1. To ensure C^0 continuity of the finite element expansion, these values must be shared by neighboring elements with common vertices. The corresponding modal functions, ζ_i^v, where $i = 1, 2, \dots, 8$, are identical to the linear nodal interpolation functions given in (6.4.3).

Edge modes

The coefficients c_i^{pq} in (6.4.13) are associated with twelve families of edge interpolation functions ζ_i^{pq}, where $i = 1, 2, \ldots, m_1 - 1$, or $i = 1, 2, \ldots, m_2 - 1$, or $i = 1, 2, \ldots, m_3 - 1$. To ensure C^0 continuity of the finite element expansion, these coefficients must be shared by neighboring elements at the common edges. The edge modes take the value of zero along all edges, except for the superscripted edge. Working as in Section 4.5, we find

$$
\begin{aligned}
\zeta_i^{12} &= \frac{1-\xi}{2}\frac{1+\xi}{2}\frac{1-\eta}{2} Lo_{i-1}(\xi) \frac{1-\zeta}{2}, \\
\zeta_i^{34} &= \frac{1-\xi}{2}\frac{1+\xi}{2}\frac{1+\eta}{2} Lo_{i-1}(\xi) \frac{1-\zeta}{2}, \\
\zeta_i^{56} &= \frac{1-\xi}{2}\frac{1+\xi}{2}\frac{1-\eta}{2} Lo_{i-1}(\xi) \frac{1+\zeta}{2}, \\
\zeta_i^{78} &= \frac{1-\xi}{2}\frac{1+\xi}{2}\frac{1+\eta}{2} Lo_{i-1}(\xi) \frac{1+\zeta}{2},
\end{aligned}
\tag{6.4.20}
$$

for $i = 1, 2, \ldots, m_1 - 1$,

$$
\begin{aligned}
\zeta_i^{13} &= \frac{1-\xi}{2}\frac{1-\eta}{2}\frac{1+\eta}{2} Lo_{i-1}(\eta) \frac{1-\zeta}{2}, \\
\zeta_i^{24} &= \frac{1+\xi}{2}\frac{1-\eta}{2}\frac{1+\eta}{2} Lo_{i-1}(\eta) \frac{1-\zeta}{2}, \\
\zeta_i^{57} &= \frac{1-\xi}{2}\frac{1-\eta}{2}\frac{1+\eta}{2} Lo_{i-1}(\eta) \frac{1+\zeta}{2}, \\
\zeta_i^{68} &= \frac{1+\xi}{2}\frac{1-\eta}{2}\frac{1+\eta}{2} Lo_{i-1}(\eta) \frac{1+\zeta}{2},
\end{aligned}
\tag{6.4.21}
$$

for $i = 1, 2, \ldots, m_2 - 1$, and

$$
\begin{aligned}
\zeta_i^{15} &= \frac{1-\xi}{2}\frac{1-\eta}{2}\frac{1-\zeta}{2}\frac{1+\zeta}{2} Lo_{i-1}(\zeta), \\
\zeta_i^{26} &= \frac{1+\xi}{2}\frac{1-\eta}{2}\frac{1-\zeta}{2}\frac{1+\zeta}{2} Lo_{i-1}(\zeta), \\
\zeta_i^{37} &= \frac{1-\xi}{2}\frac{1+\eta}{2}\frac{1-\zeta}{2}\frac{1+\zeta}{2} Lo_{i-1}(\zeta), \\
\zeta_i^{48} &= \frac{1+\xi}{2}\frac{1+\eta}{2}\frac{1-\zeta}{2}\frac{1+\zeta}{2} Lo_{i-1}(\zeta),
\end{aligned}
\tag{6.4.22}
$$

for $i = 1, 2, \ldots, m_3 - 1$.

Face modes

The coefficients c_{ij}^{pqrs} in (6.4.14) are associated with six families of face modes represented by the functions ζ_{ij}^{pqrs}, where i and j range over the limits stated in the sums. To ensure C^0 continuity of the finite element expansion, these coefficients must be shared by neighboring elements at the common faces. The face modes take the value of zero over all faces, except for the superscripted face. Requiring partial orthogonality, we find

$$\zeta_{ij}^{1234} = \frac{1-\xi}{2} \frac{1+\xi}{2} Lo_{i-1}(\xi) \frac{1-\eta}{2} \frac{1+\eta}{2} Lo_{j-1}(\eta) \frac{1-\zeta}{2},$$

$$(6.4.23)$$

$$\zeta_{ij}^{5678} = \frac{1-\xi}{2} \frac{1+\xi}{2} Lo_{i-1}(\xi) \frac{1-\eta}{2} \frac{1+\eta}{2} Lo_{j-1}(\eta) \frac{1+\zeta}{2},$$

for $i = 1, 2, \ldots, m_1 - 1$ and $j = 1, 2, \ldots, m_2 - 1$,

$$\zeta_{ij}^{1256} = \frac{1-\xi}{2} \frac{1+\xi}{2} Lo_{i-1}(\xi), \frac{1-\eta}{2} \frac{1-\zeta}{2} \frac{1+z\eta}{2} Lo_{j-1}(\zeta),$$

$$(6.4.24)$$

$$\zeta_{ij}^{3478} = \frac{1-\xi}{2} \frac{1+\xi}{2} Lo_{i-1}(\xi) \frac{1+\eta}{2} \frac{1-\zeta}{2} \frac{1+\zeta}{2} Lo_{j-1}(\zeta),$$

for $i = 1, 2, \ldots, m_1 - 1$ and $j = 1, 2, \ldots, m_3 - 1$, and

$$\zeta_{ij}^{1357} = \frac{1-\xi}{2} \frac{1-\eta}{2} \frac{1+\eta}{2} Lo_{i-1}(\eta) \frac{1-\zeta}{2} \frac{1+\zeta}{2} Lo_{j-1}(\zeta),$$

$$(6.4.25)$$

$$\zeta_{ij}^{2468} = \frac{1+\xi}{2} \frac{1-\eta}{2} \frac{1+\eta}{2} Lo_{i-1}(\eta) \frac{1-\zeta}{2} \frac{1+\zeta}{2} Lo_{j-1}(\zeta),$$

for $i = 1, 2, \ldots, m_2 - 1$ and $j = 1, 2, \ldots, m_3 - 1$.

Interior modes

The interior or bubble modes are zero over all faces and non-zero only inside the element. Requiring orthogonality, we find

$$\zeta_{ijk} = \frac{1-\xi}{2} \frac{1+\xi}{2} Lo_{i-1}(\xi) \frac{1-\eta}{2} \frac{1+\eta}{2} Lo_{j-1}(\eta) \frac{1-\zeta}{2} \frac{1+\eta}{2} Lo_{k-1}(\zeta)$$

$$= Lo_{i-1}(\xi) \, Lo_{j-1}(\eta) \, Lo_{k-1}(\zeta) \frac{1-\xi}{2} \frac{1+\xi}{2} \frac{1-\eta}{2} \left(\frac{1+\eta}{2}\right)^2 \frac{1-\zeta}{2},$$

$$(6.4.26)$$

for $i = 1, 2, \ldots, m_1 - 1$, $j = 1, 2, \ldots, m_2 - 1$, and $k = 1, 2, \ldots, m_3 - 1$.

Implementation of the modal expansion

The Galerkin finite element method for the modal expansion is implemented as discussed in Section 6.3 for the tetrahedral elements. The element mass and diffusion matrices, consisting of integrals of pairwise products of the modal functions and their derivatives, are computed numerically by the application of Gauss integration rules.

PROBLEMS

6.4.1 *Computation of an integral over a hexahedral element.*

Write a MATLAB function that uses the triple Lobatto quadrature to evaluate the integral of a given function, $f(\xi, \eta, \zeta)$, over a hexahedral element. The input to this function should include the position of the eight vertices, $\mathbf{x}_i$, for $i = 1, 2, \ldots, 8$. Run the function for $f(\xi, \eta, \zeta) = 1$, and confirm that the answer is equal to the volume of the hexahedron.

6.4.2 *Modal expansion.*

Write out specific expressions for the twelve modes corresponding to the expansion orders $m_1 = 1$, $m_2 = 1$, and $m_3 = 2$.

Appendices

Function interpolation

Given the values of a function, $f(x)$, at $N+1$ data points with abscissas located at x_i, where $i = 1, 2, \ldots, N+1$, we want to interpolate at an intermediate point, that is, we want to compute the value of the function at an arbitrary point, x. To carry out the interpolation, we replace the unknown function, $f(x)$, with a standard smooth function, $g(x)$, whose graph passes through the data points, as illustrated in Figure A.1. This means that the interpolating function $g(x)$ is required to satisfy the $N + 1$ interpolation or matching conditions

$$g(x_i) = f(x_i), \qquad (A.1)$$

for $i = 1, 2, \ldots, N + 1$, with no additional stipulations or constraints. Once the interpolating function $g(x)$ is available, it can be differentiated or integrated to yield approximations to derivatives or to a definite integral of the interpolated function, $f(x)$, in terms of the data, $f(x_i)$. Differentiation produces finite-difference approximations, and integration produces numerical integration rules and quadratures.

In this Appendix, the basic theory of polynomial interpolation is summarized. The interested reader will find further discussion in dedicated monographs and texts on numerical methods (i.e., [13, 43]).

A.1 The interpolating polynomial

The most popular choice of an interpolating function, $g(x)$, is the Nth degree polynomial,

$$P_N(x) = a_1 \, x^N + a_2 \, x^{N-1} + a_3 \, x^{N-2} + \ldots + a_N \, x + a_{N+1}. \qquad (A.1.1)$$

The interpolation condition (A.1) requires that the graph of the polynomial passes through the $N + 1$ data points,

$$P_N(x_i) = f(x_i), \qquad (A.1.2)$$

for $i = 1, 2, \ldots, N + 1$. The problem is reduced to computing the $N + 1$ coefficients, $a_1, a_2, \ldots, a_{N+1}$, so that $P_N(x)$ satisfies the $N + 1$ interpolation conditions (A.1.2). Having completed this computation, we can evaluate $P_N(x)$ for any specified value of the independent variable, x, and thus obtain an approximation to the interpolated function, $f(x)$.

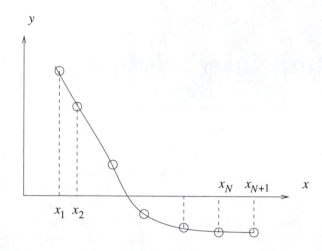

Figure A.1 Interpolation of a function through a set of $N + 1$ data points. In polynomial interpolation, the function is approximated with an Nth-degree polynomial whose graph passes through the data points.

A.1.1 Vandermonde matrix

The simplest method of computing the polynomial coefficients is by enforcing the constraints (A.1.2) at the $N + 1$ data points to derive a system of $N + 1$ linear equations for the polynomial coefficients,

$$a_1 x_1^N + a_2 x_1^{N-1} + a_3 x_1^{N-2} + \ldots + a_N x_1 + a_{N+1} = f(x_1),$$

$$a_1 x_2^N + a_2 x_2^{N-1} + a_3 x_2^{N-2} + \ldots + a_N x_2 + a_{N+1} = f(x_2),$$

$$\ldots \qquad\qquad \ldots \qquad (A.1.3)$$

$$a_1 x_N^N + a_2 x_N^{N-1} + a_3 x_N^{N-2} + \ldots + a_N x_N + a_{N+1} = f(x_N),$$

$$a_1 x_{N+1}^N + a_2 x_{N+1}^{N-1} + a_3 x_{N+1}^{N-2} + \ldots + a_N x_{N+1} + a_{N+1} = f(x_{N+1}).$$

In terms of the polynomial coefficient vector, $\mathbf{a}$, and data vector, $\mathbf{f}$, defined, respectively, as

$$\mathbf{a} \equiv \begin{bmatrix} a_{N+1} \\ a_N \\ \vdots \\ a_2 \\ a_1 \end{bmatrix}, \qquad \mathbf{f} \equiv \begin{bmatrix} f(x_1) \\ f(x_2) \\ \vdots \\ f(x_N) \\ f(x_{N+1}) \end{bmatrix}, \qquad (A.1.4)$$

the linear system takes the form

$$\mathbf{V}^T \cdot \mathbf{a} = \mathbf{f}, \tag{A.1.5}$$

where the superscript T denotes the matrix transpose, and

$$\mathbf{V}(x_1, x_2, \ldots, x_{N+1}) \equiv \begin{bmatrix} 1 & 1 & 1 & \ldots & 1 & 1 \\ x_1 & x_2 & x_3 & \ldots & x_N & x_{N+1} \\ \ldots & \ldots & \ldots & \ldots & \ldots & \ldots \\ x_1^{N-1} & x_2^{N-1} & x_3^{N-1} & \ldots & x_N^{N-1} & x_{N+1}^{N-1} \\ x_1^N & x_2^N & x_3^N & \ldots & x_N^N & x_{N+1}^N \end{bmatrix}$$

$$\tag{A.1.6}$$

is the $(N+1) \times (N+1)$ *Vandermonde matrix*. Note that the abscissas, x_i, do not have to be arranged in any particular order, for example, in ascending or descending order, and can be arbitrarily juxtaposed.

Determinant of the Vandermonde matrix

The determinant of the Vandermonde matrix is equal to the product of all possible differences between two distinct abscissas,

$$\mathrm{Det}[\mathbf{V}] = \prod_{i=1}^{N} \prod_{j=i+1}^{N+1} (x_j - x_i), \tag{A.1.7}$$

where Π denotes the product. Note that the right-hand side of (A.1.7) consists of

$$N + \ldots + 2 + 1 = \frac{1}{2}N(N+1) \tag{A.1.8}$$

multiplicative factors. When $N = 1$, $\mathrm{Det}[\mathbf{V}] = x_2 - x_1$, and when $N = 2$, $\mathrm{Det}[\mathbf{V}] = (x_2 - x_1)(x_3 - x_1)(x_3 - x_2)$. Thus, so long as the abscissas are distinct, $x_i \neq x_j$, the determinant of the Vandermonde matrix is nonzero, the matrix is nonsingular, the solution of the linear system (A.1.5) is unique, and the interpolation problem is well posed.

To prove (A.1.7), we observe that $\mathrm{Det}[\mathbf{V}(x_1, x_2, \ldots, x_N, x_{N+1})]$ is an Nth-degree polynomial in x_{N+1}. Recalling that, if two columns of a matrix are identical the determinant is equal to zero, we find that the roots of this polynomial are equal to $x_1, x_2, \ldots, x_N$. This observation motivates the recursion relation

$$\mathrm{Det}[\mathbf{V}(x_1, \ldots, x_N, x_{N+1})]$$
$$= (x_{N+1} - x_1)(x_{N+1} - x_2) \ldots (x_{N+1} - x_N) \, \mathrm{Det}[\mathbf{V}(x_1, \ldots, x_N)], \tag{A.1.9}$$

which emerges by expanding the determinant of the Vandermonde matrix by the minors of the last column. Repeated application of the recursion formula leads to (A.1.7).

Without loss of generality, we may arrange the abscissas in ascending order, so that

$$x_1 < x_2 < x_3 < \ldots < x_N < x_{N+1}, \tag{A.1.10}$$

and rescale the x axis so that $x_1 = 0$ and $x_{N+1} = 1$. Let us then hold the position of the first and last points fixed, and start increasing the polynomial order, N, from the value of one, corresponding to two data points, to higher values. Since the distances between consecutive data points are being reduced, the factors on the left-hand side of (A.1.7) tend to become smaller. In practice, the determinant of the Vandermonde matrix becomes exceedingly small, and the linear system (A.1.5) becomes nearly singular even at moderate values of N. This behavior places a serious limitation on the usefulness of the Vandermonde matrix approach.

Fekete points

An intriguing property of the Vandermonde matrix relates to the zeros of the Lobatto polynomials discussed in Appendix B. To demonstrate this connection, we hold the first and last interpolation points fixed at the standard values $x_1 = -1$ and $x_{N+1} = 1$, and search for the positions of the $N - 1$ intermediate points that maximize the magnitude of the determinant of the Vandermonde matrix. By definition, these are the Fekete points for a finite interval of the x axis. In Section A.2.5, we shall show that the Fekete points coincide with the well-known zeros of the $(N-1)$-degree Lobatto polynomial $Lo_{N-1}(x)$, discussed in Section B.6 of Appendix B. Corresponding Fekete points can be defined in higher dimensions for specified geometrical domains, as discussed in Chapters 4 and 6.

A.1.2 Generalized Vandermonde matrix

Alternatively, the Nth-degree interpolating polynomial can be expanded in a series of basis functions,

$$P_N(x) = b_1 \, \phi_{N+1}(x) + b_2 \, \phi_N(x) \tag{A.1.11}$$
$$+ \ldots + b_N \, \phi_2(x) + b_{N+1} \, \phi_1(x),$$

where $\phi_i(x)$, $i = 1, 2, \ldots, N+1$ is a set of Nth-degree polynomials that constitute a complete base for all Nth-degree polynomials. Completeness guarantees that any Nth-degree polynomial can be expressed as shown in (A.1.11), with appropriate coefficients, b_i. The standard representation (A.1.1) corresponds to the monomial base $\phi_i(x) = x^{i-1}$. Other possible bases are provided by the orthogonal polynomials discussed in Appendix B.

To compute the expansion coefficients, b_i, we require the interpolation conditions (A.1.2), and obtain the linear system

$$\mathbf{V}_\phi^T \cdot \mathbf{b} = \mathbf{f}, \tag{A.1.12}$$

where

$$
\mathbf{V}_\phi \equiv
\begin{bmatrix}
\phi_1(x_1) & \phi_1(x_2) & \cdots & \phi_1(x_N) & \phi_1(x_{N+1}) \\
\phi_2(x_1) & \phi_2(x_2) & \cdots & \phi_2(x_N) & \phi_2(x_{N+1}) \\
\cdots & \cdots & \cdots & \cdots & \cdots \\
\phi_N(x_1) & \phi_N(x_2) & \cdots & \phi_N(x_N) & \phi_N(x_{N+1}) \\
\phi_{N+1}(x_1) & \phi_{N+1}(x_2) & \cdots & \phi_{N+1}(x_N) & \phi_{N+1}(x_{N+1})
\end{bmatrix}
$$

$$(A.1.13)$$

is the $(N+1) \times (N+1)$ generalized Vandermonde matrix, and

$$
\mathbf{b} \equiv
\begin{bmatrix}
b_{N+1} \\
b_N \\
\vdots \\
b_2 \\
b_1
\end{bmatrix},
\qquad
\mathbf{f} \equiv
\begin{bmatrix}
f(x_1) \\
f(x_2) \\
\vdots \\
f(x_N) \\
f(x_{N+1})
\end{bmatrix},
\qquad (A.1.14)
$$

are the coefficient vector, and data vector. The monomial choice, $\phi_i(x) = x^{i-1}$, reduces the generalized Vandermonde matrix to the standard Vandermonde matrix displayed in (A.1.6). With a suitable choice of the polynomial base provided by the families of orthogonal polynomials discussed in Appendix B, the generalized Vandermonde system is well-conditioned even for high polynomial orders, N.

A.1.3 Newton interpolation

The aforementioned difficulty concerning the possibly nearly singular nature of the Vandermonde matrix does not imply that the problem of polynomial interpolation is ill-posed, in the sense that the results are hopelessly sensitive to the numerical round-off error. We must simply find another way of computing the interpolating polynomial, which can be done using several viable alternatives. When a numerical method does not work, we must not immediately blame the problem.

In Newton's interpolation method, the interpolating polynomial is expressed as a linear combination of a family of factorized triangular polynomials defined with reference to the first N interpolation nodes, x_i, for $i = 1, 2, \ldots, N$, as

$$
\begin{aligned}
P_N(x) = c_0 & \\
+ c_1 & (x - x_1) \\
+ c_2 & (x - x_1)(x - x_2) \\
& \cdots \\
+ c_N & (x - x_1)(x - x_2) \ldots (x - x_N),
\end{aligned}
\qquad (A.1.15)
$$

where c_i, for $i = 0, 1, \ldots, N$ are $N + 1$ unknown coefficients. Applying the interpolation condition (A.1.2) for $i = 1$, we obtain

$$
c_0 = f(x_1). \qquad (A.1.16)
$$

Applying next (A.1.2) for $i = 2$, and rearranging, we obtain

$$c_1 = \frac{f(x_2) - c_0}{x_2 - x_1}. \tag{A.1.17}$$

The rest of the coefficients can be computed in a similar manner based on the recursive formula

$$c_i = \frac{f(x_{i+1}) - c_0 - \sum_{j=1}^{i-1} c_j \left(\prod_{k=1}^{j} (x_{i+1} - x_k) \right)}{\prod_{k=1}^{i} (x_{i+1} - x_k)}. \tag{A.1.18}$$

In practice, the coefficients are computed using an algorithmic version of Newton's divided-difference table (e.g., [43], p. 272).

A.2   Lagrange interpolation

Lagrange interpolation produces the interpolating polynomial in a manner that circumvents the explicit computation of the polynomial coefficients.

To implement the method, we introduce a family of Nth-degree Lagrange polynomials associated with the abscissas. The Lagrange polynomial associated with the ith abscissa, denoted by $l_{N,i}(x)$, is defined in terms of the abscissas of all data points, x_j, where $j = 1, 2, \ldots, N + 1$. By construction, $l_{N,i}(x)$ is equal to zero at all data points, except at the ith data point where it is equal to unity, that is,

$$l_{N,i}(x_j) = \delta_{ij}, \tag{A.2.1}$$

where δ_{ij} is Kronecker's delta. One may readily verify that the polynomial

$$l_{N,i}(x) = \frac{(x - x_1)(x - x_2) \ldots (x - x_{i-1})(x - x_{i+1}) \ldots (x - x_{N+1})}{(x_i - x_1)(x_i - x_2) \ldots (x_i - x_{i-1})(x_i - x_{i+1}) \ldots (x_i - x_{N+1})} \tag{A.2.2}$$

satisfies this cardinal condition. As a mnemonic aid, we note that the factor $(x - x_i)$ is missing from the numerator, and the corresponding troublesome null factor $(x_i - x_i)$ is missing from the denominator. Moreover, the denominator is a constant, whereas the numerator is an Nth-degree polynomial in x.

We may now express the interpolating polynomial in terms of the Lagrange polynomials as

$$P_N(x) = \sum_{i=1}^{N+1} f(x_i) \, l_{N,i}(x). \tag{A.2.3}$$

Property (A.2.1) ensures that the right-hand side of (A.2.3) satisfies the matching conditions (A.1.2), and is thus the desired interpolating polynomial. Indeed,

$$P_N(x_j) = \sum_{i=1}^{N+1} f(x_i) \, l_{N,i}(x_j) = \sum_{i=1}^{N+1} f(x_i) \, \delta_{ij} = f(x_j), \tag{A.2.4}$$

as required.

A.2.1 Cauchy relations

If the interpolated function $f(x)$ is an Nth- or lower-degree polynomial, then the interpolating polynomial given in (A.2.3) is exact by construction.

Choosing $f(x) = P_N(x) = (x - a)^m$, where a is a constant, and $m = 0, 1, \ldots, N$, we obtain the identity

$$(x - a)^m = \sum_{i=1}^{N+1} (x_i - a)^m \, l_{N,i}(x), \tag{A.2.5}$$

for $0 \leq m \leq N$. Setting $a = x$, we derive the Cauchy relations

$$\sum_{i=1}^{N+1} l_{N,i}(x) = 1, \tag{A.2.6}$$

and

$$\sum_{i=1}^{N+1} (x_i - x)^m \, l_{N,i}(x) = 0, \tag{A.2.7}$$

for $m = 1, 2, \ldots, N$. Relation (A.2.6) ensures that, if $f(x_i)$ are constant, then $P_N(x)$ is also a constant.

A.2.2 Representation in terms of the Lagrange generating polynomial

An alternative representation of the Lagrange polynomials emerges by introducing the $(N + 1)$-degree Lagrange generating polynomial

$$\Phi_{N+1}(x) = (x - x_1)(x - x_2) \ldots (x - x_N)(x - x_{N+1}), \tag{A.2.8}$$

which is zero at *all* data points. Straightforward product differentiation shows that

$$\Phi'_{N+1}(x_i) = (x_i - x_1) \ldots (x_i - x_{i-1})(x_i - x_{i+1}) \ldots (x_i - x_{N+1}), \tag{A.2.9}$$

where a prime denotes a derivative with respect to x. Accordingly, the ith Lagrange interpolation polynomial can be expressed in the form

$$l_{N,i}(x) = \frac{1}{\Phi'_{N+1}(x_i)} \frac{\Phi_{N+1}(x)}{x - x_i}. \tag{A.2.10}$$

In terms of the *barycentric weights*,

$$c_i \equiv \frac{1}{\Phi'_{N+1}(x_i)}, \tag{A.2.11}$$

we obtain

$$l_{N,i}(x) = c_i \, \frac{\Phi_{N+1}(x)}{x - x_i}. \tag{A.2.12}$$

Substituting this expression in the Cauchy formula (A.2.6) and rearranging, we find

$$\frac{1}{\Phi_{N+1}(x)} = \sum_{i=1}^{N+1} \frac{c_i}{x - x_i}, \tag{A.2.13}$$

which shows that the barycentric weights arise by expanding the inverse of the generating function into a sum of partial fractions.

The interpolating polynomial can be computed using the computationally efficient form

$$P_N(x) = \Phi_{N+1}(x) \sum_{i=1}^{N+1} f(x_i) \, \frac{c_i}{x - x_i}. \tag{A.2.14}$$

Using (A.2.13), we obtain the alternative form

$$P_N(x) = \frac{\sum_{i=1}^{N+1} f(x_i) \frac{c_i}{x - x_i}}{\sum_{i=1}^{N+1} \frac{c_i}{x - x_i}}. \tag{A.2.15}$$

A.2.3   First derivative and the node differentiation matrix

Differentiating (A.2.3), we derive an expression for the first derivative of the interpolating polynomial in terms of the data,

$$\frac{\mathrm{d}P_N(x)}{\mathrm{d}x} = \sum_{i=1}^{N+1} f(x_i) \, \frac{\mathrm{d}l_{N,i}(x)}{\mathrm{d}x}. \tag{A.2.16}$$

Differentiating (A.2.2) using the rules of product differentiation, we find

$$\frac{\mathrm{d}l_{N,i}(x)}{\mathrm{d}x} = \frac{1}{\Phi'_{N+1}(x_i)} \, \frac{\Phi_{N+1}(x)}{x - x_i} \, S_i(x) = l_{N,i}(x) \, S_i(x), \tag{A.2.17}$$

where

$$S_i(x) \equiv \frac{1}{x - x_1} + \ldots + \frac{1}{x - x_{i-1}} + \frac{1}{x - x_{i+1}} + \ldots + \frac{1}{x - x_{N+1}}. \tag{A.2.18}$$

Thus, the Lagrange polynomials satisfy the differential equation

$$\frac{\mathrm{d}\ln l_{N,i}}{\mathrm{d}x} = S_i(x). \tag{A.2.19}$$

Alternatively, differentiating expression (A.2.10) using the rules of quotient differentiation, we derive the expression

$$\frac{\mathrm{d}l_{N,i}(x)}{\mathrm{d}x} = \frac{\Phi'_{N+1}(x)\,(x-x_i) - \Phi_{N+1}(x)}{(x-x_i)^2\,\Phi'_{N+1}(x_i)}. \tag{A.2.20}$$

Evaluating the derivatives on the right-hand side at the interpolation nodes, and observing that $\Phi_{N+1}(x_j) = 0$, we obtain the *node differentiation matrix*

$$d_{ij} \equiv \left(\frac{\mathrm{d}l_{N,i}}{\mathrm{d}x}\right)_{x=x_j} = \begin{cases} \dfrac{\Phi'_{N+1}(x_j)}{(x_j-x_i)\,\Phi'_{N+1}(x_i)} & \text{if } i \neq j \\[3mm] \dfrac{\Phi''_{N+1}(x_i)}{2\,\Phi'_{N+1}(x_i)} & \text{if } i = j \end{cases}. \tag{A.2.21}$$

The second expression for $i = j$ arises by using the l'Hôpital rule to evaluate the right-hand side of (A.2.20). Substituting the definition of the generating function, $\Phi_{N+1}(x)$, and carrying out straightforward differentiations, we obtain the explicit formulas

$$d_{ij} = \frac{1}{x_j - x_i}\,\frac{(x_j-x_1)\ldots(x_j-x_{j-1})(x_j-x_{j+1})\ldots(x_j-x_{N+1})}{(x_i-x_1)\ldots(x_i-x_{i-1})(x_i-x_{i+1})\ldots(x_i-x_{N+1})} \tag{A.2.22}$$

for $i \neq j$, and

$$d_{ii} \equiv \left(\frac{\mathrm{d}l_{N,i}}{\mathrm{d}x}\right)_{x=x_i} = \frac{\Phi''_{N+1}(x_i)}{2\,\Phi'_{N+1}(x_i)} \tag{A.2.23}$$

$$= S_i(x_i) = \frac{1}{x_i - x_1} + \ldots + \frac{1}{x_i - x_{i-1}} + \frac{1}{x_i - x_{i+1}} + \ldots + \frac{1}{x_i - x_{N+1}},$$

which also arises simply by evaluating (A.2.17) at $x = x_i$, For example, when $N = 1$, we find $d_{11} = d_{12} = 1/(x_1 - x_2)$, and $d_{21} = d_{22} = 1/(x_2 - x_1)$.

Point vortex motion

The sum representation of the diagonal component of the node differentiation matrix shown in the second line of (A.2.23) suggests a connection with a class of problems in mathematical physics involving multi-particle and multi-body interactions, including vortex motion in an ideal fluid [3].

Consider the mutually induced motion of a collection of $N + 1$ point vortices with identical strength, κ, initially situated along the x axis, as shown in Figure A.2.1. Using the equations of inviscid hydrodynamics, we find that the y velocity component of the ith point vortex is given

$$v_i = \frac{\kappa}{2\pi}\,S_i(x_i) = \frac{\kappa}{2\pi}\,\frac{\Phi''_{N+1}(x_i)}{2\,\Phi'_{N+1}(x_i)} \tag{A.2.24}$$

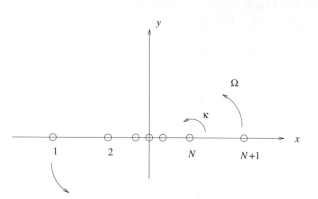

Figure A.2.1 A collection of $N + 1$ point vortices with identical strengths located at the zeros of the $(N + 1)$-degree Hermite polynomial rotates as a whole with angular velocity Ω.

(e.g., [42]). If the point vortex array rotates as a whole around the origin with angular velocity Ω, then

$$v_i = \Omega\, x_i, \tag{A.2.25}$$

and thus

$$\frac{\kappa}{2\pi}\, \frac{\Phi''_{N+1}(x_i)}{2\, \Phi'_{N+1}(x_i)} = \Omega\, x_i, \tag{A.2.26}$$

for $i = 1, 2 \dots, N + 1$, which shows that x_i are the roots of the $(N + 1)$-degree polynomial

$$\Pi_{N+1}(x) \equiv \Phi''_{N+1}(x) - \frac{4\pi\,\Omega}{\kappa}\, x\, \Phi'_{N+1}(x). \tag{A.2.27}$$

Accordingly, we can write

$$\Pi_{N+1}(x) = c\, \Phi_{N+1}(x), \tag{A.2.28}$$

where c is a constant. To ensure that the coefficients of the highest power of x on the right-hand sides of the last two expressions are the same, we set $c = -4\pi\Omega(N + 1)/\kappa$, and derive the differential equation

$$\Phi''_{N+1}(x) - \frac{4\pi\,\Omega}{\kappa}\, x\, \Phi'_{N+1}(x) + \frac{4\pi\Omega}{\kappa}\, (N + 1)\, \Phi_{N+1}(x) = 0. \tag{A.2.29}$$

In terms of the dimensionless position,

$$\widehat{x} \equiv x\, \sqrt{\frac{2\pi\Omega}{\kappa}}, \tag{A.2.30}$$

equation (A.2.29) reduces to the Hermite equation

$$\Phi''_{N+1}(\widehat{x}) - 2\,\widehat{x}\,\Phi'_{N+1}(\widehat{x}) + 2\,(N+1)\,\Phi_{N+1}(\widehat{x}) = 0, \qquad (A.2.31)$$

where a prime now designates a derivative with respect to $\widehat{x}$. The solution of this equation is the $(N+1)$-degree Hermite polynomial, whose roots, corresponding to the positions of the point vortex array, are available in analytical or tabular form (e.g., [43]).

A.2.4 Representation in terms of the Vandermonde matrix

Applying identity (A.2.5) with $a = 0$, we find

$$x^m = \sum_{i=1}^{N+1} x_i^m\, l_{N,i}(x), \qquad (A.2.32)$$

for $m = 0, 1, \ldots, N$, which can be rearranged into the linear system

$$\mathbf{V} \cdot \begin{bmatrix} l_{N,1}(x) \\ l_{N,2}(x) \\ \vdots \\ l_{N,N}(x) \\ l_{N,N+1}(x) \end{bmatrix} = \begin{bmatrix} 1 \\ x \\ \vdots \\ x^{N-1} \\ x^N \end{bmatrix}, \qquad (A.2.33)$$

where $\mathbf{V}$ is the Vandermonde matrix defined in (A.1.6). Computing the solution by Cramer's rule, we find

$$l_{N,i}(x) = \frac{\mathrm{Det}[\mathbf{V}(x_1, x_2, \ldots, x_{i-1}, x, x_{i+1}, \ldots, x_{N+1})]}{\mathrm{Det}[\mathbf{V}(x_1, x_2, \ldots, x_{i-1}, x_i, x_{i+1}, \ldots, x_{N+1})]}. \qquad (A.2.34)$$

Note that the argument x_i has been replaced by the independent variable x in the numerator on the right-hand side to give an Nth-degree polynomial. Thus, the ith Lagrange polynomial can be expressed as the ratio of the determinants of two $(N+1) \times (N+1)$ Vandermonde matrices. For brevity, we define

$$\mathcal{V}_i(x) \equiv \mathrm{Det}[\mathbf{V}(x_1, x_2, \ldots, x_{i-1}, x, x_{i+1}, \ldots, x_{N+1})], \qquad (A.2.35)$$

and write

$$l_{N,i}(x) = \frac{\mathcal{V}_i(x)}{\mathcal{V}}, \qquad (A.2.36)$$

where $\mathcal{V} = \mathcal{V}_i(x_i)$ is the determinant of the Vandermonde matrix, independent of i. To confirm that (A.2.34) is an acceptable representation, we make three key observations:

1. The numerator in (A.2.34) is an Nth-degree polynomial in x. This can be readily demonstrated by writing the Laplace expansion of the determinant with respect to the column involving x.

2. The numerator in (A.2.34) is zero when $x = x_j$ for $j \neq i$, as two columns of the Vandermonde matrix become identical.

3. The right-hand side of (A.2.34) is equal to unity when $x = x_i$, as the numerator becomes equal to the denominator.

Differentiating (A.2.36) with respect to x and rearranging, we obtain

$$\frac{\mathrm{d}\mathcal{V}_i(x)}{\mathrm{d}x} = \mathcal{V}\,\frac{\mathrm{d}l_{N,i}(x)}{\mathrm{d}x}. \qquad (A.2.37)$$

Evaluating this expression at $x = x_i$, we obtain

$$d_{ii} = \frac{1}{\mathcal{V}}\left(\frac{\mathrm{d}\mathcal{V}_i(x)}{\mathrm{d}x}\right)_{x=x_i} = \frac{1}{\mathcal{V}}\frac{\partial \mathcal{V}}{\partial x_i}, \qquad (A.2.38)$$

where d_{ii} is the diagonal component of the node differentiation matrix defined in (A.2.21) and given in (A.2.23).

Generalized Vandermonde matrix

Consider the representation of the interpolating polynomial in terms of a polynomial base, ϕ_i, as shown in (A.1.11). Working as previously in this section, we find that the Lagrange interpolating polynomials satisfy the counterpart of the linear system (A.2.33),

$$\mathbf{V}_\phi \cdot \begin{bmatrix} l_{N,1}(x) \\ l_{N,2}(x) \\ \vdots \\ l_{N,N}(x) \\ l_{N,N+1}(x) \end{bmatrix} = \begin{bmatrix} \phi_1(x) \\ \phi_2(x) \\ \vdots \\ \phi_N(x) \\ \phi_{N+1}(x) \end{bmatrix}, \qquad (A.2.39)$$

where $\mathbf{V}_\phi$ is the generalized Vandermonde matrix defined in (A.1.13). Computing the solution by Cramer's rule, we find

$$l_{N,i}(x) = \frac{\mathrm{Det}[\mathbf{V}_\phi(x_1, x_2, \ldots, x_{i-1}, x, x_{i+1}, \ldots, x_{N+1})]}{\mathrm{Det}[\mathbf{V}_\phi(x_1, x_2, \ldots, x_{i-1}, x_i, x_{i+1}, \ldots, x_{N+1})]}, \qquad (A.2.40)$$

which is a generalization of (A.2.34).

Next, we denote

$$\mathcal{V}_\phi \equiv \mathrm{Det}[\mathbf{V}_\phi(x_1, x_2, \ldots, x_{j-1}, x_j, x_{j+1}, \ldots, x_{N+1})], \qquad (A.2.41)$$

and apply the chain rule of differentiation to find

$$\frac{\partial \mathrm{Det}(\mathcal{V}_\phi)}{\partial x_j} = \sum_{i=1}^{N+1} \frac{\partial \mathrm{Det}(\mathcal{V}_\phi)}{\partial \phi_i(x_j)}\,\frac{\partial \phi_i(x_j)}{\partial x_j}. \qquad (A.2.42)$$

Using the Laplace expansion, we find that the derivative of the determinant of the generalized Vandermonde matrix with respect to the ij matrix element is given by

$$\frac{\partial \text{Det}(\mathcal{V}_\varphi)}{\partial \phi_i(x_j)} = (-1)^{i+j} \, \text{Det}(M_{ij}), \tag{A.2.43}$$

where M_{ij} is the element minor. Accordingly,

$$\frac{\partial \text{Det}(\mathcal{V}_\phi)}{\partial x_j} = \sum_{i=1}^{N+1} (-1)^{i+j} \, \text{Det}(M_{ij}) \frac{\partial \phi_i(x_j)}{\partial x_j}. \tag{A.2.44}$$

If we identify the polynomial base functions, $\phi_i(x)$, with the Lagrange interpolation functions, $l_{N,i}(x)$, the generalized Vandermonde matrix reduces to the identity matrix. Consideration of (A.2.43) then shows that the matrix of partial derivatives of the generalized Vandermonde matrix with respect to any matrix element also reduces to the unit matrix [63], and (A.2.42) simplifies to

$$\frac{\partial \text{Det}(\mathcal{V})}{\partial x_j} = \frac{\partial l_{N,j}(x_j)}{\partial x_j}. \tag{A.2.45}$$

Fekete points

If the interpolation abscissas, x_i, are distributed such that the magnitude of the associated Lagrange polynomials is less than unity over a specified interpolation domain,

$$|l_{N,i}(x)| \leq 1, \tag{A.2.46}$$

then the numerator in (A.2.34) reaches the maximum of unity when $x = x_i$, whereupon it becomes equal to the denominator. Stated differently, the function $\mathcal{V}_i(x)$ reaches an extreme value when $x = x_i$, whereupon

$$\left(\frac{\partial \mathcal{V}_i(x)}{\partial x} \right)_{x=x_i} = \frac{\partial \mathcal{V}}{\partial x_i} = 0. \tag{A.2.47}$$

Formula (A.2.38) then shows that the corresponding diagonal components of the node differentiation matrix are zero,

$$d_{ii} = 0. \tag{A.2.48}$$

The set of points that maximizes the magnitude of the determinant of the Vandermonde matrix are the Fekete points in one dimension. In Section A.5, we shall see that, when $x_1 = -1$ and $x_{N+1} = 1$, the $N - 1$ intermediate Fekete points x_i, $i = 2, 3, \ldots, N$, are the well-known roots of the $(N-1)$-degree Lobatto polynomial, $Lo_{N-1}(x)$, discussed in Section B.6 of Appendix B.

A.2.5   Lagrange polynomials corresponding to polynomial roots

Let us assume that the $(N + 1)$-degree polynomial

$$Q_{N+1}(x) = a_1 \, x^{N+1} + a_2 \, x^N + \ldots + a_{N+1} \, x + a_{N+2}, \qquad (\text{A.2.49})$$

has $N + 1$ distinct roots, x_i, where $i = 1, 2, \ldots, N + 1$, and a_i are specified polynomial coefficients. For reasons that will become evident shortly, we express this polynomial in the form

$$Q_{N+1}(x) = a_1 \, x^{N+1} - P_N(x), \qquad (\text{A.2.50})$$

where

$$P_N(x) \equiv -a_2 \, x^N - \ldots - a_{N+1} \, x - a_{N+2}, \qquad (\text{A.2.51})$$

is an Nth-degree polynomial.

Next, we formulate the Lagrange polynomials $l_{N,i}(x)$ associated with the roots, x_i, where $i = 1, 2, \ldots, N + 1$, and write the *exact* representation

$$P_N(x) = \sum_{i=1}^{N+1} P_N(x_i) \, l_{N_i}(x). \qquad (\text{A.2.52})$$

However, by definition,

$$Q_{N+1}(x_i) = a_1 \, x_i^{N+1} - P_N(x_i) = 0, \qquad (\text{A.2.53})$$

or

$$P_N(x_i) = a_1 \, x_i^{N+1}, \qquad (\text{A.2.54})$$

and thus

$$P_N(x) = a_1 \sum_{i=1}^{N+1} x_i^{N+1} \, l_{N,i}(x). \qquad (\text{A.2.55})$$

Substituting this result in (A.2.50), we derive a representation in terms of the leading-power coefficient and roots,

$$Q_{N+1}(x) = a_1 \left(x^{N+1} - \sum_{i=1}^{N+1} x_i^{N+1} \, l_{N,i}(x) \right), \qquad (\text{A.2.56})$$

which complements the more familiar representation

$$Q_{N+1}(x) = a_1 \prod_{i=1}^{N+1} (x - x_i). \qquad (\text{A.2.57})$$

A.2.6 Lagrange polynomials for Hermite interpolation

The theory of Lagrange polynomials discussed previously in this appendix fails when the abscissa of a certain point, x_j, tends to the abscissa of another point, x_i. In this limit, the interpolating polynomial continues to exist only if $f(x_j)$ also tends to $f(x_i)$. This may occur if the graph of the function $f(x)$ tends to become horizontal at the point x_i, which means that $f'(x_i) = 0$.

A generalization of these circumstances leads us to a special case of the Hermite interpolation problem, which can be stated as follows: find an Nth-degree polynomial, $P_N(x)$, so that

$$P_N(x_i) = f(x_i), \tag{A.2.58}$$

and

$$\left(\frac{\mathrm{d}P_N}{\mathrm{d}x}\right)_{x_i} = 0, \quad \ldots, \quad \left(\frac{\mathrm{d}^{m_i-1}P_N}{\mathrm{d}x^{m_i-1}}\right)_{x_i} = 0, \tag{A.2.59}$$

for $i = 1, 2, \ldots, r$, where r is the number of distinct abscissas, m_i is the grade of the ith abscissa, and

$$m_1 + m_2 + \ldots + m_r = N + 1. \tag{A.2.60}$$

When $r = N + 1$ and $m_i = 1$ for all i, we recover the standard interpolation problem discussed previously in this appendix.

The requisite interpolating polynomial can be expressed in terms of the generalized Lagrange polynomials, $\hat{l}_{N,i}$, as

$$P_N(x) = \sum_{i=1}^{r} f(x_i)\,\hat{l}_{N,i}(x), \tag{A.2.61}$$

where the generalized Lagrange polynomials are required to satisfy the super-cardinal, Hermite-like properties

$$\hat{l}_{N,i}(x_j) = \delta_{ij}, \tag{A.2.62}$$

and

$$\left(\frac{\mathrm{d}\hat{l}_{N,i}}{\mathrm{d}x}\right)_{x_j} = 0, \quad \ldots, \quad \left(\frac{\mathrm{d}^{m_j-1}\hat{l}_{N,i}}{\mathrm{d}x^{m_j-1}}\right)_{x_j} = 0, \tag{A.2.63}$$

for $i, j = 1, 2, \ldots, r$. If the interpolated function is constant, the interpolation is exact by construction for any r and m_i. This observation justifies the generalized Cauchy relation

$$\sum_{i=1}^{r} \hat{l}_{N,i}(x) = 1. \tag{A.2.64}$$

As for the standard Lagrange polynomials, we may introduce the generating function

$$\Phi_{N+1}(x) = (x - x_1)^{m_1}(x - x_2)^{m_2} \dots (x - x_r)^{m_r}, \qquad (A.2.65)$$

and express the generalized Lagrange polynomials in the form

$$\hat{l}_{N,i}(x) = \chi_{m_i-1,i}(x)\, \frac{\Phi_{N+1}(x)}{(x - x_i)^{m_i}}, \qquad (A.2.66)$$

where $\chi_{m_i-1,i}(x)$ are $(m_i - 1)$-degree polynomials arising from the partial fraction decomposition of the inverse of the generating function,

$$\frac{1}{\Phi_{N+1}(x)} = \frac{\chi_{m_1-1,1}(x)}{(x - x_1)^{m_1}} + \dots + \frac{\chi_{m_r-1,r}(x)}{(x - x_r)^{m_r}}. \qquad (A.2.67)$$

If $m_i = 1$, then

$$\chi_{0,i} \equiv c_i = \frac{1}{\Phi'_{N+1}(x_i)} \qquad (A.2.68)$$

$$= \frac{1}{(x_i - x_1)^{m_1} \dots (x_i - x_{i-1})^{m_{i-1}}(x_i - x_{i+1})^{m_{i+1}} \dots \dots (x_i - x_r)^{m_r}},$$

is a constant.

As an example, we consider interpolation with $r = 2$ distinct abscissas of grades $m_1 = 2$ and $m_2 = 1$, corresponding to $N + 1 = 3$. The generating polynomial is given by

$$\Phi_3(x) = (x - x_1)^2 (x - x_2). \qquad (A.2.69)$$

Straightforward algebra produces the partial fraction decomposition

$$\frac{1}{\Phi_3(x)} = \frac{1}{(x_2 - x_1)^2} \left(\frac{-x + (2x_1 - x_2)}{(x - x_1)^2} + \frac{1}{x - x_2} \right), \qquad (A.2.70)$$

which shows that

$$\chi_{1,1}(x) = \frac{-x + (2x_1 - x_2)}{(x_2 - x_1)^2}, \qquad \chi_{0,2} = \frac{1}{(x_2 - x_1)^2}. \qquad (A.2.71)$$

Accordingly, the generalized Lagrange polynomials are given by

$$\hat{l}_{2,1}(x) = \frac{-x + (2x_1 - x_2)}{(x_2 - x_1)^2} (x - x_2),$$

$$\qquad (A.2.72)$$

$$\hat{l}_{2,2}(x) = \frac{(x - x_1)^2}{(x_2 - x_1)^2}.$$

A.3   Error in polynomial interpolation

Unless the interpolated function, $f(x)$, happens to be a polynomial of degree equal to N or less, $P_N(x)$ and $f(x)$ will not necessarily agree between the data points. The difference between the approximate and exact value is the interpolation error,

$$e(x) \equiv P_N(x) - f(x). \tag{A.3.1}$$

The mandatory satisfaction of the interpolation conditions (A.1.2) ensures that

$$e(x_i) = 0 \tag{A.3.2}$$

for $i = 1, 2, \ldots N + 1$, that is, the function $e(x)$ has at least $N + 1$ zeros in the interpolation domain. However, in general, $e(x) \neq 0$ when $x \neq x_i$.

We shall show that, when the function $f(x)$ is sufficiently smooth, the error incurred by the polynomial interpolation is given by

$$e(x) = -\frac{f^{(N+1)}(\xi)}{(N+1)!} \, \Phi_{N+1}(x), \tag{A.3.3}$$

where:

- $(N+1)! \equiv 1 \cdot 2 \cdot \ldots \cdot N \cdot (N+1)$ is the factorial.

- The $(N+1)$-degree generating polynomial $\Phi_{N+1}(x)$ is defined in (A.2.8).

- $f^{(N+1)}(\xi)$ is the $(N+1)$-derivative of the function f evaluated at a certain point ξ that lies somewhere inside the smallest interval that contains the abscissas of all data points. Without loss of generality, we may assume that the abscissas are ordered, so that $x_1 \leq \xi \leq x_{N+1}$. Since the precise location of ξ depends on the value of x, the quantity $f^{(N+1)}(\xi)$ is truly an implicit function of x.

To prove the error formula, we hold the value of x fixed, introduce a new function of a new independent variable, z, defined as

$$Q(z) \equiv \Phi_{N+1}(z) \, e(x) - \Phi_{N+1}(x) \, e(z), \tag{A.3.4}$$

and observe that

$$Q(x_i) = 0 \tag{A.3.5}$$

for $i = 1, 2, \ldots, N + 1$, and also

$$Q(x) = 0. \tag{A.3.6}$$

Because the function $Q(z)$ has at least $N+2$ zeros, it must attain at least $N+1$ local maxima and minima inside the interval $[x_1, x_{N+1}]$. Accordingly, the first derivative, $Q'(z)$, must have at least $N + 1$ zeros, and the second derivative, $Q''(z)$, must have at least N zeros between x_1 and x_{N+1}. Continuing this deduction, we find that the $N+1$ derivative, $Q^{(N+1)}(z)$, must have at least one zero between x_1 and x_{N+1}; one of these zeros is assumed to occur at $z = \xi$. By definition then,

$$Q^{(N+1)}(\xi) = \Phi_{N+1}^{(N+1)}(\xi)\, e(x) - \Phi_{N+1}^{(N+1)}(x)\, e^{(N+1)}(\xi) = 0. \qquad (A.3.7)$$

Differentiating the right-hand side of (A.3.1) $N + 1$ times with respect to x, noting that $P_N^{(N+1)}(x) = 0$, and evaluating the resulting expression at $x = \xi$, we find

$$e^{(N+1)}(\xi) = -f^{(N+1)}(\xi). \qquad (A.3.8)$$

The $N + 1$ derivative of the Lagrange generating function is a constant,

$$\Phi_{N+1}^{(N+1)}(\xi) = (N+1)\,!\,. \qquad (A.3.9)$$

Substituting the last two expressions in (A.3.7), and solving for $e(x)$, we obtain the error formula (A.3.3).

A.3.1 Convergence and the Lebesgue constant

The max norm of a function $g(x)$, denoted by $||g(x)||$, is defined as the maximum of the absolute value of $g(x)$ over a specified interpolation interval, $a \leq x \leq b$. The preceding discussion suggests that the norm of the interpolation error, $||e(x)||$, will depend on the location of the interpolation points. Of all Nth-degree polynomials approximating the function $f(x)$, there is an optimal polynomial, denoted $P_N^{opt}(x)$, that exhibits the minimum error $||e(x)||$, called the *minimax error* and denoted by $\rho_N[f(x)]$. This optimal polynomial is not necessarily an interpolating polynomial, that is, it does not necessarily agree with the interpolated function at $N + 1$ data points.

Consider the set of all functions $f(t)$ with unit max norm, $||f(t)|| = 1$. The corresponding norm of the interpolation error is

$$||e(x)|| \equiv ||P_N(x) - f(x)|| = ||\Big(P_N(x) - P_N^{opt}(x)\Big) + \Big(P_N^{opt}(x) - f(x)\Big)||$$

$$\leq ||P_N(x) - P_N^{opt}(x)|| + ||P_N^{opt}(x) - f(x)||$$

$$= ||P_N(x) - P_N^{opt}(x)|| + \rho_N[f(x)]. \qquad (A.3.10)$$

To emphasize that the polynomial $P_N(x)$ approximates the function $f(x)$, we write it as $P_N(x, f)$. Next, we apply the Lebesgue lemma, and derive the inequality

$$||P_N(x, f) - P_N^{opt}(x)|| = ||P_N(x, f) - P_N(x, P_N^{opt})||$$

$$\leq ||P_N|| \cdot ||f(x) - P_N^{opt}(x)||, \qquad (A.3.11)$$

where

$$||P_N|| \equiv \text{Max}\Big(||P_N(f(x))||\Big), \tag{A.3.12}$$

and the maximum is computed over all permissible functions, $f(x)$. Accordingly,

$$||e(x)|| \leq (1 + ||P_N||)\, \rho_N[f(x)]. \tag{A.3.13}$$

To derive a bound for the norm $||P_N||$, we express the interpolating polynomial in terms of the Lagrange polynomials using (A.2.61). Recalling the requirement $||f(x)|| = 1$, we write

$$||P_N|| \equiv \text{Max}\Big(|\sum_{i=1}^{N+1} f(x_i)\, l_{N,i}(x)|\Big) \tag{A.3.14}$$

$$\leq \text{Max}\Big(\sum_{i=1}^{N+1} |f(x_i)|\, |l_{N,i}(x)|\Big) \leq \text{Max}\Big(\mathcal{L}_N(x)\Big),$$

where

$$\mathcal{L}_N(x) \equiv \sum_{i=1}^{N+1} |l_{N,i}(x)| \tag{A.3.15}$$

is the Lebesgue function. The maximum value of the Lebesgue function is the Lebesgue constant,

$$\Lambda_N \equiv \text{Max}\Big(\mathcal{L}_N(x)\Big). \tag{A.3.16}$$

Combining this definition with (A.3.13), we find

$$||e(x)|| \leq (1 + \Lambda_N)\, \rho_N[f(x)], \tag{A.3.17}$$

which shows that the convergence properties of the interpolation will depend on the functional dependence of the minimax error and Lebesgue constant on the polynomial order, N.

Behavior of the minimax error and Jackson's theorems

Jackson's first theorem places an upper bound on the minimax error, $\rho_N[f(x)]$. When interpolating a continuous function $f(x)$ inside the canonical interval $[-1, 1]$, the theorem states that

$$\rho_N[f(x)] \leq \Big(1 + \frac{\pi^2}{2}\Big) w\Big(\frac{1}{N}\Big), \tag{A.3.18}$$

where $w(\delta)$ is the *modulus of continuity* of $f(x)$, defined as

$$w(\delta) \equiv \text{Max}_{|x_1 - x_2| \leq \delta} |f(x_1) - f(x_2)| \tag{A.3.19}$$

([13], p. 338). For a function that satisfies the Lipschitz condition

$$|f(x_1) - f(x_2)| \leq A\,|x_1 - x_2|^\alpha, \tag{A.3.20}$$

where $\alpha > 0$ is a positive exponent, A is a positive constant, $-1 \leq x_1, x_2 \leq 1$, $w(\delta) \leq A\,\delta^\alpha$, and thus

$$\rho_N[f(x)] \leq A\left(1 + \frac{\pi^2}{2}\right)\frac{1}{N^\alpha}. \tag{A.3.21}$$

Jackson's second theorem states that, if the function $f(x)$ has a kth derivative in $[-1, 1]$, that is, the kth derivative does not become infinite in $[-1, 1]$, then

$$\rho_N[f(x)] \leq \left(1 + \frac{\pi^2}{2}\right)^{k+1}\frac{e^k}{(1+k)}\frac{1}{N^k}\,w\left(\frac{1}{N-k}\right), \tag{A.3.22}$$

for $N > k$ ([48], p. 23). The first theorem is a special case of the second theorem for $k = 0$.

Behavior of the Lebesgue constant

The Lebesgue constant is known to grow as N tends to infinity. Erdös's theorem places a lower bound on the slowest possible growth,

$$\Lambda_N > \frac{2}{\pi}\ln N + 1 - c, \tag{A.3.23}$$

where c is a positive constant. Thus, the Lebesgue constant grows at least as fast as $\ln N$.

In the event of evenly spaced points over a specified interval, the Lebesgue constant is known to grow rapidly with N, exhibiting the asymptotic behavior

$$\Lambda_N \sim \frac{2^N}{N\log N}. \tag{A.3.24}$$

In fact, numerical computations show that Λ_N increases from the value of 1 when $N = 1$, to 4.55 when $N = 6$, to 89.3 when $N = 12$, to 3171.1 when $N = 18$, to 137852.1 when $N = 24$ (e.g., [26]). This rapid growth is responsible for the deleterious Runge effect discussed in Section 2.1.

In the next two sections, we will see that, when the interpolation nodes are deployed at the zeros of orthogonal polynomials, Λ_N increases much slower, at the nearly optimal logarithmic rate.

A.4   Chebyshev interpolation

Suppose that we have the luxury of distributing the $N + 1$ nodes in any way we desire over a certain closed interval of the x axis, $[a, b]$. Is there an optimal distribution that minimizes the interpolation error in some sense? Moreover, is there a systematic way of distributing the data points so that, as N is increased, the error will decrease uniformly and at the fastest possible rate?

To address these questions, we introduce the triangular families of orthogonal polynomials discussed in Appendix B. Briefly, each family consists of an infinite sequence of polynomials $p_i(t)$, $i = 0, 1, 2, \ldots$, where p_0 is a constant, $p_1(t)$ is a first-degree polynomial, and $p_i(t)$ is an ith-degree polynomial of the independent variable t, defined over a certain interval $[c, d]$. By definition, the orthogonal polynomials of a certain family satisfy the orthogonality condition

$$(p_i, p_j) \equiv \int_c^d p_i(t)\, p_j(t)\, w(t)\, \mathrm{d}t = \begin{cases} \mathcal{G}_i & \text{if } i = j \\ 0 & \text{if } i \neq j \end{cases}, \tag{A.4.1}$$

where $w(t)$ is a weighting function, and $\mathcal{G}_i$ are constants. Any mth-degree polynomial can be expressed as a linear combination of the first $m+1$ orthogonal polynomials of any chosen family, $p_0(t), p_1(t), \ldots, p_m(t)$.

We are in a position now to tackle the issue of optimal positioning of the abscissas. As a preliminary, we introduce the independent variable t that is related to x by the linear transformation

$$x = q(t) = \frac{b+a}{2} + \frac{b-a}{2} \frac{2t-c-d}{d-c}. \tag{A.4.2}$$

As t increases from c to d, x increases from a to b. Having made this transformation, we regard the function f as a function of t, writing

$$f(x) = f[q(t)] \equiv h(t). \tag{A.4.3}$$

Using formula (A.3.3), we find that the error incurred by the polynomial interpolation of $h(t)$ with $N + 1$ abscissas, t_i, corresponding to x_i, is given by

$$e(t) \equiv P_N(t) - h(t)$$

$$\tag{A.4.4}$$

$$= -\frac{h^{(N+1)}(\xi)}{(N+1)!} (t - t_1)(t - t_2) \ldots (t - t_N)(t - t_{N+1}),$$

where ξ lies somewhere within the interval $[c, d]$.

Next, we identify the nodes $t_1, t_2, \ldots, t_{N+1}$ with the roots of an $(N+1)$-degree polynomial defined in the interval $[c, d]$ whose coefficient of the leading power is equal to unity, denoted by

$$Q_{N+1}(t) = t^{N+1} + \ldots, \tag{A.4.5}$$

and obtain

$$e(t) = -\frac{h^{(N+1)}(t)}{(N+1)!} Q_{N+1}(t). \tag{A.4.6}$$

To standardize the problem, we place without loss of generality the end-points of the interpolation interval at $c = -1$ and $d = 1$, and search for the polynomial $Q_{N+1}(t)$ with the smallest maximum (minimax) norm. The answer follows directly from the minimax property of the Chebyshev polynomials stated in Section B.7 of Appendix B. The requisite optimal polynomial is

$$Q_{N+1}(t) = 2^{-N} T_{N+1}(t), \qquad (A.4.7)$$

where $T_{N+1}(t)$ is the $(N + 1)$-degree Chebyshev polynomial,

$$T_{N+1}(t) = 2^N (t - t_1)(t - t_2) \ldots (t - t_{N+1}), \qquad (A.4.8)$$

and t_i are the zeros of the Chebyshev polynomial located at

$$t_i = \cos\left[\left(i - \frac{1}{2}\right) \frac{\pi}{N + 1} \right], \qquad (A.4.9)$$

for $i = 1, 2, \ldots, N + 1$. Substituting (A.4.7) in (A.4.6), we find

$$e(t) = -\frac{h^{(N+1)}(\xi)}{2^N (N + 1)!} T_{N+1}(t). \qquad (A.4.10)$$

Since the magnitude of the Chebyshev polynomials is less than unity for any value of t in the interval [-1, 1], the magnitude of the interpolation error is at most on the order of the fraction on the right-hand side of (A.4.10). Unless the function $h(t)$ is badly behaved, the error $e(t)$ is a rapidly decreasing function of the polynomial order, N.

More precisely, it can be shown that the error incurred by the Chebyshev interpolation is not much worse than that incurred by the minimax approximation,

$$\text{Max}[e(t)] \leq \left(\frac{2}{\pi} \log(N + 1) + 2\right) \rho_N[h(t)], \qquad (A.4.11)$$

where $\rho_N[h(t)]$ is the minimax error discussed in Section A.3 (e.g., [13, 49]). This formula shows that the Lebesgue constant increases at the optimal logarithmic rate with respect to the polynomial order, N.

A.5   Lobatto interpolation

The linear transformation

$$x = \frac{x_{N+1} + x_1}{2} + \frac{x_{N+1} - x_1}{2} t, \qquad (A.5.1)$$

maps the interpolation interval with respect to the variable x to a canonical interpolation interval with respect to the variable t, so that the first and last interpolation points, x_1 and x_{N+1}, are mapped to $t_1 = -1$ and $t_{N+1} = 1$. We may then carry the interpolation with respect to t in terms of the corresponding Nth degree Lagrange polynomials, $l_{N,i}(t)$, where $i = 1, 2, \ldots, N + 1$.

For reasons that will become evident shortly, we require that, in addition to satisfying the $N + 1$ *mandatory* cardinal interpolation conditions

$$l_{N,i}(t_j) = \delta_{ij}, \qquad (A.5.2)$$

for $i, j = 1, 2, \ldots, N + 1$, the node distribution is such that the Lagrange polynomials also satisfy the $N - 1$ *optional* conditions

$$l'_{N,i}(t_i) = 0, \qquad (A.5.3)$$

for $i = 2, 3, \ldots, N$, where a prime denotes a derivative with respect to t. Property (A.5.3) ensures that the Lagrange polynomials reach the local maximum value of unity at the $N - 1$ intermediate nodes.

Using the node differentiation matrix (A.2.21) for $i = j$, we find that (A.5.3) implies

$$\Phi''_{N+1}(t_i) = 0, \qquad (A.5.4)$$

for $i = 2, 3, \ldots, N$. Since $\Phi''_{N+1}(t)$ is an $(N - 1)$-degree polynomial, we can write

$$\Phi''_{N+1}(t) = c \, \frac{\Phi_{N+1}(t)}{(t - 1)(t + 1)} = -c \, \frac{\Phi_{N+1}(t)}{1 - t^2}, \qquad (A.5.5)$$

where c is a constant. Because the coefficient of the highest power of t in $\Phi_{N+1}(t)$ is equal to unity, we must have

$$c = N \, (N + 1). \qquad (A.5.6)$$

Consequently, the $(N+1)$-degree polynomial $\Phi_{N+1}(t)$ satisfies the second-order differential equation

$$(1 - t^2) \, \Phi''_{N+1}(t) + N(N + 1) \, \Phi_{N+1}(t) = 0. \qquad (A.5.7)$$

To this end, we recall the differential equation satisfied by the Legendre polynomials, shown in equation (B.5.11) of Appendix B, and find that the solution of (A.5.7) is

$$\Phi_{N+1}(t) = \alpha_{N+1} \, (t^2 - 1) \, L'_N(t) = \alpha_{N+1} \, (t^2 - 1) \, Lo_{N-1}(t), \qquad (A.5.8)$$

where L_N is a Legendre polynomial, Lo_{N-1} is a Lobatto polynomial, and α_{N+1} is a constant designed so that the coefficient of the highest power of t in $\Phi_{N+1}(t)$ is equal to unity. Expression (A.5.8) shows that the interpolation nodes, t_i, for $i = 2, 3, \ldots, N$ are, in fact, the zeros of the $(N - 1)$-degree Lobatto polynomial.

Spectral convergence

The significance of the preceding derivation becomes evident by considering the $2N$-degree polynomial

$$G_{2N}(t) \equiv l_{N,1}^2(t) + l_{N,2}^2(t) + \ldots + l_{N,N+1}^2(t) - 1. \tag{A.5.9}$$

Since the cardinal interpolation property (A.5.2) requires

$$G_{2N}(t_i) = 0, \tag{A.5.10}$$

for $i = 1, 2, \ldots, N+1$, we may write

$$G_{2N}(t) \equiv \Phi_{N+1}(t)\, \Psi_{N-1}(t), \tag{A.5.11}$$

where $\Psi_{N-1}(t)$ is an $(N-1)$-degree polynomial. Differentiating (A.5.9), we find

$$G'_{2N}(t) = 2 \sum_{i=1}^{N+1} l_{N,i}(t)\, l'_{N,i}(t), \tag{A.5.12}$$

which, in light of (A.5.3), shows that

$$G'_{2N}(t_i) = 0, \tag{A.5.13}$$

for $i = 2, 3, \ldots, N$. Because the points t_i, for $i = 2, 3, \ldots, N$, are double roots of $G_{2N}(t)$, we may write

$$\Psi_{N-1}(t) = d\, \frac{\Phi_{N+1}(t)}{(t+1)(t-1)} = -d\, \frac{\Phi_{N+1}(t)}{1 - t^2}, \tag{A.5.14}$$

where d is a constant. Thus,

$$G_{2N}(t) = -d\, \frac{\Phi_{N+1}^2(t)}{1 - t^2} = -\beta\, (1 - t^2)\, Lo_{N-1}^2(t), \tag{A.5.15}$$

where β is a new constant.

To compute the constant, β, we evaluate (A.5.12) at $t = 1$, and use the cardinal interpolation property to find

$$G'_{2N}(t=1) = 2\, l'_{N,N+1}(t=1) = \frac{\Phi''_{N+1}(t=1)}{\Phi'_{N+1}(t=1)}, \tag{A.5.16}$$

where the second expression arises from the node differentiation matrix shown in (A.2.21). Substituting the generating function from (A.5.8), we obtain

$$G'_{2N}(t=1) = \left(\frac{[(t^2 - 1)\, Lo_{N-1}(t)]''}{[(t^2 - 1)\, Lo_{N-1}(t)]'} \right)_{t=1}. \tag{A.5.17}$$

Since the Lobatto polynomials satisfy the differential equation

$$\left[(t^2 - 1) \, Lo_{N-1}(t)\right]' = N \, (N+1) \, L_N(t), \qquad (A.5.18)$$

we find

$$G'_{2N}(t = 1) = \frac{L'_N(t = 1)}{L_N(t = 1)} = \frac{Lo_{N-1}(t = 1)}{L_N(t = 1)}, \qquad (A.5.19)$$

where L_N is a Legendre polynomial. The Legendre polynomials are normalized such that $L_N(1) = 1$. Evaluating (A.5.18) at $t = 1$, we find

$$Lo_{N-1}(t = 1) = \frac{1}{2} \, N(N+1), \qquad (A.5.20)$$

and thus

$$G'_{2N}(t = 1) = \frac{1}{2} \, N(N+1). \qquad (A.5.21)$$

To this end, we differentiate the last expression in (A.5.15), and obtain

$$G'_{2N}(t = 1) = 2\beta \, Lo^2_{N-1}(t = 1) = \beta \, \frac{1}{2} \, N^2 \, (N+1)^2. \qquad (A.5.22)$$

Combining the last two expressions, we find $\beta = 1/[N(N+1)]$, and derive the final expression

$$G_{2N}(t) = -\frac{1}{N(N+1)} \, (1 - t^2) \, Lo^2_{N-1}(t), \qquad (A.5.23)$$

or

$$G_{2N}(t) = -\frac{1}{N(N+1)} \, (1 - t^2) \, L'^2_N(t). \qquad (A.5.24)$$

In the interval of interest, $-1 \leq t \leq 1$, the binomial $1 - t^2$ is non-negative, and the right-hand side of (A.5.24) is non-positive. Accordingly,

$$G_{2N}(t) \leq 0, \qquad (A.5.25)$$

and thus

$$l^2_{N,1}(t) + l^2_{N,2}(t) + \ldots + l^2_{N,N+1}(t) \leq 1, \qquad (A.5.26)$$

which requires

$$l^2_{N,i}(t) \leq 1, \qquad (A.5.27)$$

or

$$|l_{N,i}(t)| \leq 1, \qquad (A.5.28)$$

individually for $i = 1, 2, \ldots, N + 1$ [18, 19]. The maximum value of $|l_{N,i}(t)|$ is thus achieved when $t = t_i$.

Now, applying the Cauchy-Schwartz inequality, we find

$$\left(\sum_{i=1}^{N+1} |l_{N,i}(t)| \right)^2 \leq \sum_{i=1}^{N+1} |l_{N,i}(t)|^2. \tag{A.5.29}$$

Recalling the definition of the Lebesgue constant in (A.3.16), and using (A.5.27), we find

$$\Lambda_N^2 \leq N + 1, \quad \text{or} \quad \Lambda_N \leq \sqrt{N + 1}. \tag{A.5.30}$$

In fact, numerical investigation reveals the much stricter bound

$$\Lambda_N \leq \frac{2}{\pi} \log(N + 1) + 0.685, \tag{A.5.31}$$

which allows for well-known proofs of uniform and exponential convergence of interpolation, numerical differentiation and integration [5, 26].

Fekete points

By definition, the Fekete points for a finite interval maximize the magnitude of the determinant of the Vandermonde matrix defined in (A.1.6). Since any change in the polynomial base only multiplies the determinant by a constant factor that is independent of the point locations, the Fekete points also maximize the magnitude of the generalized Vandermonde matrix defined in (A.1.13) for any polynomial base.

Combining (A.2.40) with (A.5.28), we find that, for the Lobatto point distribution,

$$\left| \frac{\text{Det}[\mathbf{V}_\phi(x_1, x_2, \ldots, x_{i-1}, x, x_{i+1}, \ldots, x_{N+1})]}{\text{Det}[\mathbf{V}_\phi(x_1, x_2, \ldots, x_{i-1}, x_i, x_{i+1}, \ldots, x_{N+1})]} \right| \leq 1, \tag{A.5.32}$$

for $i = 2, 3, \ldots, N$. Thus, the Lobatto points are identical to the Fekete points mapped to the interval [-1, 1].

A.6 Interpolation in two and higher dimensions

Given the values of a function $f(x, y)$ at N data points in the xy plane, located at (x_i, y_i), for $i = 1, 2, \ldots, N$, we want to interpolate at an intermediate point, that is, we want to compute the value of the function at an arbitrary point, (x, y). To carry out the interpolation, we introduce a set of N linearly independent functions, $\phi_i(x, y)$, $i = 1, 2, \ldots, N$, and approximate the interpolated function, $f(x, y)$, with the interpolating function $g(x, y)$,

$$f(x, y) \simeq g(x, y) = c_1 \, \phi_N(x, y) + c_2 \, \phi_{N-1}(x, y)$$
$$+ \ldots + c_{N-1} \, \phi_2(x, y) + c_N \, \phi_1(x, y), \tag{A.6.1}$$

where the coefficients, c_i, are computed to satisfy the interpolation condition

$$g(x_i, y_i) = f(x_i, y_i) \tag{A.6.2}$$

for $i = 1, 2, \ldots, N$. Enforcing the interpolation conditions, we derive the linear system

$$\mathbf{V}_\phi^T \cdot \mathbf{c} = \mathbf{f}, \tag{A.6.3}$$

where

$$\mathbf{V}_\phi \equiv \begin{bmatrix} \phi_1(x_1, y_1) & \phi_1(x_2, y_2) & \cdots & \phi_1(x_N, y_N) \\ \phi_2(x_1, y_1) & \phi_2(x_2, y_2) & \cdots & \phi_2(x_N, y_N) \\ \cdots & \cdots & \cdots & \cdots \\ \phi_{N-1}(x_1, y_1) & \phi_{N-1}(x_2, y_2) & \cdots & \phi_{N-1}(x_N, y_N) \\ \phi_N(x_1, y_1) & \phi_N(x_2, y_2) & \cdots & \phi_N(x_N, y_N) \end{bmatrix} \tag{A.6.4}$$

is the $N \times N$ generalized Vandermonde matrix, and

$$\mathbf{c} \equiv \begin{bmatrix} c_N \\ c_{N-1} \\ \vdots \\ c_2 \\ c_1 \end{bmatrix}, \qquad \mathbf{f} \equiv \begin{bmatrix} f(x_1, y_1) \\ f(x_2, y_2) \\ \vdots \\ f(x_{N-1}, y_{N-1}) \\ f(x_N, y_N) \end{bmatrix}, \tag{A.6.5}$$

are, respectively, the coefficient vector and data vector. Unfortunately, the properties of the Vandermonde system in two dimensions, including existence and uniqueness of solution, are largely unknown. An effort to maximize the determinant of the Vandermonde matrix, and thereby ensure a solution, leads us to the notion of the Fekete points in two dimensions for a specified interpolation domain.

If we identify the basis functions, $\phi_i(x, y)$, with cardinal interpolation functions satisfying the property

$$\phi_i(x_j, y_j) = \delta_{ij}, \tag{A.6.6}$$

for $i, j = 1, 2, \ldots, N$, where δ_{ij} is Kronecker's delta, then the generalized Vandermonde matrix, $\mathbf{V}_\phi$, reduces to the identity matrix, and the expansion coefficients represent the prescribed data, $c_i = f(x_i, y_i)$.

To approximate the function $f(x, y)$ with a complete mth degree polynomial in x and y, we introduce $N = \frac{1}{2}(m+1)(m+2)$ data points, and identify the basis functions, $\phi_i(x, y)$, with the set of monomial products, $x^k y^l$, to obtain

$$\begin{aligned} f(x, y) \simeq g(x, y) = a_{00} \\ + a_{10}\, x + a_{01}\, y, \\ + a_{20}\, x^2 + a_{11}\, x\, y + a_{02}\, y^2 \\ + a_{30}\, \xi^3 + a_{21}\, x^2\, y + a_{12}\, x\, y^2 + a_{03}\, y^3 \\ \cdots \quad \cdots \quad \cdots \quad \cdots \quad \cdots \quad \cdots \quad \cdots \quad \cdots \\ + a_{m,0}\, x^m + a_{m-1,1}\, x^{m-1}\, y + \ldots + a_{1,m-1}\, x\, y^{m-1} + a_{0,m}\, y^m, \end{aligned} \tag{A.6.7}$$

where a_{ij} is a new set of coefficients that can be mapped to the previous coefficients, c_i. Other choices of basis functions include the Appell and Proriol polynomials discussed in Section 4.3.

Interpolation in three and higher dimensions is carried out by similar methods using appropriate sets of basis functions, as discussed in Section 6.3.

Orthogonal polynomials

Orthogonal polynomials play a prominent role in the theory and practice of function interpolation and approximation, numerical integration, and numerical solution of differential equations by orthogonal collocation and spectral expansions. In this Appendix, we summarize the basic theory of orthogonal polynomials in one dimension, and review certain fundamental properties pertinent to finite and spectral element methods. Extensive discussions can be found in mathematical handbooks and in the monographs by Sansone [52], Szegö [62], Krylov [31], Stroud and Secrest [59], Freud [20], Dahlquist and Björck [12], and Chihara [9].

B.1 Definitions and basic properties

Consider a triangular family of polynomials, $p_n(t)$, for $n = 0, 1, 2, \ldots$, where $p_n(t)$ is an nth-degree polynomial of the independent variable t, defined over a certain interval $[c, d]$, with leading-order coefficient equal to A_n. Explicitly, the family takes the form

$$
\begin{aligned}
p_0(t) &= A_0, \\
p_1(t) &= A_1\, t + A_{1,0}, \\
p_2(t) &= A_2\, t^2 + A_{2,1}\, t + A_{2,0}, \\
&\cdots \\
p_n(t) &= A_n\, t^n + A_{n,n-1}\, t^{n-1} + A_{n,n-2}\, t^{n-2} + \ldots + A_{n,0},
\end{aligned}
\tag{B.1.1}
$$

where $A_{i,j}$ are polynomial coefficients; to simplify the notation, we have denoted $A_n \equiv A_{n,n}$.

If the mutual weighted projection of any pair of these polynomials satisfies the orthogonality condition

$$
(p_i, p_j) \equiv \int_c^d p_i(t)\, p_j(t)\, w(t)\, \mathrm{d}t = \begin{cases} \mathcal{G}_i & \text{if } i = j \\ 0 & \text{if } i \neq j \end{cases},
\tag{B.1.2}
$$

where $w(t)$ is a positive weighting function and $\mathcal{G}_i$ are constants, then the triangular family is orthogonal. If, in addition, $\mathcal{G}_i = 1$ for any value of i, the family is *orthonormal*.

It can be shown that, for every weighting function $w(t)$ that takes non-negative values within a specified interval $[c, d]$, there is a unique corresponding family of orthonormal polynomials defined over this interval. In practice, the members of the family can be computed using the Gram-Schmidt orthogonalization method discussed later in this section.

Orthogonality against lower-degree polynomials

Equations (B.1.1) allow us to express the monomials, t^n, for $n = 0, 1, \ldots$, in terms of the orthogonal polynomials of a certain family, as

$$1 = \frac{p_0}{A_0},$$

$$t = \frac{p_1 - A_1}{A_{1,0}} = \frac{1}{A_{1,0}} \left(p_1 - \frac{A_1}{A_0} p_0 \right),$$

$$\ldots \tag{B.1.3}$$

It is evident then that we may write

$$1 = B_0 \, p_0(t),$$
$$t = B_1 \, p_1(t) + B_{1,0} \, p_0(t), \tag{B.1.4}$$

$$\ldots$$

$$t^n = B_n \, p_n(t) + B_{n,n-1} \, p_{n-1}(t) + B_{n,n-2} \, p_{n-2}(2) + \ldots + B_{n,0},$$

where $B_{i,j}$ are expansion coefficients related to the polynomial coefficients, $A_{i,j}$; for simplicity, we have denoted $B_n \equiv B_{n,n}$.

Expressions (B.1.4) can be used to uniquely express any mth-degree polynomial, $Q_m(t)$, as a linear combination of the first $m + 1$ orthogonal polynomials of any family,

$$p_0(t), \quad p_1(t), \quad \ldots, \quad p_m(t), \tag{B.1.5}$$

as

$$Q_m(t) = c_0 \, p_0(t) + c_1 \, p_1(t) + \ldots + c_m \, p_m(t), \tag{B.1.6}$$

where c_i are expansion coefficients. Using the orthogonality property, we find

$$c_i = \frac{1}{\mathcal{G}_i} \int_c^d Q_m(t) \, p_i(t) \, w(t) \, dt \equiv \frac{1}{\mathcal{G}_i} \, (p_i, Q_m), \tag{B.1.7}$$

for $i = 1, 2, \ldots, m$.

Using expansion (B.1.6), we find that the polynomial, $p_i(t)$, is orthogonal to any mth-degree polynomial, $Q_m(t)$ with $m < i$, that is,

$$(p_i, Q_m) \equiv \int_c^d p_i(t) \, Q_m(t) \, w(t) \, dt = 0. \tag{B.1.8}$$

This property suggests a method of computing the coefficients of the monomial powers in $p_{m+1}(t)$: assign an arbitrary value to the leading-order coefficient, A_{m+1}; apply (B.1.8) for a set of $m+1$ linearly independent test polynomials, $Q_m(t)$; and finally solve the linear system of equations thus derived for the rest of the coefficients.

As an example, we consider the case $m = 0$, express $p_1(t)$ as shown in (B.1.1) with $A_1 = 1$, choose $w(t) = 1$, $c = -1$, and $d = 1$, and select the test polynomial $Q_0(t) = 1$. Applying (B.1.8), we find

$$(p_1, Q_0) \equiv \int_{-1}^{1} (t + A_{1,0}) \, dt = 0, \tag{B.1.9}$$

which requires that $A_{1,0} = 0$, yielding the first-degree Legendre polynomial, $p_1(t) = t$.

Roots of orthogonal polynomials

An mth-degree orthogonal polynomial has m real distinct roots that lie within the domain of definition, $[c, d]$. Thus, the graph of the polynomial, $p_m(t)$, does not cross the t axis outside the interval $[c, d]$.

The roots of an mth-degree orthogonal polynomial interleave those of the $(m\text{-}1)$-degree polynomial in the same family. That is, there is exactly one root of the former polynomial between any two consecutive roots of the latter.

Discrete orthogonality

For every orthogonal polynomial, $p_l(t)$, we can find a set of l points t_i, where $i = 1, 2, \ldots l$, so that the following discrete orthogonality holds,

$$\sum_{i=1}^{l} p_n(t_i) \, p_m(t_i) = 0 \quad \text{if} \quad n \neq m, \tag{B.1.10}$$

where $n < l$ and $m < l$.

Gram polynomials

We can allow the weighting function $w(t)$ to be a sum of Dirac's one-dimensional delta functions, δ, centered at $N+1$ points $t_l, l = 1, 2, \ldots N+1$, distributed in some fashion over the interval $[c, d]$. That is, we can set

$$w(t) = \sum_{l=1}^{N+1} w_l \, \delta(t - t_l), \tag{B.1.11}$$

where w_l are weighting coefficients. The orthogonality condition (B.1.2) then becomes

$$(p_i, p_j) \equiv \sum_{l=1}^{N+1} p_i(t_l) \, p_j(t_l) \, w_l = \begin{cases} \mathcal{G}_i & \text{if } i = j \\ 0 & \text{if } i \neq j \end{cases}. \tag{B.1.12}$$

It can be shown that the corresponding family contains only $N + 1$ polynomials, $p_0(t), p_1(t), \ldots, p_N(t)$, while higher-order polynomials are equal to zero. An example is the family of Gram polynomials.

Recursion relation

Any family of orthogonal polynomials satisfies the recursion relation

$$
\begin{aligned}
&p_0(t) = A_0, \\
&p_1(t) = \alpha_0 \, (t - \beta_0) \, p_0(t), \\
&\ldots \\
&p_{i+1}(t) = \alpha_i \, (t - \beta_i) \, p_i(t) - \gamma_i \, p_{i-1}(t),
\end{aligned}
\tag{B.1.13}
$$

$$\ldots$$

for $i = 1, 2, \ldots$. where:

- The coefficients α_i are given by

$$
\alpha_i = \frac{A_{i+1}}{A_i},
\tag{B.1.14}
$$

  for $i = 0, 1, \ldots$, and the leading-term coefficients, A_i, defined in (B.1.1), are arbitrarily specified.

- The coefficients β_i and γ_i are given by

$$
\beta_i = \frac{1}{\mathcal{G}_i} \, (p_i, t \, p_i),
\tag{B.1.15}
$$

  for $i = 0, 1, \ldots$, and

$$
\gamma_i = \frac{\alpha_i}{\mathcal{G}_{i-1}} \, (p_i, t \, p_{i-1}) = \frac{\alpha_i \, \mathcal{G}_i}{\alpha_{i-1} \, \mathcal{G}_{i-1}},
\tag{B.1.16}
$$

  for $i = 1, 2, \ldots$, where the constants, $\mathcal{G}_i$, and the projection operator represented by the parentheses are defined in equation (B.1.2). If the graph of $w(t)$ is symmetric about the point $t = \beta$, then $\beta_i = \beta$ for all i.

As an example, we choose the flat weighting function, $w(t) = 1$, set $c = -1$, and $d = 1$, specify $p_0(t) = A_0 = 1$, whereupon $\mathcal{G}_0 = 2$, and stipulate $A_1 = 1$. Applying (B.1.14) and (B.1.15), we find $\alpha_0 = A_1/A_0 = 1$, and $\beta_0 = \frac{1}{2} \, (1, t) = \frac{1}{2} \int_{-1}^{1} t \, dt = 0$, yielding $p_1(t) = t$, which is the first-degree Legendre polynomial discussed later in this appendix.

To derive the aforementioned recursion relations, we note that the right-hand side of the last equation shown in (B.1.13) is an $(i+1)$-degree polynomial in t. Next, we introduce the expansion

$$
\begin{aligned}
\alpha_i \, (t - \beta_i) \, p_i(t) &- \gamma_i \, p_{i-1}(t) \\
&= c_0 \, p_0(t) + c_1 \, p_1(t) + \ldots + c_{i+1} \, p_{i+1}(t),
\end{aligned}
\tag{B.1.17}
$$

and require that $c_{i+1} = 1$, while all other coefficients, $c_i, c_{i-1}, \ldots, c_0$, are zero. Using (B.1.7), and exploiting the orthogonality property, we find

$$c_{i+1} = \frac{1}{\mathcal{G}_{i+1}} \int_c^d \left(\alpha_i \, (t - \beta_i) \, p_i(t) - \gamma_i \, p_{i-1}(t) \right) p_{i+1}(t) \, w(x) \, dt$$

$$= \frac{\alpha_i}{\mathcal{G}_{i+1}} \int_c^d t \, p_i(t) \, p_{i+1}(t) \, w(x) \, dt \equiv \frac{\alpha_i}{\mathcal{G}_{i+1}} \, (t \, p_i, p_{i+1}). \quad \text{(B.1.18)}$$

Setting $c_{i+1} = 1$, we obtain

$$(t \, p_i, p_{i+1}) = \frac{\mathcal{G}_{i+1}}{\alpha_i}. \quad \text{(B.1.19)}$$

Working in a similar fashion, we find

$$c_i = \frac{1}{\mathcal{G}_i} \int_c^d \left(\alpha_i \, (t - \beta_i) \, p_i(t) - \gamma_i \, p_{i-1}(t) \right) p_i(t) \, w(x) \, dt$$

$$= \frac{\alpha_i}{\mathcal{G}_i} \left(\int_c^d t \, p_i^2(t) \, dt - \beta_i \mathcal{G}_i \right) \equiv \frac{\alpha_i}{\mathcal{G}_i} \left((p_i, t \, p_i) - \beta_i \, \mathcal{G}_i \right). \quad \text{(B.1.20)}$$

Setting $c_i = 0$, we obtain the first relation in (B.1.15). The next coefficient is given by

$$c_{i-1} = \frac{1}{\mathcal{G}_{i-1}} \int_c^d \left(\alpha_i \, (t - \beta_i) \, p_i(t) - \gamma_i \, p_{i-1}(t) \right) p_{i-1}(t) \, w(x) \, dt$$

$$= \frac{1}{\mathcal{G}_{i-1}} \left(\alpha_i \, (p_i, t \, p_{i-1}) - \gamma_i \, \mathcal{G}_{i-1} \right). \quad \text{(B.1.21)}$$

Setting $c_{i-1} = 0$, and using (B.1.19), we derive the second relation in (B.1.15). Continuing in this manner, we find that the remaining coefficients on the right-hand side of (B.1.17), $c_{i-2}, c_{i-3}, \ldots, c_0$ are zero, as required.

Evaluation as the determinant of a tridiagonal matrix.

The recursion relation (B.1.13) suggests that the polynomial $p_{n+1}(t)$ can be evaluated in terms of the determinant of an $(n+1) \times (n+1)$ tridiagonal matrix, denoted by $\mathbf{T}$, as

$$p_{n+1}(t) = A_0 \, \text{Det}[\mathbf{T}(t)]. \quad \text{(B.1.22)}$$

The components of this matrix satisfy the relations

$$T_{i,i} = \alpha_{i-1} \, (t - \beta_{i-1}), \quad \text{(B.1.23)}$$

for $i = 1, 2, \ldots, n+1$, and

$$T_{i,i-1} \, T_{i-1,i} = \gamma_{i-1}, \quad \text{(B.1.24)}$$

$$d_N = c_N$$
$$d_{N-1} = (t - b_{N-1}) \, d_N + c_{N-1}$$

Do $k = N - 2, N - 1, \ldots, 0$
$$d_k = \alpha_k \, (t - \beta_k) \, d_{k+1} - \gamma_{k+1} \, d_{k+2} + c_{N-1}$$
End Do

$$Q_N(t) = A_0 \, d_0$$

Algorithm B.1.1 Clenshaw's algorithm for evaluating an expansion of orthogonal polynomials.

for $i = 2, 3, \ldots, n+1$. Setting for convenience the super-diagonal elements equal to unity, $T_{i-1,i} = 1$, we obtain the matrix

$$\mathbf{T} \equiv \begin{bmatrix} \alpha_0 \, (t - \beta_0) & 1 & 0 & 0 & \cdots \\ \gamma_1 & \alpha_1 \, (t - \beta_1) & 1 & 0 & \\ 0 & \gamma_2 & \alpha_2 \, (t - \beta_2) & 1 & \cdots \\ \cdots & \cdots & \cdots & \cdots & \cdots \\ 0 & 0 & \cdots & \cdots & \cdots \\ 0 & 0 & \cdots & \cdots & \cdots \\ 0 & 0 & \cdots & \cdots & \cdots \end{bmatrix} \tag{B.1.25}$$

$$\begin{bmatrix} \cdots & 0 & 0 & 0 & 0 & 0 \\ \cdots & 0 & 0 & 0 & 0 & 0 \\ \cdots & 0 & 0 & 0 & 0 & 0 \\ \cdots & \cdots & \cdots & \cdots & \cdots & \cdots \\ \cdots & 0 & \gamma_{n-2} & \alpha_{n-2} \, (t - \beta_{n-2}) & 1 & 0 \\ \cdots & 0 & 0 & \gamma_{n-1} & \alpha_{n-1} \, (t - \beta_{n-1}) & 1 \\ \cdots & 0 & 0 & 0 & \gamma_n & \alpha_n \, (t - \beta_n) \end{bmatrix}.$$

Clenshaw's algorithm

The fastest way of evaluating the polynomial expansion

$$Q_N(t) = \sum_{i=0}^{N} c_i \, p_i(t), \tag{B.1.26}$$

where c_i are given coefficients, is by Clenshaw's Algorithm B.1.1. The coefficients α_i, β_i and γ_i are defined in equations (B.1.13).

$$\mathbf{u}_1 = \mathbf{v}_1$$

Do $j = 2, 3, \ldots, N$

   Do $i = 1, 2, \ldots, j - 1$

   $$\alpha_{i,j} = \frac{\mathbf{u}_i \cdot \mathbf{v}_j}{\mathbf{u}_i \cdot \mathbf{u}_i}$$

   End Do

   $$\mathbf{u}_j = \mathbf{v}_j - \sum_{i=1}^{j-1} \alpha_{i,j}\, \mathbf{u}_i$$

End Do

Algorithm B.1.2 Gram-Schmidt orthogonalization of a set of linearly independent vectors, $\mathbf{v}_i$, producing a set of mutually orthogonal vectors, $\mathbf{u}_i$.

Gram-Schmidt orthogonalization

The coefficients of a chosen family of orthogonal polynomials can be computed using the Gram-Schmidt orthogonalization process, best known for generating a family of N mutually orthogonal vectors, $\mathbf{u}_i$, from an arbitrary set of N linearly independent vectors, $\mathbf{v}_i$, for $i = 1, 2, \ldots, N$. The orthogonal set of vectors is constructed working as follows:

1. Begin by setting $\mathbf{u}_1 = \mathbf{v}_1$.

2. Require that $\mathbf{u}_2$ lies in the plane of $\mathbf{u}_1$ and $\mathbf{v}_2$, and is orthogonal to $\mathbf{u}_1$.

3. Require that $\mathbf{u}_3$ lies in the space of $\mathbf{u}_1$, $\mathbf{u}_2$, and $\mathbf{v}_3$, and is orthogonal to $\mathbf{u}_1$ and $\mathbf{u}_2$.

4. Continue in this manner until the desired set is complete, following the steps of Algorithm B.1.2.

To generate orthogonal polynomials, we introduce the family of monomials,

$$q_0 = 1, \quad q_1 = x, \quad q_2 = x^2, \quad q_3 = x^3, \quad \ldots, \tag{B.1.27}$$

playing the role of the linearly independent vectors, $\mathbf{v}_i$, and then construct the orthogonal polynomials as

$$p_0 = q_0,$$
$$p_1 = q_1 - \alpha_{1,0}\, p_0,$$

$$p_2 = q_2 - \alpha_{2,1}\, p_1 - \alpha_{2,0}\, p_0,$$

$$\ldots,\tag{B.1.28}$$

where the coefficients, $\alpha_{n,m}$, are such that the orthogonality condition (B.1.2) is fulfilled. This can be done by applying Algorithm B.1.2, where the inner vector product designated by the centered dot is replaced by the projection integral defined in (B.1.2). The polynomial coefficients, $A_{i,j}$, defined in (B.1.1) are then computed as

$$A_i = 1, \qquad A_{i,j} = \sum_{l=0}^{i-1} \alpha_{i,l}\, A_{l,j},\tag{B.1.29}$$

for $i > 0$ and $j = 0, 1, \ldots, i-1$, and may be subsequently reduced to observe a desired normalization.

Orthonormal polynomials

The members of a family of orthogonal polynomials can be normalized to yield a corresponding family of orthonormal polynomials, designated by a hat,

$$\hat{p}_i(t) \equiv \frac{1}{\sqrt{\mathcal{G}_i}}\, p_i(t),\tag{B.1.30}$$

satisfying the orthonormality condition

$$(\hat{p}_i, \hat{p}_j) \equiv \int_c^d \hat{p}_i(t)\, \hat{p}_j(t)\, w(t)\, dt = \begin{cases} 1 & \text{if } i = j \\ 0 & \text{if } i \neq j \end{cases}.\tag{B.1.31}$$

By analogy with (B.1.1), we express the nth-degree orthonormal polynomial in the form

$$\hat{p}_n(t) = \hat{A}_n\, t^n + \hat{A}_{n,n-1}\, t^{n-1} + \hat{A}_{n,n-2}\, t^{n-2} + \ldots + \hat{A}_{n,0},\tag{B.1.32}$$

where $\hat{A}_{i,j}$ are polynomial coefficients; to simplify the notation, we have denoted $\hat{A}_n \equiv \hat{A}_{n,n}$. By definition then,

$$\hat{A}_n = \frac{A_n}{\sqrt{\mathcal{G}_n}}.\tag{B.1.33}$$

Christoffel-Darboux formula

Let us introduce a new independent variable, τ, and multiply the last recursion relation in (B.1.13) by $p_i(\tau)$ to obtain

$$p_{i+1}(t)\, p_i(\tau) = \alpha_i\, (t - \beta_i)\, p_i(t)\, p_i(\tau) - \gamma_i\, p_{i-1}(t)\, p_i(\tau).\tag{B.1.34}$$

Interchanging the roles of t and τ, we find

$$p_{i+1}(\tau)\, p_i(t) = \alpha_i\, (\tau - \beta_i)\, p_i(\tau)\, p_i(t) - \gamma_i\, p_{i-1}(\tau)\, p_i(t).\tag{B.1.35}$$

Forming the difference between these equations and rearranging, we obtain

$$p_{i+1}(t)\, p_i(\tau) - p_{i+1}(\tau)\, p_i(t) = \alpha_i\, (t - \tau)\, p_i(t)\, p_i(\tau)$$
$$+ \gamma_i \left(p_i(t)\, p_{i-1}(\tau) - p_i(\tau)\, p_{i-1}(t) \right), \qquad \text{(B.1.36)}$$

which can be recast into the compact form

$$\alpha_i\, (t - \tau)\, p_i(t)\, p_i(\tau) = \Pi_i(t, \tau) - \gamma_i\, \Pi_{i-1}(t, \tau), \qquad \text{(B.1.37)}$$

where

$$\Pi_i(t, \tau) \equiv p_{i+1}(t)\, p_i(\tau) - p_{i+1}(\tau)\, p_i(t). \qquad \text{(B.1.38)}$$

Now, substituting for γ_i the second expression in (B.1.15) and rearranging, we obtain

$$\frac{1}{\mathcal{G}_i}\, (t - \tau)\, p_i(t)\, p_i(\tau) = \frac{\Pi_i(t, \tau)}{\alpha_i \mathcal{G}_i} - \frac{\Pi_{i-1}(t, \tau)}{\alpha_{i-1} \mathcal{G}_{i-1}}. \qquad \text{(B.1.39)}$$

Applying this relation for $i = 0, 1, \ldots, n$, with the understanding that $\Pi_{-1} = 0$, adding the equations thus obtained, simplifying, and recalling that $\alpha_n \equiv A_{n+1}/A_n$, we derive the Christoffel-Darboux formula,

$$(t - \tau) \sum_{i=0}^{n} \frac{1}{\mathcal{G}_i}\, p_i(t)\, p_i(\tau) = \frac{A_n}{A_{n+1}\mathcal{G}_n} \Pi_n(t, \tau). \qquad \text{(B.1.40)}$$

Replacing the orthogonal polynomials and their leading-order coefficients with their orthonormal counterparts defined in (B.1.30) and (B.1.33), we derive the simpler expression

$$(t - \tau) \sum_{i=0}^{n} \hat{p}_i(t)\, \hat{p}_i(\tau) = \frac{\hat{A}_n}{\hat{A}_{n+1}}\, \hat{\Pi}_n(t, \tau), \qquad \text{(B.1.41)}$$

where

$$\hat{\Pi}_i(t, \tau) \equiv \hat{p}_{i+1}(t)\, \hat{p}_i(\tau) - \hat{p}_{i+1}(\tau)\, \hat{p}_i(t). \qquad \text{(B.1.42)}$$

Summary of orthogonal families

Table B.1.1 summarizes the defining and distinguishing properties of the most common families of orthogonal polynomials. Additional properties of the Legendre, Lobatto, Chebyshev, and Jacobi polynomials are discussed in Sections B.5–8.

Family	Symbol	$[c, d]$	$w(t)$	Standard Normalization	$\mathcal{G}_i$
Chebyshev	$T_i(t)$	[-1, 1]	$\frac{1}{\sqrt{1-t^2}}$	$T_i(1) = 1$	†
Chebyshev second kind	$\mathcal{T}_i(t)$	[-1, 1]	$\sqrt{1-t^2}$	$\mathcal{T}_i(1) = i + 1$	$\frac{\pi}{2}$
Hermite	$H_i(t)$	$(-\infty, \infty)$	$\exp(-t^2)$	$A_i = 2^i$	$\sqrt{\pi}\, 2^i\, i!$
Jacobi ††	$J_i^{(\alpha,\beta)}(t)$ $\alpha, \beta > -1$	[-1, 1]	$(1-t)^\alpha$ $\times (1+t)^\beta$	†††	††††
Laguerre	$\mathcal{L}_i(t)$	$[0, \infty)$	$\exp(-t)$	$A_i = (-1)^i$	$(i!)^2$
Legendre	$L_i(t)$	[-1, 1]	1	$L_i(1) = 1$	$\frac{2}{2i+1}$
Lobatto	$Lo_i(t)$	[-1, 1]	$1 - t^2$	$Lo_i(t) = L'_{i+1}$	$\frac{2(i+1)(i+2)}{2i+3}$
Radau	$R_i(t)$	[-1, 1]	$1 + t$	$R_i(1) = 1$	$\frac{2}{i+1}$

† $\mathcal{G}_0 = \pi$, and $\mathcal{G}_i = \frac{\pi}{2}, i = 1, 2, \ldots$

††   The Legendre polynomials arise from the Jacobi polynomials for $\alpha = 0$ and $\beta = 0$, $L_i(t) = J_i^{(0,0)}(t)$. The Lobatto polynomials arise from the Jacobi polynomials for $\alpha = 1$ and $\beta = 1$, as shown in (B.6.5). The Chebyshev polynomials arise from the Jacobi polynomials for $\alpha = -1/2$ and $\beta = -1/2$, as shown in (B.7.17).

††† $J_i^{\alpha,\beta}(1) = \frac{\Gamma(i+1+\alpha)}{\Gamma(i+1)\Gamma(\alpha+1)}$

†††† $\mathcal{G}_i = \frac{2^{\alpha+\beta+1}}{i!} \frac{\Gamma(i+1+\alpha)\,\Gamma(i+1+\beta)}{(2i+1+\alpha+\beta)\,\Gamma(i+1+\alpha+\beta)},$

Table B.1.1 Distinguishing properties of common families of orthogonal polynomials. In the expressions for the Jacobi polynomials, Γ is the Gamma function; if m is a positive integer, $\Gamma(m+1) = m!$ (e.g., [1]).

B.2   Gaussian integration quadratures

Consider the weighted definite integral of a $(2N+1)$-degree polynomial, $Q_{2N+1}(t)$, between the two integration limits, c and d,

$$\int_c^d Q_{2N+1}(t)\, w(t)\, \mathrm{d}t, \tag{B.2.1}$$

where $w(t)$ is a specified weighting function. Without loss of generality, we can select a group of $N+1$ interpolation nodes, $t_1, t_2, \ldots, t_{N+1}$, and express $Q_{2N+1}(t)$ in the *exact* form

$$Q_{2N+1}(t) = P_N(t) + \Xi_N(t)\, (t - t_1)(t - t_2) \ldots (t - t_{N+1}), \tag{B.2.2}$$

where $P_N(t)$ is the Nth-degree interpolating polynomial discussed in Appendix A, and $\Xi_N(t)$ is an Nth-degree polynomial. Note that the second term on the right-hand side of (B.2.2) is a $(2N+1)$-degree polynomial.

The integration error incurred by replacing Q_{2N+1} with the lower-degree interpolating polynomial, P_N, is

$$E = \int_c^d \Xi_N(t)\, (t - t_1)(t - t_2) \ldots (t - t_{N+1})\, w(t)\, \mathrm{d}t. \tag{B.2.3}$$

This error will be zero, provided that the $N+1$ nodes, t_i, are identified with the zeros of the $(N+1)$-degree polynomial, $p_{N+1}(t)$, that belongs to the family of orthogonal polynomials defined over the interval $[c, d]$, with weighting function $w(t)$. To prove this assertion, we observe that the product of the $N+1$ monomials, $(t - t_i)$, is proportional to $p_{N+1}(t)$, and invoke the orthogonality of $p_{N+1}(t)$ against lower-degree polynomials, as discussed in Section B.1. Consequently, the definite integral of the $(2N+1)$-degree polynomial, $Q_{2N+1}(t)$, is exactly equal to the definite integral of the Nth-degree interpolating polynomial, $P_N(t)$.

To formalize the numerical method, we consider the weighted definite integral of a non-singular function, $f(t)$,

$$\int_c^d f(t)\, w(t)\, \mathrm{d}t, \tag{B.2.4}$$

and express the Nth-degree interpolating polynomial of $f(t)$ in terms of the Lagrange interpolating polynomials, as

$$P_N(t) = \sum_{i=1}^{N+1} f(t_i)\, l_{N,i}(t), \tag{B.2.5}$$

where

$$l_{N,i}(t) = \frac{(t - t_1)(t - t_2) \ldots (t - t_{i-1})(t - t_{i+1}) \ldots (t - t_{N+1})}{(t_i - t_1)(t_i - t_2) \ldots (t_i - t_{i-1})(t_i - t_{i+1}) \ldots (t_i - t_{N+1})}. \tag{B.2.6}$$

Integrating, we derive the Gaussian quadrature formula

$$\int_c^d f(t)\, w(t)\, dt \simeq \int_c^d P_N(t)\, w(t)\, dt = \sum_{i=1}^{N+1} f(t_i)\, w_i, \tag{B.2.7}$$

where

$$w_i = \int_c^d l_{N,i}(t)\, w(t)\, dt \tag{B.2.8}$$

are the integration weights. The quadrature is exact if $f(t)$ is a $(2N+1)$-degree polynomial.

Evaluation of the integration weights

Using the representation of the Lagrange interpolation functions in terms of the generating function, as shown in (A.2.10), we write

$$l_{N,i}(t) = \frac{p_{N+1}(t)}{(t - t_i)\, p'_{N+1}(t_i)}, \tag{B.2.9}$$

which shows that the integration weights are given by

$$w_i = \frac{1}{p'_{N+1}(t_i)} \int_c^d \frac{p_{N+1}(t)}{t - t_i}\, w(t)\, dt, \tag{B.2.10}$$

for $i = 1, 2, \ldots, N+1$. To compute the integral on the right-hand side, we apply the Christoffel-Darboux formula (B.1.40) for $n = N + 1$ and $\tau = t_i$, and find

$$(t - t_i) \sum_{j=0}^{N} \frac{1}{\mathcal{G}_j}\, p_j(t)\, p_j(t_i) = \frac{A_{N+1}}{A_{N+2}\, \mathcal{G}_{N+1}} \Pi_{N+1}(t, t_i). \tag{B.2.11}$$

Observing that $p_{N+1}(t_i) = 0$, and thus

$$\Pi_{N+1}(t, t_i) \equiv p_{N+2}(t)\, p_{N+1}(t_i) - p_{N+2}(t_i)\, p_{N+1}(t)$$
$$= -p_{N+2}(t_i)\, p_{N+1}(t), \tag{B.2.12}$$

we find

$$\sum_{j=0}^{N} \frac{1}{\mathcal{G}_j}\, p_j(t)\, p_j(t_i) = -\frac{A_{N+1}\, p_{N+2}(t_i)}{A_{N+2}\, \mathcal{G}_{N+1}} \frac{p_{N+1}(t)}{t - t_i}. \tag{B.2.13}$$

Next, we multiply both sides of this equation by the weighting function, $w(t)$, and integrate over the domain of definition of the orthogonal polynomials to obtain

$$\int_c^d \frac{p_{N+1}(t)}{t - t_i}\, dt \tag{B.2.14}$$

$$= -\frac{A_{N+2}\, \mathcal{G}_{N+1}}{A_{N+1}\, p_{N+2}(t_i)} \sum_{j=0}^{N} \frac{1}{\mathcal{G}_j}\, p_j(t_i) \int_c^d p_j(t)\, w(t)\, dt.$$

Because of orthogonality, only the first term in the sum corresponding $j = 0$ makes a non-zero contribution, which is equal to unity. Thus,

$$\int_c^d \frac{p_{N+1}(t)}{t - t_i} \, dt = -\frac{A_{N+2} \, \mathcal{G}_{N+1}}{A_{N+1} \, p_{N+2}(t_i)}, \tag{B.2.15}$$

and

$$w_i = -\frac{A_{N+2} \, \mathcal{G}_{N+1}}{A_{N+1}} \frac{1}{p'_{N+1}(t_i) \, p_{N+2}(t_i)}, \tag{B.2.16}$$

for $i = 1, 2, \ldots, N + 1$. In terms of the associated orthonormal polynomials,

$$w_i = -\frac{\hat{A}_{N+2}}{\hat{A}_{N+1}} \frac{1}{\hat{p}'_{N+1}(t_i) \, \hat{p}_{N+2}(t_i)}. \tag{B.2.17}$$

Standard Gaussian quadratures

Different Gaussian integration quadratures arise from different families of orthogonal polynomials corresponding to different weighting functions and integration intervals, including the following:

Gauss-Legendre quadrature:

This quadrature applies to the integration interval $[-1, 1]$, with a flat weighting function, $w(t) = 1$, yielding

$$\int_{-1}^1 f(t) \, dt \simeq \sum_{i=1}^{N+1} f(t_i) \, w_i. \tag{B.2.18}$$

The zeros of the $(N+1)$-degree Legendre polynomial, t_i, and corresponding weights, w_i, are listed in mathematical handbooks [1, 43] (see also Section B.5).

Gauss-Chebyshev quadrature:

This quadrature applies to the integration interval $[-1, 1]$, with a singular but integrable weighting function, $w(t) = 1/\sqrt{1 - t^2}$, yielding

$$\int_{-1}^1 f(t) \, \frac{1}{\sqrt{1 - t^2}} \, dt = \sum_{i=1}^{N+1} f(t_i) \, w_i. \tag{B.2.19}$$

The zeros of the $(N + 1)$-degree Chebyshev polynomial, t_i, are given by

$$t_i = \cos\left[\frac{\pi}{N+1} \left(i - \frac{1}{2}\right)\right], \tag{B.2.20}$$

for $i = 1, 2, \ldots, N + 1$, and the corresponding weights are given by

$$w_i = \frac{\pi}{N + 1} \tag{B.2.21}$$

(see also Section B.7).

B.3   Lobatto integration quadrature

Consider the following definite integral of the $(2N - 1)$-degree polynomial, $Q_{2N-1}(t)$,

$$\int_{-1}^{1} Q_{2N-1}(t)\, dt. \tag{B.3.1}$$

To evaluate this integral, we select a group of $N + 1$ interpolation nodes, $t_1, t_2, \ldots, t_{N+1}$, stipulate that $t_1 = -1$ and $t_{N+1} = 1$, and express $Q_{2N-2}(t)$ in the *exact* form

$$Q_{2N-1}(t) = P_N(t) + \Xi_{N-2}(t)\,(1 - t^2)\,(t - t_2)\ldots(t - t_N), \tag{B.3.2}$$

where $P_N(t)$ is the Nth-degree interpolating polynomial discussed in Appendix A, and $\Xi_{N-2}(t)$ is an $(N - 2)$-degree polynomial. Note that the product of the $(N - 2)$-degree polynomial and the $(N + 1)$-degree polynomial in the second term on the right-hand side of (B.3.2) produces a $(2N - 1)$-degree polynomial. The stipulations $t_1 = -1$ and $t_{N+1} = 1$ distinguish the Lobatto integration quadrature presently considered from the Gauss-Lobatto integration quadrature falling under the auspices of the standard Gaussian quadrature, as discussed in Section B.2.

The integration error incurred by replacing $Q_{2N-1}(t)$ with the interpolating polynomial, $P_N(t)$, is

$$E = \int_{c}^{d} \Xi_{N-2}(t)\,(1 - t^2)(t - t_2)\ldots(t - t_N)\, dt. \tag{B.3.3}$$

Following the discussion of Section B.2, we conclude that this error will be zero, provided that the $N - 1$ nodes, t_i, $i = 2, 3, \ldots, N$ are identified with the zeros of the $(N - 1)$-degree Lobatto polynomial, $Lo_{N-1}(t)$.

To formalize the numerical method, we consider the definite integral of a non-singular function, $f(t)$,

$$\int_{-1}^{1} f(t)\, dt, \tag{B.3.4}$$

and express the Nth-degree interpolating polynomial of $f(t)$ as shown in (B.2.5), where the Lagrange interpolating polynomials are given in (B.2.6), and the abscissas, t_i, are the zeros of the $(N + 1)$-degree completed Lobatto polynomial,

$$Lo_{N+1}^c(t) \equiv (1 - t^2)\, Lo_{N-1}(t). \tag{B.3.5}$$

Integrating, we derive the Lobatto quadrature formula

$$\int_{-1}^{1} f(t)\, dt \simeq \int_{-1}^{1} P_N(t)\, dt = \sum_{i=1}^{N+1} f(t_i)\, w_i, \tag{B.3.6}$$

where

$$w_i = \int_{-1}^{1} l_{N,i}(t)\, dt \qquad (B.3.7)$$

are the integration weights. The quadrature is exact if $f(t)$ is a $(2N-1)$-degree polynomial.

To compute the integration weights, we express the Lagrange polynomials as shown in (2.3.4),

$$l_{N,i}(t) = \frac{1}{N\,(N+1)\,L_N(t_i)} \frac{(t^2 - 1)\,Lo_{N-1}(t)}{t - t_i}, \qquad (B.3.8)$$

where $L_N(t)$ is a Legendre polynomial. The first integration weight can be computed as

$$w_1 = \int_{-1}^{1} l_{N,1}(t)\, dt = \frac{1}{N\,(N+1)\,L_N(-1)} \int_{-1}^{1} (t-1)\,Lo_{N-1}(t)\, dt$$

$$= \frac{1}{N\,(N+1)\,L_N(-1)} \int_{-1}^{1} (t-1)\,L_N'(t)\, dt \qquad (B.3.9)$$

$$= \frac{1}{N\,(N+1)\,L_N(-1)} \left\{ \left[(t-1)\,L_N(t)\right]_{-1}^{1} - \int_{-1}^{1} L_N(t)\, dt \right\} = \frac{2}{N(N+1)}.$$

The last integration weight, w_{N+1} can be computed in a similar fashion, and the intermediate integration weights can be computed using the Christoffel-Darboux formula, as discussed in Section B.2. In summary, the integration weights are given by

$$w_1 = w_{N+1} = \frac{2}{N(N+1)},$$

$$\qquad (B.3.10)$$

$$w_i = \frac{2}{N(N+1)} \frac{1}{L_N^2(t_i)}, \quad \text{for} \quad i = 2, 3, \ldots, N,$$

where L_N is a Legendre polynomial. Numerical values for the base points, t_i, and weights are given in Table 2.2.2.

B.4 Chebyshev integration quadrature

Consider the following *weighted* definite integral of the $(2N-1)$-degree polynomial, $Q_{2N-1}(t)$,

$$\int_{-1}^{1} Q_{2N-1}(t)\, \frac{1}{\sqrt{1-t^2}}\, dt. \qquad (B.4.1)$$

To evaluate this integral, we select a set of $N + 1$ interpolation nodes, t_1, t_2, ..., t_{N+1}, stipulate that $t_1 = -1$ and $t_{N+1} = 1$, and express $Q_{2N-1}(t)$ in the exact form

$$Q_{2N-1}(t) = P_N(t) + \Xi_{N-2}(t)\,(1 - t^2)\,(t - t_2)\ldots(t - t_N), \qquad (\text{B.4.2})$$

where $P_N(t)$ is the Nth-degree interpolating polynomial discussed in Appendix A, and $\Xi_{N-2}(t)$ is an $(N - 2)$-degree polynomial. The stipulations $t_1 = -1$ and $t_{N+1} = 1$ distinguish the Chebyshev integration quadrature presently considered from the Gauss-Chebyshev integration quadrature discussed in Section B.2.

By definition, the integration error incurred by replacing $Q_{2N-1}(t)$ with $P_N(t)$ is

$$E = \int_{-1}^{1} \Xi_{N-2}(t)\,(1 - t^2)(t - t_2)\ldots(t - t_N)\,\frac{1}{\sqrt{1 - t^2}}\,dt. \qquad (\text{B.4.3})$$

We shall show that, if the $N - 1$ unspecified nodes, $t_2, \ldots, t_N$, are identified with the zeros of the $(N - 1)$-degree Chebyshev polynomial of the second kind, $\mathcal{T}_{N-1}(t)$, then the integration error is precisely equal to zero. To prove this assertion, we restate the error in the form

$$E = \frac{1}{2^N} \int_{-1}^{1} \Xi_{N-2}(t)\,(1 - t^2)\,\mathcal{T}_{N-1}(t)\,\frac{1}{\sqrt{1 - t^2}}\,dt$$

$$= \frac{1}{2^N} \int_{-1}^{1} \Xi_{N-2}(t)\,\mathcal{T}_{N-1}(t)\,\sqrt{1 - t^2}\,dt, \qquad (\text{B.4.4})$$

which is zero in light of the orthogonality of the $(N - 1)$-degree Chebyshev polynomial of the second kind against any lower-degree polynomial, with weighting function $w(t) = \sqrt{1 - t^2}$.

The emerging Chebyshev integration quadrature takes the form

$$\int_{-1}^{1} f(t)\,\frac{1}{\sqrt{1 - t^2}}\,dt \simeq \sum_{i=1}^{N+1} f(t_i)\,w_i, \qquad (\text{B.4.5})$$

where $f(t)$ is a non-singular function, and

$$t_i = \cos\left[\frac{(i - 1)\pi}{N}\right], \qquad (\text{B.4.6})$$

$i = 1, 2, \ldots, N + 1$, are the zeros of the completed Chebyshev polynomial of the second kind,

$$\mathcal{T}_{N+1}^c(t) \equiv (1 - t^2)\,\mathcal{T}_{N-1}(t). \qquad (\text{B.4.7})$$

The integration weights are found to be

$$w_1 = w_{N+1} = \frac{\pi}{2N},$$

$$(\text{B.4.8})$$

$$w_i = \frac{\pi}{N},$$

for $i = 2, 3, \ldots, N$. The quadrature is exact if $f(t)$ is a $(2N - 1)$-degree polynomial.

B.5 Legendre polynomials

In this section, we summarize the salient properties of the Legendre polynomials.

- *Domain of definition:* [-1,1] .

- *Members:*

  The first few members are listed in Table 2.2.3.

- *Explicit formula (Rodrigues):*

$$L_i(t) = \frac{1}{2^i i!} \frac{d^i (t^2 - 1)^i}{dt^i} = \frac{(2i)!}{2^i (i!)^2} t^i + \ldots$$

$$= \frac{1}{2^i} \sum_{m=0}^{[i/2]} (-1)^m \binom{i}{m} \binom{2(i-m)}{i} t^{i-2m}, \qquad (B.5.1)$$

where $[i/2]$ denotes the integral part, and the tall parentheses denote the combinatorial:

$$\binom{i}{m} = \frac{i!}{m! \, (i-m)!}. \qquad (B.5.2)$$

- *Leading-power coefficient:*

$$A_i = \frac{(2i)!}{2^i (i!)^2}. \qquad (B.5.3)$$

- *Generating function:*

$$\frac{1}{\sqrt{1 - 2\, t\, \eta + \eta^2}} = \sum_{i=0}^{\infty} L_i(t) \, \eta^i, \qquad (B.5.4)$$

  for $-1 < t < 1$ and $|\eta| < 1$.

- *Standard normalization:*

$$L_i(1) = 1. \qquad (B.5.5)$$

- *Orthogonality:*

$$\int_{-1}^{1} L_i(t) \, L_j(t) \, dt = \frac{2}{2i + 1} \, \delta_{ij}, \qquad (B.5.6)$$

  where δ_{ij} is Kronecker's delta.

- *Recursion relation:*

$$L_{i+1}(t) = \frac{2i+1}{i+1} \, t \, L_i(t) - \frac{i}{i+1} \, L_{i-1}(t). \tag{B.5.7}$$

With reference to (B.1.13),

$$A_0 = 1,$$
$$\alpha_i = \frac{2i+1}{i+1} \quad \text{for } i = 0, 1, \ldots,$$
$$\beta_i = 0 \qquad \text{for } i = 0, 1, \ldots, \tag{B.5.8}$$
$$\gamma_i = \frac{i}{i+1}, \quad \text{for } i = 1, 2, \ldots.$$

- *Recursion relations for the first derivative:*

$$\begin{aligned}
(1 - t^2) \, L'_{i+1} &= (i+1) \, (-t L_{i+1} + L_i) \\
&= (i+2) \, (-L_{i+2} + t L_{i+1}) \\
&= \frac{(i+1)(i+2)}{2i+3} \, (L_i - L_{i+2}),
\end{aligned} \tag{B.5.9}$$

and

$$L'_{i+1} - L'_{i-1} = (2i+1) \, L_i. \tag{B.5.10}$$

- *Differential equations:*

$$\left[(t^2 - 1) \, L'_i(t) \right]' = i \, (i+1) \, L_i(t), \tag{B.5.11}$$

and

$$(1 - t^2) \, L''_i(t) - 2t \, L'_i(t) + i(i+1) L_i(t) = 0. \tag{B.5.12}$$

- *Zero-mean property:*

Applying the orthogonality property with $j = 0$, we find

$$\int_{-1}^{1} L_i(t) \, dt = 0, \quad \text{for} \quad i = 1, 2, \ldots. \tag{B.5.13}$$

- *Range of variation:*

$$|L_i(t)| \le 1, \quad -1 \le t \le 1. \tag{B.5.14}$$

- *Relation to the Jacobi polynomials:*

The Legendre polynomials are related to the Jacobi polynomials by

$$L_i(t) = J_i^{(0,0)}(t). \tag{B.5.15}$$

- *Gauss-Legendre integration weights:*

$$w_i = \frac{1}{L'_m(t_i)} \int_{-1}^{1} \frac{L_m(t)}{t - t_i} \, dt = \frac{2}{1 - t_i^2} \frac{1}{L'^2_m(t_i)}$$

$$= -\frac{2}{m+1} \frac{1}{L'_m(t_i) \, L_{m+1}(t_i)} = \frac{2}{m} \frac{1}{L'_m(t_i) \, L_{m-1}(t_i)}$$

$$= \frac{2}{(m+1)^2} \frac{1 - t_i^2}{L_{m+1}^2(t_i)}, \tag{B.5.16}$$

where t_i are the zeros of $L_m(t)$, for $i = 1, 2, \ldots, m$.

B.6 Lobatto polynomials

Since the Lobatto polynomials, Lo_i, are the derivatives of the Legendre polynomials, L_i,

$$Lo_i(t) \equiv L'_{i+1}(t), \tag{B.6.1}$$

many of their properties derive from those listed in Section B.5 for the Legendre polynomials. A summary and further properties include the following:

- *Domain of definition:* [-1,1] .

- *Members:*

  The first few members are listed in Table 2.2.1.

- *Orthogonality:*

$$\int_{-1}^{1} Lo_i(t) \, Lo_j(t) \, (1 - t^2) \, dt = \frac{2(i+1)(i+2)}{2i+3} \delta_{ij}, \tag{B.6.2}$$

  where δ_{ij} is Kronecker's delta.

- *Zero-mean property:*

  Applying the orthogonality property with $j = 0$, we find

$$\int_{-1}^{1} Lo_i(t) \, (1 - t^2) \, dt = 0, \quad \text{for} \quad i = 1, 2, \ldots. \tag{B.6.3}$$

- *Lobatto integration weights:*

$$w_i = \frac{2}{m(m+1)} \frac{1}{L_m^2(t_i)}, \tag{B.6.4}$$

  where t_i are the zeros of $Lo_{m-1}(t)$, for $i = 1, 2, \ldots, m - 1$.

- *Relation to the Jacobi polynomials:*

The Lobatto polynomials are related to the Jacobi polynomials by

$$Lo_i(t) = \frac{i+2}{2} J_i^{(1,1)}(t). \tag{B.6.5}$$

As an exercise, we demonstrate the orthogonality property (B.6.2) based on the properties of the Legendre polynomials. From the definition of the Lobatto polynomials, we write

$$(Lo_i, Lo_j) \equiv \int_{-1}^{1} Lo_i(t)\, Lo_j(t)\, (1 - t^2)\, \mathrm{d}t$$

$$= \int_{-1}^{1} L'_{i+1}(t)\, L'_{j+1}(t)\, (1 - t^2)\, \mathrm{d}t. \tag{B.6.6}$$

Integrating by parts, we find

$$(Lo_i, Lo_j) = \Big[L_{i+1}(t)\, L'_{j+1}(t)\, (1 - t^2) \Big]_{-1}^{1}$$

$$- \int_{-1}^{1} L_{i+1}(t)\, \Big(L'_{j+1}(t)\, (1 - t^2) \Big)'\, \mathrm{d}t. \tag{B.6.7}$$

The first term on the right-hand side is zero. Using (B.5.11), we find that the second term can be simplified to give

$$(Lo_i, Lo_j) = (i + 1)(i + 2) \int_{-1}^{1} L_{i+1}(t)\, L_{j+1}(t)\, \mathrm{d}t. \tag{B.6.8}$$

Finally, we use the orthogonality property (B.5.6) to obtain (B.6.2).

B.7 Chebyshev polynomials

In this section, we summarize the salient properties of the Chebyshev polynomials of the first kind, $T_i(t)$, simply called the Chebyshev polynomials.

- *Domain of definition:* [-1,1] .

- *Members:*

The first few members are:

$$T_0 = 1,$$
$$T_1 = t,$$
$$T_2 = 2\,t^2 - 1,$$
$$T_3 = 4\,t^3 - 3\,t,$$

$$T_4 = 8\,t^4 - 8\,t^2 + 1,$$
$$T_5 = 16\,t^5 - 20\,t^3 + 5\,t,$$
$$\ldots$$
$$T_i = 2^{i-1}\,t^i - i\,2^{i-3}\,t^{i-2} + \ldots . \tag{B.7.1}$$

- *Explicit formula (Rodrigues):*

$$T_i(t) = \cos(i\arccos t)$$

$$= \frac{\sqrt{\pi}}{2^{i+1}\,\Gamma(i+\tfrac{1}{2})}\,\sqrt{1-t^2}\,\frac{\mathrm{d}^i(t^2-1)^{i-\frac{1}{2}}}{\mathrm{d}t^i} \tag{B.7.2}$$

$$= \frac{i}{2}\sum_{m=0}^{[i/2]}(-1)^m\,\frac{(i-m-1)!}{m!\,(i-2m)!}\,(2\,t)^{i-2m}, \tag{B.7.3}$$

where $[i/2]$ denotes the integral part.

- *Generating function:*

$$\frac{1-t\eta}{1-2\,t\,\eta+\eta^2} = \sum_{i=0}^{\infty}T_i(t)\,\eta^i, \tag{B.7.4}$$

for $-1 < t < 1$ and $|\eta| < 1$.

- *Second explicit formula:*

$$T_i = \frac{1}{2}\left(z^i + \frac{1}{z^i}\right), \tag{B.7.5}$$

where z satisfies

$$z^2 - 2\,tz + 1 = 0. \tag{B.7.6}$$

- *Standard normalization:*

$$T_i(1) = 1. \tag{B.7.7}$$

- *Orthogonality:*

$$\int_{-1}^{1}T_i(t)\,T_j(t)\,\frac{1}{\sqrt{1-t^2}}\,\mathrm{d}t = \begin{cases} \pi & \text{if } i=j=0 \\ \frac{\pi}{2} & \text{if } i=j\neq 0 \\ 0 & \text{if } i\neq j \end{cases}. \tag{B.7.8}$$

- *Recursion relation:*

$$T_{i+1}(t) = 2\,t\,T_i(t) - T_{i-1}(t). \tag{B.7.9}$$

With reference to (B.1.13),

$$\begin{aligned}
A_0 &= 1, \\
\alpha_0 &= 1, \\
\alpha_i &= 2 \quad \text{for } i = 1, 2, \ldots, \\
\beta_i &= 0 \quad \text{for } i = 0, 1, \ldots, \\
\gamma_i &= 1 \quad \text{for } i = 1, 2, \ldots.
\end{aligned}$$
(B.7.10)

- *Recursion relations for the first derivative:*

$$(1 - t^2) \, T'_{i+1} = (i+1) \, (-t \, T_{i+1} + T_i),$$
(B.7.11)

and

$$\frac{T'_{i+1}}{i+1} - \frac{T'_{i-1}}{i-1} = 2 \, T_i.$$
(B.7.12)

- *Zeros:*

$T_i(t)$ has i zeros in the interval $[-1, 1]$, given by the Chebyshev abscissas:

$$t_k = \cos\left[\frac{\pi}{i} \left(k - \frac{1}{2}\right)\right]$$
(B.7.13)

for $k = 1, 2, \ldots, i$.

- *Discrete orthogonality:*

If x_k, for $k = 1, 2, \ldots, m+1$ are the zeros of $T_{m+1}(t)$, then

$$\sum_{k=1}^{m+1} T_i(x_k) \, T_j(x_k) = \begin{cases} m+1 & \text{if } i = j = 0 \\ \frac{m+1}{2} & \text{if } i = j \neq 0 \;, \\ 0 & \text{if } i \neq j \end{cases}$$
(B.7.14)

for $0 \leq i, j \leq m$.

- *Differential equation:*

$$(1 - t^2) \, T''_i(t) - t \, T'_i(t) + i^2 \, T_i(t) = 0.$$
(B.7.15)

- *Range of variation:*

$$|T_i(t)| \leq 1, \quad -1 \leq t \leq 1.$$
(B.7.16)

- *Minimax property:*

Of all nth-degree polynomials with leading-power coefficient equal to 1, the scaled Chebyshev polynomial $2^{1-n} T_n(t)$ has the smallest maximum norm in the interval $[-1, 1]$.

- *Relation to the Jacobi polynomials:*

  The Chebyshev polynomials are related to the Jacobi polynomials by

  $$T_i(t) = \frac{n!\sqrt{\pi}}{\Gamma(i + \frac{1}{2})} J_i^{(-1/2,-1/2)}(t). \tag{B.7.17}$$

B.8 Jacobi polynomials

- *Standard notation:*

  $J_i^{(\alpha,\beta)}$, where $\alpha, \beta > -1$.

- *Domain of definition:* [-1,1] .

- *Members:*

  The first few members are:

  $$J_0^{(\alpha,\beta)}(t) = 1,$$
  $$J_1^{(\alpha,\beta)}(t) = \frac{1}{2}(\alpha + \beta + 2)\, t + \frac{1}{2}(\alpha - \beta),$$

  $$J_2^{(\alpha,\beta)}(t) = \frac{1}{8}(\alpha + \beta + 3)(\alpha + \beta + 4)\, t^2$$
  $$+ \frac{1}{4}(\alpha - \beta)(\alpha + \beta + 3)\, t$$
  $$+ \frac{1}{8}\left[\,(\alpha - \beta)^2 - (\alpha + \beta + 4)\,\right],$$

  . . .

- *Explicit formula (Rodrigues):*

  $$J_i^{(\alpha,\beta)}(t) = \frac{(-1)^i}{2^i i!}(1 - t)^{-\alpha}(1 + t)^{-\beta}\frac{d^i[\,(1 - t)^{i+\alpha}(1 + t)^{i+\beta}\,]}{dt^i}. \tag{B.8.1}$$

- *Reciprocal relation:*

  $$J_i^{(\alpha,\beta)}(t) = (-1)^i\, J_i^{(\beta,\alpha)}(-t). \tag{B.8.2}$$

- *Standard normalization:*

  $$J_i^{(\alpha,\beta)}(1) = \frac{\Gamma(i + \alpha + 1)}{\Gamma(i + 1)\,\Gamma(\alpha + 1)} = \binom{i + \alpha}{i},$$

  $$\tag{B.8.3}$$

  $$J_i^{(\alpha,\beta)}(-1) = (-1)^i\,\frac{\Gamma(i + \beta + 1)}{\Gamma(i + 1)\,\Gamma(\beta + 1)} = (-1)^i\binom{i + \beta}{i}.$$

- *Orthogonality:*

$$\int_{-1}^{1} J_i^{(\alpha,\beta)}(t) \, J_j^{(\alpha,\beta)}(t) \, (1-t)^\alpha \, (1+t)^\beta \, dt$$

$$= \frac{2^{\alpha+\beta+1}}{2i+\alpha+\beta+1} \, \frac{\Gamma(i+\alpha+1)\,\Gamma(i+\beta+1)}{i!\,\Gamma(i+\alpha+\beta+1)} \, \delta_{ij},$$

$$(B.8.4)$$

where δ_{ij} is Kronecker's delta.

- *Recursion relation:*

With reference to (B.1.13),

$$A_0 = 1,$$

$$\alpha_i = \frac{(2i+\alpha+\beta+1)(2i+\alpha+\beta+2)}{2(i+1)(i+\alpha+\beta+1)}, \quad \text{for } i = 0, 1, \ldots,$$

$$\beta_i = \frac{\beta^2 - \alpha^2}{(2i+\alpha+\beta)(2i+\alpha+\beta+2)}, \quad \text{for } i = 0, 1, \ldots,$$

$$\gamma_i = \frac{(i+\alpha)(i+\beta)(2i+\alpha+\beta+2)}{(i+1)(i+\alpha+\beta+1)(2i+\alpha+\beta)}, \quad \text{for } i = 1, 2, \ldots \quad .$$

$$(B.8.5)$$

- *Differential equations:*

$$\left[(1-t)^{1+\alpha}(1+t)^{1+\beta} \, y'(t) \right]' = -i \, (i+\alpha+\beta+1) \, y(t),$$

$$(B.8.6)$$

$$(1-t^2) \, y''(t) - \left[\alpha - \beta + (\alpha+\beta+2) \, t \right] y'(t)$$
$$+i \, (i+\alpha+\beta+1) \, y(t) = 0,$$

where $y = J_i^{(\alpha,\beta)}(t)$.

- *Generating function:*

$$\frac{2^{\alpha+\beta}}{R\,(1-\eta+R)^\alpha\,(1+\eta+R)^\beta} = \sum_{i=0}^{\infty} J_i^{(\alpha,\beta)}(t) \, \eta^i, \qquad (B.8.7)$$

for $-1 < t < 1$ and $|\eta| < 1$, where

$$R = \sqrt{1 - 2t\,\eta + \eta^2}. \qquad (B.8.8)$$

Linear solvers

$$C$$

In the finite element codes discussed in the text, the systems of linear algebraic equations arising from the implementation of the Galerkin finite element method were solved either by the Thomas algorithm for tridiagonal and pentadiagonal systems, or by invoking a MATLAB function implemented by a vector-by-matrix division.

In practice, linear systems arising in finite element applications can have large or exorbitant dimensions, on the order of 10^4 or even higher. To solve such large systems, we employ general-purpose or custom-made algorithms that exploit the sparsity and possible symmetry of the coefficient matrix to reduce the memory storage requirements and number of floating point operations per second (flops). In scientific computing, hardware performance is measured in units of megaflops, (Mflops) which is equal to 10^6 flops, gigaflops (Gflops), which is equal to 10^9 flops, and teraflops (Tflops), which is equal to 10^{12} flops. An efficient desktop computer can perform 2 floating point operations per clock cycle, which amounts to a few Gflops. At the time of this writing, the fastest supercomputer can perform 10.72 Tflops.

In the first part of this appendix, we present a general overview of selected methods for solving systems of linear equations, and demonstrate their MATLAB implementation. In the concluding section, we present a summary of topics concerning matrix storage and manipulation in finite element applications.

C.1 Gauss elimination

Gauss elimination is the most popular *direct* method for solving systems of equations of small and moderate size. Consider a system of N linear algebraic equations for the N unknowns, $x_1, x_2, \ldots, x_N$,

$$
\begin{aligned}
A_{1,1}\, x_1 + A_{1,2}\, x_2 + \ldots + A_{1,N-1}\, x_{N-1} + A_{1,N}\, x_N &= b_1, \\
A_{2,1}\, x_1 + A_{2,2}\, x_2 + \ldots + A_{2,N-1}\, x_{N-1} + A_{2,N}\, x_N &= b_2, \\
\ldots\ldots\ldots\ldots\ldots\ldots & \\
A_{N,1}\, x_1 + A_{N,2}\, x_2 + \ldots + A_{N,N-1}\, x_{N-1} + A_{N,N}\, x_N &= b_N,
\end{aligned}
\tag{C.1.1}
$$

where $A_{i,j}$ are given coefficients and b_i are given constants, for $i, j = 1, 2, \ldots, N$, In matrix notation, the system (C.1.1) takes the compact form

$$\mathbf{A} \cdot \mathbf{x} = \mathbf{b}, \tag{C.1.2}$$

where $\mathbf{A}$ is the $N \times N$ coefficient matrix

$$\mathbf{A} = \begin{bmatrix} A_{1,1} & A_{1,2} & \cdots & A_{1,N-1} & A_{1,N} \\ A_{2,1} & A_{2,2} & \cdots & A_{2,N-1} & A_{2,N} \\ \cdots & \cdots & \cdots & \cdots & \cdots \\ A_{N-1,1} & A_{N-1,2} & \cdots & A_{N-1,N-1} & A_{N-1,N} \\ A_{N,1} & A_{N,2} & \cdots & A_{N,N-1} & A_{N,N} \end{bmatrix}, \tag{C.1.3}$$

and $\mathbf{b}$ is the N-dimensional vector

$$\mathbf{b} = \begin{bmatrix} b_1 \\ b_2 \\ \vdots \\ b_{N-1} \\ b_N \end{bmatrix}. \tag{C.1.4}$$

Gauss elimination reduces the original system (C.1.2) into the upper triangular system

$$\mathbf{U} \cdot \mathbf{x} = \mathbf{y}, \tag{C.1.5}$$

where $\mathbf{U}$ is an upper triangular matrix,

$$\mathbf{U} = \begin{bmatrix} U_{1,1} & U_{1,2} & \cdots & U_{1,N-1} & U_{1,N} \\ 0 & U_{2,2} & \cdots & U_{2,N-1} & U_{2,N} \\ \cdots & \cdots & \cdots & \cdots & \cdots \\ 0 & 0 & \cdots & U_{N-1,N-1} & U_{N-1,N} \\ 0 & 0 & \cdots & 0 & U_{N,N} \end{bmatrix}, \tag{C.1.6}$$

and $\mathbf{y}$ is a properly constructed right-hand side. The elimination is followed by back-substitution, which involves solving the last equation of (C.1.5) for x_N, and substituting its value in all previous equations to eliminate this unknown. Next, we solve the penultimate equation for x_{N-1}, and substitute its value in all previous equations. Moving backward in this fashion, we compute all unknowns, all the way up to the first unknown, x_1.

As a bonus, Gauss elimination simultaneously produces a lower diagonal matrix, $\mathbf{L}$, with ones along the diagonal,

$$\mathbf{L} = \begin{bmatrix} 1 & 0 & \cdots & 0 & 0 \\ L_{2,1} & 1 & \cdots & 0 & 0 \\ \cdots & \cdots & \cdots & \cdots & \cdots \\ L_{N-1,1} & L_{N-1,2} & \cdots & 1 & 0 \\ L_{N,1} & L_{N,2} & \cdots & 0 & 1 \end{bmatrix}, \tag{C.1.7}$$

so that

$$\mathbf{A} = \mathbf{L} \cdot \mathbf{U}. \tag{C.1.8}$$

Thus, Gauss elimination performs the Doolittle **LU** decomposition of the coefficient matrix **A**, and can be used exclusively for that purpose and without reference to a system of linear equations. When the option of row pivoting is enabled, as discussed in the next section, the method performs the **LU** decomposition of a matrix that arises by interchanging rows of **A**.

The basic idea behind Gauss elimination is to solve the first equation in (C.1.1) for the first unknown, x_1, and use the expression for x_1 thus obtained to eliminate this unknown from all subsequent equations. We then retain the first equation as is, and replace all subsequent equations with their descendants that do not contain x_1. In the second stage, we solve the second equation for the second unknown, x_2, and use the expression for x_2 thus obtained to eliminate this unknown from all subsequent equations. We then retain the first and second equations, and replace all subsequent equations with their descendants that do not contain x_1 or x_2. Continuing in this manner, we arrive at the last equation, which contains only the last unknown, x_N, and this concludes the process of elimination.

C.1.1 Pivoting

Immediately before the mth equation has been solved for the mth unknown, where $m = 1, 2, \ldots, N - 1$, the linear system appears as

$$
\begin{bmatrix}
A_{1,1}^{(m)} & A_{1,2}^{(m)} & \cdots & & \cdots & A_{1,N}^{(m)} \\
0 & A_{2,2}^{(m)} & \cdots & & \cdots & A_{2,N}^{(m)} \\
0 & 0 & \cdots & & \cdots & \cdots \\
0 & 0 & A_{m-1,m-1}^{(m)} & A_{m-1,m}^{(m)} & \cdots & A_{m-1,N}^{(m)} \\
0 & \cdots & 0 & A_{m,m}^{(m)} & \cdots & A_{m,N}^{(m)} \\
0 & \cdots & 0 & \cdots & \cdots & \cdots \\
0 & \cdots & 0 & A_{N,m}^{(m)} & \cdots & A_{N,N}^{(m)}
\end{bmatrix}
\cdot
\begin{bmatrix}
x_1 \\ x_2 \\ x_3 \\ \vdots \\ x_{N-1} \\ x_N
\end{bmatrix}
=
\begin{bmatrix}
b_1^{(m)} \\ b_2^{(m)} \\ b_3^{(m)} \\ \vdots \\ b_{N-1}^{(m)} \\ b_N^{(m)}
\end{bmatrix},
$$

$$\tag{C.1.9}$$

where $A_{i,j}^{(m)}$ are intermediate coefficients, and $b_i^{(m)}$ are intermediate right-hand sides. The first equation in (C.1.9) is identical to the first equation in (C.1.1), for any value of m. Subsequent equations are different, except at the first step corresponding to $m = 1$.

An apparent failure occurs when the diagonal element $A_{m,m}^{(m)}$ is nearly or precisely equal to zero, as we may no longer solve the mth equation in (C.1.9) for x_m, as required. However, the breakdown of the algorithm, does *not* necessarily mean that the system of equations does not have a solution. To circumvent this difficulty, we simply rearrange the equations or relabel the unknowns to bring the mth unknown to the mth equation, using the method of *pivoting*. If there is no way we can make this happen, the matrix $\mathbf{A}$ is singular, and the linear system has either no solution or an infinite number of solutions.

In the method of *row pivoting*, potential difficulties are bypassed by switching the mth equation in the system (C.1.9) with the subsequent kth equation, where $k > m$, and the value of k is chosen so that $|A_{k,m}^{(m)}|$ is the maximum value of the elements in the mth column below the diagonal, $A_{i,m}^{(m)}$, for $i \geq m$. If $A_{i,m}^{(m)} = 0$ for all $i \geq m$, then the system under consideration does not have a unique solution, which means that the matrix $\mathbf{A}$ is singular. Certain important facts about pivoting are the following:

- Pivoting is not necessary for systems with diagonally dominant coefficient matrices. By definition, the magnitude of each diagonal element of a diagonally dominant matrix is larger than the sum of the magnitudes of the remaining elements in the corresponding row.

- Pivoting is not necessary for systems with symmetric and positive-definite coefficient matrices (see Section C.3.1).

- It appears that pivoting should be enabled only when the magnitude of the intermediate element, $A_{m,m}^{(m)}$, is smaller than a specified threshold. However, in practice, we want to make $|A_{m,m}^{(m)}|$ as large as possible in order to reduce the round-off error associated with the floating point representation, and thus enable pivoting even when $|A_{m,m}^{(m)}|$ is not necessarily small.

- Pivoting prohibits the parallelization of the computations.

C.1.2 Implementation

To implement the method of Gauss elimination with row pivoting, we proceed according to the following steps:

Setting up:

Formulate the $N \times (N + 1)$ partitioned augmented matrix

$$\mathbf{C}^{(1)} \equiv \left[\, \mathbf{A} \, \middle| \, \mathbf{b} \, \right], \qquad (\text{C.1.10})$$

and introduce the $N \times N$ matrix, $\mathbf{L}$, whose elements are initialized to zero.

First pass:

1. Find the location of the element with the maximum norm in the first column of $\mathbf{C}^{(1)}$. That is, search for the maximum norm of the elements $|C_{i,1}^{(1)}|$, for $i = 1, 2, \ldots, N$. Assume that this is equal to $|C_{k,1}^{(1)}|$, corresponding to the kth row.

2. Interchange the first row with the kth row of $\mathbf{C}^{(1)}$; repeat for the matrix $\mathbf{L}$. If $k = 1$, skip this step.

3. Compute the first column of $\mathbf{L}$ below the diagonal by setting $L_{i,1} = C^{(1)}_{i,1}/C^{(1)}_{1,1}$, for $i = 2, 3, \ldots, N$.

4. Subtract from the ith row of $\mathbf{C}^{(1)}$ the first row multiplied by $L_{i,1}$, for $i = 2, 3, \ldots, N$, to obtain a new augmented matrix,

$$\mathbf{C}^{(2)} \equiv [\mathbf{A}^{(2)}|\mathbf{b}^{(2)}]. \tag{C.1.11}$$

Second pass:

1. Find the location of the element with the maximum norm in the second column of $\mathbf{C}^{(2)}$, below the diagonal. That is, search for the maximum norm of the elements $|C^{(2)}_{i,2}|$, for $i = 2, 3, \ldots, N$. Assume that this is equal to $|C^{(2)}_{k,2}|$, corresponding to the kth row.

2. Interchange the second row with the kth row of $\mathbf{C}^{(2)}$; repeat for the matrix $\mathbf{L}$. If $k = 2$, skip this step.

3. Compute the second column of $\mathbf{L}$ below the diagonal, by setting $L_{i,2} = C^{(2)}_{i,2}/C^{(2)}_{2,2}$, for $i = 3, 4, \ldots, N$.

4. Subtract from the ith row of $\mathbf{C}^{(2)}$ the second row multiplied by $L_{i,2}$, for $i = 3, 4, \ldots, N$, to obtain the new augmented matrix

$$\mathbf{C}^{(3)} \equiv [\mathbf{A}^{(3)}|\mathbf{b}^{(3)}]. \tag{C.1.12}$$

...

mth **pass:**

1. Find the location of the element with the maximum norm in the mth column of $\mathbf{C}^{(m)}$, below the diagonal. That is, search for the maximum norm of the elements $|C^{(m)}_{i,m}|$, for $i = m, m + 1, \ldots, N$. Assume that this is equal to $|C^{(m)}_{k,m}|$, corresponding to the kth row.

2. Interchange the mth row with the kth row of $\mathbf{C}^{(m)}$; repeat for the matrix $\mathbf{L}$. If $k = m$, skip this step.

3. Compute the mth column of $\mathbf{L}$ below the diagonal, by setting $L_{i,m} = C^{(m)}_{i,m}/C^{(m)}_{m,m}$, for $i = m + 1, m + 2, \ldots, N$.

4. Subtract from the ith row of $\mathbf{C}^{(m)}$ the mth row multiplied by $L_{i,m}$, for $i = m + 1, \ldots, N$, to obtain the new augmented matrix

$$\mathbf{C}^{(m+1)} \equiv [\mathbf{A}^{(m+1)}|\mathbf{b}^{(m+1)}]. \tag{C.1.13}$$

...

(N-1) pass:

At the end of the $N-1$ pass, corresponding to $m = N-1$, the augmented matrix $\mathbf{C}^{(N)}$ has the form

$$\mathbf{C}^{(N)} = \left[\mathbf{A}^{(N)} \mid \mathbf{b}^{(N)}\right], \qquad (\text{C.1.14})$$

where $\mathbf{A}^{(N)} \equiv \mathbf{U}$ is the upper triangular matrix shown in (C.1.5).

Backward substitution:

Finally, solve the upper triangular system

$$\mathbf{U} \cdot \mathbf{x} = \mathbf{b}^{(N)}, \qquad (\text{C.1.15})$$

by backward substitution to extract the solution of the original system of equations $\mathbf{A} \cdot \mathbf{x} = \mathbf{b}$. This is done by solving the last equation in (C.1.15) for the last unknown, x_N; once this is available we solve the penultimate equation for x_{N-1}; continuing backward in this manner, we finally compute x_1.

Complete the matrix L (optional):

Set the diagonal elements of the matrix $\mathbf{L}$ equal to 1.

It can be shown by straightforward algebraic manipulations that the matrices $\mathbf{L}$ and $\mathbf{U}$ provide us with the $\mathbf{LU}$ decomposition of the matrix $\mathbf{A}$, that is,

$$\mathbf{L} \cdot \mathbf{U} = \mathbf{A}^{Mod}, \qquad (\text{C.1.16})$$

where the matrix $\mathbf{A}^{Mod}$ is identical to $\mathbf{A}$, except that the rows may have been interchanged due to pivoting. If pivoting is disabled, $\mathbf{A}^{Mod} = \mathbf{A}$.

Compute the determinant (optional):

The determinant of the matrix $\mathbf{A}$ follows from

$$\begin{aligned}
\text{Det}(\mathbf{A}) &= \pm\text{Det}(\mathbf{A}^{Mod}) \\
&= \pm\text{Det}(\mathbf{L}) \cdot \text{Det}(\mathbf{U}) = \pm U_{1,1}\, U_{2,2} \ldots U_{N,N}, \quad (\text{C.1.17})
\end{aligned}$$

where the plus sign applies when an even number of row interchanges have been done due to pivoting, and the minus sign otherwise.

If the original matrix $\mathbf{A}$ is symmetric and pivoting is not enabled at the risk of having to divide by zero or foster the growth of round-off error, the lower square diagonal $(N - m + 1) \times (M - m + 1)$ block of the matrix displayed in equation (C.1.9) will remain symmetric for any value of m. This observation can be exploited for storing, and working only with the upper or lower triangular parts of the evolving coefficient matrix, thereby economizing the computations.

Specifically, for any value of m, we may work only with the elements of the augmented matrix on or above the diagonal, and set the values of the elements below the diagonal equal to their symmetric counterparts, as required. This modification reduces the number of operations nearly by a factor of two. When **A** is positive definite, pivoting is not required, and the economization can be implemented without a risk.

The cost of Gauss elimination scales with $N^3/3$, where N is the system size. This means that, if the system size is increased by a factor of 2, the computational cost is raised by a factor of 8. For systems of large size, Gauss elimination requires a prohibitive amount of computational time. For example, when $N = 10^5$, a desktop computer that performs 1 Gflops requires a CPU time on the order of $\frac{1}{3} 10^{15} \times 10^{-9}$s $\sim$ 92h.

C.1.3   Gauss elimination code

FSELIB function `gel`, listed in the text implements the Gauss elimination algorithm with optional row pivoting, and also offers the possibility for an expedited solution in the case of a symmetric coefficient matrix. The following driver script `gel_dr` reads the coefficient matrix and right-hand side from file *mat_vec.dat*, and calls `gel` to compute the solution:

```
%===============================
% Driver for Gauss Elimination
%===============================

file1 = fopen('mat_vec.dat');
 N   = fscanf(file1,'%f',[1,1]);
 A   = fscanf(file1,'%f',[N,N]);
 rhs = fscanf(file1,'%f',[1,N]);
fclose(file1);

A = A';  % because A was read columnwise,
         % replace with the transpose

Iwlpvt=1; % pivoting enabled (0 to disable)
Isym=0; % system is not symmetric

[x,l,u,det,Istop] = gel (N,A,rhs,Iwlpvt,Isym);

disp ('Solution:'); x
```

To verify the solution, we type at the MATLAB prompt:

```
>> A*x'-rhs'
```

and confirm that the answer is a very small residual vector whose magnitude is comparable to the round-off error.

C.1.4 Multiple right-hand sides

A simple modification of the basic algorithm allows us to solve, at once, multiple systems of equations with the same coefficient matrix but different right-hand sides,

$$\mathbf{A} \cdot \mathbf{x}^{(j)} = \mathbf{b}^{(j)}, \tag{C.1.18}$$

where $j = 1, 2, \ldots, p$. The standard method of Gauss elimination reduces the primary system $\mathbf{A} \cdot \mathbf{x}^{(j)} = \mathbf{b}^{(j)}$ to the modified system $\mathbf{U} \cdot \mathbf{x}^{(j)} = \mathbf{c}^{(j)}$, which is then solved by backward substitution. A key observation is that the computation of the vectors $\mathbf{c}^{(j)}$ can be done simultaneously, working with the $N \times (N+p)$ partitioned augmented matrix

$$\mathbf{C}^{(1)} \equiv \left[\mathbf{A} | \mathbf{b}^{(1)} | \mathbf{b}^{(2)} | \cdots | \mathbf{b}^{(p)} | \right]. \tag{C.1.19}$$

At the end of the $N-1$ pass, the augmented matrix $\mathbf{C}^{(N)}$ will have the partially block-triangular form

$$\mathbf{C}^{(N)} = \left[\mathbf{A}^{(N)} | \mathbf{D}^{(N)} \right], \tag{C.1.20}$$

where $\mathbf{A}^{(N)} \equiv \mathbf{U}$ is an upper triangular matrix, and the p columns of the matrix $\mathbf{D}^{(N)}$ contain the evolved right-hand sides, $\mathbf{c}^{(j)}$, where

$$\mathbf{U} \cdot \mathbf{x}^{(j)} = \mathbf{c}^{(j)}. \tag{C.1.21}$$

The algorithm concludes with p back substitutions.

C.1.5 Computation of the inverse

The jth column of the inverse of a square matrix, $\mathbf{A}$, denoted by $\mathbf{x}^{(j)}$, satisfies the linear system

$$\mathbf{A} \cdot \mathbf{x}^{(j)} = \mathbf{e}^{(j)}, \tag{C.1.22}$$

where $j = 1, 2, \ldots N$, and all components of the vector $\mathbf{e}^{(j)}$ are equal to zero, except for the jth component that is equal to unity. The computation of the vectors $\mathbf{x}^{(j)}$ can be done in a compact manner, working with the $N \times 2N$ augmented matrix

$$\mathbf{C}^{(1)} \equiv \left[\mathbf{A} | \mathbf{I} \right], \tag{C.1.23}$$

where $\mathbf{I}$ is the $N \times N$ identity matrix. At the end of the $N-1$ pass, the augmented matrix $\mathbf{C}^{(N)}$ will have the form shown in (C.1.20). The algorithm concludes with p back substitutions.

```
function [x, ...
          l,u,det, ...
          Istop ] = gel (n,a,rhs,Iwlpvt,Isym)

%==========================================
% Solution of the nxn linear system:
%    a x = rhs
% by Gauss elimination with row pivoting
%
% If Iwlpvt = 1, row pivoting is enabled
% If Isym   = 1, matrix a is symmetric
%
% c: extended coefficient matrix
% l: lower triangular matrix
% u: upper triangular matrix
%
% det: determinant of a
%==========================================

%-----------
% initialize
%-----------

Istop  = 0;             % error flag
Icount = 0;             % counts row interchanges in pivoting
eps = 0.00000000001;   % tolerance

%-----------
% pivoting is not done for a symmetric system
%-----------

if(Isym==1)
 disp(' gel: system is symmetric; pivoting is disabled')
 Iwlpvt = 0;
end

%--------
% prepare
%--------

na = n-1;
n1 = n+1;

%-------------------
% initialize l and c
%-------------------

for i=1:n
  for j=1:n
    l(i,j) = 0.0;
    c(i,j) = a(i,j);
  end
  c(i,n1) = rhs(i);
end
```

Function gel: Continuing $\longrightarrow$

```
%--------------------
% begin row reductions
%--------------------

for m=1:na    % outer loop for working row

   ma = m-1; m1 = m+1;

%------------------------
% pivoting module:
%
% search the ith column
% for the largest element
%------------------------

if(Iwlpvt==1)    % pivoting will be done if Iwlpvt=1

  Ipv = m; pivot = abs(c(m,m));

    for j=m1:n
      if(abs(c(j,m)) > pivot)
        Ipv = j; pivot = abs(c(j,m));
      end
    end

    if(pivot < eps)
       disp ('gel: trouble in station 1')
       Istop = 1;
       return
    end

%-----------------------------------
% switch the working row with the row
% containing the pivot element;
% also switch rows in l
%-----------------------------------

    if(Ipv ~= m)

     for j=m:n1
       save = c(m,j);
       c(m,j) = c(Ipv,j); c(Ipv,j) = save;
     end

     for j=1:ma
       save = l(m,j);
       l(m,j) = l(Ipv,j); l(Ipv,j) = save;
     end

     Icount = Icount+1;    % increase the pivoting counter

    end
end    % end of pivoting module
```

Function gel: $\longrightarrow$ Continuing $\longrightarrow$

```
%----------------------------------------
% reduce column i beneath element c(m,m)
%----------------------------------------

  for i=m1:n

    l(i,m) = c(i,m)/c(m,m);

    if(Isym==1)
      l(i,m) = c(m,i)/c(m,m); ilow = i;
    else
      l(i,m) = c(i,m)/c(m,m); ilow = m1;
    end

    c(i,m) = 0.0;

    for j=ilow:n1
      c(i,j) = c(i,j)-l(i,m)*c(m,j);
    end

  end

%---
% end of outer loop for working row:
%---

end

%---------------------------------
% check the last diagonal element
% for a singularity
%---------------------------------

if(abs(c(n,n)) < eps)

    disp('gel: trouble in station 2')
    Istop = 1;
    return;

end

%----------------------
% complete the matrix l
%----------------------

for i=1:n
  l(i,i) = 1.0;
end

%--------------------
% define the matrix u
%--------------------
```

Function gel: $\longrightarrow$ Continuing $\longrightarrow$

```
for i=1:n
  for j=1:n
    u(i,j) = c(i,j);
  end
end

%------------------------------------
% perform back-substitution to solve
% the reduced system
% using the upper triangular matrix c
%------------------------------------

x(n) = c(n,n1)/c(n,n);

for i=na:-1:1

  sum = c(i,n1);
  for j=i+1:n
    sum = sum-c(i,j)*x(j);
  end
  x(i) = sum/c(i,i);

end

%---------------------------
% compute the determinant as:
%
% det(a) = (+-) det(l)*det(u)
%---------------------------

det = 1.0;

for i=1:n
  det = det*c(i,i);
end

if(Iwlpvt == 1)
  for i=1:Icount
    det = -det;
  end
end

%-----
% done
%-----

return
```

Function gel: ($\longrightarrow$ Continued.) An FSELIB function for solving a linear system of equations by the method of Gauss elimination with optional row pivoting, also performing the *LU* decomposition and producing the determinant of the coefficient matrix. The function offers an option for an expeditious calculation in the case of a symmetric coefficient matrix.

C.1.6 Gauss-Jordan reduction

The method Gauss-Jordan reduction is a first cousin once removed of the method of Gauss elimination, involving the following steps:

1. Divide the first equation by $A_{1,1}^{(1)}$, solve it for the first unknown, x_1, and use the expression thus obtained to eliminate x_1 from *all* subsequent equations.

2. Divide the second equation by $A_{2,2}^{(2)}$, solve it for the second unknown, x_2, and use the expression thus obtained to eliminate x_2 from the *first and all subsequent equations.*

3. Continue in this manner, until the last equation contains only the last unknown. At that point, the evolved coefficient matrix will be diagonal, with all diagonal elements equal to unity. Consequently, the solution will be displayed on the right-hand side.

To implement the method, we formulate the augmented matrix

$$\mathbf{C}^{(1)} \equiv \left[\, \mathbf{A} \,\middle|\, \mathbf{b} \,\right], \tag{C.1.24}$$

and proceed as in Gauss elimination, with straightforward modifications. At the end of the $N - 1$ pass, corresponding to $m = N - 1$, the evolved matrix $C^{(N)}$ will have the form

$$\mathbf{C}^{(N)} = \left[\, \mathbf{I} \,\middle|\, \mathbf{b}^{(N)} \,\right], \tag{C.1.25}$$

where $\mathbf{b}^{(N)}$ is the required solution. Row pivoting is implemented as in Gauss elimination.

It might appear that the method of Gauss-Jordan elimination is competitive with, if not preferable over, the method of Gauss elimination. However, counting the number of operations shows that the method is slower than the standard Gauss elimination, roughly by a factor of three. Nevertheless, the robustness of the Gauss-Jordan reduction makes it attractive as a benchmark for solving systems of small and moderate size.

C.2 Iterative methods based on matrix splitting

The need to solve systems of large size has motivated the development of a host of powerful methods for general-purpose and specialized applications. Iterative solution procedures are typically used for sparse systems arising in finite difference and finite element implementations.

In one class of iterative methods, the coefficient matrix $\mathbf{A}$ is split into two matrices,

$$\mathbf{A} = \mathbf{B} - \mathbf{C}, \tag{C.2.1}$$

and the system $\mathbf{A} \cdot \mathbf{x} = \mathbf{b}$ is recast into the form

$$\mathbf{B} \cdot \mathbf{x} = \mathbf{C} \cdot \mathbf{x} + \mathbf{b}. \tag{C.2.2}$$

The algorithm involves guessing the solution, $\mathbf{x}$, computing the right-hand side of (C.2.2), and solving for the $\mathbf{x}$ on the left-hand side. The computation is repeated until the vector $\mathbf{x}$ used to compute the right-hand side of (C.2.2) is virtually identical to that arising from solving the linear system.

The calculations are conducted based on successive substitutions, carried out according to the formula

$$\mathbf{B} \cdot \mathbf{x}^{(k+1)} = \mathbf{C} \cdot \mathbf{x}^{(k)} + \mathbf{b}, \tag{C.2.3}$$

where the superscript (k) denotes the kth iteration, for $k = 0, 1, \ldots$, and $\mathbf{x}^{(0)}$ is the initial guess. A more explicit interpretation of the iterative method emerges by recasting (C.2.3) into the form

$$\mathbf{B} \cdot \mathbf{c}^{(k)} = -\mathbf{A} \cdot \mathbf{x}^{(k)} + \mathbf{b}, \tag{C.2.4}$$

where $\mathbf{c}^{(k)} \equiv \mathbf{x}^{(k+1)} - \mathbf{x}^{(k)}$ is the correction. If $\mathbf{x}^{(k)}$ is the desired solution, the right-hand side is zero, and so is the correction.

The main advantage of the iterative approach is that, if the splitting (C.2.1) is done craftily, solving (C.2.3) for $\mathbf{x}^{(k+1)}$ is much easier than solving (C.1.2) for $\mathbf{x}$ on the left-hand side. Thus, even though multiple iterations are carried out, the recursive method can be significantly more efficient than the direct method.

To study the convergence of the iterations, we recast (C.2.3) into the form

$$\mathbf{x}^{(k+1)} = \mathbf{P} \cdot \mathbf{x}^{(k)} + \mathbf{B}^{-1} \cdot \mathbf{b}, \tag{C.2.5}$$

where

$$\mathbf{P} \equiv \mathbf{B}^{-1} \cdot \mathbf{C} \tag{C.2.6}$$

is the *projection matrix*. Theoretical analysis shows that the iterations will converge for any initial guess, provided that the spectral radius of the projection matrix is less than unity, that is, the magnitude of each real or complex eigenvalue of $\mathbf{P}$ is less than unity (e.g., [43]). If the dominant eigenvalue of the projection matrix is available, the rate of convergence of the iterations can be improved by applying Wielandt's method of spectrum deflation (e.g., [43], p. 168).

C.2.1 Jacobi's method

In this method, the matrix $\mathbf{A}$ is split into the diagonal part, $\mathbf{D}$, where $D_{ii} = A_{ii}$ and $D_{ij} = 0$ for $i \neq j$, and the remainder with zero diagonals,

$$\mathbf{B} = \mathbf{D}, \qquad \mathbf{C} = \mathbf{D} - \mathbf{A}. \tag{C.2.7}$$

The iterations are based on the formula

$$x_i^{(k+1)} = \frac{1}{A_{ii}} \left(b_i - \sum_{j=1}^{N} A_{ij} \, x_j^{(k)} \right).$$ (C.2.8)

The underlying projection matrix is

$$\mathbf{P} = \mathbf{I} - \mathbf{D}^{-1} \cdot \mathbf{A},$$ (C.2.9)

where the inverse diagonal matrix, $\mathbf{D}^{-1}$, contains the inverses of the diagonals.

C.2.2 Gauss-Seidel method

This method differs from Jacobi's method in that the newly updated values of the solution replace the old values as soon as they are available. Specifically, the iterations are based on the formula

$$x_i^{(k+1)} = \frac{1}{A_{ii}} \left(b_i - \sum_{j=1}^{i-1} A_{ij} \, x_j^{(k+1)} - \sum_{j=i+1}^{N} A_{ij} \, x_j^{(k)} \right).$$ (C.2.10)

Cursory inspection reveals that the coefficient matrix is split into

$$\mathbf{B} = \mathbf{L} + \mathbf{D} \qquad \mathbf{C} = \mathbf{L} + \mathbf{D} - \mathbf{A},$$ (C.2.11)

where:

- $\mathbf{D}$ is the diagonal part of $\mathbf{A}$.

- $\mathbf{L}$ is the strictly lower triangular part of $\mathbf{A}$, with zeros along the diagonal.

- $\mathbf{C}$ is the negative of the strictly upper triangular part of $\mathbf{A}$, with zeros along the diagonal, $\mathbf{A} = \mathbf{L} + \mathbf{D} - \mathbf{C}$.

The underlying projection matrix is

$$\mathbf{P} = -(\mathbf{L} + \mathbf{D})^{-1} \cdot \mathbf{C} = \mathbf{I} - (\mathbf{L} + \mathbf{D})^{-1} \cdot \mathbf{A},$$ (C.2.12)

where the inverse of the lower triangular matrix $(\mathbf{L} + \mathbf{D})^{-1}$ is also lower triangular.

C.2.3 Successive over-relaxation method (SOR)

The successive over-relaxation method is based on the following modification of the Gauss-Seidel splitting matrices shown in (C.2.11),

$$\mathbf{B} = \omega \, \mathbf{L} + \mathbf{D}, \qquad \mathbf{C} = \omega \, \mathbf{L} + \mathbf{D} - \mathbf{A} = (\omega - 1) \, \mathbf{L} - \mathbf{U}, \quad (C.2.13)$$

where ω is an adjustable parameter. When $\omega = 0$, we recover Jacobi's method; when $\omega = 1$, we recover the Gauss-Seidel method. The iterations are based on the formula

$$x_i^{(k+1)} = (1 - \omega) \, x_i^{(k)} + \frac{\omega}{A_{ii}} \left(b_i - \sum_{j=1}^{i-1} A_{ij} \, x_j^{(k+1)} - \sum_{j=i+1}^{N} A_{ij} \, x_j^{(k)} \right).$$

$$(\text{C.2.14})$$

It can be shown that a necessary condition for the iterations to converge is $0 \le \omega \le 2$ (e.g., [43], p. 167).

C.2.4   Operator- and grid-based splitting

In practical applications, the decomposition shown in (C.2.1) can be dictated by the splitting of differential or integral operators, or else motivated by the specifics of the numerical implementation. For example, Jacobi's method can be generalized into a block-diagonal splitting method, where the diagonal blocks correspond to different parts of the solution domain. In other applications, the coefficient matrix can be split into a tridiagonal part and the remainder, and the iterations can be carried out using the Thomas algorithm. Creativity and imagination go a long way for the efficient solution of linear systems in physical applications.

C.3   Iterative methods based on path search

A powerful class of iterative methods search for the solution vector, $\mathbf{x}$, by making steps in the N-dimensional space towards carefully selected or even optimal directions. The general strategy is to select or dynamically compose a set of search directions expressed by a finite or infinite collection of vectors,

$$\mathbf{p}^{(1)}, \qquad \mathbf{p}^{(2)}, \qquad \mathbf{p}^{(3)}, \qquad \dots, \qquad (\text{C.3.1})$$

and then advance a guessed solution stepwise, according to the algorithm

$$\mathbf{x}^{(k)} = \mathbf{x}^{(k-1)} + \alpha_k \, \mathbf{p}^{(k)}, \qquad (\text{C.3.2})$$

for $k = 1, 2, \dots$, where α_k are appropriate coefficients expressing the length of each step. The evolved solution at the end of the kth step is

$$\mathbf{x}^{(k)} = \mathbf{x}^{(0)} + \sum_{i=1}^{k} \alpha_i \, \mathbf{p}^{(i)}, \qquad (\text{C.3.3})$$

where $\mathbf{x}^{(0)}$ is the initial guess.

C.3.1 Symmetric and positive-definite matrices

Consider a square $N \times N$ matrix $\mathbf{A}$, and choose an N-dimensional vector, $\mathbf{x}$. If the scalar number $\mathbf{x} \cdot \mathbf{A} \cdot \mathbf{x}$ is positive for any non-null $\mathbf{x}$, then the matrix $\mathbf{A}$ is called positive definite.

Consider now the linear system $\mathbf{A} \cdot \mathbf{x} = \mathbf{b}$. We will show that, if the coefficient matrix $\mathbf{A}$ is symmetric and positive definite, computing the solution is equivalent to finding a vector $\mathbf{X}$ that minimizes the scalar quadratic form

$$\mathcal{F}(\mathbf{x}) = \frac{1}{2} \mathbf{x} \cdot \mathbf{A} \cdot \mathbf{x} - \mathbf{b} \cdot \mathbf{x}. \tag{C.3.4}$$

The equivalence is evident for a single equation, $a\,x = b$, where the quadratic form reduces to $\mathcal{F}(x) = x\,(\frac{1}{2}a\,x - b)$. The minimum value of $\mathcal{F}(x)$ clearly occurs at the point $x = X$ where $\partial\mathcal{F}/\partial x = aX - b = 0$.

To demonstrate the equivalence for a higher number of equations, we compute the gradient of $\mathcal{F}(\mathbf{x})$, consisting of the partial derivatives with respect to x_i, where $i = 1, 2, \ldots, N$. Taking advantage of the symmetry of $\mathbf{A}$, we write

$$\frac{\partial\mathcal{F}}{\partial x_i} = \frac{1}{2} A_{ki}\,x_k + \frac{1}{2} A_{ik}\,x_k - b_i = A_{ik}\,x_k - b_i, \tag{C.3.5}$$

and note that all derivatives are zero when $\mathbf{x} = \mathbf{X}$, where $\mathbf{A} \cdot \mathbf{X} = \mathbf{b}$. Thus, $\mathbf{X}$ is a minimum, a maximum, or a saddle point of the quadratic form, $\mathcal{F}(\mathbf{x})$. To see which one it is, we expand $\mathcal{F}(\mathbf{x})$ in a Taylor series about the critical point $\mathbf{X}$, observe that all but the constant and quadratic terms vanish, and obtain the exact representation

$$\mathcal{F}(\mathbf{x}) = \frac{1}{2}\left(\hat{\mathbf{x}} \cdot \mathbf{A} \cdot \hat{\mathbf{x}} - \mathbf{b} \cdot \mathbf{X}\right), \tag{C.3.6}$$

where $\hat{\mathbf{x}} \equiv \mathbf{x} - \mathbf{X}$ is the distance from the critical point. Because the matrix $\mathbf{A}$ has been assumed positive definite, the first term on the right-hand side is positive for any $\hat{\mathbf{x}}$, and this guarantees that $\mathcal{F}(\mathbf{x})$ attains the minimum value when $\mathbf{x} = \mathbf{X}$.

In summary, we have reduced the problem of computing the solution of the equation $\mathbf{A} \cdot \mathbf{x} = \mathbf{b}$ to the problem of finding the minimum of a quadratic form. For a certain class of matrices, $\mathbf{A}$, the solution of the minimization problem can be found with a much lower effort than that required by direct or iterative methods discussed in previous sections.

Steepest-descent search

In this method, we make an initial guess $\mathbf{x}^{(0)}$, and then improve it by making a step in the direction where the quadratic form $\mathcal{F}(\mathbf{x})$ changes most rapidly. This steepest-descent direction is aligned with the residual vector

$$\mathbf{r} = -\nabla\mathcal{F} = -\mathbf{A} \cdot \mathbf{x} + \mathbf{b}, \tag{C.3.7}$$

evaluated at $\mathbf{x}^{(0)}$, denoted by

$$\mathbf{r}^{(0)} \equiv -\mathbf{A} \cdot \mathbf{x}^{(0)} + \mathbf{b}. \tag{C.3.8}$$

Thus, the first search direction is $\mathbf{p}^{(1)} = \mathbf{r}^{(0)}$.

The question of how long a distance we should travel must now be addressed. To answer this question, we note that, as we travel in the direction of $\mathbf{r}^{(0)}$, the vector $\mathbf{x}$ is described by $\mathbf{x} = \mathbf{x}^{(0)} + \alpha\, \mathbf{r}^{(0)}$, where α is a scalar parameter, and the value of the quadratic form along this path is

$$\mathcal{F}(\mathbf{x}) = \frac{1}{2} \left(\mathbf{x}^{(0)} + \alpha\, \mathbf{r}^{(0)} \right) \cdot \mathbf{A} \cdot \left(\mathbf{x}^{(0)} + \alpha\, \mathbf{r}^{(0)} \right)$$
$$-\mathbf{b} \cdot \left(\mathbf{x}^{(0)} + \alpha\, \mathbf{r}^{(0)} \right), \tag{C.3.9}$$

which is a quadratic function of α. We want to stop traveling when $\mathcal{F}$ has reached a minimum, that is, at the point where $\partial \mathcal{F} / \partial \alpha = 0$. Taking the partial derivative of the right-hand side of (C.3.9) with respect to α, and setting the resulting expression equal to zero, we find the optimal value

$$\alpha_1 = \frac{\mathbf{r}^{(0)} \cdot \mathbf{r}^{(0)}}{\mathbf{r}^{(0)} \cdot \mathbf{A} \cdot \mathbf{r}^{(0)}}, \tag{C.3.10}$$

which yields the improved position

$$\mathbf{x}^{(1)} = \mathbf{x}^{(0)} + \alpha_1\, \mathbf{r}^{(0)}. \tag{C.3.11}$$

The minimization process is subsequently repeated in the search direction, $\mathbf{p}^{(k)} = \mathbf{r}^{(k-1)}$, with

$$\alpha_k = \frac{\mathbf{p}^{(k)} \cdot \mathbf{p}^{(k)}}{\mathbf{p}^{(k)} \cdot \mathbf{A} \cdot \mathbf{p}^{(k)}}, \tag{C.3.12}$$

until the minimum has been reached within a specified tolerance.

It is instructive to test the performance of the method for a single equation $a\,x = b$. Applying the preceding formulas, we readily find $r^{(0)} = -a\,x^{(0)} + b$ and $\alpha_1 = 1/a$, which produces the exact solution $x^{(1)} = b/a$ in a single step! Unfortunately, this excellent performance does not hold for two or more equations. In general, while the method is guaranteed to converge, the rate of convergence can be prohibitively slow. Physically, when the graph of the quadratic form has narrow valleys, successive approximations bounce off opposite sides, slowly approaching the trough.

Method of conjugate gradients

Without loss of generality, we may begin the search from the origin, that is, set the initial guess for the solution at $\mathbf{x}^{(0)} = \mathbf{0}$. Our goal is to compute the search directions so that the exact solution, $\mathbf{X}$, is found exactly after N steps, that is,

$$\mathbf{X} = \alpha_1\, \mathbf{p}^{(1)} + \alpha_2\, \mathbf{p}^{(2)} + \ldots + \alpha_N\, \mathbf{p}^{(N)}, \tag{C.3.13}$$

where N is the system size. We shall see that this is an ambitious yet achievable goal.

To address the question of how long a distance we should travel at the kth step, we note that, as we travel in the direction of $\mathbf{p}^{(k)}$, the vector $\mathbf{x}$ is described by $\mathbf{x} = \mathbf{x}^{(k-1)} + \alpha_k \, \mathbf{p}^{(k)}$, and the value of the quadratic form along this path is given by

$$
\mathcal{F}(\mathbf{x}) = \frac{1}{2} \left(\mathbf{x}^{(k-1)} + \alpha_k \, \mathbf{p}^{(k)} \right) \cdot \mathbf{A} \cdot \left(\mathbf{x}^{(k-1)} + \alpha_k \, \mathbf{p}^{(k)} \right)
$$
$$
- \mathbf{b} \cdot \left(\mathbf{x}^{(k-1)} + \alpha_k \, \mathbf{p}^{(k)} \right), \tag{C.3.14}
$$

which is a quadratic function of α_k. Setting $\partial \mathcal{F}/\partial \alpha_k = 0$ to obtain the optimal stopping point, we find

$$
\alpha_k = \frac{\mathbf{p}^{(k)} \cdot \mathbf{r}^{(k-1)}}{\mathbf{p}^{(k)} \cdot \mathbf{A} \cdot \mathbf{p}^{(k)}}, \tag{C.3.15}
$$

which is inclusive of (C.3.12).

The distinguishing feature of the method of conjugate gradients is that the N search directions are required to be A-conjugate with one another, which means that

$$
\mathbf{p}^{(i)} \cdot \mathbf{A} \cdot \mathbf{p}^{(j)} = 0 \tag{C.3.16}
$$

for $i \neq j$. We shall see shortly this constraint arises in a natural manner during the minimization of the quadratic form with respect to the search directions at every stage. It is important to note that requiring the conjugation constraint (C.3.16) does not uniquely determine the search directions, but rather imposes restrictions among them.

Assuming that relation (C.3.16) is fulfilled, we take the inner product of both sides of (C.3.13) with the vector $\mathbf{p}^{(k)} \cdot \mathbf{A}$, and find

$$
\alpha_k = \frac{\mathbf{p}^{(k)} \cdot \mathbf{A} \cdot \mathbf{X}}{\mathbf{p}^{(k)} \cdot \mathbf{A} \cdot \mathbf{p}^{(k)}} = \frac{\mathbf{p}^{(k)} \cdot \mathbf{b}}{\mathbf{p}^{(k)} \cdot \mathbf{A} \cdot \mathbf{p}^{(k)}}, \tag{C.3.17}
$$

which appears to be different from (C.3.12) and its generalization. However, the two formulas are, in fact, equivalent. To show this, we introduce the residual at the kth step,

$$
\mathbf{r}^{(k)} \equiv -\mathbf{A} \cdot \mathbf{x}^{(k)} + \mathbf{b} = -\mathbf{A} \cdot \sum_{j=1}^{k} \alpha_j \, \mathbf{p}^{(j)} + \mathbf{b}. \tag{C.3.18}
$$

Taking the inner product of both sides with the vector $\mathbf{p}^{(i)}$, where $i \leq k$, and using the A-conjugation condition and (C.3.17), we find

$$
\mathbf{p}^{(i)} \cdot \mathbf{r}^{(k)} = 0, \tag{C.3.19}
$$

for $i \leq k$, which shows that any vector that can be expressed as a linear combination of $\mathbf{p}^{(i)}$, with $i \leq k$, is orthogonal to the residual, $\mathbf{r}^{(k)}$. We proceed by expressing the kth residual as

$$\mathbf{r}^{(k)} = \mathbf{r}^{(k-1)} - \alpha_k \, \mathbf{A} \cdot \mathbf{p}^{(k)}, \qquad (\text{C.3.20})$$

take the inner product of both sides with $\mathbf{p}^{(k)}$, note that $\mathbf{p}^{(k)} \cdot \mathbf{r}^{(k)} = 0$, and thus find

$$\mathbf{p}^{(k)} \cdot \mathbf{r}^{(k-1)} = \alpha_k \, \mathbf{p}^{(k)} \cdot \mathbf{A} \cdot \mathbf{p}^{(k)}, \qquad (\text{C.3.21})$$

which reproduces (C.3.15).

It remains to show that the exact solution will be found after N steps. To prove this, all we have to do is ensure that, of all vectors that can be written as linear combinations of $\mathbf{p}^{(i)}$ with $i \leq k$, the vector

$$\mathbf{x}^{(k)} = \alpha_1 \, \mathbf{p}^{(1)} + \ldots + \alpha_k \, \mathbf{p}^{(k)} \qquad (\text{C.3.22})$$

with coefficients computed from (C.3.17) minimizes the quadratic form (C.3.4). To show that this is true, we consider the perturbed iterant $\mathbf{x}^{(k)} + \mathbf{d}$, where the vector $\mathbf{d}$ can be expressed as a linear combination of the set $\mathbf{p}^{(i)}$ with $i \leq k$, and compute

$$\begin{aligned}
\mathcal{F}(\mathbf{x}^{(k)} + \mathbf{d}) &= \frac{1}{2} \left(\mathbf{x}^{(k)} + \mathbf{d} \right) \cdot \mathbf{A} \cdot \left(\mathbf{x}^{(k)} + \mathbf{d} \right) - \mathbf{b} \cdot \left(\mathbf{x}^{(k)} + \mathbf{d} \right), \\
&= \frac{1}{2} \mathbf{x}^{(k)} \cdot \mathbf{A} \cdot \mathbf{x}^{(k)} - \mathbf{b} \cdot \mathbf{x}^{(k)} + \frac{1}{2} \mathbf{d} \cdot \mathbf{A} \cdot \mathbf{d} + \mathbf{d} \cdot (\mathbf{A} \cdot \mathbf{x}^{(k)} - \mathbf{b}).
\end{aligned}$$
$$(\text{C.3.23})$$

The last term is zero because of property (C.3.19), and the penultimate term is non-negative because $\mathbf{A}$ is positive definite. Consequently, $\mathcal{F}$ reaches its minimum value when $\mathbf{d} = \mathbf{0}$.

The issue of choosing the search directions is still pending. In the original method of conjugate gradients developed by Hestenes and Stiefel [25], the current direction $\mathbf{p}^{(k)}$ is aligned as much as possible with the current residual $\mathbf{r}^{(k-1)}$, subject to the A-conjugation constraint. The objective is to make the magnitude of the numerator in (C.3.15) as large as possible, and thereby move towards the minimum of the quadratic form at the fastest possible rate. In that case, it can be shown that the residual vectors $\mathbf{r}^{(k)}$ for a non-singular matrix $\mathbf{A}$ form an orthogonal basis of the Krylov space spanned by the vectors

$$\mathbf{r}^{(0)}, \qquad \mathbf{A} \cdot \mathbf{r}^{(0)}, \qquad \mathbf{A}^2 \cdot \mathbf{r}^{(0)}, \qquad \ldots, \qquad (\text{C.3.24})$$

that is,

$$\mathbf{r}^{(i)} \cdot \mathbf{r}^{(k)} = 0, \qquad (\text{C.3.25})$$

for $i \neq k$.

$\mathbf{x}^{(0)} = \mathbf{0}$

Do $k = 1, \ldots, N$

    If $k = 1$
$$\mathbf{p}^{(1)} = \mathbf{r}^{(0)} = \mathbf{b}$$
    Else If $k > 1$
$$\beta_k = \frac{\mathbf{r}^{(k-1)} \cdot \mathbf{r}^{(k-1)}}{\mathbf{r}^{(k-2)} \cdot \mathbf{r}^{(k-2)}}$$

$$\mathbf{p}^{(k)} = \mathbf{r}^{(k-1)} + \beta_k \, \mathbf{p}^{(k-1)}$$
    End if

$$\alpha_k = \frac{\mathbf{r}^{(k-1)} \cdot \mathbf{r}^{(k-1)}}{\mathbf{p}^{(k)} \cdot \mathbf{A} \cdot \mathbf{p}^{(k)}}$$

$$\mathbf{x}^{(k)} = \mathbf{x}^{(k-1)} + \alpha_k \, \mathbf{p}^{(k)}$$

$$\mathbf{r}^{(k)} = \mathbf{r}^{(k-1)} - \alpha_k \, \mathbf{A} \cdot \mathbf{p}^{(k)}$$

End Do

Algorithm C.3.1 A conjugate-gradients algorithm for solving the linear system $\mathbf{A} \cdot \mathbf{x} = \mathbf{b}$ with a real, symmetric, and positive definite coefficient matrix $\mathbf{A}$.

The method is implemented according to Algorithm C.3.1, which is programmed in the FSELIB function cg, listed in the text. The following driver script cg_dr reads the coefficient matrix and right-hand side from the file *mat_s_vec.dat*, and calls cg to compute the solution:

```
file1 = fopen('mat_s_vec.dat');
  N   = fscanf(file1,'%f',[1,1]);
  A   = fscanf(file1,'%f',[N,N]);
  rhs = fscanf(file1,'%f',[1,N]);
fclose(file1);
[sln] = cg (N, A, rhs);
disp ('Solution:'); sln
```

```
function [sln] = cg (n, a, rhs)

%========================================================
%  Solution of a linear symmetric positive-definite
%  system by the method of conjugate gradients
%
%  The search vectors are chosen by the method of
%  Hestenes and Steifel (1952)
%
%  SYMBOLS:
%
%  a .... symmetric positive definite matrix
%  n .... size (rows/columns) of matrix a
%  rhs .. right hand side vector (e.g. b, as in Ax=b)
%  x .... evolving solution vector
%  sln .. final solution vector
%  p .... search directions
%  r .... residual vectors
%  alpha. scale parameter for solution update
%  beta.. scale parameter for search direction
%
%========================================================

%----------------------------------------------------------
%  set the initial values of the vectors x and r (step 0)
%  set the first-step value of p
%----------------------------------------------------------

for i=1:n
  x0(i) = 0.0;
  r0(i) = rhs(i);
  p(1,i) = rhs(i);
end

%--------------------------------
%  compute the sums used in alpha
%--------------------------------

alpha_num = 0.0; alpha_den = 0.0;

for i=1:n

  alpha_num = alpha_num + r0(i)*r0(i);
  for j=1:n
    alpha_den = alpha_den + p(1,i)*a(i,j)*p(1,j);
  end

end

alpha = alpha_num/alpha_den;

%----------------------------------------------------------
%  set first step values of alpha and vectors x and r
%----------------------------------------------------------
```

Function cg: Continuing $\longrightarrow$

```
for i=1:n
  x(1,i) = x0(i)+alpha*p(1,i);
  r(1,i) = r0(i);
  for j=1:n
    r(1,i) = r(1,i)-alpha*a(i,j)*p(1,j);
  end
end

%---------------------------------------------
% loop through the remaining search vectors
% 2 to n, and compute alpha, beta,
% and the vectors p, x, and r
%---------------------------------------------

for k=2:n          %  outer loop over search directions

%---
% sums used in beta
%---

beta_num = 0.0; beta_den = 0.0D0;

for i=1:n

  beta_num = beta_num + r(k-1,i)^2;
  if(k==2)
    beta_den = beta_den + r0(i)^2;
  else
    beta_den = beta_den + r(k-2,i)^2;
  end

end

beta = beta_num/beta_den;

for i=1:n
  p(k,i) = r(k-1,i)+beta*p(k-1,i);
end

%---
% compute the sums used in alpha
%---

alpha_num = beta_num;
alpha_den = 0.0;

for i=1:n
  for j=1:n
    alpha_den = alpha_den + p(k,i)*a(i,j)*p(k,j);
  end
end

alpha = alpha_num/alpha_den;
```

Function cg: $\longrightarrow$ Continuing $\longrightarrow$

```
%---
% compute the k'th iterations of vectors x and r
%---

for i=1:n
   x(k,i) = x(k-1,i)+alpha*p(k,i);
   r(k,i) = r(k-1,i);
   for j=1:n
      r(k,i)=r(k,i)-alpha*a(i,j)*p(k,j);
   end
end

end                      % end of outer loop

%--------------------
% extract the solution
%--------------------

for i=1:n
   sln(i) = x(n,i);
end

%-----
% done
%-----

return
```

Function cg: ($\longrightarrow$ Continued.) Solution of a symmetric and positive definite system by the method of conjugate gradients.

The conjugate-gradients method produces a sequence of vectors that approximate the solution of a linear system with increasing accuracy. However, the reduction in error may be uneven through the iterations; some steps may improve the solution a little, and others a great deal. In practice, we want to make a number of steps that is substantially less than the system size, and yet obtain a good approximation to the solution. If large corrections are made at the beginning, we are fortunate; but if large corrections are made at the end, we are unfortunate. The first scenario occurs when the condition number of the matrix, identified as the magnitude of the ratio of the largest to the smallest eigenvalue, is sufficiently small.

To reduce the condition number, we precondition the linear system before applying the numerical method. This is done by multiplying both sides with the inverse of a preconditioning matrix, $\mathbf{C}^{-1}$, to obtain the preconditioned system

$$\mathbf{A}^P \cdot \mathbf{y} = \mathbf{b}^P. \tag{C.3.26}$$

where

$$\mathbf{A}^P = \mathbf{C}^{-1} \cdot \mathbf{A} \cdot \mathbf{C}^{-1}, \qquad \mathbf{y} = \mathbf{C} \cdot \mathbf{x}, \qquad \mathbf{b}^P = \mathbf{C}^{-1} \cdot \mathbf{b}. \tag{C.3.27}$$

The algorithm involves solving system (C.3.26) by the method of conjugate gradients for $\mathbf{y}$, and then recovering the solution, $\mathbf{x}$, by inverting the second equation in (C.3.27), which is done by a matrix-vector multiplication, $\mathbf{x} = \mathbf{C}^{-1} \cdot \mathbf{y}$. Note that the matrix $\mathbf{C}$ does not have to be available.

C.3.2   General methods

The method of conjugate gradients can be generalized into the method of biconjugate gradients, which is applicable to non-symmetric and non-positive-definite systems (e.g., [43]). A popular alternative is the method of Generalized Minimal Residuals (GMRES) [50]. In the method of conjugate gradients, the residuals form an orthogonal set of the Krylov space. In the GMRES method, an orthonormal basis consisting of the vectors $\mathbf{v}^{(i)}$ is generated explicitly at every step using the Gram-Schmidt orthogonalization process (e.g., [43]). Specifically, the GMRES sequence is constructed according to the formula

$$\mathbf{x}^{(k)} = \mathbf{x}^{(0)} + \alpha_1 \, \mathbf{v}^{(1)} + \alpha_2 \, \mathbf{v}^{(2)} + \ldots + \alpha_k \, \mathbf{v}^{(k)}, \qquad (\text{C.3.28})$$

where the coefficients α_i are chosen at every step to minimize the norm of the residual $|\mathbf{A} \cdot \mathbf{x}^{(k)} - \mathbf{b}|$.

C.4   Finite element system solvers

Finite element implementations usually culminate in linear systems with banded coefficient matrices of large or excessive dimensions. In some cases, the coefficient matrices are symmetric and positive-definite and the solution can be found reliably, efficiently, and economically using compact implementations of the method of Gauss elimination, LU decomposition without pivoting, and the iterative methods discussed previously in this section. For systems of large size, economizing storage and ensuring computational efficiency are of primary concern.

Significant gains in efficiency can be achieved with the implementation of bandwidth and skyline storage strategies and frontal solution algorithms (e.g., [24, 29]). In skyline storage, a dedicated skyline index matrix is introduced containing the indices of the highest and lowest elements in each column of the coefficient matrix. The matrix components are then stored in a partitioned one-dimensional skyline array. If the coefficient matrix is symmetric, the skyline index matrix contains the location of the diagonal elements in the skyline array. Methods for solving linear equations arising in finite element applications are reviewed by Schwarz ([54], Chapter 4).

Mathematical supplement D

In this Appendix, we present a brief summary of fundamental mathematical concepts and definitions arising in finite element analysis.

D.1   Index notation

In index notation, a vector $\mathbf{u}$ in the Nth-dimensional space is denoted as u_i, where $i = 1, 2, \ldots, N$, a two-dimensional matrix $\mathbf{A}$ is denoted as A_{ij}, where the indices run over appropriate ranges, and a three-dimensional matrix $\mathbf{T}$ is denoted as T_{ijk}. Similar notation is used for higher-dimensional matrices.

Einstein's repeated-index summation convention states that, if a subscript appears twice in an expression involving products, then summation over that subscript is implied in its range. Under this convention,

$$u_i\, v_i \equiv u_1\, v_1 + u_2\, v_2 + \ldots + u_N\, v_N,$$

$$(D.1.1)$$

$$A_{ii} \equiv A_{11} + A_{22} + \ldots + A_{NN},$$

where N is the maximum value of the index i. The first expression defines the inner product of the pair of vectors $\mathbf{u}$ and $\mathbf{v}$.

D.2   Kronecker's delta

Kronecker's delta δ_{ij} represents the identity or unit matrix: $\delta_{ij} = 1$ if $i = j$, and $\delta_{ij} = 0$ if $i \neq j$. Using this definition, we find

$$u_i\, \delta_{ij} = u_j, \qquad A_{ij}\, \delta_{jk} = A_{ik}, \qquad \delta_{ij}\, A_{jk}\, \delta_{kl} = A_{il},$$

$$(D.2.1)$$

$$\delta_{ij}\, \delta_{jk}\, \delta_{kl} = \delta_{il}, \qquad \delta_{ii} = N,$$

where N is the maximum value of the index i.

If x_i is a set of N independent variables, then

$$\frac{\partial x_i}{\partial x_j} = \delta_{ij}, \qquad \frac{\partial x_i}{\partial x_i} = N. \qquad (D.2.2)$$

D.3 Alternating tensor

The alternating tensor ϵ_{ijk}, where all three indices range over 1, 2, and 3, is defined such that $\epsilon_{ijk} = 0$ if any two indices have the same value, $\epsilon_{ijk} = 1$ if the indices are arranged in one of the cyclic orders 123, 312, or 231, and $\epsilon_{ijk} = -1$ otherwise. Thus,

$$\epsilon_{132} = -1, \qquad \epsilon_{122} = 0, \qquad \epsilon_{ijj} = 0,$$

$$\epsilon_{ijk} = \epsilon_{jki} = \epsilon_{kij}. \tag{D.3.1}$$

Two useful properties of the alternating matrix are:

$$\epsilon_{ijk}\,\epsilon_{ijl} = 2\,\delta_{kl},$$

$$\epsilon_{ijk}\,\epsilon_{ilm} = \delta_{jl}\,\delta_{km} - \delta_{jm}\,\delta_{kl}. \tag{D.3.2}$$

D.4 Two- and three-dimensional vectors

The inner product of a pair of two- or three-dimensional vectors $\mathbf{a}$ and $\mathbf{b}$ is a scalar defined as $s \equiv \mathbf{a} \cdot \mathbf{b} = a_i\,b_i$. If the inner product is zero, then the two vectors are orthogonal. The inner product is equal to the product of the lengths of the two vectors and the cosine of the angle subtended between the vectors in their plane.

The outer product of a pair of two- or three-dimensional vectors $\mathbf{a}$ and $\mathbf{b}$ is a new vector $\mathbf{c} \equiv \mathbf{a} \times \mathbf{b}$ given by the determinant of a matrix,

$$\mathbf{c} \equiv \begin{bmatrix} \mathbf{e}_1 & \mathbf{e}_2 & \mathbf{e}_3 \\ a_1 & a_2 & a_3 \\ b_1 & b_2 & b_3 \end{bmatrix}, \tag{D.4.1}$$

where $\mathbf{e}_i$ are unit vectors in the subscripted directions. If the outer product vanishes, then the two vectors are parallel. The magnitude of the outer product is equal to the product of the lengths of the vectors $\mathbf{a}$ and $\mathbf{b}$, and the sine of the angle subtended between these vectors in their plane. The outer product is oriented normal to the plane of $\mathbf{a}$ and $\mathbf{b}$, and its direction is determined by the right-handed rule applied to the ordered triplet $\mathbf{a}, \mathbf{b}$, and $\mathbf{c}$.

The triple scalar product of the ordered triplet of vectors $\mathbf{a}$, $\mathbf{b}$, and $\mathbf{c}$ is the scalar $s \equiv (\mathbf{a} \times \mathbf{b}) \cdot \mathbf{c} = (\mathbf{c} \times \mathbf{a}) \cdot \mathbf{b} = (\mathbf{b} \times \mathbf{c}) \cdot \mathbf{a}$, which is equal to the determinant of the matrix

$$\begin{bmatrix} a_1 & a_2 & a_3 \\ b_1 & b_2 & b_3 \\ c_1 & c_2 & c_3 \end{bmatrix}. \tag{D.4.2}$$

D.5   Del or nabla operator

The Cartesian components of the *del* or *nabla* operator, ∇, are the partial derivatives with respect to the corresponding coordinates,

$$\nabla \equiv \left(\frac{\partial}{\partial x}, \frac{\partial}{\partial y}, \frac{\partial}{\partial z} \right). \tag{D.5.1}$$

In two dimensions, only the x and y derivatives appear.

D.6   Gradient and divergence

If f is a scalar function of position, $\mathbf{x}$, then its gradient, ∇f, is a vector defined as

$$\nabla f = \mathbf{e}_x \frac{\partial f}{\partial x} + \mathbf{e}_y \frac{\partial f}{\partial y} + \mathbf{e}_z \frac{\partial f}{\partial z}, \tag{D.6.1}$$

where $\mathbf{e}_i$ are unit vectors in the directions of the subscripted axes.

The Laplacian of the scalar function f is a scalar defined as

$$\nabla \cdot (\nabla f) \equiv \nabla^2 f = \frac{\partial^2 f}{\partial x^2} + \frac{\partial^2 f}{\partial y^2} + \frac{\partial^2 f}{\partial z^2}. \tag{D.6.2}$$

The Laplacian is equal to the divergence of the gradient. The divergence of a vector function of position, $\mathbf{F} = (F_x, F_y, F_z)$, is a scalar defined as

$$\nabla \cdot \mathbf{F} = \frac{\partial F_i}{\partial x_i} = \frac{\partial F_x}{\partial x} + \frac{\partial F_y}{\partial y} + \frac{\partial F_z}{\partial z}. \tag{D.6.3}$$

If $\nabla \cdot \mathbf{F}$ vanishes at every point, then the vector field $\mathbf{F}$ is called *solenoidal*.

The gradient of the vector field $\mathbf{F}$, denoted by $\mathbf{U} \equiv \nabla \mathbf{F}$, is a two-dimensional matrix with elements

$$U_{ij} = \frac{\partial F_j}{\partial x_i}. \tag{D.6.4}$$

The divergence of $\mathbf{F}$ is equal to the trace of $\mathbf{U}$, which is equal to the sum of the diagonal elements.

The curl of the vector field $\mathbf{F}$, denoted as, $\nabla \times \mathbf{F}$, is another vector field computed according to the usual rules of the outer vector product, treating the del operator as a regular vector, yielding

$$\nabla \times \mathbf{F} = \mathbf{e}_x \left(\frac{\partial F_z}{\partial y} - \frac{\partial F_y}{\partial z} \right) + \mathbf{e}_y \left(\frac{\partial F_x}{\partial z} - \frac{\partial F_z}{\partial x} \right) + \mathbf{e}_z \left(\frac{\partial F_y}{\partial x} - \frac{\partial F_x}{\partial y} \right). \tag{D.6.5}$$

D.7   Vector identities

If f is a scalar function and $\mathbf{F}$ and $\mathbf{G}$ are two vector functions, we can show by working in index notation that

$$\nabla \cdot (f\,\mathbf{F}) = f\,\nabla \cdot \mathbf{F} + \mathbf{F} \cdot \nabla f, \tag{D.7.1}$$

$$\nabla\,(\mathbf{F} \cdot \mathbf{G}) = \mathbf{F} \cdot \nabla \mathbf{G} + \mathbf{G} \cdot \nabla \mathbf{F} + \mathbf{F} \times (\nabla \times \mathbf{G}) + \mathbf{G} \times (\nabla \times \mathbf{F}), \tag{D.7.2}$$

$$\nabla \cdot (\mathbf{F} \times \mathbf{G}) = \mathbf{G} \cdot \nabla \times \mathbf{F} - \mathbf{F} \cdot \nabla \times \mathbf{G}, \tag{D.7.3}$$

$$\nabla \times (\mathbf{F} \times \mathbf{G}) = \mathbf{F}\,\nabla \cdot \mathbf{G} - \mathbf{G}\,\nabla \cdot \mathbf{F} + \mathbf{G} \cdot \nabla \mathbf{F} - \mathbf{F} \cdot \nabla \mathbf{G}, \tag{D.7.4}$$

$$\nabla \times (\nabla f) = \mathbf{0}, \tag{D.7.5}$$

$$\nabla \cdot (\nabla \times \mathbf{F}) = 0, \tag{D.7.6}$$

$$\nabla \times (\nabla \times \mathbf{F}) = \nabla\,(\nabla \cdot \mathbf{F}) - \nabla^2 \mathbf{F}. \tag{D.7.7}$$

D.8   Divergence theorem in three dimensions

Let V_c be an arbitrary control volume bounded by the closed surface D, and $\mathbf{n}$ be the unit vector that is normal to D pointing *outward*. The Gauss divergence theorem states that the volume integral of the divergence of any differentiable vector function $\mathbf{F} = (F_x, F_y, F_z)$ over V_c is equal to the flow rate of $\mathbf{F}$ across D,

$$\iiint_{V_c} \nabla \cdot \mathbf{F}\,\mathrm{d}V = \iint_D \mathbf{F} \cdot \mathbf{n}\,\mathrm{d}S. \tag{D.8.1}$$

Making the three sequential selections $\mathbf{F} = (f, 0, 0)$, $\mathbf{F} = (0, f, 0)$, and $\mathbf{F} = (0, 0, f)$, where f is a differentiable scalar function, we obtain the vector form of the divergence theorem

$$\iiint_{V_c} \nabla f\,\mathrm{d}V = \iint_D f\,\mathbf{n}\,\mathrm{d}S. \tag{D.8.2}$$

The particular choices $f = x$, $f = y$, or $f = z$ yield the volume of V_c in terms of a surface integral of the x, y, or z component of the normal vector.

Setting $\mathbf{F} = \mathbf{a} \times \mathbf{G}$, where $\mathbf{a}$ is a constant vector and $\mathbf{G}$ is a differentiable function, and then discarding the arbitrary constant $\mathbf{a}$, we obtain the new identity

$$\iiint_{V_c} \nabla \times \mathbf{G}\,\mathrm{d}V = \iint_D \mathbf{n} \times \mathbf{G}\,\mathrm{d}S. \tag{D.8.3}$$

D.9   Divergence theorem in two dimensions

Let A_c be an arbitrary control area in the xy plane that is bounded by the closed contour C, and $\mathbf{n}$ be the unit vector that is normal to C pointing *outward*. The Gauss divergence theorem states that the areal integral of the divergence of any two-dimensional differentiable vector function $\mathbf{F} = (F_x, F_y)$ over A_c is equal to the flow rate of $\mathbf{F}$ across C,

$$\iint_{A_c} \nabla \cdot \mathbf{F}\, \mathrm{d}A = \oint_C \mathbf{F} \cdot \mathbf{n}\, \mathrm{d}l, \tag{D.9.1}$$

where l is the arc length along C.

Making the sequential choices $\mathbf{F} = (f, 0)$ and $\mathbf{F} = (0, f)$, where f is a differentiable scalar function, we obtain the vector form of the divergence theorem

$$\iint_{A_c} \nabla f\, \mathrm{d}A = \oint_C f \mathbf{n}\, \mathrm{d}l. \tag{D.9.2}$$

The particular choices $f = x$ or $f = y$ yield the area of A_c in terms of a line integral of the x or y component of the normal vector.

D.10   Stokes' theorem

Let C be an arbitrary closed loop with unit tangent vector $\mathbf{t}$, D be an arbitrary surface bounded by C, and $\mathbf{n}$ be the unit vector that is normal to D and is oriented according to the right-handed rule with respect to $\mathbf{t}$ and with reference to a designated side of D. As we look at the designated side of D, the normal vector points towards us and the tangent vector describes a counterclockwise path.

Stokes' theorem states that the circulation of a differentiable vector function $\mathbf{F}$ along C is equal to the flow rate of the curl of $\mathbf{F}$ across D,

$$\oint_C \mathbf{F} \cdot \mathbf{t}\, \mathrm{d}l = \iint_D (\nabla \times \mathbf{F}) \cdot \mathbf{n}\, \mathrm{d}S, \tag{D.10.1}$$

where l is the arc length along C.

Setting $F = \mathbf{a} \times \mathbf{G}$, where $\mathbf{a}$ is a constant vector and $\mathbf{G}$ is a differentiable function, expanding out the integrand on the right-hand side of (A.10.1), and then discarding the arbitrary constant $\mathbf{a}$, we obtain the new identity

$$\oint_C \mathbf{G} \times \mathbf{t}\, \mathrm{d}l = \iint_D [\mathbf{n}\, \nabla \cdot \mathbf{G} - (\nabla \mathbf{G}) \cdot \mathbf{n}]\, \mathrm{d}S. \tag{D.10.2}$$

Element grid generation

Discretizing the solution domain into a collection of elements by the process of grid generation or tessellation is an important aspect of a finite or spectral element implementation. In fact, grid generation for complicated geometries can be the most demanding and expensive module of a code. Grids can be broadly classified as *structured*, meaning that the element vertices and edges can be indexed by simple algorithms and nearly all nodes have the same number of neighbors, and *unstructured*, meaning that their manipulation relies heavily on a non-obvious connectivity matrix for element and node identification. Structured grids are more restrictive in that transposing the labels of two elements results in an unstructured grid, but not *vice versa*.

One-dimensional grid generation

Element discretization in one dimension is straightforward compared to its two- and three-dimensional counterparts. In adaptive discretization, the algorithm is designed so that the more rapidly the solution varies in a certain interval, or the higher the magnitude of the local numerical error quantified in some sense, the smaller the local element size and the denser the element distribution. For problems governed by one-dimensional equations defined over a curved, planar or three-dimensional line, the element size is adjusted according to the local curvature using appropriate criteria to ensure adequate spatial resolution (e.g., [32]).

Two- and three-dimensional grid generation

Several algorithms are available for the automated discretization of two- and three-dimensional domains, including the method of successive subdivision, the method of Delaunay triangulation (both discussed in Chapter 3), the advancing front method (AFM), and their hybrid implementations.

In the advancing front method in two dimensions, the boundary contour is discretized into one-dimensional elements defining the initial front (e.g., [32]). Triangles are then added inward, so that each element has at least one edge at the initial front. When the initial front has been depleted, the process is repeated with a redefined new front until the whole of the discretization domain has been tiled. A similar method can be implemented in three dimensions with

an initial surface front described by triangles, using tetrahedral elements to
tessellate the solution domain.

Resources on the Internet

Internet sites with numerous links to grid generation pages can be found at:

Mesh generation and grid generation on the web:

http://www-users.informatik.rwth-aachen.de/~roberts/meshgeneration.html

Mesh generators:

http://www.engr.usask.ca/~macphed/finite/fe_resources/mesh.html

Internet finite element resources:

http://www.engr.usask.ca/~macphed/finite/fe_resources/fe_resources.html

Meshing research corner:

http://www.andrew.cmu.edu/user/sowen/mesh.html

A number of high-quality grid generation codes listed in these sites are available
in the public domain, subject to the terms of the GNU license agreement. Two
examples are the DistMesh and the triangle.

DistMesh

This is a MATLAB code for generating unstructured triangular meshes with
triangular and tetrahedral elements [39]. The source code is available from the
Internet site:

> http://www-math.mit.edu/~persson/mesh/

For the actual mesh generation, DISTMESH uses the Delaunay triangulation
routine of MATLAB , and seeks to optimize the node locations by a force-based
smoothing procedure, while the topology is regularly updated by delaunay.
The boundary points are only allowed to move tangentially by projection using
a distance function. The iterative procedure typically results in well-shaped
meshes.

triangle

This award-winning discretization code, written in C, generates exact Delaunay triangulations, constrained Delaunay triangulations, Voronoi diagrams, and
quality conforming Delaunay triangulations [58]. The source code is available
from the Internet site:

> http://www-2.cs.cmu.edu/~quake/triangle.html

Glossary

F

This glossary is meant to complement the subject index by providing further definitions and references on specialized aspects of the finite element method.

Arbitrary Lagrangian-Eulerian formulation (ALE): A methodology for unsteady problems with moving finite elements, in which the nodal velocities are set arbitrarily yet judiciously depending on the physics of the problem under consideration (see also "Lagrangian formulation" and "Eulerian formulation" in this Appendix), For example, nodes lying in a material interface may move with the normal velocity alone to track the motion of the interface while avoiding tangential accumulation [15].

Babuška–Brezzi condition: A condition for the convergence of the finite element solution for the Stokes flow problem, in which a C^0 expansion is used for the velocity, and a lower-order discontinuous expansion is used for the pressure (e.g., [68]).

Boundary-fitted mesh: The edges and faces of the finite elements are physical boundaries of the solution domain where boundary conditions are prescribed. Troublesome non-boundary-fitted meshes are sometimes called non-aligned.

Conformal mesh: The nodes, edges and sides of neighboring elements are perfectly matched. Non-conformal meshes include hanging nodes and overlapping zones.

Discontinuous Galerkin method (DG): A modification of the standard Galerkin method for convection problems governed by conservation laws, where the finite element solution is allowed to be discontinuous across the element boundaries. This formulation is a hybrid of the standard Galerkin method and the finite volume method (FVM).

Eulerian formulation: The element nodes are stationary marker points embedded in a convected field, such as a fluid flow.

Free-surface problem: The solution domain consists of two distinct media whose interface is either stationary or evolves in time. A free-surface problem is an interfacial problem in which one of the media is inactive, meaning that the requisite solution and its interfacial distribution have simple forms.

hp Spectral element method: Adaptive finite element method in which the mesh size, h, and the order of the polynomial expansion over each element, p, are adjusted to produce a high-accuracy solution, while ensuring an adequate spatial resolution [30, 53].

Infinite element: Not surprisingly, an infinite element extends to infinity. When the asymptotic behavior of the solution is known, the element interpolation functions are craftily designed to capture the far-field behavior in terms of *a priori* unknown coefficients.

Lagrangian formulation: The finite element nodes are material (Lagrangian) point particles moving with the medium velocity, and the finite element grid is convected with the instantaneous flow. In practice, this may result in severe grid distortion that can be prevented either by re-griding or by resorting to the arbitrary Lagrangian–Eulerian formulation (ALE).

Least-squares finite element method (LSFEM): An alternative formulation in which algebraic and ordinary differential equations for the nodal values are produced by minimizing a least-squares functional instead of the Rayleigh functional associated with the Galerkin finite element method (GFEM).

Mortar finite element method: The domain of solution is divided into different regimes, and the individual solutions are matched at generally nonconforming elements by means of mortar functions.

Moving-boundary problems: In this class of problems, the boundary of the solution domain is moving due to convection, accumulation, or dissolution, and boundary nodes are convected to describe the boundary motion.

Patch test: A patch is composed of the union of all elements attached to a node, called the patch node. More generally, a patch is a well-defined substructure of a finite element grid that retains its identity as the grid is refined. The patch test is used to ensure that the solution of the problem under consideration over the participating elements, subject to a stipulated condition around the patch boundary, is in agreement with expectation. In solid mechanics, the displacement, the traction, or both, are prescribed as boundary conditions.

Singular finite element method (SFEM): When the solution is expected to exhibit a singular behavior at a point or along a line, special singular elements are employed to capture the singularity. The coefficient of the singular term is included in the finite element expansion in lieu of a nodal value.

Streamwise Upwind Petrov-Galerkin method (SUPG): In this variation of the Galerkin method for the convection equation, the governing equation is projected onto functions that are biased in the flow direction in lieu of upwind differencing.

Matlab primer

G

MATLAB is a software product developed by the commercial company *Mathworks*, designed to run on a variety of Operating Systems (OS), including UNIX and LINUX. Initially, MATLAB was developed as a virtual laboratory for matrix calculus and linear algebra. Today, MATLAB can be described both as a *programming language* and a *computing environment*.

As a programming language, MATLAB is roughly equivalent, in some ways superior and in some ways inferior to traditional upper-level languages such as FORTRAN, PASCAL, C, and C++. An attractive feature of MATLAB is that it incorporates a broad range of utility commands and intrinsic functions, most notably, graphics. A simplifying feature of MATLAB is that the dimensions of vectors and matrices used in the calculations are automatically assigned and can be changed in the course of a session, thereby circumventing the need for memory declaration and allocation. Although the graphics component makes MATLAB dependent on the graphics display library that accompanies the Operating System (OS), this is nearly transparent to the user, and becomes evident only when newer versions of MATLAB fail to run properly, or when an installed version of MATLAB stops working properly after the operating system has been upgraded.

As a computing environment, MATLAB is able to run indefinitely in its own workspace. Thus, a session defined by the values of all initialized variables and graphical objects can be saved in a file and reinstated at a later time. In this sense, MATLAB is an operating system running inside the operating system empowering the computer.

To invoke MATLAB in a UNIX or LINUX operating system endowed with a window manager, we open a new terminal command line, and issue the command: `matlab`. Assuming that the application is in the shell path, this initiates MATLAB in some graphics or line command form. To invoke MATLAB in an operating system that does not offer a terminal command line, we launch the application (executable MATLAB binary file) by double-clicking on the MATLAB icon.

An alternative to MATLAB is the public domain application SCILAB, which is freely available from the Internet site:

`http://scilabsoft.inria.fr`

SCILAB includes a wealth of mathematical functions and allows for interfacing with various programming languages, including C and FORTRAN. SCILAB works on most personal computers, UNIX and LINUX workstations.

G.1 Programming in Matlab

Only elementary knowledge of computer programming is required to under-stand and write MATLAB code. The code is written in one file or a collection of files, called the *source* or *program* files, using a standard file editor, such as the *vi* editor on UNIX and LINUX. The source code contains the main program, sometimes unfortunately called a script, and the necessary user-defined func-tions. The names of the program files must be endowed with the suffix .m . Execution begins by typing the name of the file containing the main program, in the MATLAB environment. Alternatively, the code can be typed one line at a time followed by the RETURN keystroke in the MATLAB environment.

MATLAB is an interpreted language, which means that the instructions are translated into machine language and executed in real time, one at a time. In contrast, a source code written in FORTRAN or C must first be compiled to produce the object files. These are then linked together with the necessary system libraries to produce the executable image, sometimes called the binary file or application. In fact, the MATLAB distribution is an application residing in an appropriate directory. Intrinsic MATLAB functions are precompiled and do not have to be interpreted in the course of the execution.

G.1.1 Grammar and syntax

Following is a list of general rules regarding the general grammar and syntax of MATLAB. When confronted with an error after issuing a command or during execution, this list should serve as a first check point:

- MATLAB *variables are (lower and upper) case sensitive:*
  The variable `echidna` is different than the variable `echiDna`. Similarly, the MATLAB command `return` is not equivalent to the erroneous command `Return`; the latter will not be recognized by the interpreter.

- MATLAB *variables must start with a letter:*
  A variable name is described by a string of up to 31 characters including letters, digits, and the underscore; punctuation marks are not allowed.

- MATLAB *string variables are enclosed by a single quote:*
  For example, we may define the string variable:
  `thinker_764 = 'Thucydides'`

- *Beginning and end of a command line:*
  A MATLAB command may begin at any position in a line, and may con-tinue practically indefinitely in the same line.

- *Line continuation:*
  To continue a command onto the next line, we put three dots at the end of the line.

- *Multiple commands in a line:*
  Two or more commands may be placed in the same line, provided that they are separated with a semi-colon (;).

- *Display:*
  When a command is executed directly or by running a MATLAB code, MATLAB displays the numerical value assignment or the result of a calculation. To suppress the output, we put a semi-colon (;) at the end of the command.

- *Blank spaces:*
  More than one empty space between words are ignored by the compiler. However, numbers cannot be split in sections separated by blank spaces.

- *Range of indices:*
  Vectors and arrays must have positive and non-zero indices; the vector entry v(-3) is unacceptable in MATLAB. This annoying restriction can be circumvented in clever ways by redefining the indices.

- *Comments:*
  A line beginning with the % character, or the tail-end of a line after the % character, is a comment, and is ignored by the MATLAB interpreter.

- *Mathematical symbols and special characters:*
  Table G.1.1 lists mathematical symbols and special characters used in MATLAB interactive dialog and programming.

- *Logical control flow commands:*
  Table G.1.2 lists the basic logical control flow commands.

- *Input/Output commands:*
  Table G.1.3 lists basic Input/Output (I/O) commands, functions, and format. Once the output format is set, it remains in effect until changed.

G.1.2 Precision

MATLAB stores all numbers in the long format of the floating-point representation. This means that real numbers have a finite precision of roughly sixteen significant digits, and a range of definition roughly varying between 10^{-308} and 10^{+308} in absolute value. Numbers smaller than 10^{-308} or larger than 10^{+308} in absolute value cannot be accommodated.

MATLAB performs all computations in double precision. However, this should not be confused with the ability to view and print numbers with a specified number of significant figures using the commands listed at the bottom of Table G.1.3.

+	Plus
-	Minus
*	Matrix multiplication
.*	Array multiplication
^	Matrix power
.^	Array power
kron	Kronecker tensor product
\	Backslash or left division
/	Slash or right division
./	Array division
:	Colon
()	Parentheses
[]	Brackets
.	Decimal point
..	Parent directory
...	Line continuation
,	Comma
;	Semicolon, used to suppress the screen display
%	Indicates that the rest of the line is a comment
!	Exclamation point
'	Matrix transpose
"	Quote
.'	Nonconjugated transpose
=	Set equal to
==	Equal
~=	Not equal
<	Less than
<=	Less than or equal to
>	Greater than
>=	Greater than or equal to
&	Logical *and*
|	Logical *or*
~	Logical *not*
xor	Logical *exclusive or*
i, j	Imaginary unit
pi	number $\pi = 3.14159265358\ldots$

Table G.1.1 MATLAB operators, symbols, special characters, and constants.

Control flow commands:

break	Terminate the execution
else	Use with the **if** statement
elseif	Use with the **if** statement
end	Terminate a **for** loop, a **while** loop, or an **if** block
error	Display a message and abort
for	Loop over commands a specific number of times
if	Conditionally execute commands
pause	Wait for user's response
return	Return to the MATLAB environment, invoking program or function
while	Repeat statements an indefinite number of times until a specified condition is met

Table G.1.2 MATLAB logical control flow commands and logical construct components.

G.1.3   Matlab commands

Once invoked, MATLAB responds interactively to various commands, statements, and definitions issued by the user in its window. These are implemented by typing the corresponding name, single- or multi-line syntax, and then pressing the ENTER key. Table G.1.4 lists general utility and interactive-input MATLAB commands. Issuing the command *demos* initiates various demonstrations and illustrative examples of MATLAB code, worthy of exploration.

To obtain a full explanation of a MATLAB command, statement, operator or function, use the MATLAB **help** facility, which is the counterpart of the UNIX **man** facility. For example, issuing the command: **help break** in the MATLAB environment, produces the description:

```
BREAK Terminate execution of WHILE or FOR loop.
   BREAK terminates the execution of FOR and WHILE loops.
   In nested loops, BREAK exits from the innermost loop only.
   If you use BREAK outside of a FOR or WHILE loop in a MATLAB
   script or function, it terminates the script or function at
   that point.  If BREAK is executed in an IF, SWITCH-CASE, or
   TRY-CATCH statement, it terminates the statement at that point.
```

The command **clear** is especially important, as it resets all variables to the "uninitialized" status, and thereby prevents the use of improper values defined or produced in a previous calculation. A detailed explanation of this command can be obtained by typing: **help clear** .

I/O commands:

disp	Display numerical values or text
	Use as: `disp` `disp()` `disp('text')`
fclose	Close a file
fopen	Open a file
fread	Read binary data from a file
fwrite	Write binary data to a file
fgetl	Read a line from a file, discard newline character
fgets	Read a line from a file, keep newline character
fprintf	Write formatted data to a file using C language conventions
fscanf	Read formatted data from a file
feof	Test for end-of-file (EOF)
ferror	Inquire the I/O error status of a file
frewind	Rewind a file
fseek	Set file position indicator
ftell	Get file position indicator
sprintf	Write formatted data to string
sscanf	Read formatted string from file
csvread	Read from a file values separated by commas
csvwrite	Write into file values separated by commas
uigetfile	Retrieve the name of a file to open through dialog box
uiputfile	Retrieve the name of a file to write through dialog box

Interactive input:

input	Prompt for user input
keyboard	Invoke keyboard as though it were a script file
menu	Generate menu of choices for user input

Output format:

format short	fixed point with 4 decimal places (default)
format long	fixed point with 14 decimal places
format short e	scientific notation with 4 decimal places
format long e	scientific notation with 15 decimal places
format hex	hexadecimal format

Table G.1.3 MATLAB Input/Output (I/O) commands, functions, and format.

clear	Clear variables and functions from memory
demo	Run demos
exit	Terminate a MATLAB session
help	On-line documentation
load	Retrieve variables from a specified directory
save	Save workspace variables to a specified directory
saveas	Save figure or model using a specified format
size	Reveal the size of matrix
who	List current variables
quit	Terminate a MATLAB session

Table G.1.4 General utility MATLAB commands.

G.1.4 Elementary examples

In the following examples, the interactive use of MATLAB is demonstrated by elementary examples. In these sessions, a line that begins with two "greater than" signs (>>) denotes the MATLAB command line where we type a definition or issue a command. Unless stated otherwise, a line that does not begin with >> is MATLAB output. Recall that the command `clear` clears the memory content from previous definitions to prevent misappropriation.

• Numerical value assignment and addition:

```
>> a = 1
a =
      1
>> b = 2
b =
      2
>> c = a + b
c =
      3
```

• Numerical value assignment and subtraction:

```
>> clear
>> a=1; b=-3; c=a-b
c =
      4
```

- Multiplication of numbers:

```
>> clear
>> a = 2.0; b=-3.5; c=a*b;
>> c
c =
    -7
```

Note that typing the variable c displays its current value, in this case -7.

- Vector definition:

```
>> clear
>> v = [2 1]
v =
    2    1

>> v(1)
ans =
    2

>> v' % transpose
ans =
    2
    1
```

Note that typing v(1) produces the first component of the vector v as an answer. The comment "transpose" is ignored since it is preceded by the comment delimiter "%." The answer **ans** is, in fact, a variable evaluated by MATLAB.

- Vector addition:

```
>> v = [1 2]; u = [-1, -2]; u+v
ans =
    0    0
```

- Matrix definition, addition, and multiplication:

```
>> a = [1 2; 3 4]
a =
    1    2
    3    4

>> b = [ [1 2]' [2 4]' ]
b =
    1    2
    2    4
```

```
>> a+b
ans =
        2       4
        5       8
>> c = a*b
c =
        5      10
       11      22
```

- Multiply a complex matrix by a complex vector:

```
>> a = [1+2i 2+3i; -1-i 1+i]
a =
   1.0000 + 2.0000i   2.0000 + 3.0000i
  -1.0000 - 1.0000i   1.0000 + 1.0000i

>> v = [1+i 1-i]
v =
   1.0000 + 1.0000i   1.0000 - 1.0000i

>> c = a*v'
c =
   2.0000 + 6.0000i
  -2.0000 + 2.0000i
```

Note that, by taking its transpose indicated by a prime, the row vector, v, becomes a column vector that is conformable with the square matrix, a.

- Print π:

```
>> format long
>> pi
ans =
   3.14159265358979
```

- for loop:

```
>> for j=-1:1
     j
   end
j =
     -1
j =
      0
j =
      1
```

In this example, the first three lines are entered by the user.

- if statement:

```
>> j=0;
>> i=1;
>> if i==j+1, disp 'case 1', end
case 1
```

- for loop:

```
>> n=3;
>> for i=n:-1:2
disp 'i='; disp (i), end
i=
     3
i=
     2
```

The loop is executed backward, starting at n, with step of -1.

- if loop:

```
>> i=1; j=2;
>> if i==j+1;    disp 'case 1'
elseif i==j; disp 'case2'
else;         disp 'case3'
end
case3
```

In this example, all but the last line are entered by the user.

- while loop:

```
>> i=0;
>> while i<2, i=i+1; disp(i), end
    1

    2
```

The four statements in the while loop could be typed in separate lines; that is, the commas can be replaced by the ENTER keystroke.

G.2 Matlab functions

In scientific computing, a function is an evaluation procedure that receives single or multiple input, and produces single or multiple output. MATLAB comes with an extensive library of intrinsic or embedded functions for numerical computation and data visualization.

Table G.2.1 lists general and specialized MATLAB mathematical functions. MATLAB performs these functions based on algorithms, procedures, and approximations discussed in mathematical handbooks and texts of numerical methods (e.g., [43]). To obtain specific information on the proper usage of a function, use the MATLAB help facility. If you are unsure about the proper syntax or reliability of a function, it is best to write your own code from first principles. It is both rewarding and instructive to duplicate a MATLAB function and create a personal library of user-defined functions based on control-flow commands.

User-defined functions

In MATLAB, a user-defined function is written in a file whose name is the same as the name of the function. The file name must be suffixed with the MATLAB identifier: .m . Thus, a function named *koumbaros* must reside in a file named koumbaros.m, whose general structure is:

```
function [output1, output2, ...] = koumbaros(input1, input2,...)
      ......
   return
```

The three dots indicate additional input and output arguments separated by commas; the six dots indicate additional lines of code. The output string, output1, output2, ..., consists of numbers, vectors, matrices, and string variables evaluated by the function by performing operations involving the input string, input, input2,.... To execute this function in the MATLAB environment or invoke it from a program file, we issue the command:

```
[evaluate1, evaluate2, ...] = koumbaros(parameter1, parameter2,...)
```

After the function has been successfully executed, evaluate1 takes the value of output1, evaluate2 takes the value of output2, and the rest of the output variables take corresponding values.

If a function evaluates only one number, vector, matrix, character string, entity or object, then the function statement and corresponding function can be simplified to:

```
function evaluate =  koumbaros(input1, input2,...)
    ...
   return
```

An example of a simple function residing in the function file bajanakis.m is:

Common:

abs	Absolute value
acos	Inverse cosine
acosh	Inverse hyperbolic cosine
acot	Inverse cotangent
acoth	Inverse hyperbolic cotangent
acsc	Inverse cosecant
acsch	Inverse hyperbolic cosecant
angle	Phase angle
asec	Inverse secant
asech	Inverse hyperbolic secant
asin	Inverse sine
asinh	Inverse hyperbolic sine
atan	Inverse tangent
atan2	Four quadrant inverse tangent
atanh	Inverse hyperbolic tangent
cart2pol	Cartesian to polar coordinate conversion
cart2sph	Cartesian to spherical coordinate conversion
conj	Complex conjugate
cos	Cosine
cosh	Hyperbolic cosine
cot	Cotangent
coth	Hyperbolic cotangent
csc	Cosecant
csch	Hyperbolic cosecant
exp	Exponential
expm	Matrix exponential
fix	Round towards zero
floor	Round towards minus infinity
gcd	Greatest common divisor
imag	Complex imaginary part
lcm	Least common multiple
log	Natural logarithm
log10	Common logarithm
pol2cart	Polar to Cartesian coordinate conversion

Table G.2.1 Continuing $\longrightarrow$

real	Complex real part
rem	Remainder after division
round	Round towards nearest integer
sec	Secant
sech	Hyperbolic secant
sign	Signum function
sin	Sine
sinh	Hyperbolic sine
sph2cart	Polar to Cartesian coordinate conversion
sqrt	Square root
tan	Tangent
tanh	Hyperbolic tangent

Specialized:

bessel	Bessel functions
besseli	Modified Bessel functions of the first kind
besselj	Bessel functions of the first kind
besselk	Modified Bessel functions of the second kind
bessely	Bessel functions of the second kind
beta	Beta function
betainc	Incomplete beta function
betaln	Logarithm of the beta function
ellipj	Jacobi elliptic functions
ellipke	Complete elliptic integral
erf	Error function
erfc	Complementary error function
erfcx	Scaled complementary error function
erfinv	Inverse error function
expint	Exponential integral
gamma	Gamma function
gammainc	Incomplete gamma function
gammaln	Logarithm of gamma function
legendre	Associated Legendre functions
log2	Dissect floating point numbers
pow2	Scale floating point numbers

Table G.2.1 $\longrightarrow$ Continuing $\longrightarrow$

rat	Rational approximation
rats	Rational output

Matrix and vector initialization:

eye	Identity matrix
ones	Matrix of ones
rand	Uniformly distributed random numbers and arrays
randn	Normally distributed random numbers and arrays
zeros	Matrix of zeros

Table G.2.1 ($\longrightarrow$ Continued.) Common and specialized MATLAB mathematical functions.

```
function bname = bajanakis(isel)

  if(isel == 1)
   bname = 'phaedrus';
  elseif(isel == 2)
   bname = 'phaethon';
  else
   bname = 'alkiviadis';
  end

 return
```

Numerous examples of user-defined functions can be found in the directories of FSELIB listed in the text.

G.3 Numerical methods

MATLAB includes a general-purpose numerical methods library whose functions perform numerical linear algebra, solve algebraic equations, carry out function differentiation and integration, solve differential equations, and execute a variety of other tasks. Special-purpose libraries of interest to a particular discipline are accommodated in *toolboxes*. The theory underlying the numerical methods is discussed in texts on numerical methods and scientific computing (e.g., [43]). Table G.3.1 lists selected MATLAB numerical methods functions. While these functions are generally robust and reliable, the reader should always work under the premises of the Arabic proverb: "Trust in Allah but always tie your camel."

cat	Concatenate arrays
cond	Condition number of a matrix
det	Matrix determinant
eig	Matrix eigenvalues and eigenvectors
inv	Matrix inverse
lu	LU-decomposition of a matrix
ode23	Solution of ordinary differential equations by the second/third order Runge-Kutta method
ode45	Solution of ordinary differential equations by the fourth/fifth order Runge-Kutta-Fehlberg method
qr	QR-decomposition of a matrix, where Q is an orthogonal matrix, and R is an upper triangular (right) matrix
poly	Produces the characteristic polynomial of a matrix
quad	Function integration by Simpson's rule
root	Polynomial root finder
svd	Singular-value decomposition
trapz	Function integration by the trapezoidal rule
x = A\b	Solves the linear system $\mathbf{A} \cdot \mathbf{x} = \mathbf{b}$, where $\mathbf{A}$ is an $N \times N$ matrix, and $\mathbf{b}, \mathbf{x}$ are N-dimensional column vectors
x = b/A	Solves the linear system $\mathbf{x} \cdot \mathbf{A} = \mathbf{b}$, where $\mathbf{A}$ is an $N \times N$ matrix, and $\mathbf{b}, \mathbf{x}$ are N-dimensional row vectors

Table G.3.1 Examples of general-purpose numerical methods MATLAB functions.

G.4 Matlab Graphics

A powerful feature of MATLAB is the ability to produce professional-quality graphics, including animation. Graphics are displayed in dedicated windows appearing in response to the graphics commands. Graphics functions are listed in Tables G.4.1–G.4.3. in several categories. To obtain a detailed description of a graphics function, use the *help* facility.

Some useful tips are:

- To generate a new graphics window, use the command: `figure`

- To produce a graphics file, use the `export` option under the *file* pull-down menu in the figure window.

- To superimpose graphs, use the `hold` command.

Two-dimensional (xy) graphs:

bar	Bar graph
comet	Animated comet plot
compass	Compass plot
errorbar	Error bar plot
feather	Feather plot
fplot	Plot a function
fill	Draw filled two-dimensional polygons
hist	Histogram plot
loglog	Log-log scale plot
plot	Linear plot
polar	Polar coordinate plot
rose	Angle histogram plot
semilogx	Semi-log scale plot, x-axis logarithmic
semilogy	Semi-log scale plot, y-axis logarithmic
stairs	Stair-step plot
stem	Stem plot for discrete sequence data

Graph annotation and operations:

grid	Grid lines
gtext	Mouse placement of text
legend	Add legend to plot
text	Text annotation
title	Graph title
xlabel	x-axis label
ylabel	y-axis label
zoom	Zoom in and out of a two-dimensional plot

Table G.4.1 Elementary and specialized xy graphs, and commands for graph annotation and operations.

Line and fill commands:

fill3	Draw filled three-dimensional polygons
plot3	Plot lines and points

Two-dimensional graphs of three-dimensional data:

clabel	Contour plot elevation labels
comet3	Animated comet plot
contour	Contour plot
contour3	Three-dimensional contour plot
contourc	Contour plot computation (used by contour)
image	Display image
imagesc	Scale data and display as image
pcolor	Pseudocolor (checkerboard) plot
quiver	Quiver plot
slice	Volumetric slice plot

Surface and mesh plots:

mesh	Three-dimensional mesh surface
meshc	Combination mesh/contour plot
meshgrid	Generate x and y arrays
meshz	Three-dimensional mesh with zero plane
slice	Volumetric visualization plot
surf	Three-dimensional shaded surface
surfc	Combined surf/contour plot
surfl	Shaded surface with lighting
trimesh	Triangular mess plot
trisurf	Triangular surface plot
waterfall	Waterfall plot

Table G.4.2 Continuing $\longrightarrow$

Three-dimensional objects:

cylinder	Generate a cylinder
sphere	Generate a sphere

Graph appearance:

axis	Axis scaling and appearance
caxis	Pseudocolor axis scaling
colormap	Color lookup table
hidden	Mesh hidden line removal mode
shading	Color shading mode
view	Graph viewpoint specification
viewmtx	View transformation matrices

Graph annotation:

grid	Grid lines
legend	Add legend to plot
text	Text annotation
title	Graph title
xlabel	x-axis label
ylabel	y-axis label
zlabel	z-axis label for three-dimensional plots

Table G.4.2 ($\longrightarrow$ Continued.) MATLAB functions for three-dimensional graphics.

Graphics control:

capture	Screen capture of current figure in UNIX
clf	Clear current figure
close	Abandon figure
figure	Create a figure in a new graph window
gcf	Get handle to current figure
graymon	Set default figure properties for gray-scale monitors
newplot	Determine correct axes and figure for new graph
refresh	Redraw current figure window
whitebg	Toggle figure background color

Axis control:

axes	Create axes at arbitrary position
axis	Control axis scaling and appearance
caxis	Control pseudo-color axis scaling
cla	Clear current axes
gca	Get handle to current axes
hold	Hold current graph
ishold	True if hold is on
subplot	Create axes in tiled positions

Graphics objects:

figure	Create a figure window
image	Create an image
line	Generate a line
patch	Generate a surface patch
surface	Generate a surface
text	Create text
uicontrol	Create user interface control
uimenu	Create user interface menu

Table G.4.3 Continuing $\longrightarrow$

Graphics operations:

delete	Delete object
drawnow	Flush pending graphics events
findobj	Find object with specified properties
gco	Get handle of current object
get	Get object properties
reset	Reset object properties
rotate	Rotate an object
set	Set object properties

Hard copy and storage:

orient	Set paper orientation
print	Print graph or save graph to file
printopt	Configure local printer defaults

Movies and animation:

getframe	Get movie frame
movie	Play recorded movie frames
moviein	Initialize movie frame memory

Miscellaneous:

ginput	Graphical input from mouse
ishold	Return hold state
rbbox	Rubber-band box for region selection
waitforbuttonpress	Wait for key/button press over figure

Table G.4.3 $\longrightarrow$ Continuing $\longrightarrow$

Color controls:

caxis	Pseudocolor axis scaling
colormap	Color lookup table
shading	Color shading mode

Color maps:

bone	Gray-scale with a tinge of blue color map
contrast	Contrast-enhancing gray-scale color map
cool	Shades of cyan and magenta color map
copper	Linear copper-tone color map
flag	Alternating RGB and black color map
gray	Linear gray-scale color map
hsv	Hue-saturation-value color map
hot	Black-red-yellow-white color map
jet	Variation of HSV color map (no wrap)
pink	Pastel shades of pink color map
prism	Prism-color color map
white	All white monochrome color map

Color map functions:

brighten	Brighten or darken color map
colorbar	Display color map as color scale
hsv2rgb	Hue-saturation-value to RGB equivalent
rgb2hsv	RGB to hue-saturation-value conversion
rgbplot	Plot color map
spinmap	Spin color map

Lighting models:

diffuse	Diffuse reflectance
specular	Specular reflectance
surfl	Three-dimensional shaded surface with lighting
surfnorm	Surface normals

Table G.4.3 ($\longrightarrow$ Continued.) Miscellaneous MATLAB graphics functions.

In the remainder of this section, we present several graphics sessions followed
by the graphics output.

- Graph of the function: $f(x) = sin^3(\pi x)$

```
>> x=-1.0:0.01:1.0;        % define an array of abscissae
>> y = sin(pi*x).^3;       % note the .^ operator (Table F.1.1)
>> plot(x,y)
```

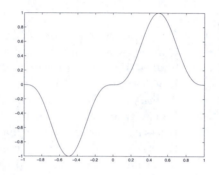

- Graph of the Gaussian function: $f(x) = e^{-x^2}$

```
>> fplot('exp(-x^2)',[-5, 5])
```

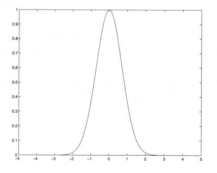

- Paint a polygon in black:

```
>> x =[0.0 1.0 1.0]; y=[0.0 0.0 1.0]; c='k';
>> fill (x,y,c)
```

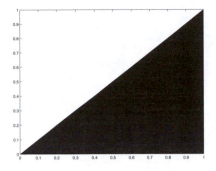

● mesh plot:

```
>> [x, y] = meshgrid(-1.0:0.10:1.0, -2.0:0.10:2.0);
>> z = sin(pi*x+pi*y);
>> mesh(z)
```

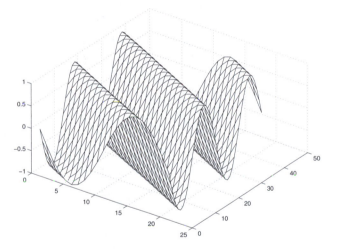

● Kelvin foam:

FSELIB script `tetrakai` (not listed in the text) generates a periodic, space-filling lattice of Kelvin's tetrakaidecahedron (14-faced polyhedron) encountered in liquid and metallic foam.

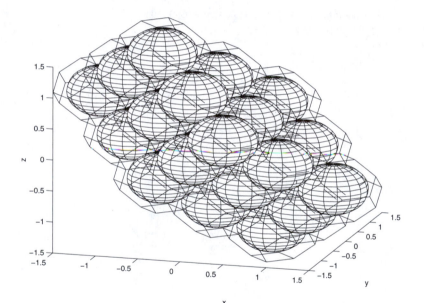

References

[1] ABRAMOWITZ, M. & STEGUN, I. A., 1970, *Handbook of Mathematical Functions*, Dover.

[2] APPELL, P., 1881, Sur des polynômes de deux variables analogues aux polynômes de Jacobi, *Arch. Math. Phys.*, **66**, 238–245.

[3] AREF, H., NEWTON, P. K., STREMLER, M. A., TOKIEDA, T. & VAINCHTEIN, D. L., 2003, Vortex crystals, *Adv. Appl. Mech.*, **39**, 1–79.

[4] BLYTH, M. G. & POZRIKIDIS, C., 2005, A Lobatto interpolation grid over the triangle, *IMA J. Appl. Math.*, in press.

[5] BOS, L., 1983, Bounding the Lebesgue function for Lagrange interpolation in a simplex, *J. Approx. Theory*, **38**, 43–59.

[6] BOS, L., TAYLOR, M. A. & WINGATE, B. A., 2000, Tensor product Gauss-Lobatto points are Fekete points for the cube, *Math. & Comput.*, **70**, 1543–1547.

[7] CHEN, Q. & BABUŠKA, I., 1995, Approximate optimal points for polynomial interpolation of real functions in an interval and in a triangle, *Comput. Meth. Appl. Mech. Engrg.*, **128**, 405–417.

[8] CHEN, Q. & BABUŠKA, I., 1996, The optimal symmetrical points for polynomial interpolation of real functions in the tetrahedron, *Comput. Meth. Appl. Mech. Engrg.*, **137**, 89–94.

[9] CHIHARA, T. S., 1978, *An Introduction to Orthogonal Polynomials*, Gordon & Breach.

[10] CIARLET, P. G., 1978, *The Finite Element Method for Elliptic Problems*, North-Holland.

[11] CLOUGH, R. W. & FELIPPA, C. A., 1968, A refined quadrilateral element for analysis of plate bending, *Matrix Methods in Structural Mechanics, Proc. Second Conf.*, 399–440.

[12] DAHLQUIST, G. & BJÖRCK, Å., 1974, *Numerical Methods*, Prentice Hall.

[13] DAVIS, P. J., 1975, *Interpolation and Approximation*, Dover.

[14] DEMMEL, J. W., 1997, *Applied Numerical Linear Algebra*, SIAM.

[15] DONEA, J. & HUERTA, A., 2003 *Finite Element Methods for Flow Problems*, Wiley.

[16] DOSSOU, K. & PIERRE, R., 2003, A Newton-GMRES approach for the analysis of the postbuckling behavior of the solutions of the von Kármán equations, *SIAM J. Sci. Comput.*, **24**, 1994–2012.

[17] DUBINER, M., 1991, Spectral methods on triangles and other domains, *J. Sci. Comput.*, **6**, 345–390.

[18] FEJÉR, L., 1932, Lagrangesche interpolation und die zugehörigen konjugierten punkte, *Mathematische Annalen*, **106**, 1–55.

[19] FEJÉR, L., 1932, Bestimmung derjenigen abszissen eines intervalles für welche die quadratsumme der grundfunktionen der Lagrangeschen interpolation im invervalle [-1, 1] ein möglichst kleines maximum besitzt, *Ann Scuola Norm. Sup. Pisa Sci. Fis. Mt. Ser. II*, **1**, 263–273.

[20] FREUD, G., 1966, *Orthogonal Polynomials*, Pergamon.

[21] GALLAGHER, R. H., 1975, *Finite Element Analysis Fundamentals*, Prentice Hall.

[22] GLOWINSKI, R., 1984, *Numerical Methods for Nonlinear Variational Problems*, Springer.

[23] GRESHO, P. M. & SANI, R. L., 1998, *Incompressible Flow and the Finite Element Method, Volume I, Advection-Diffusion*, Wiley.

[24] HARBANI, Y. & ENGELMAN, M., 1979, Out-of-core solution of linear equations with non-symmetric coefficient matrix, *Computers & Fluids*, **7**, 13–31.

[25] HESTENES, M. & STIEFEL, E., 1952, Methods of conjugate gradients for solving linear systems, *J. Res. Natl. Bur, Stand.*, **49**, 409–436.

[26] HESTHAVEN, J. S., 1998, From electrostatics to almost optimal nodal sets for polynomial interpolation in a simplex, *SIAM J. Numer. Anal.*, **35**, 655–676.

[27] HESTHAVEN, J. S. & TENG, C. H., 2000, Stable spectral methods on tetrahedral elements, *SIAM J. Sci. Comput.*, **21**, 2352–2380.

[28] HRABOK, M. M. & HRUDEY, T. M., 1984, A review and catalog of plate bending finite elements. *Computers & Structures*, **19**, 479–495.

[29] HOOD, P., 1976, Frontal solution program for unsymmetric matrices, *Int. J. Num. Meth. Eng.*, **10**, 379–399.

[30] KARNIADAKIS, G. E. & SHERWIN, S. J., 1999, *Spectral/hp Element Methods for CFD*, Oxford University Press.

[31] KRYLOV, V. I., 1962, *Approximate Calculation of Integrals*, Macmillan.

[32] KWAK, S. & POZRIKIDIS, C., 1998, Adaptive triangulation of evolving, closed or open surfaces by the advancing-front method, *J. Comp. Phys.*, **145**, 61–88.

[33] KWON, Y. W. & BANG, H., 2000, *The Finite Element Method using MATLAB*, Chapman & Hall/CRC.

[34] LUO, H. & POZRIKIDIS, C., 2005, A Lobatto interpolation grid in the tetrahedron, Submitted.

[35] MCFARLAND, D., SMITH, B. L. & BERNHART, W. D., 1972, *Analysis of Plates*, Spartan Books.

[36] MORLEY, L. S. D., 1968, The triangular equilibrium problem in the solution for plate bending problems. *Aero. Quart.*, **19**, 149–169.

[37] NOVOSHILOV, V. V., 1961, *Theory of Elasticity*, Israel Program for Scientific Translations.

[38] PATERA, A. T., 1984, A spectral element method for fluid dynamics: Laminar flow in a channel expansion, *J. Comp. Phys.*, **54**, 468–488.

[39] PERSSON, P.-O. & STRANG, G., 2004, A simple mesh generator in MATLAB, *SIAM Rev.*, **46**, 329–345.

[40] PEYRET, R., 2002, *Spectral Methods for Incompressible Viscous Flow*, Springer.

[41] POPOV, E. P., 1991, *Engineering Mechanics of Solids*, Second edition, Prentice Hall.

[42] POZRIKIDIS, C., 1997, *Introduction to Theoretical and Computational Fluid Dynamics*, Oxford University Press.

[43] POZRIKIDIS, C., 1998, *Numerical Computation in Science and Engineering*, Oxford University Press.

[44] POZRIKIDIS, C., 2002, *A Practical Guide to Boundary-Element Methods with the Software Library BEMLIB*, Chapman & Hall/CRC.

[45] POZRIKIDIS, C., 2004, A finite-element method for interfacial surfactant transport with application to the flow-induced deformation of a viscous drop. *J. Eng. Math.*, **49**, 163–180.

[46] PRORIOL, J., 1957, Sur une famille de polynomes á deux variables orthogonaux dans un triangle. *Comptes Rendus de'l Académie des Sciences Paris*, **245**, 2459–2461.

[47] REBAY, S., 1993, Efficient unstructured mesh generation by means of Delaunay triangulation and Bower-Watson algorithm. *J. Comp. Phys.*, **106**, 125–138.

[48] RIVLIN, T. J., 1969, *An Introduction to the Approximation of Functions*, Dover.

[49] RIVLIN, T. J., 1974, *The Chebyshev Polynomials*, Wiley.

[50] SAAD, Y. & SCHULTZ, M., 1986, GMRES: A generalized minimal residual algorithm for solving nonsymmetric linear systems, *SIAM J. Sci. Statist. Comput.*, **7**, 856–869.

[51] DE SAMPIO, P. A. B., 1990, A Petrov-Galerkin/modified operator formulation for convection–diffusion problems, *Int. J. Num. Meth. Eng.*, **30**, 331–347.

[52] SANSONE, G., 1959, *Orthogonal Functions*, Dover, Reprinted in 1991.

[53] SCHWAB, CH., 1998, *p and hp-Finite Element Methods*, Oxford University Press.

[54] SCHWARZ, H. R., 1988, *Finite Element Methods*, Academic Press.

[55] SHERWIN, S. J. & KARNIADAKIS, G. E., 1995, A triangular spectral element method; applications to the incompressible Navier-Stokes equations, *Comp. Meth. Appl. Mech. Eng.*, **123**, 189–229.

[56] SHERWIN, S. J. & KARNIADAKIS, G. E., 1995, A new triangular and tetrahedral basis for high-order (hp) finite element methods, *Int. J. Num. Meth. Eng.*, **38**, 3775–3802.

[57] SHERWIN, S. J. & KARNIADAKIS, G. E., 1996, Tetrahedral hp finite elements: Algorithms and flow simulations, *J. Comp. Phys*, **124**, 14–45.

[58] SHEWCHUK, J. R., 2002, Refinement Algorithms for Triangular Mesh Generation, *Computational Geometry: Theory and Applications*, **22**, 21–74.

[59] STROUD, A. H. & SECREST, D., 1966, *Gaussian Quadrature Formulas*, Prentice Hall.

[60] SUETIN, P. K., 1999, *Orthogonal Polynomials in Two Variables*, Gordon & Breach (Translation of original Russian edition published by Nauka, Moscow, 1988.)

[61] SZABÓ, B. & BABUŠKA, I., 1991, *Finite Element Analysis*, Wiley.

[62] SZEGÖ, G., 1975, *Orthogonal Polynomials*, Fourth edition, American Mathematical Society, Providence.

[63] TAYLOR, M. A., WINGATE, B. A. & VINCENT, R. E., 2000, An algorithm for computing Fekete points in the triangle, *SIAM J. Numer. Anal.*, **38**, 1707–1720.

[64] TIMOSHENKO, S. P. & GERE, J. M., 1961, *Theory of Elastic Stability*, First edition, McGraw-Hill.

[65] TIMOSHENKO, S. & WOINOWSKY-KRIEGER, S., 1959, *Theory of Plates and Shells*, McGraw-Hill.

[66] YU, C.-C. & HEINRICH, J. C., 1986, Petrov-Galerkin methods for the time-dependent convective transport equation, *Int. J. Num. Meth. Eng.*, **23**, 883–901.

[67] WILSON, E. L., 1974, The static condensation algorithm, *Int. J. Num. Meth. Eng.*, **8**, 198–203.

[68] ZHANG, S., 2004, A new family of stable mixed finite elements for the 3D Stokes equations. *Math. Comp.*, **74**,543–554.

Index

Program index

645

Subject index